# THE METROLOGY HANDBOOK

*The Measurement Quality Division, ASQ*
*Jay L. Bucher, Editor*

ASQ Quality Press
Milwaukee, Wisconsin

*The Metrology Handbook*
The Measurement Quality Division
© 2004 by ASQ
All rights reserved. Published 2004
Printed in the United States of America

12 11 10 09 08 07 06 05     5 4 3

**Library of Congress Cataloging-in-Publication Data**

The metrology handbook / The Measurement Quality Division ; Jay Bucher, editor ; [authors] Jay L. Bucher ... [et al.].-- 1st ed.
    p. cm.
 Includes bibliographical references and index.
 ISBN 0-87389-620-3 (hardcover : alk. paper)
 1. Mensuration--Handbooks, manuals, etc. 2. Calibration--Handbooks, manuals, etc. 3. Quality assurance--Handbooks, manuals, etc. I. Bucher, Jay L., 1949- II. American Society for Quality. Measurement Quality Division. III. Title.

T50.M423 2004
620'.0044--dc22

2004003464
No part of this book may be reproduced in any form or by any means, electronic, mechanical, photocopying, recording, or otherwise, without the prior written permission of the publisher.

ISBN 0-87389-620-3

The idea for the cover picture, and the photograph of the micrometer, are by Graeme C. Payne. The image of Earth, AS17-148-22727, is courtesy of Earth Sciences and Image Analysis Center, NASA Johnson Space Center. (http://eol.jsc.nasa.gov/) This image, now known as the "Blue Marble" photograph, was taken on December 7, 1972 by the crew of Apollo 17 as they were en route to the Moon for the final landing of the 20th Century.

Publisher: William A. Tony
Acquisitions Editor: Annemieke Hytinen
Project Editor: Paul O'Mara
Production Administrator: Randall Benson
Special Marketing Representative: Matt Meinholz

ASQ Mission: The American Society for Quality advances individual, organizational, and community excellence worldwide through learning, quality improvement, and knowledge exchange.

Attention Bookstores, Wholesalers, Schools and Corporations: ASQ Quality Press books, videotapes, audiotapes, and software are available at quantity discounts with bulk purchases for business, educational, or instructional use. For information, please contact ASQ Quality Press at 800-248-1946, or write to ASQ Quality Press, P.O. Box 3005, Milwaukee, WI 53201-3005.

To place orders or to request a free copy of the ASQ Quality Press Publications Catalog, including ASQ membership information, call 800-248-1946. Visit our Web site at www.asq.org or http://qualitypress.asq.org.

Printed in the United States of America

 Printed on acid-free paper

Quality Press
600 N. Plankinton Avenue
Milwaukee, Wisconsin 53203
Call toll free 800-248-1946
Fax 414-272-1734
www.asq.org
http://qualitypress.asq.org
http://standardsgroup.asq.org
E-mail: authors@asq.org

# Table of Contents

*List of Figures and Tables* . . . . . . . . . . . . . . . . . . . . . . . . . . . . . . . . . . . . . . . . . . . . . . . *vii*
*CD Contents* . . . . . . . . . . . . . . . . . . . . . . . . . . . . . . . . . . . . . . . . . . . . . . . . . . . . . . . . . . *ix*
*Foreword* . . . . . . . . . . . . . . . . . . . . . . . . . . . . . . . . . . . . . . . . . . . . . . . . . . . . . . . . . . . . *xi*
*Preface* . . . . . . . . . . . . . . . . . . . . . . . . . . . . . . . . . . . . . . . . . . . . . . . . . . . . . . . . . . . . . *xiii*
*Acknowledgements* . . . . . . . . . . . . . . . . . . . . . . . . . . . . . . . . . . . . . . . . . . . . . . . . . . . *xv*

## Part I. Background . . . . . . . . . . . . . . . . . . . . . . . . . . . . . . . . . . . . . . . . . . 1

Chapter 1   History and Philosophy of Metrology/Calibration . . . . . . . . . . . . . . . . . . . . 3

## Part II. Quality Systems . . . . . . . . . . . . . . . . . . . . . . . . . . . . . . . . . . . . . 17

Chapter 2   The Basics of a Quality System . . . . . . . . . . . . . . . . . . . . . . . . . . . . 19
Chapter 3   Quality Standards and Their Evolution . . . . . . . . . . . . . . . . . . . . . . . . . . . 25
Chapter 4   Quality Documentation . . . . . . . . . . . . . . . . . . . . . . . . . . . . . . . . . . . . . . 37
Chapter 5   Calibration Procedures and Equipment Manuals . . . . . . . . . . . . . . . . . . 41
Chapter 6   Calibration Records . . . . . . . . . . . . . . . . . . . . . . . . . . . . . . . . . . . . . . . . . 47
Chapter 7   Calibration Certificates . . . . . . . . . . . . . . . . . . . . . . . . . . . . . . . . . . . . . . 51
Chapter 8   Quality Manuals . . . . . . . . . . . . . . . . . . . . . . . . . . . . . . . . . . . . . . . . . . . 55
Chapter 9   Traceability . . . . . . . . . . . . . . . . . . . . . . . . . . . . . . . . . . . . . . . . . . . . . . . 61
Chapter 10   Calibration Intervals . . . . . . . . . . . . . . . . . . . . . . . . . . . . . . . . . . . . . . . . 67
Chapter 11   Calibration Standards . . . . . . . . . . . . . . . . . . . . . . . . . . . . . . . . . . . . . . . 71
Chapter 12   Audit Requirements . . . . . . . . . . . . . . . . . . . . . . . . . . . . . . . . . . . . . . . . 89
Chapter 13   Scheduling and Recall Systems . . . . . . . . . . . . . . . . . . . . . . . . . . . . . . . 93
Chapter 14   Labels and Equipment Status . . . . . . . . . . . . . . . . . . . . . . . . . . . . . . . . 95
Chapter 15   Training . . . . . . . . . . . . . . . . . . . . . . . . . . . . . . . . . . . . . . . . . . . . . . . . . 99
Chapter 16   Environmental Controls . . . . . . . . . . . . . . . . . . . . . . . . . . . . . . . . . . . . 103
Chapter 17   Industry-Specific Requirements . . . . . . . . . . . . . . . . . . . . . . . . . . . . . . 109
Chapter 18   Computers and Automation . . . . . . . . . . . . . . . . . . . . . . . . . . . . . . . . . 137

## Part III. Metrology Concepts . . . . . . . . . . . . . . . . . . . . . . . . . . . . . . . . . 147

Chapter 19   A General Understanding of Metrology . . . . . . . . . . . . . . . . . . . . . . . . 149
Chapter 20   Measurement Methods, Systems, Capabilities, and Data . . . . . . . . . . . . 157
Chapter 21   Specifications . . . . . . . . . . . . . . . . . . . . . . . . . . . . . . . . . . . . . . . . . . . . 177
Chapter 22   Substituting Calibration Standards . . . . . . . . . . . . . . . . . . . . . . . . . . . . 195
Chapter 23   Proficiency Testing, Measurement Assurance Programs,
            and Laboratory Intercomparisons . . . . . . . . . . . . . . . . . . . . . . . . . . . . . 203

## Part IV. Mathematics and Statistics: Their Use in Measurement ............................. 223

    Chapter 24  Number Formatting ................................................. 225
    Chapter 25  Unit Conversions ................................................... 239
    Chapter 26  Ratios ................................................................ 259
    Chapter 27  Statistics ............................................................ 265
    Chapter 28  Mensuration, Volume, and Surface Areas ............................ 285

## Part V. Uncertainty in Measurement ............................. 303

    Chapter 29  Uncertainty in Measurement ........................................ 305

## Part VI. Measurement Parameters ............................. 321

    Chapter 30  Introduction to Measurement Parameters ........................... 323
    Chapter 31  DC and Low Frequency ............................................. 325
    Chapter 32.  Radio Frequency and Microwave .................................. 347
    Chapter 33  Mass and Weight .................................................. 373
    Chapter 34  Dimensional and Mechanical Parameters ........................... 387
    Chapter 35  Other Parameters: Chemical, Analytical,
                     Electro-Optical, and Radiation .................................... 397

## Part VII. Managing a Metrology Department or Calibration Laboratory ............................. 411

    Chapter 36  Getting Started ..................................................... 413
    Chapter 37  Best Practices ...................................................... 417
    Chapter 38  Personnel Organizational Responsibilities ......................... 429
    Chapter 39  Process Workflow .................................................. 435
    Chapter 40  Budgeting and Resource Management .............................. 441
    Chapter 41  Vendors and Suppliers ............................................. 443
    Chapter 42  Housekeeping and Safety .......................................... 447

## Appendix A  Professional Associations ....................... 451

## Appendix B  ASQ and Certification .......................... 455

## Appendix C  Acronyms and Abbreviations .................... 463

## Appendix D  Glossary of Terms ............................. 471

## Appendix E  Common Conversions ........................... 487

*Bibliography* ................................................................. 513
*Author biographies* ......................................................... 523
*Index* ........................................................................ 527

# List of Figures and Tables

| | | |
|---|---|---|
| Figure 1.1 | Early dimensional standard at the Royal Observatory, Greenwich, England. | 8 |
| Figure 5.1 | Sample calibration procedures with specification tables. | 43 |
| Figure 5.2 | Example of specific instructions in a calibration procedures. | 44 |
| Figure 7.1 | A sample certificate of calibration | 52 |
| Figure 8.1 | A sample table of contents for an ISO/IEC 17025-based quality manual | 57 |
| Figure 9.1 | Traceability pyramid—example 1. | 63 |
| Figure 9.2 | Traceability pyramid—example 2. | 64 |
| Table 11.1 | SI derived units. | 72 |
| Table 11.2 | SI prefixes. | 72 |
| Table 11.3 | Definitions of various types of standards. | 86 |
| Table 11.4 | Realization techniques of SI base units. | 88 |
| Table 16.1 | General-purpose calibration laboratories. | 105 |
| Table 19.1 | Frequently used constants. | 150 |
| Table 19.2 | Common measurement parameters. | 151 |
| Table 19.3 | Common measurands and equipment used to generate them. | 153 |
| Table 19.4 | Common measurands and some of their associated formulas. | 154 |
| Figure 20.1 | Normal frequency distribution curve—1 standard deviation. | 161 |
| Table 20.1 | X-bar and R-bar control chart calculations. | 165 |
| Table 20.2 | Table for calculating the control limits. | 166 |
| Figure 20.2 | X-bar chart for example of time interval measurements. | 167 |
| Figure 20.3 | R-bar chart for the example of time interval measurement. | 167 |
| Table 20.3 | Range R&R study. | 169 |
| Figure 20.4 | Example of appraiser variations. | 171 |
| Table 21.1 | Specifications of meters A & B. | 189 |
| Table 21.2 | Measurement unit values at 28 V AC. | 190 |
| Table 22.1 | Measurement standards in original procedures. | 196 |
| Table 22.2 | Measurement requirements. | 198 |
| Table 22.3 | Equivalent standards to substitute. | 199 |
| Figure 23.1 | Measurement comparison scheme. | 204 |
| Table 23.1 | Assigning a reference value. | 205 |
| Table 23.2 | Measurement comparison scheme: raw data and calculations. | 206 |
| Figure 23.2 | Measurement comparison scheme: mean ± 3 standard deviations. | 207 |

| | | |
|---|---|---|
| **Figure 23.3** | Measurement comparison scheme: mean ± U (k=2). | 207 |
| **Figure 23.4** | Uncertainty overlap plot. | 208 |
| **Table 23.3** | Interlaboratory testing comparison data. | 209 |
| **Figure 23.5** | Interlaboratory testing comparison data. | 209 |
| **Table 23.4** | Derivation of consensus or reference value. | 210 |
| **Table 23.5** | Split-sample data analysis. | 211 |
| **Figure 23.6** | Analysis of variance. | 211 |
| **Figure 23.7** | Interpretation of one-way ANOVA data. | 212 |
| **Table 23.6** | Control chart constants. | 215 |
| **Table 23.7** | Data for individual measurements. | 216 |
| **Figure 23.8** | Individual measurements chart. | 218 |
| **Figure 23.9** | Moving range chart. | 218 |
| **Table 23.8** | Multiple measurement data. | 219 |
| **Figure 23.10** | X-bar chart. | 221 |
| **Figure 23.11** | Range chart. | 221 |
| **Table 25.1** | SI units derived from base. | 242 |
| **Table 25.2** | Derived units with specialized names and symbols. | 242 |
| **Table 25.3** | Other derived units. | 243 |
| **Table 25.4** | Currently recognized SI unit prefixes. | 245 |
| **Table 25.5** | SI-named units that are multiples of SI base or derived units. | 246 |
| **Table 25.6** | Units not to be used within SI system of units. | 246 |
| **Table 25.7** | SI units arranged by unit category. | 248 |
| **Table 25.8** | Conversion matrix: plane angle units. | 252 |
| **Table 25.9** | Frequently used constants. | 254 |
| **Table 27.1** | Bimodal distribution. | 268 |
| **Table 29.1** | Individual data. | 307 |
| **Table 29.2** | Standard deviation of the mean calculation. | 309 |
| **Figure 29.1** | An example of an uncertainty budget. | 311 |
| **Figure 29.2** | A sample uncertainty report. | 314 |
| **Table 30.1** | The five major measurement parameters | 324 |
| **Table 31.1** | Direct voltage parameters. | 327 |
| **Table 31.2** | Thermoelectric effects from connector materials. | 329 |
| **Table 31.3** | Direct current parameters. | 330 |
| **Table 31.4** | Resistance parameters. | 331 |
| **Table 31.5** | Alternating voltage parameters. | 332 |
| **Table 31.6** | Alternating current parameters. | 334 |
| **Table 31.7** | Capacitance parameters. | 335 |
| **Table 31.8** | Inductance parameters. | 336 |
| **Table 31.9** | Time interval and frequency parameters. | 337 |
| **Table 31.10** | Phase angle parameters. | 341 |
| **Table 31.11** | Electrical power parameters. | 343 |
| **Table 32.1** | Common frequency bands and names. | 349 |
| **Table 32.2** | Common wavelength bands. | 350 |
| **Table 32.3** | RF power parameters. | 351 |
| **Figure 32.1** | Power sensor calibration system—block diagram. | 356 |
| **Table 32.4** | Attenuation or insertion loss parameters. | 357 |
| **Table 32.5** | Reflection coefficient, standing wave ratio parameters. | 359 |

| | | |
|---|---|---:|
| **Figure 32.2** | Smith chart. | 360 |
| **Table 32.6** | RF voltage parameters. | 361 |
| **Table 32.7** | Modulation parameters. | 362 |
| **Table 32.8** | Noise figure, excess noise ratio parameters. | 363 |
| **Figure 32.3** | A scalar network analyzer | 366 |
| **Figure 32.4** | A vector network analyzer | 366 |
| **Figure 32.5** | A general two-port network. | 367 |
| **Figure 32.6** | S-parameter flowgraph of a two-port network. | 369 |
| **Table 33.1** | Most-used conversion factors… | 376 |
| **Figure 33.1** | Platform scale schematic. | 382 |
| **Table 34.1** | Conversion factors for plane angle units. | 390 |
| **Figure 35.1** | Converging (positive) lens. | 397 |
| **Figure 35.2** | Diverging (negative) lens. | 398 |
| **Figure 35.3** | Concave mirrors reflect light… | 398 |
| **Figure 35.4** | Refracting telescope. | 398 |
| **Figure 35.5** | Reflecting telescope. | 398 |
| **Figure 37.1** | Scheduled forecast of calibrations due | 422 |
| **Figure 39.1** | A sample business process interaction diagram | 439 |

# CD Contents

| | |
|---|---|
| **Appendix I** | Acronyms.pdf |
| **Appendix II** | Chapter 23 example worksheet.xls |
| **Appendix III** | Chapter 29 examples of uncertainty.xls |
| **Appendix IV-a** | Statistical tables, tools, and formula.xls |
| **Appendix IV-b** | Metrology, math, statistics, and engineering formulas.xls |
| **Appendix IV-c** | Exponential interpolation—Linear and nonlinear.xls |
| **Appendix V** | Units—Fundamental and conversions.xls |
| **Appendix VI** | Internet resources and links.xls |
| **Appendix VII** | Tolerance calculator version 4.0.exe |
| **Appendix VIII** | Uncertainty calculator version 3.2.exe |

# Foreword

My introduction to real metrology was quite rude. I was applying for a job as an analytical chemist, and the laboratory director interviewing me asked how to run an infrared spectrum. After rather proudly reciting what I had been taught in college, I was immediately deflated when he asked me how accurate the measurement result would be. As I stammered my non-answer, I realized that job was not going to happen. Oh, if I only had *The Metrology Handbook* at that time.

Over the course of the last few decades, metrology has changed significantly. Concepts have been defined more rigorously. Electronics, computers, micro-technology, lasers and many more technological developments have pushed measurement capability to almost unbelievable accuracy. The industrial world has realized that measurement is fundamental to product and process quality. Consumers demand product characteristics and functionality realizable only with today's measurement accuracy. It is the job of the metrologist to meet the need for greater measurement accuracy by using all tools now available.

The American Society for Quality has assembled in *The Metrology Handbook* the basic components of the practice of metrology as it is known today. For those who want to be part of the ever growing metrology community, it introduces the fundamental concepts in a clear and precise way. For those who are already metrology professionals, it is an ideal companion to supplement and expand your knowledge. And for those who work in or manage metrology and calibration laboratories—either today or tomorrow—*The Metrology Handbook* covers the essentials of operating a respected laboratory.

The practice of metrology is both constant and changing. Metrology is constant in that it relies on a strong sense of the fundamental. Good measurement is based on understanding what must be controlled, what must be reported and what must be done to ensure that repeated measurements continue to maintain accuracy. Metrology is always changing as scientists, engineers, and technicians learn more about how to make good measurements. In the latter part of the last century, concepts such as uncertainty, quality systems, statistics, and good metrology laboratory management underwent significant changes, resulting in better metrological practice. All this and more is covered here.

It is essential that every metrology professional be familiar with these new approaches and use them daily in his or her work. A metrology professional must continue to expand his or her knowledge through courses, conferences and study. Through the work of organizations such as the American Society for Quality and handbooks such as *The Metrology Handbook*, the task of keeping up with advances has been made much simpler.

Dr. John Rumble, Jr.<br>
Gaithersburg, Maryland

# Preface

Metrology, in one form or another, has been around from the early days of *Homo-sapiens* when they had to hunt for survival. Back then, traceable standards were not available, and unbroken chains of comparison did not exist. In an article in *Quality in Manufacturing*, Nathalie Mitard states that metrology only became important when people started making tools from metal instead of stone, bone, and wood.[1] Be that as it may, the science of measurement was alive and well, and made itself apparent with the dawning of each new day.

In order for the hunter to kill game with a weapon, what felt good or worked effectively was replicated to gain the desired results over and over again. If a bow was not of the correct length, arrows did not have enough force to penetrate fur, bone, and muscle of wild game; or it was too difficult to draw the bow for the strength and length of the arm. Through trial and error, early humans became the hunters instead of the hunted. This happened because they remembered what worked the best and disregarded what didn't! With today's sophisticated machines and technology, reproducing bows and arrows to match your specific measurements, use, and function is as easy as ordering online. And, hopefully, you don't have to worry about a saber-toothed tiger attacking you while waiting for your new weapon.

Designing and/or manufacturing has made giant leaps throughout history due to improvements in measurement within agriculture, building construction, tool making, clothing, food, and transportation, to name a few (see Chapter 1). From sowing wild oats to cultivating genetically modified organisms; living in a cave to building the world's tallest structures that are designed to be earthquake resistant; from using the original hammer, a rock, to crafting specialty tools that work in deep space and at the bottom of the sea; going from the fig leaf to donning fire retardant clothes, from walking to space shuttles: Metrology affects everyone on a daily basis, whether we realize it or not. These, of course, are only general examples in the long history of *metrology*, the science of measurement.

Our purpose in writing this handbook was to develop a practical metrology reference for calibration professionals. We have intentionally focused on information for the vast majority of practicing professionals providing calibration/testing services, realizing that to do justice to the immense volumes of graduate and postgraduate level published metrology work would not be practical in a single handbook.

Whether you're changing disciplines in your career field; helping to becoming certified to a new or different standard; accepting more responsibilities as a supervisor or manager; training your fellow calibration practitioners; or using it to prepare for ASQ's

Certified Calibration Technician (CCT) exam . . . we hope this handbook provides the information, guidance, and/or knowledge to help you achieve your goals.

## Endnote

1. Mitard, Nathalie. 2001. From the cubit to optical inspection. *Quality in Manufacturing* (September/October).
   www.manufacturingcenter.com/qm/archives/0901/0901gaging_suppl.asp

# Acknowledgments

This handbook is a synthesis of the education, training, experience, and hard work of several individuals. The authors acknowledge and are grateful for the support of our faiths, our families, and our employers during this project. This book has touched the lives of friends, coworkers, supervisors, clients, and families. We would be remiss if we did not acknowledge their part in allowing us to do our part. Their patience and understanding made this task a great deal easier and, in some cases, turned an insurmountable job into just a bump in the road. To all of you, we collectively say, "Thank you from the bottom of our hearts. We could not have done this without your understanding and support."

To Annemieke Hytinen of ASQ, our acquisitions editor . . . the English language does not have enough words to express our gratitude. Your patience, understanding, and continual support during the easy and hard times have provided a bright light at the end of that long tunnel the team has traveled for the past year. We hope you will think kindly of calibration practitioners, consultants, and metrologists.

To Ann Benjamin of ASQ, for providing valuable collaborative resources and timely support in helping to make the handbook come to fruition.

To the metrology community . . . this book was written with your needs and requirements in mind. On behalf of ASQ's Measurement Quality Division, we solicit your observations, comments, and suggestions for making this a better book in the future. Calibration and metrology are living entities that continue to grow and evolve. What we do would not be much fun if they didn't.

And finally, to this unique team of coauthors—thank you. Two simple words that say it all. This book is the result of cooperation, understanding, and devotion to the calibration/metrology community and a burning desire to share what we have learned, experienced, and accomplished in our individual careers. I thank you, the rest of the team thanks you, and the metrology community surely will thank you.

Jay L. Bucher, MSgt, USAF (Ret), ASQ CCT  
Editor, *The Metrology Handbook*

Primary authorship responsibility for *The Metrology Handbook* is as follows:

| | |
|---|---|
| Keith Bennett: | Chapter 1 |
| Hershal Brewer: | Chapters 8 and 17 |
| David Brown: | Chapter 24, 25, 26, 27, 28, and Appendices E, IV, and V |
| Jay L. Bucher: | Chapters 2, 4, 5, 6, 7, 9, 10, 11, 12, 13, 14, 15, 16, 17, 18, 35, 36, 37, 39, 40, 41, 42, and Appendix A |
| Christopher Grachanen: | Chapter 19, 20, 38, and Appendices A, B, VII, and VIII |
| Emil Hazarian: | Chapters 33 and 34 |
| Graeme C. Payne: | Chapter 3, 17, 21, 22, 30, 31, 32, and Appendices A, C, D, I, II, and VI |
| Dilip Shah: | Chapters 23 and 29, and Appendix III |

The following individuals in addition to their contribution as coauthors also provided invaluable review of *The Metrology Handbook* contents: Jay L. Bucher, Christopher Grachanen, Graeme C. Payne, and Dilip Shah.

---

## A CALIBRATION TECHNICIAN . . .

. . . tests, calibrates, and maintains electronic, electrical, mechanical, electromechanical, analytical, physical inspection measuring and test equipment (IM&TE) using calibration standards and calibration procedures based on accepted metrological practices derived from fundamental theories of physics, chemistry, mathematics, and so on. In some organizations calibration technicians perform IM&TE repairs as well as maintaining calibration standards.

A Calibration Technician typically:
- has a working knowledge of applied mathematics, basic algebra, and basic statistics
- has in-depth knowledge of at least one field of metrology and at least basic familiarity with others
- is a graduate from a technical trade school, military training school, or has at least a two year associates degree
- possesses good written and oral communication skills
- has basic computer skills

# Part I
## Background

**Chapter 1**  History and Philosophy of Metrology/Calibration

# Chapter 1
# History and Philosophy of Metrology/Calibration

*Weights and measures may be ranked among the necessaries of life to every individual of human society. They enter into the economical arrangements and daily concerns of every family. They are necessary to every occupation of human industry; to the distribution and security of every species of property; to every transaction of trade and commerce; to the labors of the husbandman; to the ingenuity of the artificer; to the studies of the philosopher; to the researches of the antiquarian; to the navigation of the mariner; and the marches of the soldier; to all the exchanges of peace, and all the operations of war. The knowledge of them, as in established use, is among the first elements of education, and is often learned by those who learn nothing else, not even to read and write. This knowledge is riveted in the memory by the habitual application of it to the employments of men throughout life.*

John Quincy Adams, Report to the Congress, 1821

## ANCIENT MEASUREMENT

Weights and measures were some of the earliest tools invented. From the beginning of mankind there were needs to standardize measurements used in everyday life, such as construction of weapons used for hunting and protection, gathering and trading of food and clothing, and territorial divisions. The units of specific measurements, such as length, were defined as the length of an individual's arm (other parts of the human anatomy were also used). Weight probably would have been defined as the amount a man could lift or the weight of a stone the size of a hand. Time was defined by the length of a day and days between the cycles of the moon. Cycles of the moon were used to determine seasons. Although definitions of these measurements were rudimentary, they were sufficient to meet local requirements. Little is known about the details of any of these measurements, but artifacts over 20,000 years old indicate some form of timekeeping.

The earliest Egyptian calendar was based on the moon's cycles, but later the Egyptians realized that the Dog Star in Canis Major, which we call Sirius, rose next to the sun every 365 days, about when the annual inundation of the Nile began. Based on this knowledge, they devised a 365-day calendar that seems to have begun in 4236 B.C., which appears to be one of the earliest years recorded in history.

As civilizations grew they became more sophisticated and better definitions for measurement were required. The history of Egypt begins at approximately 3000 B.C. This is when Upper and Lower Egypt became unified under one ruler, Menes, and when the first pyramids were being built. At this time trade of goods was common as well as levies or taxes on them. Definitions for liquid and dry measures were important. Length was also an important measure as well as time. Some key Babylonian units were the *Kush* (cubit) for length, *Sar* (garden-plot) for area and volume, *Sila* for capacity, and *Mana* for weight. At the base of the system is the barleycorn (*She*), used for the smallest unit in length, area, volume, and weight.

## Length

The smallest unit of length is the *She* (barleycorn), which equals about $\frac{1}{360}$ meter.

| | | |
|---|---|---|
| 6 *She* | = | 1 *Shu-Si* (finger) |
| 30 *Shu-Si* | = | 1 *Kush* (cubit—about ½ m.) |
| 6 *Kush* | = | 1 *Gi / Ganu* (reed) |
| 12 *Kush* | = | 1 *Nindan / GAR* (rod—6 m.) |
| 10 *Nindan* | = | 1 *Eshe* (rope) |
| 60 *Nindan* | = | 1 *USH* (360 m.) |
| 30 *USH* | = | 1 *Beru* (10.8 km.) |

## Area and Volume

The basic area unit is the *Sar*, an area of 1 square *Nindan*, or about 36 square meters. The area *She* and *Gin* are used as generalized fractions of this basic unit.

| | | |
|---|---|---|
| 180 *She* | = | 1 *Gin* |
| 60 *Gin* | = | 1 *Sar* (garden plot 1 sq. *Nindan*—36 sq. m.) |
| 50 *Sar* | = | 1 *Ubu* |
| 100 *Sar* | = | 1 *Iku* (1 sq. *eshe*—0.9 acre, 0.36 *Ha*.) |
| 6 *Iku* | = | 1 *Eshe* |
| 18 *Iku* | = | 1 *Bur* |

## Capacity

These units were used for measuring volumes of grain, oil, beer, and so on. The basic unit is the *Sila*, about 1 liter. The semistandard Old Babylonian system used in mathematical texts is derived from the ferociously complex mensuration systems used in the Sumerian period.

| | | |
|---|---|---|
| 180 *She* | = | 1 *Gin* |
| 60 *Gin* | = | 1 *Sila* (1 liter) |
| 10 *Sila* | = | 1 *Ban* |
| 6 *Ban* | = | 1 *Bariga* |
| 5 *Bariga* | = | 1 *Gur* |

## Weight

The basic unit of weight is the *Mana*, about ½ kilogram.

| | | |
|---|---|---|
| 180 *She* | = | 1 *Gin/Shiqlu* (Shekel) |
| 60 *Gin* | = | 1 *Mana* (*Mina*—500 gm.) |
| 60 *Mana* | = | 1 *Gu/Biltu* (talent, load—30 kg.) |

## Royal Cubit Stick

The *royal cubit* (524 millimeters or 20.62 inches) was subdivided in an extraordinarily complicated way. The basic subunit was the *digit*, doubtlessly a finger's breadth, of which there were 28 in the royal cubit.

- Four digits equaled a *palm*, five *a hand*.
- Twelve digits, or three palms, equaled *a small span*.
- Fourteen digits, or one-half a cubit, equaled *a large span*.
- Sixteen digits, or four palms, made *one t'ser*.
- Twenty-four digits, or six palms, were a *small cubit*.

The Egyptians studied the science of geometry to assist them in the construction of the pyramids. The royal Egyptian cubit was decreed to be equal to the length of the forearm from the bent elbow to the tip of the extended middle finger plus the width of the palm of the hand of the pharaoh or king ruling at that time.

The royal cubit master was carved out of a block of granite to endure for all times. Workers engaged in building tombs, temples, pyramids, and so on were supplied with cubits made of wood or granite. The royal architect or foreman of the construction site was responsible for maintaining and transferring the unit of length to workers instruments. They were required to bring back their cubit sticks at each full moon to be compared to the royal cubit master. Failure to do so was punishable by death. Though the punishment prescribed was severe, the Egyptians had anticipated the spirit of the present day system of legal metrology, standards, traceability, and calibration recall.

With this standardization and uniformity of length, the Egyptians achieved surprising accuracy. Thousands of workers were engaged in building the Great Pyramid of Giza. Through the use of cubit sticks, they achieved an accuracy of 0.05%. In roughly 756 feet or 9,069.4 inches, they were within 4 ½ inches.

## Digit

The digit was in turn subdivided. Reading from right to left in the upper register, the 14th digit on a cubit stick was marked off into 16 equal parts. The next digit was divided into 15 parts, and so on, to the 28th digit, which was divided into two equal parts. Thus, measurement could be made to digit fractions with any denominator from two through 16. The smallest division, $\frac{1}{16}$ of a digit, was equal to $\frac{1}{448}$ part of a royal cubit.

Although the Egyptians achieved very good standardization of length, this standardization was regional. There were multiple standards for the cubit, which varied greatly due to the standard they were based on, the length from the tip of the middle finger to the elbow. Variations of the cubit are as follows:

- Arabian (black) cubit of 21.3 inches
- Arabian (hashimi) cubit of 25.6 inches
- Assyrian cubit of 21.6 inches
- Ancient Egyptian cubit of 20.6 inches
- Ancient Israeli cubit of 17.6 inches
- Ancient Grecian cubit of 18.3 inches
- Ancient Roman cubit of 17.5 inches

Of these seven cubits the variation from the longest to the shortest was 8.1 inches, with an average value of 20.36 inches. These variations made trade difficult between different regions. As time evolved there became greater need for trade on a regional basis. The need for more sophistication and accuracy became greater.

## MEASUREMENT PROGRESS IN THE LAST 2000 YEARS

Efforts at standardizing measurement evolved around the world, not just in Egypt. English, French, and American leaders strived to bring order to their marketplaces and governments.

**732** King of Kent—The measurement of an acre is in common use.

**960** Edgar the Peaceful decree, "All measure must agree with standards kept in London and Winchester."

**1215** King John agrees to have national standards of weights and measures incorporated into the Magna Carta.

**1266** Henry III declares in an act that:

- One penny should weigh the same as 32 grains of wheat.
- There would be 20 pennies to the ounce.
- There would be 12 ounces to the pound.
- There would be eight pounds to the weight of one gallon of wine.

**1304** Edward I declares in a statute that:

- For medicines, one pound equals 12 ounces (apothecaries, still used in United States).
- For all other liquid and dry measures one pound equaled 15 ounces.
- One ounce still equals 20 pennies.

**1585** In his book *The Tenth*, Simon Stevin suggests that a decimal system should be used for weights and measures, coinage, and divisions of the degree of arc.

**1670** Authorities give credit for originating the metric system to Gabriel Mouton, a French vicar.

**1790** Thomas Jefferson proposes a decimal-based measurement system for the United States. France's Louis XVI authorizes scientific investigations aimed at a reform of French weights and measures. These investigations lead to the development of the first metric system.

**1792** The U.S. Mint is formed to produce the world's first decimal currency (the U.S. dollar consisting of 100 cents).

**1795** France officially adopts the metric system.

**1812** Napoleon temporarily suspends the compulsory provisions of the 1795 metric system adoption.

**1824** George IV, in a Weights and Measures Act (5 GEO IV c 74) establishes the "Imperial System of Weights and Measures," which is still used.

**1840** The metric system is reinstated as the compulsory system in France.

**1866** The use of the metric system is made legal (but not mandatory) in the United States by the (Kasson) Metric Act of 1866. This law also makes it unlawful to refuse to trade or deal in metric quantities.

## STANDARDS, COMMERCE, AND METROLOGY

Standards of measurement before the 1700s were local and often arbitrary, making trade between countries—and even cities—difficult. The need for standardization as an aid to commerce became apparent during the Industrial Revolution. Early standardization and metrology needs were based on military requirements, especially those of large maritime powers such as Great Britain and the United States. A major task of navies in the eighteenth and nineteenth centuries was protection of their country's international trade, much of which was carried by merchant ships. Warships would sail with groups of merchant ships to give protection from pirates, privateers, and ships of enemy nations; or they would sail independently to "show the flag" and enforce the right of free passage on the seas. A typical ship is the frigate USS *Constitution*. She was launched in 1797 and armed with 34 24-pound cannon and 20 32-pound cannon. (The size of cannon in that era was determined by the mass, and therefore the diameter, of the spherical cast-iron shot that would fit in the bore.) For reasons related to accuracy, efficiency, and economy, the bores of any particular size of cannon all had to be the same diameter. Likewise, the iron shot had to be the same size. If a cannonball was too large, it would not fit into the muzzle; too small, and it would follow an unpredictable trajectory when fired. The requirements of ensuring that dimensions were the same led to the early stages of a modern metrology system, with master gages, transfer standards, and regular comparisons. Figure 1.1 is an example of a length standard that probably dates from this era. This one, mounted on the wall of the Royal Observatory at Greenwich, England, is one of several that were placed in public places by the British government for the purpose of standardizing dimensional measurements. It has standards for three and six inches, one and two feet, and one yard. In use, a transfer standard would be placed on the supports, and presumably it should fit snugly between the flats of the large posts. The actual age of this standard is now unknown.

**Figure 1.1** Early dimensional standard at the Royal Observatory, Greenwich, England.
(Photo by Graeme C. Payne.)

Over time, as measurements became standardized within countries, the need arose to standardize measurements between countries. A significant milestone in this effort was the adoption of the Convention of the Metre treaty in 1875. This treaty set the framework for and still governs the international system of weights and measures. It can be viewed as one of the first voluntary standards with international acceptance, and possibly the most important to science, industry, and commerce. The United States was one of the first nations to adopt the Metre Convention.

**1875** The Convention of the Metre is signed in Paris by 18 nations, including the United States. The Meter Convention, often called the Treaty of the Meter in the United States, provides for improved metric weights and measures and the establishment of the General Conference on Weights and Measures (CGPM) devoted to international agreement on matters of weights and measures.

**1878** Queen Victoria declares the Troy pound illegal. Commercial weights could only be of the quantity of 56 lb, 28 lb, 14 lb, 7 lb, 4 lb, 2 lb, 1 lb, 8 oz, 4 oz, 2 oz, and so on.

**1889** As a result of the Meter Convention, the United States receives a prototype meter and kilogram to be used as measurement standards.

**1916** The Metric Association is formed as a nonprofit organization advocating adoption of the metric system in U.S. commerce and education. The organizational name started as the American Metric Association and was changed to the U.S. Metric Association (USMA) in 1974.

**1954** The International System of Units (SI) begins its development at the tenth CGPM. Six of the new metric base units are adopted.

**1958** A conference of English-speaking nations agrees to unify their standards of length and mass, and define them in terms of metric measures. The American yard was shortened and the imperial yard was lengthened as a result. The new conversion factors are announced in 1959 in the *Federal Register*.

## HISTORY OF QUALITY STANDARDS

In some respects, the concept we call *quality* has been with humankind through the ages. Originally aspects of quality were passed on by word of mouth, from parent to child, and later from craftsman to apprentice. As the growth of agriculture progressed, people started settling in villages and resources were available to support people who were skilled at crafts other than farming and hunting. These craftsmen and artisans would improve their skill by doing the same work repeatedly and make improvements based on feedback from customers. The main pressure for quality was social because the communities were small and trade was local. A person's welfare was rooted in his or her reputation as an honest person who delivered a quality product—and the customers were all neighbors, friends, or family.

The importance and growth of quality and measurement probably increased with the development of the first cities and towns, on the order of 6000 to 7000 years ago. Astronomy, mathematics, and surveying were important trades. Standard weights and measures were all developed as needed. All of these were important for at least three activities: commerce, construction, and taxation. The systems of measuring also represented an important advance in human thought because it is an advance from simple counting to representing units that can be subdivided at least to the resolution available to the unaided eye. In Egypt, systems of measurement were good enough 5000 years ago to survey and construct the pyramids at Giza with dimensional inaccuracies on the order of 0.05%. This was made possible by the use of regular calibration and traceability. Remember, the cubit rules used by the builders were required to be periodically compared to the Pharaoh's granite master cubit. Juran mentions evidence of written specifications for products about 3500 years ago, and Bernard Grun notes regulations about the sale of products (beer, in this case) about the same time. Standardization of measurements increased gradually as well, reaching a peak in the later years of the Roman Empire. (It is often said, humorously, that the standard-gauge spacing of railway tracks in Europe and North America is directly traceable to the wheel spacing of Roman war chariots.)

Measurement science and quality experienced a resurgence as Europe emerged from the Dark Ages. Again, some of the driving forces were commerce, construction, and taxation. Added to these were military requirements and the needs of the emerging fields of science. Many of the quality aspects were assumed by trade and craft guilds. The guilds set specifications and standards for their trades, and developed a training system that persists to this day: the system of apprentices, journeymen, and master craftsmen. Guilds exercised quality control through inspection, but often stifled quality improvement and product innovation.

The Industrial Revolution accelerated the growth of both quality and measurement. Quantities of manufactured items increased, but each part was still essentially custom made. Even while referring to a master template, it was difficult to construct parts that could be randomly selected and assembled into a functioning device. By the mid-1700s the capability of a craftsman to produce substantially identical parts was demonstrated in France. In 1789, Eli Whitney used this capability in the United States when he won a government contract to provide 10,000 muskets with interchangeable parts, but it took him 10 years to fill the contract. By 1850 it was possible for a skilled machinist to make hundreds of repeated parts with dimensional uncertainties of no more than ±0.002 inch. Those parts were still largely handmade, one at a time.

## MEASUREMENT AND THE INDUSTRIAL REVOLUTION

The Industrial Revolution began around 1750. Technology began to progress quickly with materials, energy, time, architecture, and man's relationship with the earth. Industry began to quickly evolve. With a growing population the need for clothing, transportation, medicines, and food drove industry to find better, more efficient methods to support this need. Technology had evolved sufficiently to support this growth.

During this time there was tremendous growth of discoveries in quantum mechanics and molecular, atomic, nuclear, and particle physics. These discoveries laid the groundwork for much of the seven base units of the current International System of Units (SI). These seven units are well-defined and dimensionally independent. These units are the meter (m), kilogram (kg), second (s), ampere (A), kelvin (K), mole (mol), and candela (cd).

During the middle 1800s there was a lot of progress with temperature and thermodynamics. Note the speed of discovery.

**1714** Mercury and alcohol thermometer are invented by Daniel Gabriel Fahrenheit.

**1821** Thermocouples are invented by Thomas Johann Seebeck. The discovery that metals have a positive temperature coefficient of resistance, which led to the use of platinum as a temperature indicator (PRT) is made by Humphrey Davy.

**1834** Lord Kelvin formulates of the Second Law of Thermodynamics.

**1843** Discovers the mechanical equivalent of heat.

**1848** Lord Kelvin discovers the absolute zero point of temperature (0 K).

**1889** Platinum thermometers are defined by many different freezing points and boiling points of ultra pure substances such as the freezing point of $H_2O$ at 0°C, the boiling point of $H_2O$ at 100°C, and the boiling point of sulfur at 444.5°C.

**1900** The Blackbody radiation law is formulated by Max Planck.

**1927** The Seventh CGPM adopted the International Temperature Scale of 1927.

There were many more discoveries during this period that covered electromagnetic emissions, radioactivity, nuclear chain reactions, superconductivity, and others. The following is a list of some of the properties and quantities of electricity that were defined.

**1752** Benjamin Franklin proves that lightning and the spark from amber are the same thing.

**1792** Alessandro Volta shows that when moisture comes between two different metals electricity is created. This led him to invent the first electric battery, the voltaic pile, which he made from thin sheets of copper and zinc separated by moist pasteboard. The unit of electrical potential, the *volt*, is named after Volta.

**1826** George Simon Ohm states Ohm's law of electrical resistance.

**1831** Michael Faraday discovers the first method of generating electricity by means of motion in a magnetic field.

**1832** Faraday discovers the laws of electrolysis.

**1882** A New York street is lit by electric lamps.

**1887** The photoelectric effect is discovered by Heinrich R. Hertz.

**1897** J. J. Thomson discovers the electrons.

**1930** Paul A. M. Dirac introduces the electron hole theory.

With the advancement in the discoveries of these physical phenomena, technology took advantage and developed products around these discoveries. With these advancements better measurements were needed. To build a single machine, an inventor could probably get by without much standardization of his measurements, but to build multiple machines, using parts from multiple suppliers, measurements had to be standardized to a common entity.

Metrology is keeping up with physics and industry. Recent successes in the industry that wouldn't be possible without metrology include the following: Fission and fusion are being refined for both weapons and as a source of energy. Semiconductors are being developed, refined, and applied on a larger scale with the invention of the integrated circuit, solar cells, light-emitting diodes, and liquid crystal displays. Communication technology has developed through application of satellites and fiber optics. Lasers have been invented and applied in useful technology to include communication, medicine, and industrial applications. The ability to place a humanmade object outside the Earth's atmosphere started the space race and has led to placing a man on the moon, satellites used for multiple purposes, probes to other planets, the space shuttle, and space stations. A few spacecraft have passed beyond the farthest known planets and are headed into interstellar space. One thing is certain: as technology continues to grow, measurement challenges will grow proportionally.

## MILESTONES IN U.S. FOOD AND DRUG LAW HISTORY

From the beginnings of civilization people have been concerned about the quality and safety of foods and medicines. In 1202, King John of England proclaimed the first English food law, the Assize of Bread, which prohibited adulteration of bread with such ingredients as ground peas or beans. Regulation of food in the United States dates from early colonial times. Federal controls over the drug supply began with inspection of imported drugs in 1848. The following chronology describes some of the milestones in the history of food and drug regulation in the United States.

**1820** Eleven physicians meet in Washington, D.C., to establish the U.S. Pharmacopeia, the first compendium of standard drugs for the United States.

**1848** Drug Importation Act passed by Congress requires U.S. Customs Service inspection to stop entry of adulterated drugs from overseas.

**1862** President Lincoln appoints a chemist, Charles M. Wetherill, to serve in the new Department of Agriculture. This was the beginning of the Bureau of Chemistry, the predecessor of the Food and Drug Administration.

**1880** Peter Collier, chief chemist, U.S. Department of Agriculture, recommends passage of a national food and drug law, following his own food adulteration investigations. The bill was defeated, but during the next 25 years more than 100 food and drug bills were introduced in Congress.

**1883** Dr. Harvey W. Wiley becomes chief chemist, expanding the Bureau of Chemistry's food adulteration studies. Campaigning for a federal law, Wiley is called the Crusading Chemist and Father of the Pure Food and Drugs Act. He retired from government service in 1912 and died in 1930.

**1902** The Biologics Control Act is passed to ensure purity and safety of serums, vaccines, and similar products used to prevent or treat diseases in humans.

Congress appropriates $5,000 to the Bureau of Chemistry to study Chemical Perservatives and Colors and their effects on digestion and health. Wiley's studies draw widespread attention to the problem of food adulteration. Public support for passage of a federal food and drug law grows.

**1906** The original Food and Drugs Act is passed by Congress on June 30 and signed by President Theodore Roosevelt. It prohibits interstate commerce in misbranded and adulterated foods, drinks and drugs. The Meat Inspection Act is passed the same day. Shocking disclosures of unsanitary conditions in meat packing plants, the use of poisonous preservatives and dyes in foods, and cure-all claims for worthless and dangerous patent medicines were the major problems leading to the enactment of these laws.

**1927** The Bureau of Chemistry is reorganized into two separate entities. Regulatory functions are located in the Food, Drug, and Insecticide Administration, and nonregulatory research is located in the Bureau of Chemistry and Soils.

**1930** The name of the Food, Drug, and Insecticide Administration is shortened to Food and Drug Administration (FDA) under an agricultural appropriations act.

**1933** FDA recommends a complete revision of the obsolete 1906 Food and Drugs Act. The first bill is introduced into the Senate, launching a five-year legislative battle.

**1937** Elixir of Sulfanilamide, containing the poisonous solvent diethylene glycol, kills 107 persons, many of whom are children, dramatizing the need to establish drug safety before marketing and to enact the pending food and drug law.

**1938** The Federal Food, Drug, and Cosmetic (FDC) Act of 1938 is passed by Congress, containing new provisions:

- Extending control to cosmetics and therapeutic devices
- Requiring new drugs to be shown safe before marketing—starting a new system of drug regulation
- Eliminating the Sherley Amendment requirement to prove intent to defraud in drug misbranding cases
- Providing that safe tolerances be set for unavoidable poisonous substances
- Authorizing standards of identity, quality, and fill-of-container for foods
- Authorizing factory inspection
- Adding the remedy of court injunctions to the previous penalties of seizures and prosecutions

Under the Wheeler-Lea Act, the Federal Trade Commission is charged with overseeing advertising associated with products otherwise regulated by FDA, with the exception of prescription drugs.

**1943** In *U.S. v. Dotterweich*, the Supreme Court rules that the responsible officials of a corporation, as well as the corporation itself, may be prosecuted for violations. It need not be proven that the officials intended, or even knew of, the violations.

**1949** FDA publishes Guidance to Industry for the first time. This guidance, "Procedures for the Appraisal of the Toxicity of Chemicals in Food," came to be known as the black book.

**1951** Durham-Humphrey Amendment defines the kinds of drugs that cannot be safely used without medical supervision and restricts their sale to prescription by a licensed practitioner.

**1958** Food Additives Amendment enacted, requiring manufacturers of new food additives to establish safety. The Delaney proviso prohibits the approval of any food additive shown to induce cancer in humans or animals.

FDA publishes in the Federal Register the first list of Substances generally recognized as safe (GRAS). The list contains nearly 200 substances.

**1959** U.S. Cranberry crop recalled three weeks before Thanksgiving for FDA tests to check for aminotriazole, a weed killer found to cause cancer in laboratory animals. Cleared berries were allowed a label stating that they had been tested and had passed FDA inspection, the only such endorsement ever allowed by FDA on a food product.

**1962** Thalidomide, a new sleeping pill, is found to have caused birth defects in thousands of babies born in western Europe. News reports on the role of

Dr. Frances Kelsey, FDA medical officer, in keeping the drug off the U.S. market, arouse public support for stronger drug regulation.

Kefauver-Harris Drug Amendments passed to ensure drug efficacy and greater drug safety. For the first time, drug manufacturers are required to prove to FDA the effectiveness of their products before marketing them. The new law also exempts from the Delaney proviso animal drugs and animal feed additives shown to induce cancer, but which leave no detectable levels of residue in the human food supply.

Consumer Bill of Rights is proclaimed by President John F. Kennedy in a message to Congress. Included are the right to safety, the right to be informed, the right to choose, and the right to be heard.

**1972** Over-the-counter drug review begun to enhance the safety, effectiveness and appropriate labeling of drugs sold without prescription.

Regulation of biologics—including serums, vaccines, and blood products-is transferred from NIH to FDA.

**1976** Medical Device Amendments passed to ensure safety and effectiveness of medical devices, including diagnostic products. The amendments require manufacturers to register with FDA and follow quality control procedures. Some products must have premarket approval by FDA; others must meet performance standards before marketing.

Vitamins and Minerals Amendments (Proxmire Amendments) stop FDA from establishing standards limiting potency of vitamins and minerals in food supplements or regulating them as drugs based solely on potency.

**1978** Good manufacturing practices become effective.

**1979** Good laboratory practices become effective.

**1983** Orphan Drug Act passed, enabling FDA to promote research and marketing of drugs needed for treating rare diseases.

**1984** Fines Enhancement Laws of 1984 and 1987 amend the U.S. Code to greatly increase penalties for all federal offenses. The maximum fine for individuals is now $100,000 for each offense and $250,000 if the violation is a felony or causes death. For corporations, the amounts are doubled.

**1988** Food and Drug Administration Act of 1988 officially establishes FDA as an agency of the Department of Health and Human Services with a Commissioner of Food and Drugs appointed by the president with the advice and consent of the Senate and broadly spells out the responsibilities of the secretary and the commissioner for research, enforcement, education, and information.

The Prescription Drug Marketing Act bans the diversion of prescription drugs from legitimate commercial channels. Congress finds that the resale of such drugs leads to the distribution of mislabeled, adulterated, subpotent, and counterfeit drugs to the public. The new law requires drug wholesalers to be licensed by the states; restricts reimportation from other countries; and bans sale, trade, or purchase of drug samples, and traffic or counterfeiting of redeemable drug coupons.

**1990** Safe Medical Devices Act is passed, requiring nursing homes, hospitals, and other facilities that use medical devices to report to FDA incidents that suggest that a medical device probably caused or contributed to the death, serious illness, or serious injury of a patient. Manufacturers are required to conduct postmarket surveillance on permanently implanted devices whose failure might cause serious harm or death, and to establish methods for tracing and locating patients depending on such devices. The act authorizes FDA to order device product recalls and other actions.

**1995** FDA declares Cigarettes to be drug delivery devices. Restrictions are proposed on marketing and sales to reduce smoking by young people.

**1996** Federal Tea Tasters Repeal Act repeals the Tea Importation Act of 1897 to eliminate the Board of Tea Experts and user fees for FDA's testing of all imported tea. Tea itself is still regulated by FDA.

**1997** Food and Drug Administration Modernization Act reauthorizes the Prescription Drug User Fee Act of 1992 and mandates the most wide-ranging reforms in agency practices since 1938. Provisions include measures to accelerate review of devices, regulate advertising of unapproved uses of approved drugs and devices, and regulate health claims for foods.

**1998** First phase to Consolidate FDA Laboratories nationwide from 19 facilities to nine by 2014 includes dedication of the first of five new regional laboratories.

## References

BIPM (International Bureau of Weights and Measures). 2000. The Convention of the Metre. www.bipm.fr/enus/1_Convention/foreword.html

Bertermann, Ralph E. 2002. *Understanding current regulations and international standards; Calibration compliance in FDA regulated companies.* Mt. Prospect, IL: Lighthouse Training Group.

Columbia Encyclopedia. 2003. "Measurement" Expanded Columbia Electronic Encyclopedia, Columbia University Press. Accessed through the History Channel, www.historychannel.com/perl/print_book.pl?ID-100597

Evans, James R. and William M. Lindsay. 1993. *The management and control of quality*, 2nd ed. St. Paul, MN: West Publishing Company.

Juran, Joseph M. 1997. Early SQC: A historical supplement. *Quality Progress*, 30, (September): 73–81.

Juran, Joseph M. 1999. *Juran's quality handbook.* Wilton, CT: McGraw-Hill.

NBS (National Bureau of Standards). 1975. *The International Bureau of Weights and Measures 1875–1975.* NBS Special Publication 420. Washington, DC: U.S. Government Printing Office.

NCSL International (NCSLI). 2003. "The Royal Egyptian Cubit." www.ncsli.org/misc/cubit.cfm

NIST (National Institute of Standards and Technology). 2003. Frequently asked questions about the Malcolm Baldrige National Quality Award. www.nist.gov/public_affairs/factsheet/baldfaqs.htm

Swade, Doron. 2000. *The Difference Engine: Charles Babbage and the quest to build the first computer.* London: Viking Penguin Books.

U. S. Navy. 1966. *Why calibrate?* Washington, DC: Department of the Navy. Training film number MN-10105.

U. S. Navy. 2003. "United States Ship Constitution." Washington, DC: Department of the Navy www.ussconstitution.navy.mil/statpage.htm

# Part II
## Quality Systems

| | |
|---|---|
| **Chapter 2** | The Basics of a Quality System |
| **Chapter 3** | Quality Standards and Their Evolution |
| **Chapter 4** | Quality Documentation |
| **Chapter 5** | Calibration Procedures and Equipment Manuals |
| **Chapter 6** | Calibration Records |
| **Chapter 7** | Calibration Certificates |
| **Chapter 8** | Quality Manuals |
| **Chapter 9** | Traceability |
| **Chapter 10** | Calibration Intervals |
| **Chapter 11** | Calibration Standards |
| **Chapter 12** | Audit Requirements |
| **Chapter 13** | Scheduling and Recall Systems |
| **Chapter 14** | Labels and Equipment Status |
| **Chapter 15** | Training |
| **Chapter 16** | Environmental Controls |
| **Chapter 17** | Industry-Specific Requirements |
| **Chapter 18** | Computers and Automation |

# Chapter 2
# The Basics of a Quality System

What is a quality system? Who says a laboratory or department has to have one, and why? Let us answer the second question first. According to ANSI/ISO 17025-1999, chapter 4.2.1: "The laboratory shall establish, implement and maintain a quality system . . ." Q9001-2000, chapter 4.1 says, "The organization shall establish, document, implement and maintain a quality management system . . ." Z540 says in chapter 5.1: "The laboratory shall establish and maintain a quality system . . ." RP-6 states in chapter 5: "A calibration control system should include . . ." And M1-1996 states in chapter 3.3: "The organization and operation of the calibration system shall be consistent with ANSI Standard Z1.15, *Generic Guidelines for Quality Systems* . . ." All of these quality organizations, which have major roles in metrology systems, require that a quality system *shall*, not *should*, be maintained.

Now to answer the first question. The basic premise and foundation of a good quality system is to *Say what you do, do what you say, record what you did, check the results, and act on the difference*. In simple terms: *Say what you do* means write, in detail, how to do your job. This includes calibration procedures, standard operating procedures (SOPs), protocols, work instruction, work cards, and so on. *Do what you say* means follow the documented procedures or instructions every time you calibrate, validate, or perform a function that follows specific written instructions. *Record what you did* means precisely record the results of your measurements and adjustments, including what your standard(s) read or indicated both before and after adjustment. *Check the results* means make certain the inspection, measuring, and test equipment (IM&TE) meets the tolerances, accuracies, or upper/lower limits specified in your procedures or instructions. *Act on the difference* means if the IM&TE is out of tolerance, does not meet the specified accuracies, or exceeds the upper/lower test limits written in your procedures, you're required to inform the user because he or she may have to reevaluate manufactured goods, change a process, or recall a product and/or previously calibrated equipment that used that particular standard.

To help ensure that all operations throughout a metrology department, calibration laboratory, or work area where calibrations are accomplished occur in a stable manner, one needs to establish a quality management system. The effective operation of such a system should result in stable processes and, therefore, in a consistent output from those processes. Once stability and consistency are achieved, then it's possible to initiate improvements. Each calibration technician must follow the calibration procedure as it is written, collect the data as it is found, and document the results each and every time. Then trends can be evaluated, intervals changed, and improvements to the processes and procedures implemented. This is the basic foundation of a quality system. Although these steps check your process for correctness, it never validates your process, which you define, for correctness relative to accepted practices in that specific field. For example, if a quality system states, "Calibration on reference standards will be performed every five years," as long as these reference standards are calibrated every five years, this will meet the requirements of the quality standard. It will not meet accepted metrology practices of defining calibration intervals based on performance of the standard in question, unless you use the historical data and check the results and act on the difference when compiling your information. Trends will either support the calibration interval, or give sufficient data to increase or decrease those intervals.

Some of the quality tools that help determine the status of information, data, intervals, and so on are check sheets, Pareto charts, flowcharts, cause-and-effect diagrams (also called fish bone diagrams), histograms, scatter diagrams, and control charts.

## QUALITY TOOLS

Various quality tools are available to assist a company, department, or laboratory to continually improve, update, and adapt its programs, policies, and procedures to maintain a business advantage, identify problems before they affect the bottom line, and help keep the quality system alive and healthy.

### Checksheet

The function of a checksheet is to present information in an efficient, graphical format. This may be accomplished with a simple listing of items, however, the utility of the checksheet may be significantly enhanced, in some instances, by incorporating a depiction of the system under analysis into the form.

### Pareto Chart

Pareto charts are extremely useful because they can be used to identify those factors that have the greatest cumulative effect on the system and thus screen out the less significant factors in an analysis. Ideally, this allows the user to focus attention on a few important factors in a process. Pareto charts are created by plotting the cumulative frequencies of the relative frequency data (event count data), in descending order. When this is done, the most essential factors for the analysis are graphically apparent in an orderly format.

### Flowchart

Flowcharts are pictorial representations of a process. By breaking the process down into its constituent steps, flowcharts can be useful in identifying where errors are likely to be found in the system.

## Cause-and-Effect Diagram

This diagram, also called an Ishikawa diagram (or fish bone diagram), is used to associate multiple possible causes with a single effect. Thus, given a particular effect, the diagram is constructed to identify and organize possible causes for it.

The primary branch represents the effect (the quality characteristic that is intended to be improved and controlled) and is typically labeled on the right side of the diagram. Each major branch of the diagram corresponds to a major cause (or class of causes) that directly relates to the effect. Minor branches correspond to more detailed causal factors. This type of diagram is useful in any analysis, as it illustrates the relationship between cause and effect in a rational manner.

## Histogram

Histograms provide a simple, graphical view of accumulated data, including their dispersion and central tendency. Histograms are easy to construct and provide the easiest way to evaluate the distribution of data.

## Scatter Diagram

Scatter diagrams are graphical tools that attempt to depict the influence that one variable has on another. A common scatter diagram usually displays points representing the observed value of one variable corresponding to the value of another variable.

## Control Chart

The control chart is the fundamental tool of statistical process control, as it indicates the range of variability that is built into a system (known as *common cause* or *random variation*). Thus, it helps determine whether or not a process is operating consistently or if a special cause or nonrandom event has occurred to change the process mean or variance. The process control chart may also be called a *process description chart*.

The bounds of the control chart are marked by upper and lower control limits that are calculated by applying statistical formulas to data from the process. Data points that fall outside these bounds represent variations due to special causes, which can typically be found and eliminated. On the other hand, improvements in common cause variation require fundamental changes in the process, which leads to process improvement techniques.

Various tools are available for accomplishing process improvements. Any quality system should be a living entity: constantly changing, improving, and adapting to the business environment where it is used. Without process improvement, the system becomes stagnant and will fall behind the times.

# PROCESS IMPROVEMENT TECHNIQUES

## PDCA

The plan do check act (PDCA or PDSA) cycle was originally conceived by Walter Shewhart in the 1930s and later adopted by W. Edwards Deming. The model provides a framework for the improvement of a process or system. It can be used to guide the entire improvement project or to develop specific projects once target improvement areas have been identified.

The PDCA cycle is designed to be used as a dynamic model. The completion of one turn of the cycle flows into the beginning of the next. Following in the spirit of *continuous* quality improvement, the process can always be reanalyzed and a new test of change can begin. This continual cycle of change is represented in the ramp of improvement. Using what we learn in one PDCA trial, we can begin another, more complex trial.

The first step is to plan. In this phase, analyze what you intend to improve, looking for areas that hold opportunities for change. Choose areas that offer the most return for the effort—the biggest bang for your buck. To identify these areas for change, consider using a flowchart or Pareto chart.

Next, do what is planned. Carry out the change or test, preferably on a small scale.

The check or study phase is a crucial step in the PDCA cycle. After you have implemented the change for a short time, determine how well it is working. Is it really leading to improvement in the way you had hoped? Decide on several measures with which you can monitor the level of improvement. Run charts can be helpful with this measurement.

After planning a change, implementing, and then monitoring it, you must decide whether it is worth continuing that particular change. If it consumed too much time, was difficult to adhere to, or even led to no improvement, consider aborting the change and planning a new one. If the change led to a desirable improvement or outcome, however, consider expanding the trial to a different area or slightly increasing your complexity. Act on your discovery. This sends you back into the Plan phase of the cycle.

## Brainstorming

Most problems are not solved automatically by the first idea that comes to mind. To get to the best solution it is important to consider many possible solutions. One of the best ways to do this is called *brainstorming*. Brainstorming is the act of defining a problem or idea and coming up anything related to the topic, no matter how remote a suggestion may sound. All of these ideas are recorded and evaluated only after the brainstorming is completed.

To begin brainstorming, gather a group. Select a leader and a recorder (it may be the same person). Define the problem or idea to be brainstormed. Make sure everyone is clear on the topic being explored. Set up the rules for the session. They should include: the leader is in control; everyone can contribute; no one will insult, demean, or evaluate another participant or a response; no answer is wrong; each answer will be recorded unless it is a repeat; and a time limit will be set and adhered to.

Start the brainstorming! Have the leader select members of the group to share their answers. The recorder should write down all responses, if possible so everyone can see them. Make sure not to evaluate or criticize any answers until done brainstorming.

Once the brainstorming is finished, go through the results and begin evaluating the responses. Some initial qualities to look for when examining the responses include: looking for any answers that are repeated or similar; grouping like concepts together; and eliminating responses that definitely do not fit. Now that the list is pruned, discuss the remaining responses as a group.

The mere use of the quality control tools does not necessarily constitute a quality program. Thus, to achieve lasting improvements in quality, it is essential to establish a system that will continuously promote quality in all aspects of its operation.

The following chapters will explain in detail what is required in calibration procedures, records, certificates, and a quality manual. It's one thing to say that a quality system is required, it's another to explain what it is, what needs to be in it, what the requirements are in different standards and regulations, and how to apply them in a systematic approach that can be tailored for your individual situations. No two calibration laboratories or metrology departments/groups are the same, however, they may have similar guidelines as they pertain to metrology. Each has specific requirements that need to be met, while generally providing the same basic function for the company—traceable calibration measurements.

## References

ANSI/ASQ M1-1996, *American National Standard for calibration systems*. Milwaukee: ASQC Quality Press.

ANSI/ISO 17025-1999, *American National Standard—General requirements for the competence of testing and calibration laboratories*. Milwaukee: ASQ Quality Press.

ANSI/ISO/ASQ Q9001-2000, *American National Standard—Quality management systems— Requirements*. Milwaukee: ASQ Quality Press.

ANSI/NCSL Z540-1-1994, *American National Standard for calibration—Calibration laboratories and measuring and test equipment—General requirements*. Boulder, CO: National Conference of Standards Laboratories.

BSR/ISO/ASQ Q10012:2003(E), *Measurement management systems—Requirements for measurement processes and measuring equipment*. Milwaukee: ASQ Quality Press.

Bucher, Jay L. 2000. *When your company needs a metrology program, but can't afford to build a calibration laboratory . . . What can you do?* Boulder, CO: National Conference of Standards.

www.dartmouth.edu/~ogehome/CQI/PDCA.html

deming.eng.clemson.edu/pub/tutorials/qctools/qct.htm

The Healthcare Metrology Committee. 1999. *Calibration control systems for the biomedical and pharmaceutical industry*. Recommended Practice RP-6. Boulder, CO: National Conference of Standards Laboratories.

Hirano, Hiroyuki. 1995. *5 pillars of the visual workplace: The sourcebook for 5S implementation*. Translated by Bruce Talbot. Shelton, CT: Productivity Press.

Kimothi, S. K. 2002. *The uncertainty of measurements, physical and chemical metrology impact and analysis*. Milwaukee: ASQ Quality Press.

NIST Special Publication 811, 1995 Edition, *Guide for the use of the International System of Units (SI)*. Gaithersburg, MD: National Institute of Standards and Technology.

Pinchard, Corinne. 2001. *Training a calibration technician . . . in a metrology department?* Boulder, CO: National Conference of Standards Laboratories.

projects.edtech.sandi.net/staffdev/tpss99/processguides/brainstorming.html

# Chapter 3
# Quality Standards and Their Evolution

## WHAT ARE QUALITY STANDARDS AND WHY ARE THEY IMPORTANT?

Businesses exist to produce products and provide services.[1] Products or services will be purchased if they meet the needs and requirements of the customer. Over time, certain business management practices have been observed that enable a business to consistently produce product at required quality levels. *Quality management system standards*, as we know them today, are primarily written descriptions of those practices, stated as sets of requirements. In theory, a business that meets the requirements of a quality management standard should be capable of producing its products (or providing its service) at a consistent level of quality.

> Be aware that even though the phrases *quality standard* or *quality system* are often used, they are shorthand forms of referring to the *quality management system standards* or the *quality management system* of an organization. These terms should *not* be taken to refer to the technical requirements of a product that constitute its product quality attributes.

There are two different classes of quality standards: those that are required by law or regulation and those that are voluntary. A government law, or a regulation of a government agency, may include or specify quality standard requirements. In these cases, a business *must* comply if it is either in a regulated industry or wishes to sell products or services to the government. In all other cases, a quality standard is technically voluntary. This means a business *may* follow it if it chooses to or may ignore it. The voice of the customers and the forces of competition have an effect though. In the modern business environment, relations with many customers are governed by the terms of purchase orders or other contracts. Relations between competing companies are sensitive to conditions in the overall market. As a practical matter, some quality standards are voluntary only to the extent that the organization can afford to lose business by not following them. All of this results in several forces that are driving the importance of the voluntary quality standards.

- **Customer Requirements.** On a retail level, customers will not buy if the product or service does not meet their requirements. Large customers, such as other businesses, can state specific quality requirements in a request for quote and then in a purchase order. In many cases that is done now with language stating that the supplier must provide proof that it meets the requirements of a specific quality standard. For example, the largest automobile manufacturers in North America and Europe currently require that their primary suppliers be registered to ISO/TS 16949 (although implementation is delayed in some cases).

- **Competitive Advantage.** A second force is a desire to stand out in a competitive environment. There are many industries where there is little to distinguish one supplier from all of the others. A company may decide to adopt a quality standard in order to gain a marketing advantage.

- **Response to Competition.** Another force is response to the pressures of competition. When a critical mass of businesses in an industry sector formally adopts a quality standard, this creates pressure on the others to do likewise. The alternative is often loss of business and the decline of the companies that do not respond. It is now a generally accepted axiom that if a business expects to be competitive on a national or international level, then registration to an appropriate quality standard is a minimum business necessity.[2]

- **Government Requirements.** In the United States, the federal government is a very large customer and a regulator of certain industries. As customers, many government agencies require suppliers to adhere to quality requirements. The Department of Defense (DOD), for example, now uses the ISO 9000 series quality management standards in many procurement contracts. Other government agencies have their own quality requirements that must be followed by the industries they regulate. Chief among these agencies are the Food and Drug Administration (FDA), the Nuclear Regulatory Commission (NRC), and the Federal Aviation Administration (FAA). Some agencies also apply external quality standards to themselves. In September 1999, the National Aeronautics and Space Administration (NASA) became the first government agency in the world to be fully registered to ISO 9001 at every location.

## THE ROLE OF ISO

Many standards documents are referred to as ISO standards. This indicates that the standard was developed by or in cooperation with the International Organization for Standardization (ISO), headquartered in Geneva, Switzerland.[3] ISO is a nongovernmental organization. The members are organizations: the national body "most representative of standardization in its country."[4] For example, the organization representing the United States is the American National Standards Institute (ANSI). ISO was created in 1946 as a replacement for an earlier organization that ceased operations in 1942.[5]

The purpose of ISO is to remove technical barriers to trade and to otherwise aid and improve international commerce by harmonizing existing standards and developing new voluntary standards where needed. Many of the more than 13,700 ISO standards affect people every day because they cover areas as diverse as the size of credit cards, the properties of camera film, the threads on bolts in the engine of a car, the shipping containers used for international freight, the format of the music recorded on compact disks, and business management systems for quality, environmental safety, and social responsibility. One of the most important business standards to people working in metrology is ISO/IEC 17025:1999, *Requirements for the Competence of Testing and Calibration Laboratories.*

It is important to note some differences in terminology with respect to third-party audits of quality systems. The terms used may be different due to historic language and cultural differences. In the United States, an organization is *registered* to be in conformance to a standard. In most of the rest of the world, the organization is *certified*. In this context, then, *registration* and *certification* mean the same thing—a qualified third-party assessor formally states that the organization meets the requirements of the specified standard. When the conformance standard is ISO/IEC 17025, the organization is always *accredited*. Accreditation of a testing or calibration organization always includes evaluation of technical competence to perform the work listed in the scope.

Voluntary product and service technical standards developed by ISO and other organizations—notably the International Electrotechnical Commission (IEC) and the International Telecommunications Union (ITU)—aid international commerce by specifying technical requirements. These requirements are often expressed in terms of a physical quantity such as dimension, mass, voltage, temperature, or chemical composition. The role of metrology is to ensure that measurements of the physical quantities are substantially the same no matter where the measurements are made or where the resulting products are produced or sold. The most important aid to this is that all measurements are now made with reference to a single set of measurement standards, the International System of Units (SI), that is defined and managed by the BIPM. Since the measurements mean the same thing in different countries, international commerce is much easier because these and other technical barriers to trade are removed. The result is that a product manufactured in one country can be sold and used in any other country if it is otherwise suitable for the intended use.[6]

## EVOLUTION OF QUALITY STANDARDS

Concepts of quality, and the business practices needed to achieve consistent quality, have evolved over time. The most dramatic changes have occurred since the start of the twentieth century. Then, companies set their own practices and standards, and production was all-important. Now, business has moved to the present state where quality is as important, and international standards exist to define the minimum acceptable practices.

## HISTORY OF QUALITY STANDARDS—1900 TO 1940

Quality control, quality assurance, and quality standards as we are familiar with them now have their immediate roots in the early twentieth century. This coincided with very rapid growth of mass production, early use of automation, and scientific research and development. Inspection as a means of quality control was already in use, but was inefficient and was often under control of the production departments. During the 1920s, several people started systematic studies of applying statistical analysis for the first time to improve production quality. Most of this work was pioneered at Western Electric's Hawthorne Works in Chicago, with important work done by people including Walter Shewhart, Harold Dodge and Joseph Juran. Statistical studies were used by Shewhart to provide a means of saving money by controlling a process, by Dodge for improving the inspection sampling process, and by Juran for evaluating quality improvements and educating top management. Inspection sampling systems, control charts and other quality tools invented at the Hawthorne Works form the foundations of the modern statistical quality assurance system.[7]

## HISTORY OF QUALITY STANDARDS —1940 TO THE PRESENT

When the United States entered World War II at the end of 1941, the production needs of the War Department marked the imposition of these new statistical quality control ideas on industry. This also brought the terms *military specification* and *military standard* into common use. The specifications defined what was needed, and many of the standards defined how to ensure the delivered product was acceptable. They were product- and process-based, and depended on inspection to ensure the quality of the final output.

A significant number of the people recruited by the War Department came from the Western Electric system—notably Juran, Dodge, and Shewhart; as well as a colleague of Shewhart's from the Department of Agriculture, W. Edwards Deming. They brought inspection sampling methods, control charts, and other statistical tools. All of these were developed further during the war, and some became military standards, especially the inspection sampling tables, MIL-STD-105 and MIL-STD-414. (These are now known as ANSI/ASQ Z1.4 and Z1.9, respectively.)

Effective use of this new statistical quality control (SQC) required that large numbers of engineers and other practitioners be trained in the new subject. Some of these people formed local groups to study and share experiences outside the work environment. In 1946, most of these local groups merged to create the American Society for Quality Control (ASQC) as a formal professional society.[8]

After World War II, the United States started aiding Germany and Japan in rebuilding their devastated economies. As part of the rebuilding effort in Japan, a few people were invited to teach the new SQC methods and management methods to Japanese engineers and manager. The best known were Deming and Juran. The Japanese took the lessons to heart and incorporated the new lessons into their industrial culture from the top down. The teaching of Deming was so highly regarded that the Union of Japanese Scientists and Engineers (JUSE) established an annual quality award named in honor of him—the Deming Prize.[9]

The next significant event was the appearance of another military standard, MIL-Q-9858 *Quality Program Requirements*, in 1959. This standard was the first to include most of the elements of a modern quality management system. The current ISO quality management system standards can be traced to roots in quality practices developed by the U.S. War Department during World War II. At some point, those practices were incorporated into a military specification, MIL-Q-5923, *General Quality Control Requirements*. What appears to be the major step, though, occurred when MIL-Q-9858 was released in April 1959 as a replacement for the earlier specification. Other important documents followed over the next few years.

> **April 1959** MIL-Q-9858, Quality Program Requirements. This was replaced in December 1963 by a revised version (MIL-Q-9858A) that remained in force until being canceled in 1996.

**February 1962** MIL-C-45662, Calibration System Requirements. This was replaced by MIL-STD-45662 in 1980. The final revision (MIL-STD-45662A) was canceled in 1995 and replaced by ANSI/NCSL Z540-1-1994 and ISO 10012-1.

**October 1960** MIL-H-110 (Interim), Quality Control and Reliability Handbook. This was replaced by MIL-HDBK-50 Evaluation of a Contractor's Quality Program in April 1965; this is still in force.

**June 1963** MIL-Q-21549B, Product Quality Program Requirements for Fleet Ballistic Missile Weapon System Contractors. The original version was probably introduced in 1961.[10] This standard has since been canceled.

These standards contain the roots of what are now considered to be the fundamentals of an effective quality management system. During the late 1960s, the North Atlantic Treaty Organization (NATO) incorporated these standards into various Allied Quality Assurance Procedures (AQAP) documents. In the early 1970s, there was a movement in British industrial circles to create a set of generic quality management system standards equivalent to the AQAPs but for commercial use.

This led to the submission of a first draft standard by the Society of Motor Manufacturers and Traders to the British Standards Institution. This was circulated for public comment in 1973 . . . (and led) to the publication of the three part BS 5179 series of guidance standards in 1974:

> BS 5179: *Guide to the operation and evaluation of quality assurance systems.*
>
> Part 1: *Final inspection system*
>
> Part 2: *Comprehensive inspection system*
>
> Part 3: *Comprehensive quality control system*[11]

BS 5179 was re-numbered to BS 5750 in 1979. When ISO/TC176 started work on what was to become the first edition of ISO 9000, they drew on that and the related standards for background.[12]

Other elements were included in additional military standards that appeared over the next few years. As explained by Stanley Marash[13]:

> In addition to documentation and auditing, other integral features of military and aerospace standards included management responsibility . . . corrective action processes, control of purchasing, flow-down of requirements to suppliers and subcontractors, control of measuring and test equipment, identification and segregation of nonconforming product, application of statistical methods, and other requirements we now take for granted.

Use of these and other government quality management standards however, was mostly confined to companies fulfilling defense and other government contracts or companies in regulated industries where the standards had force.

Over this period, many U.S. industries had forgotten much of the quality stuff that had become mandatory during the war. They deemed production numbers and schedules more important than quality. After all, they simply had to respond to the public's increasing demand for more and better automobiles, appliances, stereos, and so on. This led to large amounts of rework and waste (again) and quality by inspection (again). It also led directly to such tragedies as the Apollo 1 fire that killed three astronauts in January 1967, which was traced directly to poor workmanship quality in the electrical wiring.[14]

Awareness of the need to improve quality management performance was growing across all industries. In 1968, ASQC introduced the Certified Quality Engineer program. The 1970s and early 1980s made American industry very aware of the competitive need for improved product quality. Since 1945, the quality lessons from wartime production had been largely forgotten or ignored and suddenly products made in Japan were a major competitive threat. Compared to similar domestic products, the imports usually had better product quality and lower cost. In the late 1970s, one of the authors was working for a large consumer electronics retailer. At one point, another company was acquired and their products were integrated into the existing system, complete with going through receiving inspection at the warehouses. The other company had carried a much higher proportion of U.S.-made products. During the product line integration period, it was not uncommon for the warehouses to reject 10 percent or more of product from U.S. manufacturers, while the reject rate from Japanese plants was less than 1 percent and getting better.

The economic pressure was not driven solely by competition from Japan and Taiwan. Inflation and resource scarcity (such as periodic shortages of petroleum) were also significant at the time. But there was evidence that things could be better and actually were in some places. In June 1980, U.S. industry got a major wake-up call when NBC News aired a 90-minute white paper program called *If Japan Can . . . Why Can't We?* Many authorities credit this television show with starting the resurgence in quality management in American business. About one-third of the program explored the management theory of Deming and the influence of his early 1950s teaching of SQC and his management theory (continual improvement of the whole process, which is a system) on Japanese industry. In addition to rediscovering Deming, American industry started exploring the work of Juran (another rediscovery), and new authorities such as Armand Fiegenbaum, Philip Crosby, Ichiro Ishikawa, and others.

Another aid in quality management improvement was the 1987 establishment of the Malcolm Baldrige National Quality Award (MBNQA) program in the U.S. Department of Commerce. Although not a conformance standard, this annual award is important because it has become a widely accepted benchmark for performance excellence of quality management system. Many organizations use the MBNQA criteria as a guide for self-assessment, and they are the basis of many state and local quality award programs. The awards themselves, and related publications by the winners, may be viewed as a store of best practices used by a wide range of businesses, and healthcare and educational institutions. The MBNQA general assessment criteria are:[15]

- **Leadership**. Examines how senior executives guide the organization and how the organization addresses its responsibilities to the public and practices good citizenship.

- **Strategic planning**. Examines how the organization sets strategic directions and how it determines key action plans.

- **Customer and market focus**. Examines how the organization determines requirements and expectations of customers and markets; builds relationships with customers; and acquires, satisfies, and retains customers.

- **Measurement, analysis, and knowledge management.** Examines the management, effective use, analysis, and improvement of data and information to support key organization processes and the organization's performance management system.

- **Human resource focus.** Examines how the organization enables its workforce to develop its full potential and how the workforce is aligned with the organization's objectives.

- **Process management.** Examines aspects of how key production/delivery and support processes are designed, managed, and improved.

- **Business results.** Examines the organization's performance and improvement in its key business areas: customer satisfaction, financial and marketplace performance, human resources, supplier and partner performance, operational performance, and governance and social responsibility. The category also examines how the organization performs relative to competitors.

## START OF ISO 9000 SERIES

When ISO Technical Committee 176 (ISO/TC176) was formed in the early 1980s to start work on an international quality management standard, it drew heavily on the earlier work of the U.S. government standards. The first edition of the ISO 9000 series appeared in 1987. The required elements of an effective quality management system were essentially the same as those identified in MIL-Q-9858 and other standards, and even the language was copied in some cases.

At the beginning of 1993, the European Economic Community (EEC, now the European Union or EU) took another step toward a unified system. As part of a group of agreements, a single set of conformity assessment laws became effective in all member countries. While mostly technical specifications for regulated products, this also included the ISO 9000 series of quality management standards. The European economy is about the same size as the economy of the United States (in 2003), and the two are major trading partners. Combined with other global trade, the impact of this change in Europe was a subject of much apprehension and fears of increased barriers to trade. This was the first time that a major economic bloc of nations had adopted ISO 9000 at the same time, and it was the first time they were adopted as legal requirements by a major global trader. In Europe, many things are legal requirements that in the U.S. and other countries are voluntary standards. Because the quality management standards were suddenly in the business news, awareness of them increased. As awareness and understanding of the requirements and benefits of ISO 9000 increased, the early fears have largely abated. Additional information on ISO 9000 can be found in Chapter 17.

In the past, ISO had three conformance standards: ISO 9001:1994, ISO 9002:1994, and ISO 9003:1994. Now there's only one conformance standard: ISO 9001:2000. The other two standards have been dropped.

If an organization is currently ISO 9002:1994 or ISO 9003:1994 certified, it will now need to become ISO 9001:2000 certified. If an organization is now ISO 9001:1994 certified, it's going to have to update the quality system in order to meet the new ISO 9001:2000 requirements.[16]

When comparing ISO 9001:1994 and ISO 9001:2000 you'll notice that ISO has abandoned the 20-clause structure of the old standard. Instead of 20 sections, the new standard now has five sections. ISO reorganized the ISO 9001 standard to create a more logical structure and to make it more compatible with the ISO 14001 environmental management standard. While this reorganization is largely a cosmetic change, it could have some rather profound implications if you've organized your current quality manual around the old 20-part structure.

In general, the new standard is more customer oriented than the old standard. While the old standard was also oriented toward meeting customer requirements and achieving customer satisfaction, the new standard addresses this in much greater detail. In addition, it expects you to communicate with customers and to measure and monitor customer satisfaction.

The new standard also emphasizes the need to make improvements. While the old standard did implicitly expect organizations to make improvements, the new standard makes this explicit. Specifically, ISO 9001 now wants an organization to evaluate the effectiveness and suitability of its quality management system and to identify and implement systemic improvements.

In the past, organizations that wished to be certified were referred to as *suppliers* because they supplied products and services to customers. Since many people were confused by this usage, ISO has decided to use the word *organization* instead. Now the ISO standards focus on the *organization*, not the supplier.

The term *supplier* now refers to the organization's supplier. The new redefined term *supplier* replaces the old term *subcontractor* (which was dropped). While this may sound a bit confusing, this new usage simply reflects the way these words are normally used. While you're probably familiar with the previous concepts, you may not have heard of the next one. ISO now uses the phrase *product realization*. While this is a rather abstract concept, it is now central to ISO's approach. In fact, ISO devotes an entire section to this new concept. So what does it mean?

In order to grasp what it means you need to recognize that a product starts out as an idea. The idea is realized or actualized by following a set of product realization processes. *Product realization* refers to the interconnected processes that are used to bring products into being. In brief, when you start out with an idea and end up with a product, you've gone through the process of product realization.

The new ISO 9001:2000 standard introduces some new requirements and modifies some old ones. These requirements are summarized in the following list. For more detail, see the associated ISO 9001:2000 clauses listed in parentheses.

- Communicate with customers (7.2.3).
- Identify customer requirements (5.2, 7.2.1).
- Meet customer requirements (5.2).
- Monitor and measure customer satisfaction (8.2.1).
- Meet regulatory requirements (5.1).
- Meet statutory requirements (5.1).
- Support internal communication (5.5.3).
- Provide quality infrastructure (6.3).

- Provide a quality work environment (6.4).
- Evaluate the effectiveness of training (6.2.2).
- Monitor and measure processes (8.2.3).
- Evaluate the suitability of quality management system (8.4).
- Evaluate the effectiveness of quality management system (8.4).
- Identify quality management system improvements (5.1, 8.4).
- Improve quality management system (5.1, 8.5).

Under the new ISO 9001:2000 standard, you may ignore or exclude some requirements. Requirements that may be ignored under special circumstances are known as *exclusions*. According to ISO, you may ignore or exclude any of the requirements found in *Section 7 Product realization* as long as you meet certain conditions.

You may exclude a Section 7 requirement if you cannot apply it. More precisely, you may exclude or ignore a requirement if you cannot apply it because of the nature of your organization, or you cannot apply it because of the nature of your products or services.

You may *not* exclude or ignore Section 7 requirements if doing so will compromise your ability or willingness to meet the requirements set by customers and regulators.

We believe that this permissible exclusion clause is a very important improvement. It makes implementation more flexible and conformance less rigid. Because of this significant innovation, you're more likely to end up with a quality management system that not only complies with ISO's standards, but also meets your organization's unique needs. This new, more flexible, approach is further demonstrated in another way. When you study the new ISO 9001 standard, you'll notice that it is less prescriptive than the old standard. In general, the new standard tells you what to do *not* how to do it. This is particularly evident when you look at how many times procedures are required. When you compare the old and the new standard, you'll notice that procedures are much less often required by the new standard. This more flexible approach gives you more freedom to decide how you're going to meet the requirements. In general, this should make it easier for you to develop a more suitable and effective quality management system.

In order to understand ISO 9001:2000 at a deeper level, you need to recognize that ISO uses a process approach to quality management. While the process approach is not new, the increased emphasis ISO now gives to it is new. It is now central to the way ISO thinks about quality management systems. An introduction to the process approach is in ISO 9000:2000, one of the guidance standards that are part of the ISO 9000 family.

According to this approach, a quality management system can be thought of as a single large process that uses many inputs to generate many outputs. This large process is, in turn, made up of many smaller processes. Each of these processes uses inputs from other processes to generate outputs, which, in turn, are used by still other processes.

A detailed analysis of the standard reveals that an ISO 9001:2000 quality management system is made up of at least 21 processes (22 if you recognize that the quality management system as a whole is also a process). These 21 processes are as follows:

1. Quality management process
2. Resource management process
3. Regulatory research process
4. Market research process
5. Product design process
6. Purchasing process
7. Production process
8. Service provision process
9. Product protection process
10. Customer needs assessment process
11. Customer communications process
12. Internal communications process
13. Document control process
14. Record-keeping process
15. Planning process
16. Training process
17. Internal audit process
18. Management review process
19. Monitoring and measuring process
20. Nonconformance management process
21. Continual improvement process

Note that, as mentioned in clause 2.7.2 of ANSI/ISO/ASQ Q9000:2000, the record-keeping process listed above can be considered part of the document control process. The standard states that records are simply "documents that provide objective evidence of activities performed or results achieved."

If your organization does not already have a documented quality management system, then in order to develop a quality management system that meets the new ISO 9001:2000 standard, you must create or modify each of those 21 processes. You must develop each process, document each process, implement each process, monitor each process, improve each process, and describe the interactions with other processes.

Each process uses inputs to generate outputs, and all of these processes are interconnected using these input–output relationships. The output from one process becomes the input for other processes. Because of this, inputs and outputs are really the same thing. If the organization already has a documented system (from an older version of ISO 9000, for example), then the main thing that needs to be done is to document how the existing system meets the requirements of the new standard. A matrix diagram is ideal for this. If the process of doing that turns up any gaps, they must be addressed. Fortunately, that is usually a small task that does not require wholesale change of the existing system.

The following is an incomplete list of some general types of inputs/outputs: products, *services* (a service is defined as a type of product), information, documents, reports, records, results, needs, data, expectations, requirements, complaints, comments, feedback, resources, measurements, authorizations, decisions, plans, ideas, solutions, proposals, and instructions.

An ISO 9001:2000 Quality Management System is made up of many processes, and these processes are glued together by means of many input–output relationships. These input-output relationships turn a simple list of processes into an *integrated system*. Without these input-output relationships, you wouldn't have a Quality Management System.

The most recent revision of the ISO 9000 series is often seen as a very revolutionary change when compared to the older versions, but is actually more evolutionary when considered alongside the work of Deming, Crosby, Juran, and others. Instead of details of manufacturing and support process tasks with an emphasis on how they should be done, the emphasis is now on the results to be achieved and the interactions of the management processes used to achieve them. The 2000 version emphasizes the view of a system of processes, quality assurance over quality control, the voice of the customer, and the Shewhart/Deming PDCA cycle.

## Endnotes

1. The concept of *product* includes services, but they are often stated separately anyway. See ISO 9000:2000 for more information.
2. Barker, 12.
3. ISO is not an acronym! When ISO was formed, the delegates realized that an acronym would be different in every language, yet they wanted something common that would reflect the work of standardization. They selected the word *iso* (from the Greek *isos*, meaning equal), so that the short form of the organization's name would be the same in every language—ISO.
4. ISO 2003a, www.iso.ch/iso/en/aboutiso/isomembers
5. International Federation of Standardizing Associations (ISA); founded in 1926.
6. ISO 2003b, www.iso.ch/iso/en/aboutiso/introduction
7. Juran 1997, 73-81.
8. ASQ, www.asq.org/join/about/history
9. Kilian, 357-376.
10. Marash, 18.
11. Bird, 3-6.
12. Bird, 3-6; ISO 2003c, www.iso.ch/iso/en/iso9000-14000/tour/wherfrom.html; and Marquardt, 21.
13. Marash, 18.
14. Pellegrino and Stoff, 91-93.
15. NIST 2003. The annual award criteria, applications and other materials are available from the NIST MBNQA web site, www.baldrige.nist.gov
16. Portions of the remainder of this chapter are based on material on the web site of Praxiom Research Group, praxiom.com

## References

ASQ. 2003. *Origins of the American Society for Quality*. www.asq/org/join/about/history

Barker, Joel Arthur. 1993. *Paradigms: The business of discovering the future*. New York: HarperCollins.

Bird, Malcolm. 2002. "A few small miracles give birth to an ISO quality management systems standard for the automotive industry." ISO Bulletin, August 2002. www.iso.ch/iso/en/commcentre/isobulletin/articles/2002/pdf/automotiveind02-08.pdf

Delaney, Helen. 2000. *Impact of conformity assessment on trade: American and European perspectives.* Proceedings of the SES 2000 conference, Baltimore.
www.ses-standards.org/library/00proceedings/delaney.pdf

Fischer, Thomas C. and David Williams. 2000. The United States, the European Union and the *"Globalization" of world trade: Allies or adversaries?* Westport, CT: Quorum Books: 24.

Gillespie, Helen. 1994. What Is ISO 9000? *Scientific Computing and Automation.* (February).
www.limsource.com/library/limszine/applica_tions/apscaiso/apwhat94.html

Grun, Bernard. 1991. *The timetables of history.* New 3rd rev. ed. New York: Simon & Schuster.

ISO (International Organization for Standardization). 2003a. "ISO Members."
www.iso.ch/iso/en/aboutiso/isomembers

ISO (International Organization for Standardization). 2003b. "Overview."
www.iso.ch/iso/en/aboutiso/introduction

ISO (International Organization for Standardization). 2003c. "Where ISO 9000 came from and who is behind it." www.iso.org/iso/en/iso9000-14000/tour/wherfrom.html

Juran, Joseph M. 1997. Early SQC: A Historical Supplement. *Quality Progress* 30 (September): 73-81.

Juran, Joseph M. 1999. *Juran's quality handbook.* Wilton, CT: McGraw-Hill.

Kilian, Cecelia S. 1992. *The world of W. Edwards Deming.* Knoxville, TN: SPC Press.

Marash, Stanley A. 2003. What's good for defense . . . how security needs of the Cold War inspired the quality revolution. *Quality Digest* 23, (June): 18.

Marquardt, Donald W. 1997. Background and development of ISO 9000 standards. *The ISO 9000 Handbook,* 3rd ed., Edited by Robert W. Peach. New York: McGraw-Hill: 14–20.

Microsoft Corporation. 2003. "Chariots." In *Microsoft Encarta 2004.* Redmond, WA: Microsoft Corporation.

NASA (National Aeronautics and Space Administration). 2003. "NASA Quality Management Systems." www.hq.nasa.gov/iso

NIST (National Institute of Standards and Technology). 2003. Frequently asked questions about the Malcolm Baldrige National Quality Award.
www.nist.gov/public_affairs/factsheet/baldfaqs.htm

Pellegrino, Charles R., and Joshua Stoff. 1985. *Chariots for Apollo.* New York: Atheneum.

Praxiom Research Group Limited, 9619 - 100A Street, Edmonton, Alberta, T5K 0V7, Canada

# Chapter 4
# Quality Documentation

We've all heard the phrase, "The job isn't done till the paperwork is complete." Within a quality system, there is more to it than that. What about changes made along the way, whether it's to the procedures, certificates, records, database, software, or quality manual? How do you control those changes, and how do you let staff know what the latest revision is to those documents? According to ANSI/ISO 17025-1999, ANSI/ISO/ASQ Q9001-2000, and ANSI/NCSL Z540-1-1994, there are specific steps that must be taken to ensure compliance within the quality system. Documentation control ensures that the latest and greatest calibration procedures, certificates, labels, forms, software, and quality records are available for use by those authorized to use them. This reinforces the organization's ability to give the customers the quality they deserve.

Quality documentation encompasses an organization's procedures, records, certificates, and quality manual. The following chapters go into specific details on each of these topics, however, each is related to the other in that none could occur without the others. You must have calibration procedures to follow in order to get the data to put on the calibration certificate (or in cases where certificates are not used, calibration records and labels). The quality manual lays the groundwork for everything in the system and is a living document that continually changes as customer and company needs dictate.

According to 17025-1999, chapter 4.3, Document control: "The laboratory shall establish and maintain procedures to control all documents . . . all documents issued . . . shall be reviewed and approved. A master list . . . identifying the current revision status and distribution of documents in the quality system shall be established. Changes to documents shall be reviewed and approved. . . ." Q9001-2000, chapter 4.2.3, Control of documents states: "Documents required by the quality management system shall be controlled. A documented procedure shall be established to define the controls needed to approve documents . . . review and update . . . ensure that changes are identified . . . that relevant versions are available . . . to prevent the unintended use of obsolete documents. . . ." And Z540 states in chapter 5.2: "The quality manual and related documentation shall also contain d) procedures for control and maintenance of documentation." The requirements for a documentation control system are easily found in many manuals and standards. How to put them into practice can be one of the more time consuming parts of a quality system. But once established, it will help keep the most up-to-date procedures in the hands of technicians, while removing errors and obsolete procedures from the quality system.

One of the most important parts of a quality system should be the document control procedures. How do you control who makes changes, how new documents and/or changes to old documents get posted, and when the users are notified or made aware that changes have been made? There are software packages available that can assist in controlling an organization's document system, but a small business may not be able to afford those packages. Here's a brief overview of what can be done with the resources an organization may already be using.

Each of the controlled documents (procedures, records, certificates, and so on) should have a unique identification number (many systems use a number often known as a *control number*) as well as a revision number. A master list with all this information should be available for anyone to see what the current documents are within their quality system. This list should also include the revision date of the document. It needs to be updated every time changes are made and approved for documents. A simple spreadsheet or word processor document can fulfill this requirement, as long as it is updated and maintained. It has to be available to the technicians who use the various procedures, records, and certificates. This isn't a difficult problem in a small group where only a few people work with the various documents, but if an organization has two or three shifts, different locations or off-site calibration responsibilities, the opportunity for using out-of-date or incorrect documents could easily become a problem. Notification of changes and training, when applicable, should be documented as a form of keeping everyone informed and up-to-date.

Part of an organization's training program should include when and how to inform and train its staff of changes to the documents. Some organizations ensure training and/or notification of changes has occurred before they allow the latest revisions to be posted. In some systems, the new revision must be posted in order for the user to have access to the documents. Another approach would be to maintain all of your quality documentation electronically via an intranet. In this process any printed documents would be invalid. This process insures only the latest and greatest procedures are available to all that they pertain to. Whichever way an organization's system works, it is vital that everyone involved be informed and trained when changes are made, and that only the latest revisions are available for their use.

As a minimum, an organization's master list should have, for each of its controlled documents a unique identification number, document name, revision or edition number, and revision date. In addition it is helpful to include the name of the approver or approval authority and something stating that only the revisions listed should be used. Archiving a copy of previous revisions can have benefits, but must be in a location where they cannot be readily accessed for use by staff. Some systems use black lines in their borders to indicate where changes have been made; others annotate the changes in a reference section at the end of the document; and still others refer to comparisons of archived documents as the only reference to changes. No matter which system or combination of systems is used, they only need to meet the quality system requirements that have been set for the organization.

## References

ANSI/ISO 17025-1999, *American National Standard—General requirements for the competence of testing and calibration laboratories*. Milwaukee: ASQ Quality Press.

ANSI/ISO/ASQ Q9001-2000, *American National Standard—Quality management systems—Requirements*. Milwaukee: ASQ Quality Press.

ANSI/NCSL Z540-1-1994, *American National Standard for calibration—calibration laboratories and measuring and test equipment—general requirements*. Boulder, CO: National Conference of Standards Laboratories.

# Chapter 5

# Calibration Procedures and Equipment Manuals

All quality systems that address calibration require written instructions for the calibration of IM&TE. Different standards have various requirements, and those will be addressed in Chapter 17, Industry-Specific Requirements, wherever they differ from the information in this chapter. Under the quality system, this is the *say what you do* portion, which means you need to write down in detail how to do your job (this includes calibration procedures, SOPs, protocols, work instructions, work cards, and so on). Why follow formal instructions or procedures? Simple . . . in order to get consistent results from a calibration, you must be able to follow step-by-step instructions each and every time you perform those calibrations.

Three different sources focus on written calibration procedures. To begin with, BSR/ISO/ASQ Q10012:2003, chapter 6.2.1 states: "Measurement management system procedures shall be documented to the extent necessary and validated to ensure the proper implementation, their consistency of application, and the validity of measurement results. New procedures or changes to documented procedures shall be authorized and controlled. Procedures shall be current, available and provided when required."

NCSLI RP-6 states, in chapter 5.9, Calibration Procedures, "Documentation should be provided containing sufficient information for the calibration of measurement equipment." The Requirements in 5.9.1 are:

- **Source.** The calibration procedure may be . . . prepared internally, by another agency, by the manufacturer, or by a composite of the three.

- **Completeness.** The procedure should contain sufficient instruction and information to enable qualified personnel to perform the calibration.

- **Approval.** All procedures should be approved and controlled, evidence should be displayed on the document.

- **Software.** When used instead of an actual procedure should follow the computer software recommendation for control.

Under Format in chapter 5.9.2, internal procedures should include:

- **Performance requirements.** This includes device description, manufacture, type or model number, environmental conditions, specifications, and so on.
- **Measurement standards.** This includes generic description of measurement standards and performance requirements, accuracy ratio and/or uncertainty, and any auxiliary tools.
- **Preliminary operations.** This includes any safety or handling requirements, cleaning prerequisites, reminders, or operational checks.
- **Calibration process.** This includes the detailed set of instructions for process verification in well-defined segments, upper and lower tolerance limits, and required further instructions.
- **Calibration results.** This is a performance-results data sheet or form to record the calibration data when required.
- **Closing operations.** This includes any labeling, calibration safeguards, and material-removal requirements to prevent contamination of product.
- **Storage and handling.** These are requirements to maintain accuracy and fitness for use.

And chapter 5.9.3, Identification, states: "For reference purposes, a system should be established for identifying calibration procedures."

ANSI/ASQC, M1-1996, chapter 4.9 states: "Documented procedures, of sufficient detail to ensure that calibrations are performed with repeatable accuracy, shall be utilized for the calibration of all ensembles." That explains it all in one sentence. (This standard describes a calibration system as an *ensemble*.)

Now that you know what the standards require for calibration procedures, where do you go from here? Some companies use original equipment manufacturer (OEM) procedures that are in their service manuals as a starting point. Keep in mind that a lot of the service manuals have procedures for adjusting the IM&TE as well as (or instead of!) the calibration (performance verification) process. Also, some OEM procedures are vague and lack specific requirements needed to insure a good calibration, such as equipment requirements, environmental conditions, and so on. Finally, in many cases the equipment manufacturer simply does not provide any calibration or service information as a matter of policy. By writing your own procedures or using prewritten calibration procedures, you might save time by eliminating the adjustment process if it is not required and/or improve the outcome of the calibration. An example of a calibration procedure can be found in Figures 5.1 and 5.2.

With regard to adjusting IM&TE, there are several schools of thought on the issue. On one end of the spectrum, some (particularly government regulatory agencies) require that an instrument be adjusted at every calibration, whether or not it is actually required. At the other end of the spectrum, some hold that any adjustment is tampering with the natural system (from Deming) and what should be done is simply record the values and make corrections to measurements.

An intermediate position is to adjust the instrument only if (a) the measurement is outside the specification limits; or (b) if the measurement is inside but near the specification

## CHAPTER 5: CALIBRATION PROCEDURES AND EQUIPMENT MANUALS

| Title: Balance and Scale Calibration Procedure | | Procedure No. SOP11C002 | Rev. No. 02 |
|---|---|---|---|
| **Submitted by:** Anna Terese Public | **Date:** 1/16/92 | **Approved by:** Ayumi Jane Deaux | |

**READ THE ENTIRE PROCEDURE BEFORE BEGINNING.**

1. PURPOSE
   This standard operating procedure (SOP) describes the responsibilities of the Metrology Department as they relate to the calibration of all balances and scales. The intent of this SOP is to give the reader an idea of how to format and structure a calibration procedure.

2. SCOPE
   This SOP applies to all balances and scales that impact the quality of goods supplied by Acme Metrology Services, LLC, Eatmorecheese, Wisconsin.

3. RESPONSIBILITIES
   3.1. It is the responsibility of all metrology technicians who calibrate balances and scales to comply with this SOP.
   3.2. The person responsible for the repair and calibration of the balance or scale will wear rubber gloves and eye protection. The balance or scale must be cleaned and/or decontaminated by the user before work can be accomplished.

4. DEFINITIONS
   4.1. **NIST** – National Institute of Standards and Technology
   4.2. **IM&TE** – Inspection, Measurement, and Test Equipment.

**IM&TE SPECIFICATIONS**

| Manuf. | P/N | Usable Range | Accuracy |
|---|---|---|---|
| Allied | 7206A | 500 mg ~ 500 g | ± 30 mg (500 mg ~ 30 g) > 30 g ± 0.1% of Rdg |
| Denver | 400 | 500 mg ~ 400 g | ± 20 mg (500 mg ~ 20 g) > 20 g ± 0.1% of Rdg |
| Ohaus | V02130 | 50 mg ~ 210 g | ± 3 mg (50 mg ~ 3 g) > 3 g ± 0.1% of Rdg |

**EQUIPMENT REQUIREMENTS (STANDARDS)**

| Weight Size | Accuracy (mg) | Class |
|---|---|---|
| 25 Kilograms | ± 2.5 grams | F |
| 5 Kilograms | ± 12.0 | 1 |
| 50 milligrams | ± 0.01 | 1 |
| 1 milligrams | ± 0.01 | 1 |

**Figure 5.1** Sample calibration procedures with specification tables.

limits, where *near* is defined by the uncertainty of the calibration standards; or (c) if a documented history of the values of the measured parameter shows that the measurement trend is likely to take it out of specification before the next calibration due date.

## 5. PROCEDURE

5.1. General inspection

    5.1.1. This is to give the reader an idea of a numbering system and a formatting scheme to use for calibration procedures.

5.2. Leveling the test instrument

    5.2.1. Be as specific as possible in your instructions. Write for the benefit of the least experienced technician, not the senior person on your staff.

5.3. Calibrating the "edges" of the weighing pan

    5.3.1. Following the example in Figure 1, place a weight equal to approximately ½ the capacity of the test instrument on edge 1 (place the single weight ½ the distance between the pan center and the usable edge). Record the reading on the calibration worksheet.

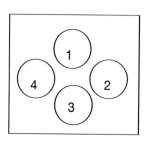 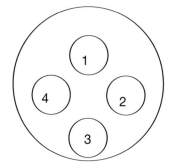

5.4. Completing the calibration worksheet or form

    5.4.1. Identify the standard weight(s) used for each weight check by its identification number(s) in the block provided.

## 6. RELATED PROCEDURES

## 7. FORMS AND RECORDS

## 8. DOCUMENT HISTORY

| Rev. # | Change Summary |
|---|---|
| 00 | New document |
| 01 | Added edge calibration verbiage. |
| 02 | Changed accuracies to be ± 0.1% of reading and instituted usable range at low and high end of all balances. |

**DISCONNECT AND SECURE ALL IM&TE**

**IM&TE CALIBRATION POINTS**

| Manufacturer | Part Number | Calibration Test Points in Grams |
|---|---|---|
| Allied | 7206A | .5, 1, 10, 50, 100, 500 |
| Denver/Fisher | 400 | .5, 10, 20, 50, 100, 400 |
| Ohaus | V02130 | .05, .5, 5, 10, 100, 210 |

**Figure 5.2** Example of specific instructions in calibration procedures.

## References

ANSI/ASQ M1-1996, *American National Standard for calibration systems*. Milwaukee: ASQC Quality Press.

BSR/ISO/ASQ Q10012:2003(E), *Measurement management systems—Requirements for measurement processes and measuring equipment*. Milwaukee: ASQ Quality Press.

The Healthcare Metrology Committee. 1999. *Calibration control systems for the biomedical and pharmaceutical industry*. Recommended Practice RP-6. Boulder, CO: National Conference of Standards Laboratories.

# Chapter 6
# Calibration Records

Some readers might ignore this chapter altogether. For them, there are no requirements to keep or maintain records of their calibrations or of the maintenance performed on the IM&TE that they support. For many of us, this would seem almost unbelievable, but it's true. If their IM&TE passes calibration, meeting the tolerances (staying within the lower and upper test limits of their procedures), they are only required to place a calibration label on the unit, and update their database with the next date due calibration. For the rest of us, this is not an option. Documentation of the As Found, and when needed, As Left readings, along with what the standard read is as natural as using calibration procedures and traceable standards.

Who needs records? They're a waste of time and energy! There is a peculiar aspect to metrology that is "old is good." This concept is not readily understood by most, but after understanding the relationship of historical data and stability of a standard, the need for good historical records is apparent. For example, a standard with 10 years of quality documentation as it relates to its performance is much more valuable than a standard with no history. With historical data you can predict the stability of a particular instrument and build on this, without the history there is nothing to build from. This is a common concept in metrology and why quality documentation/calibration records are invaluable.

Why do doctors, lawyers, dentists, and mechanics keep records? Some because it is required by law (record archiving for liability purposes have been with us for a long time), others because they need a history in order to know what has happened previously. When working with food, pharmaceuticals, cosmetics, aircraft, and many other sections of industry where IM&TE is used, previous circumstances have proven that the procedures, data, and history of the calibration and/or repair of their test equipment is critical in finding why accidents happened, defaults were accepted, or mistakes were made. It is also important in forecasting how IM&TE will function over periods of time. Without historical information, we are bound to repeat mistakes and not make the process improvements we should all be striving for. Records play a crucial part of the calibration process, and their retention and availability should be an integral part of your quality system.

Here's a short list of where to find specific requirements for documenting your calibrations using records. ANSI/ISO 17025-1999, chapter 4.12, paragraph 4.12.2.1, states:

"The laboratory shall retain records of original observations, derived data . . . to establish an audit trail . . . for a defined period. The records for each calibration shall contain sufficient information to . . . enable the test or calibration to be repeated." Paragraph 4.12.2.2 states: "Observations, data and calculations shall be recorded at the time they are made. . . ." BSR/ISO/ASQ Q10012:2003, chapter 6.2.3 states: "Records containing information required for the operation of the measurement management system shall be maintained. Documented procedures shall ensure the identification, storage, protection, retrieval, retention time and disposition of records." NCSLI RP-6, chapter 5.6, Records, states: "Records should be maintained for all IM&TE that is included in the calibration system." M1-1996, chapter 4.7, Records, states: " . . . records shall include, but not be limited to . . . a. Description of equipment and unique identification; b. Date most recent calibration was performed; c. Indication of procedure used; d. Calibration interval; e. Calibration results obtained (i.e., in or out of tolerance); f. By whom the ensemble was calibrated; and g. Standards used. Specific record requirements for other standards or regulations can be found in Chapter 17, Industry Specific Requirements.

So what do these standards ask you to do? They say to collect the IM&TE readings as they occur and record them, either on a hard copy record or electronically, if that is how your system works. Data can also be recorded using computer process equipment or software designed to perform that function. Remember that those units must be qualified and the process validated to perform those functions prior to using them to collect data for inclusion in your records.

So, for argument's sake, let's assume you have to keep a record of your calibrations. What needs to be in the record? What happens to the record when you're through with the calibration and collecting the data? Is the record saved, destroyed, archived, or used as scratch paper during the recycling process? Before we go any further, let's remember that generally speaking, there are two types of users. The first is the group that calibrates IM&TE for their company and do not have any external or commercial customers. The second group deals with external customers who are paying for their services. Both groups may require some information contained in a record, and some may be unique to each group. Some information is required because of the standard that covers your calibration activity, some due to customer requirements, and some because it just may be prudent to have in the record. So, having said that, let's answer each of these questions one at a time.

In order for a record to be valid, it must:

1. Identify which IM&TE it is recording data on. It also must have a unique identification number assigned to it, identify who owns it (when appropriate), record the IM&TE part or manufacturer's number, record the IM&TE serial number and/or asset number if applicable, and be in appropriate location. If it is for internal use only, possibly show the department, group, or cost center that owns it and the calibration interval assigned to it.

2. Contain the environmental conditions during calibration (when applicable or required).

3. List the instrument's ranges and tolerances (when applicable or required).

4. Show traceability for the standards you are using to perform the calibration by identifying your standards and when they are currently due calibration.

5. List the procedure used for that particular calibration and the revision number.

6. List your standard's tolerances and/or traceability information.

7. Have an area for recording the standard's readings, as well as what the IM&TE read (this is what is known as the As Found readings). There must also be an area for the As Left readings, those readings taken after repair, alignment, or adjustment of the unit. When required, the out-of-tolerance readings may be given in magnitude and direction in order to make an impact assessment.

8. List the next Date Due Calibration if required. More information on the actual date (day, month, year, or just month and year) can be found in Chapter 10 and Chapter 14.

9. Contain an area for comments or remarks when clarification of limits, repairs, or adjustments that occurred during the process of calibration is required.

10. Have the person performing the calibration's signature and date. The date will be when all calibration functions have been completed. Some items take more than a day to calibrate, and when this occurs, the final day of calibration is the calibration date and also the date used for calculating when it is next due for calibration. Some organizations have a requirement for a second set of eyes to audit, perform a quality assurance function, or just ensure all the data are present. Whatever the case may be, there needs to be a place for the second signature and possibly date accomplished.

Most organizations require records be maintained for a predetermined period of time and list them in their records retention policy and post this schedule for ease of reference. According to ANSI/ISO 17025-1999, chapter 4.12.2.1: "The laboratory shall retain original observations, derived data and sufficient information to establish an audit trail, calibration records, staff records and a copy of each test report or calibration certificate issued, for a defined period."

You also need to ensure that the record is safe, secure from tampering or alteration, and available when needed. If changes are made to a record, the error needs to be crossed out (not erased, made illegible, or removed/deleted) and the correct value or reading entered next to it. The person making the change needs to initial and date next to the entry. If you record or store your records electronically, equivalent measures need to be taken to avoid any loss or change in the original data. If changes are made to an electronic record, a new record needs to be added. The new record will contain the original data with the changes, as well as some way to identify who made the change and when, and a link to the old record. The old record should be unchanged, but have a flag showing it is archived and a link to the replacement record.

There are different approaches on how to store records for easy accessibility. You might consider storing by the IM&TE's unique identification number or by the location of the unit, with all items in a specific room, lab, or department having their records stored in that location or file. When storing electronically, it is not so simple. You have to be able to easily find the record during audit or review. One system that this author has found useful during electronic storage is using the unique identification number (such as a five-digit numeric system) followed by the Julian date of calibration. In this case, the Julian date is the year designated by two digits (03 for 2003) and the sequential number of the calendar year (001 for January 1, or 349 for December 15, so this example would be 03001 or 03349). Not only does this identify the unit, but readily shows the date calibrated, distinguishing it from records of previous calibrations.

## References

ANSI/ASQ M1-1996, *American National Standard for calibration systems*. Milwaukee: ASQC Quality Press.

ANSI/ISO 17025-1999, *American National Standard—General requirements for the competence of testing and calibration laboratories*. Milwaukee: ASQ Quality Press.

BSR/ISO/ASQ Q10012:2003(E), *Measurement management systems—Requirements for measurement processes and measuring equipment*. Milwaukee: ASQ Quality Press.

The Healthcare Metrology Committee. 1999. *Calibration control systems for the biomedical and pharmaceutical industry*. Recommended Practice RP-6. Boulder, CO: National Conference of Standards Laboratories.

# Chapter 7
# Calibration Certificates

If your calibration responsibilities include the generation and/or completion of calibration certificates, this chapter is for you. There are many instances, however, where calibration records are a direct substitute for certificates, such as when you only perform calibrations internally for your own customers. Then, the combination of the calibration record and calibration label attached to the IM&TE serves the same purpose. They provide all the information needed to prove the item has received traceable calibration. The information included in a calibration record/label can provide the same important information as included on a calibration certificate. Refer to Chapter 6, Calibration Records, for more information.

According to ANSI/ISO 17025-1999, paragraphs 5.10.2 and 5.10.4: "Each certificate shall include:

- A title
- Name and address of the laboratory
- Unique identification of the certificate
- Name and address of the client
- Identification of the method used
- A description and condition of the item calibrated
- The date(s) of calibration
- The calibration results and units of measurement
- The name, function and signature of the person authorizing the certificate
- The environmental conditions during calibration
- The uncertainty of measurement
- Evidence that the measurements are traceable."

The uncertainty of measurement statement on most certificates will fall between a statement of "Does not exceed a 4:1 TUR unless noted otherwise" to an elaborate uncertainty analysis that includes formulas, correction factors, confidence levels, and so on.

Calibration certificates can be in either hard copy or electronic format. They must be legible, readable, and provide the necessary information requested or required by the user. Most bear statements that they cannot be copied in part without the consent of the calibration authority issuing the certificate. See a sample calibration certificate in Figure 7.1.

---

### CERTIFICATE OF CALIBRATION

**Customer:**   **Customer Nbr:**

| | |
|---|---|
| **Cert/RA Nbr:** 1-UF75R-1-4 | **Date Received:** Nov. 22, 2002 |
| **Manufacturer:** Agilent Technologies | **Date Calibrated:** Nov 22, 2002 |
| **Description:** Digital Multimeter | **Next Calibration Due:** Nov 22, 2003 |
| **Model Nbr:** 34401 A | **Calibration Proc:** 33K8-4-1029-1 |
| **Serial Nbr:** MY41004563 | **Data Sheet Nbr:** 33-00007-4 |
| **ID Nbr:** NONE | **Item Received:** Out of Tolerance |
| **PO Nbr:** 28268 | **Item Returned:** In Tolerance |

For out of tolerance data, see Supplemental Report for RA Nbr 1 -UF75R -1-4
**Temperature:** 74°F / 23.3°C     **Relative Humidity:** 46%

Transcat Calibration Laboratories have been audited and found in compliance with ISO/IEC 17025. Accreditation calibrations performed within the Lab's Scope of Accreditation are indicated by the presence of the A2LA Logo and Certificate Number on this Certificate of Calibration. Any measurement on an accredited calibration not covered by that Lab's Scope are noted within the data and/or accompanying paperwork.

Transcat calibrations, as applicable, are performed in compliance with requirements of ISO 9002-1994, QS-9000, MIL-STD-45662A, ANSI/NCSL Z540-1994 and ISO 10012-1992. When specified contractually, the requirements of 10CFR21, 10CFR50 App. B and NQA-1 are covered.

Transcat will maintain and document the traceability of all its standards to the National Institute of Standards and Technology, NIST (formerly NBS), or the National Research Council NRC, of Canada, or to other recognized national or international standard bodies, or to measurable conditions created in our laboratory, or accepted fundamental and/or natural physical constraints, ratio type of calibration, or by comparison to consensus standards.

Complete records of work performed are maintained by Transcat and are available for inspection. Laboratory assets used in performance of this calibration are shown below.

The results in this report relate only to the item calibrated or tested.

All calibrations have been performed using processes having a test uncertainty ratio of four or more times greater than the unit calibrated, unless otherwise noted. Uncertainties have been estimated at a 95 percent confidence level (k=2). Calibration at 4:1 TUR provides reasonable confidence that the instrument is within the manufacturer's published specifications. Limitations on the uses of this instrument are detailed in the manufacturer's operating instructions.

**Notes:**

| Assets | Manufacturing | Model | Description | Cal Date | Due Date | Traceability Numbers |
|---|---|---|---|---|---|---|
| 19370 | General Resistance | RTD-100 | Simulator, RTD | 10/30/2002 | 10/30/2003 | 21639 |
| 21639 | Hewlett Packard Comp | 3458A | Digital Multimeter | 04/17/2002 | 04/17/2003 | 34527 |
| 2469 | Hewlett Packard Comp | 3324A | Function Generator | 12/19/2001 | 12/19/2002 | 1-2469-10-1 |
| 2487 | Fluke Corporation | 5700A | Calibrator | 01/31/2002 | 01/31/2003 | 105016-0 |
| 2488 | Fluke Corporation | 5725A | Amplifier | 01/31/2002 | 01/31/2003 | 10516-0 |

| Calibrated at: | Facility Responsible | |
|---|---|---|
| 300 Industrial Dr | 35 Vantage Point Dr | by: Chris DeZutter     Date |
| Grand Island, NY 14072-1294 | Rochester, NY 14624 | Onsite Representative |

**Figure 7.1** A sample certificate of calibration.

What do you do with the calibration certificate once the IM&TE and certificate are returned to you? Most just check the data to ensure they are for the correct item and that all parameters have passed, and then file them, never to be viewed again. Others, however, look for As Found and As Left data on the certificates. If your IM&TE was out of tolerance or required adjustment, there should be As Left data showing what the item indicated after being adjusted. It is the customer's (or user of the IM&TE) responsibility to assess the impact on its products or processes if its IM&TE was out of tolerance. Depending on its quality system, the customer may need to document all out-of-tolerance conditions, what impact that IM&TE had within its system, and the actions taken to preclude a reoccurrence.

Others may use the data on the certificate for their uncertainty budgets, to complete other calibration certificates, or to review or analyze for historical trends. In any case, they can be very useful and their retention and storage is very important.

## References

ANSI/ISO 17025-1999, *American National Standard—General requirements for the competence of testing and calibration laboratories*. Milwaukee: ASQ Quality Press.

# Chapter 8
# Quality Manuals

Quality manuals are the summary of documentation for the management system of an organization, not just for the quality portion of the management system. They should describe how the organization is managed, how documentation and information flows and is controlled, how the customer is serviced and supported, and how the management system monitors itself.

Quality manuals that are written for accreditation under ISO/IEC 17025 often follow the format of the standard, although that is not required. Using the format of the standard allows easier correlation of the manual to the standard. The laboratory should use the format that is most appropriate for its own needs. Regardless of the format that is used, the quality manual needs to be easy to read and use by all personnel. Some issues to avoid include (1) a lack of page numbers, (2) a lack of section headings, and (3) a lack of references to other related documents. A quality manual that lacks these items will be unclear and difficult to use. Quality manuals that are exceptionally verbose are difficult to use. Quality manuals that are difficult to use do not get used by the personnel who perform the day-to-day work of the laboratory.

Many alternative methods exist for quality manuals, including the use of flowcharts. Flowcharts are typically simpler to understand, visual, and therefore are more likely to be used by all personnel.

The quality manual is meant to be a living document. Organizations undergo change, and the quality manual should reflect the changes and should reflect the philosophy of how the organization is managed on a day-to-day basis. Quality manuals are meant to reflect the specific organization or portion of an organization, and so each manual is or, at least, should be, unique.

A quality manual may have somewhat different amounts and types of information, depending on the needs of the organization. For example, a small laboratory may have one manual containing more descriptive information, rather than having several layers of procedural documentation that a large laboratory may have. Formats and layouts will likely be different for different manuals. The key aspects of the quality manual are that the quality manual meets the needs of the laboratory, accurately documents the management system used in the laboratory, has provisions to service the customer and safeguard customer information, and addresses any external requirements such as ISO/IEC 17025.

Every organization has information flowing through it all the time. For a calibration laboratory, information is flowing in from the customer and back out to the customer. The flow of information takes many forms, from simple conversations to formal contracts or complaints. The quality manual should describe the protocols and controls that are used to direct, store, use, safeguard, and communicate the information is the manner best suited for the laboratory. The quality manual should include protocols for both physical and electronic documentation, and especially describe the protocols to safeguard confidentiality of customer information, whether it is transmitted electronically or stored in a file. Some examples of controls of information include a description of how records are maintained and how procedures and forms are controlled.

Service and support of the customer needs to be documented clearly in the quality manual. After all, the customer pays the bills! Some ways to service the customer are really communication methods. One example of servicing the customer is providing an equipment-due-for-calibration list. Documentation of the methods used to service the customer can be accomplished in different ways. Detailed descriptions can be provided in a quality manual or referred to and detailed elsewhere. The needs and size of the laboratory will dictate exactly how service to the customer is documented.

A management system needs to be monitored, even in a small laboratory. Formal reviews of the management system need to be held on some periodic basis. Audits can monitor the system. Complaints are another tool. Various business metrics can also be monitored. Business metrics can be very simple, such as volume of a particular type of equipment over some period of time to very complex and detailed statistical analysis. Whether the laboratory is large or tiny, the quality manual should describe how the management system is monitored or should point to more specific documents such as procedures that describe the monitoring.

Quality manuals have been in existence in one form or another, and by various names, for a long time. The current term *quality manual* and the more precise definition of its layout and contents is much more recent. ISO 9001:1987 was the first truly international document to give the quality manual real definition and a more consistent purpose. ISO 9001:2000 and ISO/IEC 17025:1999 develop the purpose and requirements for the quality manual more completely than previous documents. For a calibration laboratory that services customers in most fields of business, ISO/IEC 17025 is the primary document used to define the management system and the quality manual. ISO/IEC 17025 has several prescriptive requirements for management review, corrective action, training, and other aspects of the management system. Many customers also require compliance to ISO 9001:2000, so the laboratory will include any additional requirements from that standard. Some industries have prescriptive requirements that may be unique to that industry. A calibration laboratory that services customers in those industries must include the prescriptive requirements in their quality manual.

Some standards such as ISO/IEC 17025 and ISO 9001 require that a statement known as a quality policy be included in the quality manual. The quality policy is simply a summary statement of the laboratory's interpretation of its goals for providing a quality service or product, servicing its customers, and ensuring compliance to standards.

Every employee should be familiar with the quality manual and its contents. Refer to the manual whenever necessary instead of relying on memory. Interpret and apply the principles on a personal level. Management should provide open communication and refresher indoctrination to the manual on a periodic basis. Update the manual as necessary to ensure it stays current. If these steps are followed, the success of the quality manual and of the management system as a whole is assured.

Figure 8.1 is an example of a table of contents for an ISO/IEC 17025-based quality manual, with some explanation of the headings.

---

**TABLE OF CONTENTS**

**Introduction**

**Quality Policy Statement**

**1.0 Scope.** This is the technical scope of the laboratory's operations, that is, what services are provided to clients. If the laboratory is accredited, then the listing will specifically states what services are or are not under the scope of accreditation.

**2.0 References.**

**3.0 Terms and Definitions.**

**4.0 Management Requirements.**

**4.1 Organization.** This is where the organization would delineate that it has legal responsibility for its operation and are organized to operate in accordance with the requirements of ISO 17025. The company would also have an organizational chart showing staff, responsibilities, and so on.

**4.2 Quality System.** The company must document its quality system to include its quality manual and all quality documents. The quality manual is the principal document that defines the quality system for the company. Quality system procedures are used to establish and maintain continuity of each activity within a calibration facility whenever that activity affects quality. Quality procedures should be readily available to personnel for performing calibrations, repairing IM&TE, whenever used as reference material, and during the writing or updating of any quality procedure. Under a four-tier system, the quality document structure contains the company's (1) quality manual, (2) quality procedures, (3) work instructions, and (4) quality records.

**4.3 Quality Policy and Document Control.** The company must have a formal, written quality policy. It should also provide adequate procedures to satisfy all appropriate standards, requirements, or regulations that fall under the scope of the company's calibration function.

**4.4 Review of Requests, Tenders, and Contracts.** All contracts/orders should be legal, binding contracts. They also should be clear, concise, and easily understood, providing the customer with satisfactory service in a timely manner. Cost is also a major player in the willingness of customers to obtain a company's services. This system should have a formal system for record reviews on a regular basis and, as other parts of a quality system, this review is documented and saved for future reference. Whenever there is any deviation from a contract or service with a customer, it must be in writing and saved or archived.

**4.5 Subcontracting of Tests and Calibration.** If calibrations are sent to another facility, for whatever reason, the customer must be informed and all parameters of the original agreement must be met by the subcontractor. It is critical to ensure traceability of measurement for the customer that the proper documentation is received and passed to the customer.

**4.6 Purchasing Services and Supplies.** To comply with most ISO guidelines, a company is usually required to audit outside vendors or suppliers. This is especially critical when its services or parts affect the quality of service provided by your company to your clients.

**4.7 Service to the Client.** A major player in most third-party calibration facilities is being able to provide confidentiality to their customers and being able to maintain that confidentiality during outside audits and/or visits by other customers. The segregation of the customer's IM&TE may be a major part of confidentiality if it uses unique, specialized equipment that is being sent to one's company for calibration. Also, the security of records, calibration certificates, and so on, must also be considered in the support of those customers.

**4.8 Complaints.** The company must have a formal complaint system, both in how the complaints can be received and how they are managed, resolved, and documented.

---

**Figure 8.1** A sample table of contents for an ISO/IEC 17025-based quality manual. *continued*

*continued*

**4.9 Control of Nonconforming Testing and/or Calibration Work.** The company must establish and maintain a policy and procedures that are used whenever its testing and/or calibrations do not meet stated tolerances, the scope of its work, or any part of what has been agreed upon with the customer. These procedures should state that nonconforming product, services, calibrations, or any type of work performed for the customer are identified, segregated, and managed to prevent unintended use. All of this must be performed in writing, according to stated written policies, with the customer being informed of all actions taken and problem resolution solutions. Records showing how the quality system was improved to prevent further occurrences must also be documented.

**4.10 Corrective Action.** This section could be seen as being reactive to identified problems. Problems must also be thoroughly documented and a system must be in place for controlling corrective actions.

**4.11 Preventive Action.** This section could be seen as being proactive in finding solutions to problems. As in paragraph 4.10, everything is documented from the procedure to all actions taken and resolved.

**4.12 Control of Records.** Records that must be controlled include internal audit and management review records. Calibration information, including original handwritten observations are records and must be controlled also.

**4.13 Internal Audits.** Internal audits must be on a preplanned schedule. Follow-up audits are required for corrective actions.

**4.14 Management Reviews.** Management reviews must be periodic and on a prearranged schedule. Many laboratories perform annual reviews.

**5.0 Technical Requirements.**

**5.1 Technical Requirements—General.**

**5.2 Personnel.** Training plans must exist and records of training qualifications must be maintained.

**5.3 Accommodations and Environmental Conditions.** Environmental conditions must be specified for both in-laboratory and on-site environments. Acceptable requirements may vary widely for some applications, but generally will be consistent with requirements outlined by NCSLI and ASTM.

**5.4 Test and Calibration Methods and Method Validation.** Most calibration methods will come from sources such as the manufacturer or the available military procedures (for example, USAF 33K, USN 17-20 series). Calibration procedures developed by the laboratory must be validated. This typically requires predefined criteria to measure the success of the procedure and a report that describes the observed results of the use of the procedure and specific acceptance by management.

**5.5 Equipment.** Equipment must be fit for use and purpose. Equipment must be calibrated if it measures a quantity.

**5.6 Measurement Traceability.** Traceability is to be to SI units through national or international standards, whenever possible. For consensus or similar standards, traceability to SI is rarely possible and alternative methods outlined in other area of the handbook will provide more complete guidance.

**5.7 Sampling.** This is often not applicable for calibration laboratories, as most laboratories calibrate 100 percent of items sent by clients, where the item can be calibrated.

**5.8 Handling and Transportation of Test and/or Calibration Items.** The laboratory must have procedures and facilities to ensure that client equipment is kept secure and safe from damage.

**5.9 Assuring the Quality of Test and Calibration Results.** This will involve the use of proficiency testing and often other methods for internal monitoring also.

**5.10 Reporting the Results.** This section includes the prescriptive requirements for certificates and reports.

**Figure 8.1** A sample table of contents for ISO/IEC 17025-based quality manual.

## References

ANSI/ISO 17025-1999, *American National Standard—General requirements for the competence of testing and calibration laboratories.* Milwaukee: ASQ Quality Press.

ANSI/ISO/ASQ Q9001-2000, *American National Standard—Quality management systems—Requirements.* Milwaukee: ASQ Quality Press.

# Chapter 9
# Traceability

What is the definition of traceability? According to the ISO International Vocabulary of Basic and General Terms of Metrology (VIM), 1993, paragraph 6.10: "Property of the result of a measurement or the value of a standard whereby it can be related to stated references, usually national or international standards, through an unbroken chain of comparisons having stated uncertainties. Notes: 1. The concept is often expressed by the adjective traceable. 2. The unbroken chain of comparisons is called a traceability chain." By the end of this chapter, the reader should have a better understanding when the phrase *unbroken chain of comparisons* is stated in any document, customer requirement, or by a particular standard. This phrase helps form the foundation for traceable measurements in the metrology profession.

According to ANSI/ISO 17025, paragraph 5.6.2.1.1: "For calibration laboratories, the programme for calibration of equipment shall . . . ensure that calibrations and measurements . . . are traceable to the International System of Units (SI) . . . establishes traceability . . . by means of an unbroken chain of calibrations or comparisons." ANSI/NCSL Z540, NCSLI RP-6, and ANSI/ASQC M1-1996 all state the same qualification for any calibration system. In the United States, when we talk about national standards, we are usually referring to the standards kept at the National Institution of Standards and Technology (NIST) formerly known as the National Bureau of Standards (NBS). NIST is also called a national metrology institute (NMI). NMIs are the highest level of traceability for most countries. NMIs will realize a definition of an intrinsic standard set forth by the Bureau International des Poids et Mesures (BIPM). The BIPM ensures worldwide uniformity of measurements and their traceability to the SI and is located in Sèvres, France. Upon realization of the defined standard set forth by the BIPM, NMIs will disseminate these measurements to reference levels. These two acronyms, NIST and NMI, could be used interchangeably when talking about the highest level of standards in the United States.

Accreditation bodies such as American Association for Laboratory Accreditation (A2LA) have their own guidelines for traceability as defined:

Traceability is characterized by six essential elements:

1. **An unbroken chain of comparison.** Traceability begins with an unbroken chain of comparisons originating at national, international, or intrinsic standards of

measurement and ending with the working reference standards of a given metrology laboratory.

2. **Measurement uncertainty.** The measurement uncertainty for each step in the traceability chain must be calculated according to defined methods and must be stated at each step of the chain so that an overall uncertainty for the whole chain can be calculated.

3. **Documentation.** Each step in the chain must be performed according to documented and generally acknowledged procedures and the results must be documented, in a calibration or test report.

4. **Competence.** The laboratories or bodies performing one or more steps in the chain must supply evidence of technical competence, for example, by demonstrating that they are accredited by a recognized accreditation body.

5. **Reference to SI units.** Where possible, the primary national, international, or intrinsic standards must be primary standards for realization of the SI units.

6. **Recalibrations.** Calibrations must be repeated at appropriate intervals in such a manner that traceability of the standard is preserved.

What is an unbroken chain of comparisons back to a national or international standard? Does this mean all of a laboratory's standards must be sent to NIST for calibration to comply with the different requirements? The answer is both simple and complex. No, a lab does not have to send its standards to NIST for their calibrations, but it must be able to prove that an unbroken chain of comparisons exists. This way its customers can show they have traceability to a national or international standard, through their calibrations and those of any outside vendor whose services they may be using. Let's start with a standard mass, the kilogram.

An organization's working standard weight meets the tolerance specified by the vendor that calibrated it. It knows because the calibration supplier provided it with a calibration certificate showing the uncertainty of the weight, as well as the uncertainty of the weight it was compared against. The calibration supplier also documents that its reference or standard weight was calibrated by or is traceable to NIST, and lists the procedures and tolerances that were used. An organization must also make the same statements to its customer about the item it is calibrating. If the organization is a commercial calibration facility, its calibration certificates should include all the pertinent data showing traceability. If it is calibrating only for internal customers, then calibration records should have a traceability statement, as well as identify which standard was used, when it was due calibration, and its uncertainty.

Here's a cautionary note about a concept called *circular calibration*. This happens when a particular standard is used to calibrate another unit. Then, the other unit is in turn used to certify or verify the tolerances of the first standard. This is not an unbroken chain of comparisons, since there is no link to a higher standard to show traceability. In this case, traceability does not exist for this standard. One way to observe if traceability is present is to check for documentation that shows a more accurate standard was used to calibrate the unit and that this was accomplished at each stage along the traceability chain.

Why is this unbroken chain of comparisons required? Does it really benefit anyone or anything? Consider these examples. An aircraft loads its passengers and takes off from the United States. It flies to its destination somewhere outside the United States. Aircraft personnel find that it requires some repairs and needs to have some parts replaced before it can return home. If there is no traceability back to a higher standard, the parts may not fit if manufactured by a different vendor. By having everything calibrated to a traceable standard, it makes no difference where the parts are manufactured, shipped, or replaced. They will work as long as they meet the design specifications and tolerances. The torque wrenches used to secure the replacement parts require calibration, too. If the parts are not torqued to the correct specification, components could come loose in flight, either after initially being installed or after replacement.

Here's another example that happens everyday throughout the world. Scientists are looking for the cure for cancer. They experiment with new molecules and compounds to find the one silver bullet that will save millions of lives and make cancer a thing of the past. They use certain types of IM&TE every day in their experiments. Pipettes, balances, centrifuges, autoclaves, spectrophotometers, and water baths are but a few of the IM&TE used by biotechnology and pharmaceutical companies to aid in the discovery of cures and manufacture their products. Without traceable calibration, they cannot reproduce their results, both in the R&D environment and on the production line. One milliliter of product must meet the exacting tolerances no matter where in the world is it used. Using traceably calibrated IM&TE is the answer for getting repeatable, reliable readings.

The standard traceability pyramid is similar to the one in Figure 9.1. (See Figure 9.2 for another example).

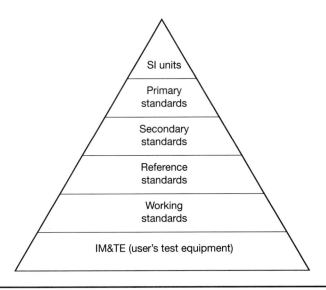

**Figure 9.1** Traceability pyramid—example 1.

64  Part II: Quality Systems

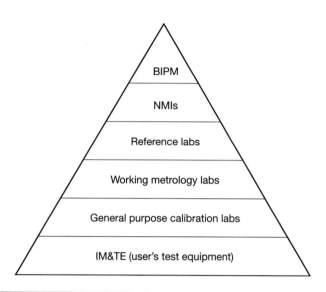

**Figure 9.2** Traceability pyramid—example 2.

As you can see from Figures 9.1 and 9.2, it's possible to skip a level either up or down the pyramid and still maintain traceability. As long as an organization can show the uncertainty up or down the chain and the items being used as standards are more accurate than those being calibrated, it will maintain its traceability. Within an organization, one may have as many as two or three levels of standards. Using the correct standard to perform a calibration could be important. If one is using the highest level standard to calibrate IM&TE that has a very large tolerance range, one could be wasting valuable resources for a small cost benefit. Using a $5,000 standard that is accurate to ± 0.025° C to calibrate a $25.00 temperature device with an accuracy of ± 5° C could be a waste of time and money. On the other hand, using an $80.00 thermometer that has an accuracy of ± 0.5° C, could still maintain traceability while maximizing resources and keeping cost at a minimum. How to allocate resources and use them to their greatest advantage is covered further in Part VII.

It is usually not feasible or desirable to have all IM&TE calibrated by NIST or any NMI. To do so would greatly increase costs while probably not giving the return on investment that would keep a profit business. Also, it would be a burden on the NMI to service all its customers when generally there would be so many overkills on accuracy as to make this a futile effort. Most IM&TE can be calibrated against a company's working standards and still maintain the desired traceability and accuracy required. Working standards should be calibrated against internal reference standards or sent to an outside vendor who can provide traceable calibration to the NMI. During each stage of calibration, the test uncertainty ratio (TUR) is calculated and uncertainty stated in a calibration certificate or record. This provides the unbroken chain of comparisons from the NMI all the way to the IM&TE used daily to find the cure for cancer or ensure safety of flight for aircraft.

Another term that should be discussed is *reverse traceability*. If it is found that the standard being used to calibrate IM&TE is out of tolerance, a determination must be made to account for all of the IM&TE that used that standard between the time it was calibrated to the time it was found to be out of tolerance. Some calibration software programs have reverse traceability reports built into them. Others do not. It can be very labor- and time-intensive to research historical records to ascertain what items were calibrated using a particular standard. But this is exactly what must occur when standards are found to be out of tolerance. An assessment of how this might affect product, processes, or manufacturing must be analyzed and a determination made as to recall of product or stoppage of production. If a reverse traceability program is available for use, it can save valuable time and resources.

Once the items that were calibrated by the out-of-tolerance standard are identified, they must be isolated to preclude any or further usage. An impact assessment of nonconformance must be made on both the IM&TE and the product or process that may have been used during any manufacturing processes or procedures. Was the customer notified in writing? Was a product recall required or initiated? Was this accomplished because of written procedures or because it made sense? Was a root-cause analysis accomplished to find the reason for all the extra work and expense? A lot of questions to be answered when one of the company's standards are found to be out of tolerance. There should be written procedures to address each of these situations, with documentation and records proving what was found, what was done, and who the approving authority was in each case. This is all part of a good quality system and cannot be overlooked.

It is important to note again two of the important phrases from the definition of *traceability*: it is a property of the result of a measurement and all of the comparisons must have stated uncertainties. Traceability applies to the measured value and its uncertainty, as a single entity. One without the other is not traceable. This also means that only the numbers that form the result of a calibration may have the property of traceability. In the metrological sense, traceability never applies to a report document, a calibration procedure, a piece of equipment, or a laboratory. It also means that traceability is not given by having a supplier's test report number on your calibration certificate. That may have some meaning for a purchase order, but it is meaningless for metrological traceability.

# References

ANSI/ISO 17025-1999, *American National Standard—General requirements for the competence of testing and calibration laboratories*. Milwaukee: ASQ Quality Press.

ISO. 1993. *International vocabulary of basic and general terms of metrology (VIM)*. Geneva: ISO.

# Chapter 10
# Calibration Intervals

IM&TE that makes a quantitative measurement requires calibration. Most people will not argue with this statement, but how often does it need to be calibrated? Is the interval between calibrations adjustable, or once it is set, is it written in stone? Who determines the various calibration intervals for different types of IM&TE and for different types of companies? Let's start the discussion with why calibration intervals are required, which should help to answer how often, or what's the calibration interval.

The dependability of any IM&TE has little to do with its accuracy, uncertainty budget, or the number of external knobs. How often an item is used, the type of environment where it is used, and to what extent the user expects it to repeat measurements plays a major role in determining how often it requires calibration. For example, in the world of aircraft, the more often a plane is flown, the more reliable it becomes. This is because some systems stay reliable if they are regularly used, other parts are rebuilt or replaced at defined intervals, and the whole airframe is regularly inspected and maintained. (An equivalent operation for your family automobile would be to have it stripped of all removable items, have all of those items inspected and repaired or replaced as needed, have the body and frame inspected and repaired, and everything put back together—every year.) With proper maintenance a modern aircraft will last indefinitely. If it sits on a ramp or in a hangar for extended periods of time, its components have a higher rate of failure than those used on a regular basis. In some instances, this could be true for IM&TE; it performs more reliably with regular use. In other areas, state-of-the-art units seem to work better with less usage. There are also differences due to the use environment. Two companies may have equal numbers of the same model of IM&TE but have different calibration intervals because one company's use is all indoors and the other company uses their its for field service in all kinds of weather. This is why most companies set its own calibration intervals, using input from the IM&TE's manufacturer, historical data collected over many calibration cycles, and/or data retrieved from outside sources that are using the same or similar types of IM&TE.

There are software solutions available that greatly reduce the number crunching required to analyze data. Others find simple methods more conducive to their applications. Whether purchased software or user-developed methods are used, there can be no doubt that calibration intervals are an important part of any calibration program. Let's examine a simple method used by a newly implemented, fictitious metrology department.

The initial calibration interval for the majority of IM&TE at the Acme Widget Company was set at 12 months. Most manufacturers recommend this for a couple of reasons. If their equipment will not hold calibration for at least a year under normal operating conditions, they probably will not stay in business very long. Second, most IM&TE is sent back to the manufacturer to be calibrated, generating income and after-sale service. It is to the benefit of the manufacturer to have the shortest interval, while ensuring its equipment continues to function properly between calibrations. The Acme Widget Co. was monitoring how often each calibration performed met or did not meet specifications. No matter how far out of tolerance the IM&TE might be, the out-of-tolerance condition was recorded. Over a set period of time, say a year or 18 months, the total number of calibrations for that particular type of IM&TE was tabulated, as well as the number of times the same type of IM&TE did not pass specification. The pass rate for that type of equipment was tallied and recorded.

This same exercise was accomplished on each general type of IM&TE. When completed, the pass rates varied between 92.9 percent and 100 percent. For items with a pass rate greater than 98 percent, their intervals were doubled. For items with a pass rate between 95 percent and 98 percent, their intervals were increased by 50 percent. Pass rates less than 95 percent were examined to see if they would be reduced, or monitored for another round of calibrations. There were instances where items exceeded 95 percent or 98 percent pass rates, but their calibration intervals were not lengthened as per the formula. Those particular items were being used in critical areas of production, on a more frequent basis, or a combination of both. In these instances, it was determined to be more prudent to exercise caution, rather than run the risk of having to recall product or remanufacture goods. Each company must make hard choices with a critical eye to the cost of doing unnecessary calibrations (too frequently for the way the items are used) versus calibrating at extended intervals and run the risk of turning out bad product and spoiling its reputation at the expense of a few dollars. By performing a thorough analysis of calibration intervals on a regular basis, an organization can get the most bang for the buck from its calibration program while reducing the risk associated with lengthening intervals only for the sake of saving time and money.

Because there is no single best practice, and there are several interval analysis methodologies, the first recommended practice (RP) developed by NCSLI was RP-1 *Establishment and Adjustment of Calibration Intervals.* Methods are categorized by their effectiveness, cost to implement, their suitability for large or small inventories, and other factors. One of the factors is the renewal, or adjustment, policy implemented by the calibration laboratory:

- **Renew Always.** An equipment management policy or practice in which IM&TE parameters are adjusted or otherwise optimized (where possible) at every calibration.

- **Renew-if-Failed.** An equipment management policy or practice in which IM&TE parameters are adjusted or otherwise optimized (if possible) only if found out of tolerance at calibration.

- **Renew-as-Needed.** An equipment management policy or practice in which IM&TE parameters are adjusted or otherwise optimized (if necessary) if found outside safe adjustment limits.

The methods listed in RP-1 are in two broad groups: those that use statistical tests and those that do not. Each has benefits and drawbacks. The nonstatistical approaches are low-cost and can be implemented easily. They are very slow to settle to a stable calibration interval, however. It is possible for an instrument to be on a different interval after each calibration, which can be havoc on the laboratory work scheduling. The statistical approaches can achieve a good calibration interval very quickly. They require a lot of historical data, however, and range from moderately to very expensive to implement. If an organization has a large inventory with large numbers of similar items (such as 300 of a single model of meter), a statistical approach may be useful. A small inventory or many different models with only a few units of each may require time periods longer than the average life of the equipment for sufficient data to be collected for a statistically valid test.

A side topic of calibration intervals is how the next due date is determined within the calibration interval program. Some calculate the next due date by day, month, and year from the day it was last calibrated. Others simply use only the month and year. Here are some pros and cons for both options.

Using only the month and year allows organizations more flexibility as to when they can schedule the next calibrations. This option also helps reduce the number of overdue calibrations that show up during audits and inspections. Any item due calibration during the month will not show up as overdue until the following month. If historical data support the reliability of the IM&TE using this system, it can also make a calibration function look better to upper management during audits. The downside to this system is encountered when working with IM&TE that have a short calibration interval of only one or two months. As an example, assume an item has a two-month calibration interval. It was calibrated on February 2, making it due calibration during the month of April. If it is next calibrated on April 30, it has gone almost three months between calibrations. If it was calibrated on February 27, and next calibrated on April 1, it was barely a month between calibrations. This is one of the problems encountered using the month/year method of stating the next calibration date.

When using the day/month/year system, the exact interval is applied to each IM&TE and allows the user to know exactly when he or she can no longer use it—at midnight of the day on the calibration label it is out of calibration. No guessing or wondering on the part of the user or calibration technician. Most software packages used in calibration systems calculate the next due date according to these conventions, and consistency throughout the industry reduces training on this topic as technicians move from company to company. If a company is using only month and year on their calibration labels, however, it might better suit its needs in calibrating all pieces of IM&TE while a production line is down for maintenance or repairs. That way nothing goes overdue during that month, no matter which day of the month it was calibrated.

Another factor to consider with this is that time between calibrations is data used by interval analysis methods. When grouping items for statistical analysis, one of the key criteria is that they all have to be on the same calibration interval, and all should have been on that interval long enough to have at least two calibrations each. Historically, many organizations have recorded intervals in a variety of units—usually weeks, months, or years—and varying it by individual item. It will be easier to manage interval analysis and adjustment if the same units are used for all items in the calibration program. It does not matter what unit is used, as long as it results in whole numbers and is the same used for all items. Many organizations use months as the units, and others use weeks. Some use years for everything, but that is practical only if all items are very durable and highly reliable.

Finally, the calibration procedure used is also a key factor in interval analysis. A requirement of statistical analysis methods is that all items in the group have to have been calibrated using the same procedure. If a calibration procedure is changed, only the data collected before or after the change can be used in the analysis.

While on the topic of calibration intervals, it should be mentioned that most organizations use a software program to trigger an effective reminder/recall for recalibration of their IM&TE, plus maintain history using that program, as opposed to a paper system. If you are still using a paper-based recall system, it is possible that many time-saving functions are being overlooked, and the chances of items falling through the cracks are significantly higher.

A laboratory—particularly an in-house lab—will often get pressure from customers to extend a calibration interval mainly for their own convenience. This should be resisted. A calibration interval should only be increased based on data that show the reliability of the instruments support it. It is much easier, and more acceptable, to shorten an interval based on limited data. For example, suppose that a company has a goal of 95 percent reliability; that is, 95 percent of all items should be in tolerance when calibrated. If a particular model of meters has failed five of the last 10 calibrations, reducing the interval is highly indicated. An increase in the interval, however, could only be justified after the next 100 calibrations have been performed with no failures. (In the case of this example, the interval was reduced *and* it was suggested to the customer it seek out a model with higher reliability for their needs.)

How often to calibrate IM&TE is a balancing act between the costs associated with time and resources (buying standards, providing adequate facilities, and hiring and training competent staff, to name a few). This is true for both in-house calibration functions as well as those that outsource all or part of their calibrations. By extending calibration intervals without enough historical data, an organization runs the risk of being on the ragged edge; compared with making its interval adjustments in a systematic, calculated fashion, putting their program on the cutting edge, in both reliability and dependability.

## References

The Healthcare Metrology Committee. 1999. *Calibration control systems for the biomedical and pharmaceutical industry*. Recommended Practice RP-6. Boulder, CO: National Conference of Standards Laboratories.

# Chapter 11
## Calibration Standards

It was a common belief among calibration technicians in U.S. Air Force Precision Measurement Equipment Laboratories (PMELs) that IM&TE calibrated at one PMEL would get the same results if calibrated at another PMEL, no matter if it was on the same base or half way around the world. The same is true in most industries. An item calibrated at one location should be able to make a like measurement anywhere within that system. This is possible because measurement systems that use standards traceable to a national or international standard will form the foundation for all measurements used within that SI. As one can see, calibration standards are the basis for all of our measurements, no matter how large or small the system, measurement, or equipment.

As mentioned in Chapter 1, Britain had a system of units using the foot, pound, and second. The other system was the meter, kilogram, and second. The use of different systems of measurement caused confusion in international trade and communications. Measurement results can be understood by all the parties involved only if the unit in which it is expressed is standardized. This was a major challenge for metrologists worldwide beginning in the nineteenth century.

The use of various systems of units affected trade between countries. The CGPM in 1954 agreed to form a uniform and comprehensive system of units based on the metric system then in use. In 1960, the CGPM established the SI for use throughout the world. This system originally had six base units, but in 1971 the unit of chemical substances, the mole, was added to it. The SI is the foundation of modern metrology and is practically the sole legitimate system of measurement units in use throughout the world today.

A *unit of measurement* has been defined as a particular quantity defined and adopted by convention, with which other quantities of the same kind are compared in order to express their magnitude relative to that quantity. The SI has been defined as the coherent system of units adopted and recommended by the GMCP. SI units consist of seven base units and two supplementary units. The base units are regarded as dimensionally independent, and all the other units are derived from the base units or from other derived units. The base and supplementary units consist of length (the *meter*, m), mass (*kilogram*, kg), time (*second*, s), electric current (*ampere*, A), thermodynamic temperature (*kelvin*, K), luminous intensity (*candela*, cd), amount of a substance (*mole*, mol), plane angle (*radian*, rad), and solid angle (*steradian*, Sr).

The SI has 19 derived units that are obtained by forming various combinations of the base units, supplementary units, and other derived units. Their names, symbols and how their values are obtained can be found in Table 11.1. Table 11.2 contains their prefixes.

**Table 11.1** SI derived units.

| Parameter | Unit (symbol) | Value |
|---|---|---|
| Frequency | Hertz (Hz) | l/s |
| Force | Newton (N) | kg·m/s² |
| Pressure, stress | Pascal (Pa) | N/m2 |
| Energy, work, quantity of heat | Joule (J) | N◊m |
| Power, radiant flux | Watt (W) | J/s |
| Electric potential difference | Volt (V) | W/A |
| Electric resistance | Ohm (Ω) | V/A |
| Electric charge | Coulomb (C) | A·s |
| Electric capacitance | Farad (F) | C/V |
| Electric conductance | Siemens (S) | A/V |
| Magnetic flux | Weber (Wb) | V·S |
| Magnetic flux density | Tesla (T) | Wb/m² |
| Inductance | Henry (H) | Wb/A |
| Celsius temperature | Degree (°C) | – |
| Luminous flux | Lumen (lm) | cd·Sr |
| Illuminance | Lux (lx) | Lm/m² |
| Activity (of a radionuclide) | Becquerel (Bq) | – |
| Absorbed dose | Gray (Gy) | J/kg |
| Dose equivalent | Sievert (Sv) | J/kg |

**Table 11.2** SI prefixes.

| Factor | Prefix | Symbol | Factor | Prefix | Symbol |
|---|---|---|---|---|---|
| $10^{24} = (10^3)^8$ | yotta | Y | $10^{-1}$ | deci | d |
| $10^{21} = (10^3)^7$ | zetta | Z | $10^{-2}$ | centi | c |
| $10^{18} = (10^3)^6$ | exa | E | $10^{-3} = (10^3)^{-1}$ | milli | m |
| $10^{15} = (10^3)^5$ | peta | P | $10^{-6} = (10^3)^{-2}$ | micro | μ |
| $10^{12} = (10^3)^4$ | tera | T | $10^{-9} = (10^3)^{-3}$ | nano | n |
| $10^9 = (10^3)^3$ | giga | G | $10^{-12} = (10^3)^{-4}$ | pico | p |
| $10^6 = (10^3)^2$ | mega | M | $10^{-15} = (10^3)^{-5}$ | femto | f |
| $10^3 = (10^3)^1$ | kilo | k | $10^{-18} = (10^3)^{-6}$ | atto | a |
| $10^2$ | hecto | h | $10^{-21} = (10^3)^{-7}$ | zepto | z |
| $10^1$ | deka | da | $10^{-24} = (10^3)^{-8}$ | yocto | y |

It is well known that SI units have been accepted internationally and are the basis of modern measurements for governments, academia, and industry. The evolution of practical national and international measurement systems is achieved in four stages.

- **Definition of the unit.** The accuracy of the definition of a unit is important, as it will be reflected in the accuracy of the measurements that can be achieved.

- **Realization of the unit.** The definition of the unit has to be realized so that it can be used as a reference for measurement. This task is carried out by NMIs. The units are realized in the form of experimental setups.

- **Representation of the unit.** The realized experimental setup of the system of the unit is the physical representation of the unit. NMIs are responsible for the maintenance of this representation. They ensure that these experimental setups continue to represent the SI and are available for reference.

- **Dissemination of the unit.** The end users of measurements are trade, industry, and calibration laboratories. They do not have access to the representations to the SI units held by NMIs. The end users also need the values of the SI units for reference. This is accomplished through the process of dissemination, wherein the units are made available to the end users of measurement results.

A harmonized measurement unit ensures that everybody concerned with a measurement result understands it the same way. For acceptability, it is also essential that the measurement results for the same parameter measured at different places and by different people be in agreement. This implies that measurement results should be correlated. To achieve this agreement in measurement results, it is essential that everybody draw his or her measurement units from a common acceptable standard. Thus, a *standard* is a physical object or a characteristic of a physical apparatus that represents the conceptual unit chosen to represent a particular measurable attribute. As mentioned previously, measurement is critical to trade, industry and government worldwide. The same can be said for measurement standards.

The following are taken from NIST Special Publication 811, sections 6 and 7. These Rules and Style Conventions could greatly assist anyone in their use and understanding of SI units.

---

**6 Rules and Style Conventions for Printing and Using Units**

**6.1 Rules and style conventions for unit symbols**
The following eight sections give rules and style conventions related to the symbols for units.

**6.1.1 Typeface**
Unit symbols are printed in roman (upright) type regardless of the type used in the surrounding text. (See also Sec. 10.2 and Secs. 10.2.1 to 10.2.4.)

**6.1.2 Capitalization**
Unit symbols are printed in lower-case letters except that:

(a) the symbol or the first letter of the symbol is an upper-case letter when the name of the unit is derived from the name of a person; and

(b) the recommended symbol for the liter in the United States is L [see Table 6, footnote (*b*)].

*Examples:*   m (meter) s (second) V (volt) Pa (pascal) lm (lumen) Wb (weber)

### 6.1.3 Plurals
Unit symbols are unaltered in the plural.

*Example:*   $l = 75$ cm *but not:* $l = 75$ cms

*Note: l* is the quantity symbol for length. (The rules and style conventions for expressing the values of quantities are discussed in detail in Chapter 7.)

### 6.1.4 Punctuation
Unit symbols are not followed by a period unless at the end of a sentence.

*Example:* '   'Its length is 75 cm." or "It is 75 cm long." *but not:* "It is 75 cm. long."

### 6.1.5 Unit symbols obtained by multiplication
Symbols for units formed from other units by multiplication are indicated by means of either a half-high (that is, centered) dot or a space. However, this *Guide*, as does Ref. [8], prefers the half-high dot because it is less likely to lead to confusion.

*Example:*   N • m or N m

*Notes:*
1. A half-high dot or space is usually imperative. For example, m • $s^{-1}$ is the symbol for the meter per second while $ms^{-1}$ is the symbol for the reciprocal millisecond ($10^3$ $s^{-1}$—see Sec. 6.2.3).

2. Reference [6: ISO 31-0] suggests that if a space is used to indicate units formed by multiplication, the space may be omitted if it does not cause confusion. This possibility is reflected in the common practice of using the symbol kWh rather than kW • h or kW h for the kilowatt hour. Nevertheless, this *Guide* takes the position that a half-high dot or a space should always be used to avoid possible confusion; and that for this same reason, only one of these two allowed forms should be used in any given manuscript.

### 6.1.6 Unit symbols obtained by division

Symbols for units formed from other units by division are indicated by means of a solidus (oblique stroke, /), a horizontal line, or negative exponents.

*Example:*   m/s, $\frac{m}{s}$, or m • $s^{-1}$

However, to avoid ambiguity, the solidus must not be repeated on the same line unless parentheses are used.

*Examples:*   m/$s^2$ or m • $s^{-2}$ *but not:* m/s/s

   m • kg/($s^3$ • A) or m • kg • $s^{-3}$ • $A^{-1}$ *but not:* m • kg/$s^3$/A

Negative exponents should be used in complicated cases.

### 6.1.7 Unacceptability of unit symbols and unit names together
Unit symbols and unit names are not used together. (See also Secs. 9.5 and 9.8.)

*Example:* C/kg, C • kg$^{-1}$, or coulomb per *but not:* coulomb/kg; coulomb per kg; kilogram C/kilogram; coulomb • kg$^{-1}$; C per kg; coulomb/kilogram

### 6.1.8 Unacceptability of abbreviations for units
Because acceptable units generally have internationally recognized symbols and names, it is not permissible to use abbreviations for their unit symbols or names, such as sec (for either s or second), sq. mm (for either mm$^2$ or square millimeter), cc (for either cm$^3$ or cubic centimeter), mins (for either min or minutes), hrs (for either h or hours), lit (for either L or liter), amps (for either A or amperes), AMU (for either u or unified atomic mass unit), or mps (for either m/s or meter per second). Although the values of quantities are normally expressed using symbols for numbers and symbols for units (see Sec. 7.6), if for some reason the name of a unit is more appropriate than the unit symbol (see Sec. 7.6, note 3), the name of the unit should be spelled out in full.

## 6.2 Rules and style conventions for SI prefixes
The following eight sections give rules and style conventions related to the SI prefixes.

### 6.2.1 Typeface and spacing
Prefix symbols are printed in roman (upright) type regardless of the type used in the surrounding text, and are attached to unit symbols without a space between the prefix symbol and the unit symbol. This last rule also applies to prefixes attached to unit names.

*Examples:* mL (milliliter) pm (picometer) GΩ (gigaohm) THz (terahertz)

### 6.2.2 Capitalization
The prefix symbols Y (yotta), Z (zetta), E (exa), P (peta), T (tera), G (giga), and M (mega) are printed in upper-case letters while all other prefix symbols are printed in lower-case letters (see Table 5). Prefixes are normally printed in lower-case letters.

### 6.2.3 Inseparability of prefix and unit
The grouping formed by a prefix symbol attached to a unit symbol constitutes a new inseparable symbol (forming a multiple or submultiple of the unit concerned) which can be raised to a positive or negative power and which can be combined with other unit symbols to form compound unit symbols.

*Examples:*
$2.3 \text{ cm}^3 = 2.3 \text{ (cm)}^3 = 2.3 \text{ (}10^{-2}\text{ m)}^3 = 2.3 \times 10^{-6} \text{ m}^3$
$1 \text{ cm}^{-1} = 1 \text{ (cm}^{-1}) = 1 \text{ (}10^{-2}\text{ m)}^{-1} = 10^2 \text{ m}^{-1}$
$5000 \text{ μs}^{-1} = 5000 \text{ (μs)}^{-1} = 5000 \text{ (}10^{-6}\text{ s)}^{-1} = 5000 \times 10^6 \text{ s}^{-1} = 5 \times 10^9 \text{ s}^{-1}$
$1 \text{ V/cm} = (1 \text{ V})/(10^{-2}\text{m}) = 10^2 \text{ V/m}$

Prefixes are also inseparable from the unit names to which they are attached. Thus, for example, millimeter, micropascal, and meganewton are single words.

### 6.2.4 Unacceptability of compound prefixes
Compound prefix symbols, that is, prefix symbols formed by the juxtaposition

of two or more prefix symbols, are not permitted. This rule also applies to compound prefixes.

*Example:* nm (nanometer) *but not:* mμm (millimicrometer)

### 6.2.5 Use of multiple prefixes
In a derived unit formed by division, the use of a prefix symbol (or a prefix) in both the numerator *and* the denominator may cause confusion. Thus, for example, 10 kV/mm is acceptable, but 10 MV/m is often considered preferable because it contains only one prefix symbol and it is in the numerator.

In a derived unit formed by multiplication, the use of more than one prefix symbol (or more than one prefix) may also cause confusion. Thus, for example, 10 MV • ms is acceptable, but 10 kV • s is often considered preferable.

*Note:* Such considerations usually do not apply if the derived unit involves the kilogram. For example, 0.13 mmol/g is *not* considered preferable to 0.13 mol/kg.

### 6.2.6 Unacceptability of stand-alone prefixes
Prefix symbols cannot stand alone and thus cannot be attached to the number 1, the symbol for the unit one. In a similar vein, prefixes cannot be attached to the name of the unit one, that is, to the word "one." (See Sec. 7.10 for a discussion of the unit one.)

*Example:* the number density of Pb atoms is $5 \times 10^6/m^3$ *but not:* the number density of Pb atoms is $5 M/m^3$

### 6.2.7 Prefixes and the kilogram
For historical reasons, the name "kilogram" for the SI base unit of mass contains the name "kilo," the SI prefix for $10^3$. Thus, because compound prefixes are unacceptable (see Sec. 6.2.4), symbols for decimal multiples and submultiples of the unit of mass are formed by attaching SI prefix symbols to g, the unit symbol for gram, and the names of such multiples and submultiples are formed by attaching SI prefixes to the name "gram."

*Example:* $10^{-6}$ kg = 1 mg (1 milligram) *but not:* $10^{-6}$ kg = 1 μkg (1 microkilogram)

### 6.2.8 Prefixes with the degree Celsius and units accepted for use with the SI
Prefix symbols may be used with the unit symbol °C and prefixes may be used with the unit name "degree Celsius." For example, 12 m°C (12 millidegrees Celsius) is acceptable. However, to avoid confusion, prefix symbols (and prefixes) are not used with the time- related unit symbols (names) min (minute), h (hour), d (day); nor with the angle-related symbols (names) ° (degree), ' (minute), and " (second) (see Table 6).

Prefix symbols (and prefixes) may be used with the unit symbols (names) L (liter), t (metric ton), eV (electronvolt), and u (unified atomic mass unit) (see Tables 6 and 7). However, although submultiples of the liter such as mL (milliliter) and dL (deciliter) are in common use, multiples of the liter such as kL (kiloliter) and ML (megaliter) are not. Similarly, although multiples of the metric ton such as kt (kilometric ton) are commonly used, submultiples such as mt (millimetric ton), which is equal to the kilogram (kg), are not. Examples of the use of prefix symbols with eV and u are 80 MeV (80 megaelectronvolts) and 15 nu (15 nanounified atomic mass units).

## 7 Rules and Style Conventions for Expressing Values of Quantities

### 7.1 Value and numerical value of a quantity

The *value* of a quantity is its magnitude expressed as the product of a number and a unit, and the number multiplying the unit is the *numerical value* of the quantity expressed in that unit.

More formally, the value of quantity $A$ can be written as $A = \{A\} [A]$, where $\{A\}$ is the numerical value of $A$ when the value of $A$ is expressed in the unit $[A]$. The numerical value can therefore be written as $\{A\} = A / [A]$, which is a convenient form for use in figures and tables. Thus, to eliminate the possibility of misunderstanding, an axis of a graph or the heading of a column of a table can be labeled "$t/°C$" instead of "$t$ (°C)" or "Temperature (°C)." Similarly, an axis or column heading can be labeled "$E/(V/m)$" instead of "$E$ (V/m)" or "Electric field strength (V/m)."

*Examples:*
1. In the SI, the value of the velocity of light in vacuum is $c$ = 299 792 458 m/s exactly. The number 299 792 458 is the numerical value of $c$ when c is expressed in the unit m/s, and equals $c$ /(m/s).

2. The ordinate of a graph is labeled $T/(10^3 K)$, where $T$ is thermodynamic temperature and K is the unit symbol for kelvin, and has scale marks at 0, 1, 2, 3, 4, and 5. If the ordinate value of a point on a curve in the graph is estimated to be 3.2, the corresponding temperature is $T/(10^3 K)$ = 3.2 or $T$ = 3200 K. Notice the lack of ambiguity in this form of labeling compared with "Temperature $(10^3 K)$."

3. An expression such as ln $(p/MPa)$, where $p$ is the quantity symbol for pressure and Mpa is the unit symbol for megapascal, is perfectly acceptable because $p/MPa$ is the numerical value of $p$ when $p$ is expressed in the unit MPa and is simply a number.

*Notes:*
1. For the conventions concerning the grouping of digits, see Sec. 10.5.3.

2. An alternative way of writing $c/(m/s)$ is $\{c\}_{m/s}$, meaning the numerical value of $c$ when $c$ is expressed in the unit m/s.

### 7.2 Space between numerical value and unit symbol

In the expression for the value of a quantity, the unit symbol is placed after the numerical value and a *space* is left between the numerical value and the unit symbol. The only exceptions to this rule are for the unit symbols for degree, minute, and second for plane angle: °, ', and ", respectively (see Table 6), in which case no space is left between the numerical value and the unit symbol.

*Example:* α= 30°22'8"

*Note:* α is a quantity symbol for plane angle.

This rule means that:

(a) The symbol °C for the degree Celsius is preceded by a space when one expresses the values of Celsius temperatures.

*Example:*   $t = 30.2$ °C *but not:* $t = 30.2$°C or $t = 30.2$° C

(b) Even when the value of a quantity is used in an adjectival sense, a space is left between the numerical value and the unit symbol. (This rule recognizes that unit symbols are not like ordinary words or abbreviations but are mathematical entities, and that the value of a quantity should be expressed in a way that is as independent of language as possible — see Secs. 7.6 and 7.10.3.)

*Examples:*   a 1 m end gauge *but not* : a 1-m end gauge
a 10 kΩ resistor *but not:* a 10-kΩ resistor

However, if there is any ambiguity, the words should be rearranged accordingly. For example, the statement "the samples were placed in 22 mL vials" should be replaced with the statement "the samples were placed in vials of volume 22 mL."

*Note:* When unit names are spelled out, the normal rules of English apply. Thus, for example, "a roll of 35-millimeter film" is acceptable (see Sec. 7.6, note 3).

### 7.3 Number of units per value of a quantity

The value of a quantity is expressed using no more than one unit.

*Example:*   $l = 10.234$ m *but not:* $l = 10$ m 23 cm 4 mm

*Note:* Expressing the values of time intervals and of plane angles are exceptions to this rule. However, it is preferable to divide the degree decimally. Thus one should write 22.20° rather than 22°12', except in fields such as cartography and astronomy.

### 7.4 Unacceptability of attaching information to units

When one gives the value of a quantity, it is incorrect to attach letters or other symbols to the unit in order to provide information about the quantity or its conditions of measurement. Instead, the letters or other symbols should be attached to the quantity.

*Example:*   $V_{max} = 1000$ V *but not:* $V = 1000$ $V_{max}$

*Note:* V is a quantity symbol for potential difference.

### 7.5 Unacceptability of mixing information with units

When one gives the value of a quantity, any information concerning the quantity or its conditions of measurement must be presented in such a way as not to be associated with the unit. This means that quantities must be defined so that they can be expressed solely in acceptable units (including the unit one — see Sec. 7.10).

*Examples:*
the Pb content is 5 ng/L *but not:* 5 ng Pb/L or 5 ng of lead/L;
the sensitivity for $NO_3$ molecules is $5 \times 10^{10}/cm^3$ *but not:* the sensitivity is
    $5 \times 10^{10}$ $NO_3$ molecules/$cm^3$;
the neutron emission rate is $5 \times 10^{10}/s$ *but not:* the emission rate is $5 \times 10^{10}$ n/s;
the number density of $O_2$ atoms is $3 \times 10^{18}/cm^3$ *but not:* the density is $3 \times 10^{18}$
    $O_2$ atoms/$cm^3$;
the resistance per square is 100 $\Omega$ *but not:* the resistance is 100 $\Omega$/ square

**7.6 Symbols for numbers and units versus spelled-out names of numbers and units**

This *Guide* takes the position that the key elements of a scientific or technical paper, particularly the results of measurements and the values of quantities that influence the measurements, should be presented in a way that is as independent of language as possible. This will allow the paper to be understood by as broad an audience as possible, including readers with limited knowledge of English. Thus, to promote the comprehension of quantitative information in general and its broad understandability in particular, values of quantities should be expressed in acceptable units using

— the Arabic symbols for numbers, that is, the Arabic numerals, *not* the spelled-out names of the Arabic numerals; and

— the symbols for the units, *not* the spelled-out names of the units.

*Examples:*
the length of the laser is 5 m *but not:* the length of the laser is five meters;
the sample was annealed at a temperature of 955 K for 12 h
    *but not* : the sample was annealed at a temperature of 955 kelvins
for 12 hours

*Notes:*

1. If the intended audience for a publication is unlikely to be familiar with a particular unit symbol, it should be defined when first used.

2. Because the use of the spelled-out name of an Arabic numeral with a unit symbol can cause confusion, such combinations must strictly be avoided. For example, one should never write "the length of the laser is five m."

3. Occasionally, a value is used in a descriptive or literary manner and it is fitting to use the spelled-out name of the unit rather than its symbol. Thus this *Guide* considers acceptable statements such as "the reading lamp was designed to take two 60-watt light bulbs," or "the rocket journeyed uneventfully across 380000 kilometers of space," or "they bought a roll of 35-millimeter film for their camera."

4. The *United States Government Printing Office Style Manual* (Ref. [4], pp. 165-171) gives the rule that symbols for numbers are always to be used when one expresses (a) the value of a quantity in terms of a unit of measurement, (b) time (including dates), and (c) an amount of money. This publication should be consulted for the rules governing the choice between the use of symbols for

numbers and the spelled-out names of numbers when numbers are dealt with in general.

### 7.7 Clarity in writing values of quantities

The value of a quantity is expressed as the product of a number and a unit (see Sec. 7.1). Thus, to avoid possible confusion, this *Guide* takes the position that values of quantities must be written so that it is completely clear to which unit symbols the numerical values of the quantities belong. Also to avoid possible confusion, this *Guide* strongly recommends that the word "to" be used to indicate a range of values for a quantity instead of a range dash (that is, a long hyphen) because the dash could be misinterpreted as a minus sign. (The first of these recommendations once again recognizes that unit symbols are not like ordinary words or abbreviations but are mathematical entities—see Sec. 7.2.)

*Examples:*

51 mm × 51 mm × 25 mm *but not:* 51 × 51 × 25 mm

225 nm to 2400 nm or (225 to 2400) nm *but not:* 225 to 2400 nm

0 °C to 100 °C or (0 to 100) °C *but not:* 0 °C – 100 °C

0 V to 5 V or (0 to 5) V *but not:* 0 – 5 V

(8.2, 9.0, 9.5, 9.8, 10.0) GHz *but not:* 8.2, 9.0, 9.5, 9.8, 10.0 GHz

63.2 m ± 0.1 m or (63.2 ± 0.1) m *but not:* 63.2 ± 0.1 m or 63.2 m ± 0.1

129 s – 3 s = 126 s or (129 – 3) s = 126 s *but not:* 129 – 3 s = 126 s

*Note:* For the conventions concerning the use of the multiplication sign, see Sec. 10.5.4.

### 7.8 Unacceptability of stand-alone unit symbols

Symbols for units are never used without numerical values or quantity symbols (they are not abbreviations).

*Examples:*

there are $10^6$ mm in 1 km *but not*: there are many mm in a km
it is sold by the cubic meter *but not*: it is sold by the $m^3$
$t$ /°C, $E$ /(V/m), $p$/MPa, and the like are perfectly acceptable (see Sec. 7.1)

### 7.9 Choosing SI prefixes

The selection of the appropriate decimal multiple or submultiple of a unit for expressing the value of a quantity, and thus the choice of SI prefix, is governed by several factors. These include:

— the need to indicate which digits of a numerical value are significant,

— the need to have numerical values that are easily understood, and

— the practice in a particular field of science or technology.

A digit is significant if it is required to express the numerical value of a quantity. In the expression $l = 1200$ m, it is not possible to tell whether the last two zeroes are significant or only indicate the magnitude of the numerical value of $l$. However, in the expression $l = 1.200$ km, which uses the SI prefix symbol for $10^3$ (kilo, symbol k), the two zeroes are assumed to be significant because if they were not, the value of $l$ would have been written $l = 1.2$ km.

It is often recommended that, for ease of understanding, prefix symbols should be chosen in such a way that numerical values are between 0.1 and 1000, and that only prefix symbols that represent the number 10 raised to a power that is a multiple of 3 should be used.

*Examples:* $3.3 \times 10^7$ Hz may be written as $33 \times 10^6$ Hz = 33 MHz

0.009 52 g may be written as $9.52 \times 10^{-3}$ g = 9.52 mg

2703 W may be written as $2.703 \times 10^3$ W = 2.703 kW

$5.8 \times 10^{-8}$ m may be written as $58 \times 10^{-9}$ m = 58 nm

However, the values of quantities do not always allow this recommendation to be followed, nor is it mandatory to try to do so.

In a table of values of the same kind of quantities or in a discussion of such values, it is usually recommended that only one prefix symbol should be used even if some of the numerical values are not between 0.1 and 1000. For example, it is often considered preferable to write "the size of the sample is 10 mm × 3 mm × 0.02 mm" rather than "the size of the sample is 1 cm × 3 mm × 20 µm."

In certain kinds of engineering drawings it is customary to express all dimensions in millimeters. This is an example of selecting a prefix based on the practice in a particular field of science or technology.

**7.10 Values of quantities expressed simply as numbers: the unit one, symbol 1**
Certain quantities, such as refractive index, relative permeability, and mass fraction, are defined as the ratio of two mutually comparable quantities and thus are of dimension one (see Sec. 7.14). The coherent SI unit for such a quantity is the ratio of two identical SI units and may be expressed by the number 1. However, the number 1 generally does not appear in the expression for the value of a quantity of dimension one. For example, the value of the refractive index of a given medium is expressed as $n = 1.51 \times 1 = 1.51$.

On the other hand, certain quantities of dimension one have units with special names and symbols which can be used or not depending on the circumstances. Plane angle and solid angle, for which the SI units are the radian (rad) and steradian (sr), respectively, are examples of such quantities (see Sec. 4.3).

**7.10.1 Decimal multiples and submultiples of the unit one**
Because SI prefix symbols cannot be attached to the unit one (see Sec. 6.2.6), powers of 10 are used to express decimal multiples and submultiples of the unit one.

*Example:* $\mu_r = 1.2 \times 1026$ *but not:* $\mu_r = 1.2$ µ

*Note:* $\mu_r$ is the quantity symbol for relative permeability.

### 7.10.2 %, percentage by, fraction

In keeping with Ref. [6: ISO 31-0], this *Guide* takes the position that it is acceptable to use the internationally recognized symbol % (percent) for the number 0.01 with the SI and thus to express the values of quantities of dimension one (see Sec. 7.14) with its aid. When it is used, a space is left between the symbol % and the number by which it is multiplied [6: ISO 31-0]. Further, in keeping with Sec. 7.6, the symbol % should be used, not the name "percent."

*Example:* $x_B = 0.0025 = 0.25$ % *but not:* $x_B = 0.0025 = 0.25$% or $x_B = 0.25$ percent

*Note:* $x_B$ is the quantity symbol for amount-of-substance fraction of B (see Sec. 8.6.2).

Because the symbol % represents simply a number, it is not meaningful to attach information to it (see Sec. 7.4). One must therefore avoid using phrases such as "percentage by weight," "percentage by mass," "percentage by volume," or "percentage by amount of substance."

Similarly, one must avoid writing, for example, "% $(m/m)$," "% (by weight)," "% $(V/V)$," "% (by volume)," or "% (mol/mol)." The preferred forms are "the mass fraction is 0.10," or "the mass fraction is 10 %," or "$w_B = 0.10$," or "$w_B = 10$ %" ($w_B$ is the quantity symbol for mass fraction of B — see Sec. 8.6.10); "the volume fraction is 0.35," or "the volume fraction is 35 %," or "$\varphi_B = 0.35$," or "$\varphi_B = 35$ %" ($\varphi_B$ is the quantity symbol for volume fraction of B —see Sec. 8.6.6); and "the amount-of-substance fraction is 0.15," or "the amount-of-substance fraction is 15 %," or "$x_B = 0.15$," or "$x_B = 15$ %." Mass fraction, volume fraction, and amount-of-substance fraction of B may also be expressed as in the following examples: $w_B = 3$ g/kg; $\varphi_B = 6.7$ mL/L; $x_B = 185$ μmol/mol. Such forms are highly recommended. (See also Sec. 7.10.3.)

In the same vein, because the symbol % represents simply the number 0.01, it is incorrect to write, for example, "where the resistances $R_1$ and $R_2$ differ by 0.05 %," or "where the resistance $R_1$ exceeds the resistance $R_2$ by 0.05 %." Instead, one should write, for example, "where $R_1 = R_2(1 + 0.05$ %)," or define a quantity $\Delta$ via the relation $\Delta = (R_1 - R_2)/R_2$ and write "where $\Delta = 0.05$ %." Alternatively, in certain cases, the word "fractional" or "relative" can be used. For example, it would be acceptable to write "the fractional increase in the resistance of the 10 kΩ reference standard in 1994 was 0.002 %."

### 7.10.3 ppm, ppb, and ppt

In keeping with Ref. [6: ISO 31-0], this *Guide* takes the position that the language dependent terms part per million, part per billion and part per trillion, and their respective abbreviations "ppm," "ppb," and "ppt" (and similar terms and abbreviations), are not acceptable for use with the SI to express the values of quantities. Forms such as those given in the following examples should be used instead.

*Example:*

a stability of 0.5 (µA/A)/min *but not:* a stability of 0.5 ppm/min

a shift of 1.1 nm/m *but not:* a shift of 1.1 ppb

a frequency change of $0.35 \times 10^{-9} f$ *but not:* a frequency change of 0.35 ppb

a sensitivity of 2 ng/kg *but not:* a sensitivity of 2 ppt

the relative expanded uncertainty of the resistance $R$ is $U_r = 3$ µΩ/Ω

or the expanded uncertainty of the resistance $R$ is $U = 3 \times 10^{-6} R$

or the relative expanded uncertainty of the resistance $R$ is $U_r = 3 \times 10^{-6}$

*but not:* the relative expanded uncertainty of the resistance $R$ is $U_r = 3$ ppm

Because the names of numbers $10^9$ and larger are not uniform worldwide, it is best that they be avoided entirely (in most countries, 1 billion = $1 \times 10^{12}$, not $1 \times 10^9$ as in the United States); the preferred way of expressing large numbers is to use powers of 10. This ambiguity in the names of numbers is one of the reasons why the use of ppm, ppb, ppt, and the like is deprecated. Another, and a more important one, is that it is inappropriate to use abbreviations that are language dependent together with internationally recognized signs and symbols, such as MPa, ln, $10^{13}$, and %, to express the values of quantities and in equations or other mathematical expressions (see also Sec. 7.6).

*Note:* This *Guide* recognizes that in certain cases the use of ppm, ppb, and the like may be required by a law or a regulation. Under these circumstances, Secs. 2.1 and 2.1.1 apply.

**7.10.4 Roman numerals**

It is unacceptable to use Roman numerals to express the values of quantities. In particular, one should not use C, M, and MM as substitutes for $10^2$, $10^3$, and $10^6$, respectively.

**7.11 Quantity equations and numerical-value equations**

A quantity equation expresses a relation among quantities. An example is $l = vt$, where $l$ is the distance a particle in uniform motion with velocity $v$ travels in the time $t$.

Because a quantity equation such as $l = vt$ is independent of the units used to express the values of the quantities that compose the equation, and because $l$, $v$, and $t$ represent quantities and not numerical values of quantities, it is incorrect to associate the equation with a statement such as "where $l$ is in meters, $v$ is in meters per second, and $t$ is in seconds."

On the other hand, a numerical value equation expresses a relation among numerical values of quantities and therefore does depend on the units used to express the values of the quantities. For example, $\{l\}_m = 3.6^{-1} \{v\}_{km/h} \{t\}_s$ expresses the relation among the numerical values of $l$, $v$, and $t$ only when the values of $l$, $v$, and $t$ are expressed in the units meter, kilometer per hour, and second, respectively. (Here $\{A\}_X$ is the numerical value of quantity $A$ when its value is expressed in the unit X — see Sec. 7.1, note 2.)

An alternative way of writing the above numerical value equation, and one that is preferred because of its simplicity and generality, is $l\,/m = 3.6^{-1} [\,v/(km/h)](t/s)$. NIST authors should consider using this preferred form instead of the more traditional form "$l = 3.6^{-1}\,vt$, where $l$ is in meters, $v$ is in kilometers per hour, and $t$ is in seconds." In fact, this form is still ambiguous because no clear distinction is made between a quantity and its numerical value. The correct statement is, for example, "$l^* = 3.6^{-1}\,v^*\,t^*$, where $l^*$ is the numerical value of the distance $l$ traveled by a particle in uniform motion when $l$ is expressed in meters, $v^*$ is the numerical value of the velocity $v$ of the particle when $v$ is expressed in kilometers per hour, and $t^*$ is the numerical value of the time of travel $t$ of the particle when $t$ is expressed in seconds." Clearly, as is done here, it is important to use different symbols for quantities and their numerical values to avoid confusion.

It is the strong recommendation of this *Guide* that because of their universality, quantity equations should be used in preference to numerical-value equations. Further, if a numerical value equation is used, it should be written in the preferred form given in the above paragraph and if at all feasible, the quantity equation from which it was obtained should be given.

*Notes:*

1. Two other examples of numerical-value equations written in the preferred form are as follows, where $E_g$ is the gap energy of a compound semiconductor and $\kappa$ is the conductivity of an electrolytic solution:

$E_g\,/eV = 1.425 - 1.337x + 0.270x^2$, $0 \le x \le 0.15$, where $x$ is an appropriately defined amount-of-substance fraction (see Sec. 8.6.2).

$\kappa/(S/cm) = 0.065\,135 + 1.7140 \times 10^{-3}(t\,/°C) + 6.4141 \times 10^{-6}(t\,/°C)^2 - 4.5028 \times 10^{-8}(t\,/°C)^3$, $0°C \le t \le 50\,°C$, where $t$ is Celsius temperature.

2. Writing numerical-value equations for quantities expressed in inch-pound units in the preferred form will simplify their conversion to numerical-value equations for the quantities expressed in units of the SI.

### 7.12 Proper names of quotient quantities

Derived quantities formed from other quantities by division are written using the words "divided by" rather than the words "per unit" in order to avoid the appearance of associating a particular unit with the derived quantity.

*Example:* pressure is force divided by area *but not:* pressure is force per unit area

### 7.13 Distinction between an object and its attribute

To avoid confusion, when discussing quantities or reporting their values, one should distinguish between a phenomenon, body, or substance, and an attribute ascribed to it. For example, one should recognize the difference between a body and its mass, a surface and its area, a capacitor and its capacitance, and a coil and its inductance. This means that although it is acceptable to say "an object of mass 1 kg was attached to a string to form a pendulum," it is not acceptable to say "a mass of 1 kg was attached to a string to form a pendulum."

### 7.14 Dimension of a quantity

Any SI derived quantity $Q$ can be expressed in terms of the SI base quantities length ($l$), mass ($m$), time ($t$), electric current ($I$), thermodynamic temperature ($T$), amount of substance ($n$), and luminous intensity ($I_v$) by an equation of the form

$$Q = l^\alpha \, m^\beta \, t^\gamma \, I^\delta \, T^\varepsilon \, n^\zeta \, I_v^\eta \sum_{k=1}^{K} a_k ,$$

where the exponents $\alpha, \beta, \gamma, \ldots$ are numbers and the factors $a_k$ are also numbers. The dimension of $Q$ is defined to be

$$\dim Q = L^\alpha M^\beta T^\gamma I^\delta \Theta^\varepsilon N^\zeta J^\eta ,$$

where L, M, T, I, Θ, N, and J are the *dimensions* of the SI base quantities length, mass, time, electric current, thermodynamic temperature, amount of substance, and luminous intensity, respectively. The exponents $\alpha, \beta, \gamma, \ldots$ are called "dimensional exponents." The SI derived unit of $Q$ is $m^\alpha \cdot kg^\beta \cdot s^\gamma \cdot A^\delta \cdot K^\varepsilon \cdot mol^\zeta \cdot cd^\eta$, which is obtained by replacing the dimensions of the SI base quantities in the dimension of $Q$ with the symbols for the corresponding base units.

*Example:* Consider a nonrelativistic particle of mass $m$ in uniform motion which travels a distance $l$ in a time $t$. Its velocity is $v = l/t$ and its kinetic energy is $E_k = mv^2/2 = l^2 m t^{-2}/2$. The dimension of $E_k$ is $\dim E_k = L^2 M T^{-2}$ and the dimensional exponents are 2, 1, and –2. The SI derived unit of $E_k$ is then $m^2 \cdot kg \cdot s^{-2}$, which is given the special name "joule" and special symbol J.

A derived quantity of dimension one, which is sometimes called a "dimensionless quantity," is one for which all of the dimensional exponents are zero: $\dim Q = 1$. It therefore follows that the derived unit for such a quantity is also the number one, symbol 1, which is sometimes called a "dimensionless derived unit."

*Example:* The mass fraction $w_B$ of a substance B in a mixture is given by $w_B = m_B/m$, where $m_B$ is the mass of B and $m$ is the mass of the mixture (see Sec. 8.6.10). The dimension of $w_B$ is $\dim w_B = M^1 M^{-1} = 1$; all of the dimensional exponents of $w_B$ are zero, and its derived unit is $kg^1 \cdot kg^{-1} = 1$ also.

A *measurement standard* has been defined as a material measure, measuring instrument, reference material, or measuring system intended to define, realize, conserve, or reproduce a unit of one or more values of a quantity to serve as a reference. The various categories of standards used throughout our industry, are given in Table 11.3.

**Table 11.3** Definitions of various types of standards.

| Type of Standard | Definition | Example |
|---|---|---|
| International | A standard recognized by international agreement to serve internationally as the basis for fixing the value of all other standards of the quantity concerned. | The prototype of the kilogram maintained at the International bureau of Weights and Measures (BIPM) is an international standard of mass. |
| National | A standard recognized by an official national decision to serve in a country as the basis for fixing the value of all other standards of the quantity concerned. Generally, the national standard in a country is also a primary standard to which other standards are traceable. | National prototypes of the kilogram, which are identical to the international prototype of the kilogram, are maintained as national standards of mass in various NMIs. |
| Primary | A standard that is designated or widely acknowledged as having the highest metrological quality and whose value is accepted without reference to other standards of the same quantity. National standards are generally primary standards. | The metrological quality of the Josephson-junction-based voltage standard is far superior to that of the standard cell. However, it could take quite some time to replace the standard cell as the national standard of voltage. Until then it remains the primary standard. |
| Secondary | A standard whose value is based on comparisons with some primary standard. Note that a secondary standard, once its value is established, can become a primary standard for some other user. | The national standard of length consists of a stabilized laser source. High-accuracy gage blocks are used as a secondary standard of length. These standards are assigned values based on their comparison with national standards. |
| Reference | A standard having the highest metrological quality available at a given location from which the measurements made at that location are derived. | In the Unites States, state legal metrology laboratories maintain NIST-calibrated kilogram standards. These serve as reference standards for them. |
| Working | A measurement standard not specifically reserved as a reference standard, which is intended to verify measuring equipment of lower accuracy. | Multifunction calibrators are used as working standards for the calibration of IM&TE to be used in the measurement of various electrical parameters. |
| Transfer | A standard that is the same as a reference standard except that it is used to transfer a measurement parameter from one organization to another for traceability purposes. | Standard cells are used as transfer standards for the transfer of voltage parameters from the national standard to other standards. |

A *reference material* has been defined as a material or substance one or more of whose property values are sufficiently homogeneous and well established to be used for the calibration of an apparatus, the assessment of a measurement method, or the assigning of values to materials. *Certified reference material* has been defined as reference material accompanied by a certificate, one or more of whose property values are traceable to a procedure that establishes traceability to an accurate realization of the unit in which property values are expressed and for which each certified value is accompanied by an uncertainty at a stated level of confidence.

The need for international compatibility of measurement results came about in the beginning of the nineteenth century. With the start of international trade, the necessity for harmonization of measurement units and standards on a global scale was realized. The first success in these efforts was achieved in 1875 when the Treaty of the Meter was signed (see Chapter 1). Under this treaty, the signatory states agreed to set up a permanent scientific organization, BIPM. Basically, the Treaty of the Meter is a diplomatic treaty. For the execution of the various tasks required to achieve the objectives of the treaty, there are two other international organizations actively engaged in metrology activities. The first is the CGPM, which meets every four years to make decisions on important matters. This is the supreme policy-making and decision-making body under the treaty. The International Committee for Weights and Measures (CIPM) is appointed by the CGPM and is responsible for planning and executing the decisions of the CGPM. The members of the CIPM are eminent scientists and metrologists of different nationalities.

The concept of national and international measurement systems has been accepted globally. Nations have given their NMIs the responsibility of realizing and maintaining national standards of measurement, which are the representations of the SI units for particular parameters. These NMIs also take part in the process of dissemination of these units to the actual users. This task is carried out by the National Institute of Standards and Technology (NIST) in the United States, the National Physical Laboratory (NPL) in the United Kingdom, Physikalisch-Technische Bundesanstalt (PTB) in Germany, and the NPL in India. The traceability of measurements is achieved through the use of standards in the calibration process.

Intrinsic standards are realized based on standard procedures. It is assumed that following the procedure correctly will generate a standard that will produce a quantity that has a low uncertainty of realization—an uncertainty within accepted limits. An intrinsic standard has been defined as: a standard recognized as having or realizing, under its prescribed conditions of use and intended application, as an assigned value, the basis of which is an inherent physical constant or an inherent and sufficiently stable physical property. Some examples of the intrinsic standards used by metrology organizations and calibration laboratories globally are:

- Josephson-junction-based voltage standards
- Quantum-Hall-effect-based resistance standards
- Cesium atomic standards for time interval and frequency
- The International Temperature Scale of 1990 (ITS-90)

NMIs establish and maintain SI units and disseminate them within the country. The realization techniques used for the base units by the various NMIs are similar. See Table 11.4 for their details.

Table 11.4 Realization techniques of SI base units.

| Base unit | Realization technique |
|---|---|
| Length (meter) | Realized through a laser source as recommended by the CIPM |
| Mass (kilogram) | Realized through a national prototype of the kilogram (Note: mass is the only SI unit that is defined by a physical artifact). |
| Time (second) | Realized through a cesium atomic clock |
| Electric current (ampere) | Realized through units of voltage (volts) and resistance (ohms) |
| Temperature (Kelvin) | Realized through the triple point of a water cell and the ITS-90 with a number of fixed points at thermal equilibrium |
| Luminous intensity | Realized through a group of incandescent lamps and a calibrated radiometer (candela) |

# Endnote

1. Bucher, 2.

# References

Bucher, Jay L. 2000. *When your company needs a metrology program, but can't afford to build a calibration laboratory . . . what can you do?* Boulder, CO: National Conference of Standards Laboratories.

Kimothi, S. K. 2002. *The uncertainty of measurements, physical and chemical metrology impact and analysis.* Milwaukee: ASQ Quality Press.

NIST Special Publication 811, 1995 Edition, *Guide for the use of the international system of units (SI).* Gaithersburg, MD: National Institute of Standards and Technology.

# Chapter 12
# Audit Requirements

This chapter could be the most important chapter in this entire book! Why? First, read this explanation of a quality system. "The basic premise and foundation of a good quality system is to say what you do, do what you say, record what you did, check the results, and act on the difference. Also, for the whole system to work, the organization needs to establish a quality management system to ensure that all operations throughout the metrology department, calibration laboratory, or work area where calibrations are accomplished, occur in a stable manner. The effective operation of such a system will result in stable processes and, therefore, in a consistent output from those processes. Once stability and consistency are achieved, then it's possible to initiate improvements.[1]"

How does the organization know that everything is working correctly or according to written instructions? How does it know where improvements can/should be made? How does it know it's giving customers the quality service they are paying for? The answer is simple . . . perform an audit.

According to a dictionary definition, an *audit* is an examination of records or accounts to check their accuracy.[2] It is not vulgar, profane, or have any unwanted calories, but mention the word *audit* to a manager, supervisor, or person in a quality position, and the normal reaction is one of disdain, disgust, despise, and derision. The fact is, nothing could or should be further from the truth. Any organization that is responsive to its customers, both internal and external, and desires to find problems before they affect their products, services, or customers will conduct audits based on its quality system on a regular basis.

An audit process asks questions; looks at how an organization is supposed to be conducting business; and checks if it's following its defined procedures. All this is done with a mind-set of helping itself, its customer base, and its bottom line! An audit is not a bad thing. It costs little in both time and money. Audits afford an organization the opportunity to correct errors, make improvements, and find areas where it can change for the better before it affects customers, quality, or certification. It is a self-policing effort.

Are audits a requirement within the various systems? Absolutely. Here are some examples of those requirements. ANSI/ISO 17025-1999, paragraph 4.13.1, reads in part, "The laboratory shall periodically . . . conduct internal audits of its activities to verify that its operations continue to comply with the requirements of the quality system and this International Standard. The internal audit programme shall address all elements of the quality system, including the testing and/or calibration activities. It is the responsibility of the quality manager to plan and organize audits as required by the schedule and requested by management. Such audits shall be carried out by trained and qualified personnel who are, wherever resources permit, independent of the activity to be audited." BSR/ISO/ASQ Q10012:2003, paragraph 8.2.3, reads, "The metrological function shall plan and conduct audits of the measurement management system to ensure its continuing effective implementation and compliance with the specified requirements. Audit results shall be reported to affected parties within the organization's management. The results of all audits . . . shall be recorded." NCSL International's RP-6, chapter 5.4, reads, "The calibration control systems should be subject to periodic audits conducted at a frequency and to a degree that will ensure compliance with all elements of the system procedures and documented requirements. It is recommended that a procedure describing the system audits and controls be available and includes:

- Function or group responsible for conducting audits
- Frequency and extent of audits to ensure compliance with procedures
- Description of the methods used to ensure that measurements and calibration have been performed with the required accuracy
- Deficiency reporting and corrective actions required and taken."

As a minimum, if there is no internal audit function requirement, a self-inspection program could go a long ways in preparing the organization for audits and inspections. By setting up a self-inspection program the organization is:

- Showing an effort to find problems
- Seeing where it is not meeting the quality system
- Demonstrating a desire to continuously improve its program through self-initiative
- Finding opportunities before they are found by others
- Making itself proactive instead of reactive to problems and solutions

One option for self-auditing is to follow the "Say what you do, do what you say, record what you did, check the results, and act on the difference" theme. Check if the organization is actually following its quality procedures. Do all its records contain the required information, and do they show a paper trail for traceability purposes? If an item was found to be out of tolerance during calibration, was action taken? Was the customer informed, and does it have data to show that it occurred? The more specific the question, the easier it is to answer. Self-inspections can be an important continuous process improvement, but it takes time, effort, and honesty at all levels. Self inspections, period equipment confidence checks, and so on, all need to be documented if they are to be used in an audit.

Generally, there are three types of audits. An *internal audit* (first-party audit), conducted by personnel from within an organization, department, or organization, examines its system and records the results for internal eyes only. Internal audits are usually performed by a person who is independent of the department or process that is being audited to avoid potential conflict of interest.

An *external audit* is conducted by a customer (second-party audit) for the purpose of evaluating its supplier for granting or continuation of business. An external audit could also be conducted by an auditing agency (third-party audit), with the results being forwarded to the management of the company, department, or organization. Most external audits are performed to see if an organization is in compliance to a specific standard, guideline, or regulation. They can be either subjective or directive in nature. For example, if an organization was audited for compliance to cGMP requirements (FDA), it would be informed of any findings by use of the FDA's Form 483, which is part of the public record. In some cases, accreditation on-site assessments typically require a demonstration of proficiency—over-the-shoulder evaluation/observation. Depending on the inspections criteria, it could include examination of the IM&TE, the technician doing the calibration, the process (calibration procedure, records, documentation, and so on), or all three areas.

How often audits are conducted might depend on who is performing the audit and for what purpose. An organization may receive an initial audit for ISO 9000:2000 compliance and then have surveillance audits every six months or yearly. Most internal audits are conducted on a yearly basis, unless problems are found, in which case more frequent audits may be performed to ensure improvements are made and conformance is met. Some requirements require a specific time period for audits, while others leave the time period to the individual company or laboratory.

Once an audit is conducted, the results need to be documented and kept on file a predetermined amount of time. How long records are maintained should be stated in the records retention policy. The important point here is that the results are saved for future reference. Follow-up audits to ensure observations, findings, and/or write-ups have been corrected also need to be filed for future needs. Are any corrective and/or preventive actions identified from the results of the audit? Are any opportunities identified to perform the process in a more efficient manor? Is the proper authority sent the final audit results? Are the people using the quality system aware of the findings and updated on any changes to the system? Is there documentation that supports all of this? If procedures are changed, are the technicians, supervisors, and manager trained in the new procedures and their training records updated accordingly? Is there an area in the audit for checking if training records are maintained properly? See Chapter 15 for more information on training and training records. It is important to assign custodial responsibility for audit discrepancies—the person assigned to follow through with the corrective action plan and the timetable for correcting discrepancies. This should be part of the quality system with timelines, the responsible party that the findings are sent to, and how long the results are maintained.

## Endnotes

1. Bucher, 2.
2. *The American Heritage Dictionary*, 141.

## References

*The American Heritage Dictionary*. 1985. Boston, MA: Houghton-Mifflin Co.

ANSI/ISO 17025-1999, *American National Standard—General requirements for the competence of testing and calibration laboratories*. Milwaukee: ASQ Quality Press.

BSR/ISO/ASQ Q10012:2003(E), *Measurement management systems—Requirements for measurement processes and measuring equipment*. Milwaukee: ASQ Quality Press.

Bucher, Jay L. 2000. *When your company needs a metrology program, but can't afford to build a calibration laboratory . . . what can you do?* Boulder, CO: National Conference of Standards.

The Healthcare Metrology Committee. 1999. *Calibration control systems for the biomedical and pharmaceutical industry*. Recommended Practice RP-6. Boulder, CO: National Conference of Standards Laboratories.

# Chapter 13
## Scheduling and Recall Systems

The past, the present, and the future . . . what do they all have in common? In the world of calibration, you can access all three through your calibration management software. The past shows not only what you *have* calibrated, but also what you *have not* calibrated—the IM&TE that is overdue calibration!

The IM&TE waiting to be calibrated could be referred to as the present. They sit on the incoming shelf waiting for time, standards, a free technician, funds from their owner, or possibly technical data, owner's manuals, parts, and so on.

The future has another name in the metrology community. It is called *the schedule*. Depending on your system, requirements, resources, or directives from upper management, your schedule could look into the future for 30 days, 60 days, 90 days, or even longer. Some of the deciding factors include if the customer is internal or external, or both. Also, the amount of staff, standards, bench space, cables/accessories, and so on, that will be available to perform the calibration.

The importance of proper scheduling of test equipment cannot be overemphasized. By anticipating what will be required of a department's staff, standards, space, and time; work assignments, use of standards, and combining or scheduling like items to be calibrated can turn what might be chaos into an orderly schedule of events. There is still the unexpected to consider, though. Staff are affected by illness or injury. Customers buy new equipment, and they don't always eliminate old items. Some new equipment in your workload may require you to suddenly purchase a new measurement standard and have some people trained on it. The laboratory may gain or lose an important customer. In truth, none of these items are really unexpected, but their probability is low enough that they are not considered as much, and they are admittedly hard to quantify for planning purposes. So the orderly schedule can be devised and is a desirable goal, but the prudent manager should be prepared for variation in it.

Some departments evaluate their workload on a weekly basis, but do so by using their latest 30-day schedule. Any updates would have already been accomplished as work was completed and should be current in their system. Any changes that might impact work should also show up on the new schedule and can be easily included. It has been found that calibrating like items (see Chapter 39 for more on calibrating like items) can reduce the normal time it takes to accomplish one item done multiple times.

Is there a system in place in case of the dreaded *R* word . . . recall? What are the ramifications of equipment found out of tolerance (your standards, customer equipment, or notification from your outside vendors)? What are the ins and outs of recalling equipment and determining the necessity through stated requirements. Are any of these problems written into your quality system in case they occur? Who is responsible, and what are they responsible for? The following suggestions might help.

Some software management systems have a reverse traceability function that allows the user to identify the standard or system used for a particular function. Then, when queried, the system can identify all items that had been calibrated using a particular standard or system. Most standards have requirements for having a system in place, in writing, that identifies how to proceed if a recall is required. This could include recall of standards, products, calibrated test equipment, and so on.

Here's how a reverse traceability problem might occur and be solved. A company sends its standard out for calibration. The vendor informs the company that the standard was out of tolerance when received and provides the As Found and As Left data. The company must make a determination if the out-of-tolerance condition of its standard had an impact on the IM&TE it was used to calibrate and if the calibrated equipment also had an impact in the production or process it was also used on. Without having the proper documentation available to trace when and where the items were used, it would be impossible to know this information. This is another reason to maintain documentation records. Also, by having the ability to generate a reverse traceability list using your software, you reduce the time it takes to find the equipment involved and can remove it from service before it impacts other product or processes.

Once the IM&TE is identified, it should be segregated from other IM&TE and labeled as such. Some companies have designated areas for this type of equipment or products. Simply identifying the problem without identifying the equipment and removing it from service is not enough. What's needed is a paper trail showing what was accomplished, when it happened, what was accomplished to preclude it from happening again, and what the ramifications of the entire process involved was, and all this should be part of the permanent record for your company. Not only will an auditor want to see these records, but they can be used for future training on what not to do and how not to do it. If we do not learn from our mistakes, we are bound to repeat them. In the world of metrology, this has proven to be very true.

# Chapter 14
# Labels and Equipment Status

What's the status of your IM&TE? Does the user know? Can an auditor tell? Do you have to go to your computer or printout to know? The proper use of calibration labels and their reflection in your calibration management system could be the answer to all these questions. Quick, simple, and easy to use—they are a requirement in most systems.

BSR/ISO/ASQ Q10012:2003(E) reflects the need for labels, as stated in paragraph 7.1.1, General, which reads, "Information relevant to the metrological confirmation status of measuring equipment shall be readily available to the operator, including any limitations or special requirements." ANSI/ASQC M1-1996 states, in paragraph 4.10, Identification of Calibration Status, "Instruments shall be labeled to indicate when the next calibration is due." And *Calibration Control Systems for the Biomedical and Pharmaceutical Industry*, RP-6 reads in paragraph 5.8, Labels, "To alert the user to the status of a piece of equipment, all equipment should be labeled or coded to indicate its status. Equipment not labeled is considered 'not calibrated.'" Sometimes equipment is of a particular size or shape that does not allow for the labels to be easily attached. In some of those cases, color codes are used to show when they are next due calibration, by either a month schedule, quarter, and so on. In those cases, the system that is followed is well documented and there is a procedure that identifies which color is for which time period, and so on.

What kind of labels are referred to in this chapter? The average calibration technician calls them *calibration stickers*, *no-calibration required stickers*, or *limited calibration stickers*. They all have the same things in common. They identify the IM&TE that they are affixed to by their unique identification number; they show when the unit was calibrated in the form of a date; they show when the unit will need to be recalibrated, usually called the *date due calibration*, and they list the name, stamp, or signature of the person who performed the calibration. In some calibration systems, if there is not a calibration label attached to the IM&TE, then it is assumed to not require calibration; in others, it is assumed to be uncalibrated. Most systems, however, require the attachment of a No Calibration Required (NCR) label to be readily apparent. When an NCR sticker is applied, it can be advantageous to place the unique identification number of the test instrument somewhere on the NCR label. This precludes a second party from removing the label and placing it on another unit. During audits and inspections, the equipment user has been known to take drastic measures to keep from getting written up on their IM&TE. This practice basically keeps honest people honest.

Some companies also make use of Do Not Use, Out of Calibration labels for items that are out of calibration, broken, or waiting for service of some type. These bright red labels easily get the attention of any user and allow for quick identification of IM&TE that can not be used.

Another label used is the Out of Calibration If Seal Is Broken sticker. It is placed over screws or devices securing covers or panels used to enclose test equipment. They are also used to cover holes or access panels that have adjustment areas, screws, or knobs that need to have access limited to authorized personnel. This type of seal (sometimes called a *tamper seal*) or equivalent is required by ISO 17025:1999 (at 5.5.12) and BSR/ISO/ASQ Q10012:2003 (at 7.1.3), as well as by many regulatory agencies and company policies. Even if not required, it is still a good practice. A tamper seal serves as a deterrent to inappropriate adjustment and is a visual indicator of the likely integrity of the calibration.

Just because there are calibration labels on your IM&TE does not mean that everything is fine with the system. The labels must match the information in the calibration system database as well as the information on the calibration record or certificate, be legible, easy to find, and up-to-date. Here are some hints on how to manage your labels:

- Use black or dark ink. (No pencil, Sharpies™, magic markers™ or crayons).

- Cover the label with tape to help preserve the data. (Some companies use chemical- or UV-resistant tape when applicable).

- Make a new label if an error is made. (Line-outs, white-out, and so on are is not acceptable).

- Never use another technician's stamp, name, or identifying mark on your work.

- Keep all labels in a secure, locked area, with limited access only to those authorized to use them.

- If the IM&TE is small or makes attaching a label difficult, use an alternate system (metal tag, color coding, or manila tag) and document how the alternate system will be managed.

Note that many laboratory database systems have provisions for printing calibration labels from the results of a calibration event, and there are a number of printer and label systems available. This type of system eliminates many of the potential problems. In particular, dates, technician identification, and equipment serial number should never have any errors because the data come from the actual record that is stored.

Other types of labels that may need to be used in different systems include limited calibration stickers, any type of chart or graph that displays data for the user's benefit, Calibration Before Use labels, radiation labels, and/or preventive maintenance labels.

Limited calibration labels have the same data as a regular calibration label with the addition of an area that identifies either the range or tolerances of the IM&TE that are limited, or those that can not be used. Charts or graphs that are attached or referred to on a limited calibration label need to have the same information on them as are on a regular calibration label: the unique identification number for that particular unit, the date calibrated, the next calibration date, and the name of the person accomplishing the calibration.

*Calibrate Before Use* (CBU) stickers generally are attached to units that require calibration before they can be placed back in service. Some examples of these would be items that see very little use over an extended period of time or units that received a calibration and revert to CBU status after a defined period of time, with a calibration sticker and CBU sticker both attached to the unit. Most items selected for CBU status have limited or specialized use and the time and money expended to keep them in service is not worth the cost.

*Standardize Before Use* stickers can be used on items such as pH meters and conductivity meters that require a standardization against buffers or solutions that have traceable qualities back to a national or international standard. This label can also be used on any other type of equipment that requires standardization, normalization, or (most commonly) self-calibration before it is used.

When a preventive maintenance inspection has been performed on IM&TE, it is sometimes advisable to place a label on the unit showing its status. The information helps the customer know that the preventive maintenance has been performed if the unit did not receive a calibration. When used for informational purposes, labels can be employed to assure customers of equipment status, dates that need to be followed, and who to contact if more information is required. This is not to say that the IM&TE should appear to look like a well-used automobile bumper with a variety of stickers, but sometimes information usable to the customer on a regular/daily basis can be displayed on a label or sticker for ready use.

Any information displayed on a label should also be recorded in the calibration record, management software database, and/or customer information folder. In some cases it is better to be redundant with information, than lose it when a label is removed or lost.

# References

ANSI/ASQ M1-1996, *American National Standard for calibration systems*. Milwaukee: ASQC Quality Press.

ANSI/ISO 17025-1999, *American National Standard—General requirements for the competence of testing and calibration laboratories*. Milwaukee: ASQ Quality Press.

BSR/ISO/ASQ Q10012:2003(E), *Measurement management systems—Requirements for measurement processes and measuring equipment*. Milwaukee: ASQ Quality Press.

The Healthcare Metrology Committee. 1999. *Calibration control systems for the biomedical and pharmaceutical industry*. Recommended Practice RP-6. Boulder, CO: National Conference of Standards Laboratories.

# Chapter 15
# Training

Everybody requires training. From our early years when potty training is completed, to higher education and beyond. Training is part of our lives, both in expanding our knowledge and experience and in learning to adapt to new ideas, concepts, and problems. No one is born knowing how to calibration IM&TE. They are taught through formal education, on-the-job training, and/or self-taught through the Internet, home study, or correspondence courses. No matter how one receives new information, training is a lifelong endeavor.

The extent of one's education, both formal and hands-on, needs to be documented for a couple of reasons. First, it will preclude having to remember facts that are easily forgotten over time. Second, it is readily available for viewing by inspectors or auditors. Third, it allows workers to easily see what they are qualified to do, what they need additional training in, and where they lack knowledge, skill, or experience. A comprehensive training record will ensure the information is available and accurate.

Here are some references that show training is required in calibration facilities. First, ANSI/ISO 17025 states in paragraph 5.2, personnel: "The laboratory . . . management shall ensure the competence of all who operate specific equipment, perform tests and/or calibrations, evaluate results, and sign test reports and calibration certificates. When using staff who are undergoing training, appropriate supervision shall be provided. Personnel performing specific tasks shall be qualified on the basis of appropriate education, training, experience and/or demonstrated skills, as required . . . shall have a policy and procedures for identifying training needs and providing training of personnel . . . shall maintain current job descriptions for managerial, technical and key support personnel involved in tests and/or calibrations . . . shall maintain records of the relevant authorization(s), competence, educational and professional qualifications, training, skills and experience of all technical personnel."

ANSI/NCSL Z540-1-1994 also addresses the requirements for training in section 6, personnel: "The calibration laboratory shall have sufficient personnel, having the necessary education, training, technical knowledge and experience for their assigned functions. The calibration laboratory shall ensure that the training of its personnel is kept up-to-date consistent with employee assignments and development. Records on the relevant qualifications, training, skills and experience of the technical personnel shall be maintained and be available to the laboratory."

BSR/ISO/ASQ Q10012:2003(E) continues this theme in section 6.1.2, Competence and training, when it states, "The management of the metrological function shall ensure that personnel involved in the measurement management system have demonstrated their ability to perform their assigned tasks . . . shall ensure that training is provided to address identified needs, records of training activities are maintained, and that the effectiveness of the training is evaluated and recorded."

In a paper she presented at the 2001 NCSL International Workshop & Symposium in Washington, D.C. Corinne Pinchard said, "How do you save $65,000 a year and get two technicians for the price of one? Easy—you provide in-house training for new technicians on how to calibrate the test equipment used by your company. It is not necessary for your entire core group to be experienced calibration technicians. One experienced person can pass on their knowledge and skills using a well-rounded training program. By providing a solid foundation of knowledge . . . you can have a solid training program in place for minimum cost and effort. A good training program can reap benefits for years to come, especially if it is continually upgraded and improved as circumstances, test equipment, and technician skills change." That was true decades ago and will continue to be true in the future.

In most cases the hiring of experienced calibration technicians, trained in the latest technology or system, will fulfill any calibration laboratory's staffing requirements. But technology never stands still. The only guarantee about change is that is will occur. Providing the necessary training to the technicians that will be using the systems and ensuring their training records reflect that knowledge or skill helps both the company and technician. The company can prove the qualification of its staff, and the technician can prove his or her qualifications when moving from section to section or job to job.

Training and training records consistently fall in the top three areas written up during any audit or inspection. Possibly one of the main reasons is because training is time-consuming (some mistakenly believe the company is making no money during training —nothing could be further from the truth). Training is costly (because personnel must be sent to an outside agency to receive qualification or certification); and training records are hard to maintain and nobody really cares (The auditor cares, and it's a requirement in many standards.). Generally, trained personnel perform their jobs right the first time (saving the company money by not doing work over); can be used as trainers for untrained personnel (saves the cost of sending everyone out for training); and puts quality at the most important place . . . where the unit is calibrated, not when product is rejected or inspected.

It is repetitive and boring to calibrate the same widget 69 times in the same week, while following the same old boring procedure. When someone's life, however, or the cure for deadly diseases is on the line, the technician better be doing it right the first time. Following procedures is the only way to ensure that is going to happen, and the only way to keep track of changes and improvements is to document them within your quality system. This is only a small part of the criticality of documenting training.

## References

ANSI/ISO 17025-1999, *American National Standard—General requirements for the competence of testing and calibration laboratories.* Milwaukee: ASQ Quality Press.

ANSI/NCSL Z540-1-1994, *American National Standard for calibration—Calibration laboratories and measuring and test equipment—General requirements.* Boulder, CO: National Conference of Standards Laboratories.

BSR/ISO/ASQ Q10012:2003(E), *Measurement management systems—Requirements for measurement processes and measuring equipment.* Milwaukee: ASQ Quality Press.

Pinchard, Corinne. 2001. *Training a calibration technician . . . in a metrology department?* Boulder, CO: National Conference of Standards Laboratories.

# Chapter 16
# Environmental Controls

The importance of the environment in which IM&TE is calibrated does not immediately come to mind for most technicians. There are a couple of reasons for this. One is that unless the equipment being calibrated is of such a high tolerance that temperature, humidity, radio frequency, vibration, or dust control might have an impact on its ability to make an accurate measurement, the environment is not a concern. Another reason could be that the vast majority of IM&TE is designed and built to be used in a wide variety of environments, but in real life they are used in stable facilities where there is no impact on their ability to make accurate measurements. All IM&TE has uncertainty, however, and the environment where it is calibrated and/or used can play a critical role in determining the item's known uncertainty. A review of what the standards require concerning environmental controls should remove any doubt as to how critical a part it plays in a company's calibration process.

According to ANSI/ISO 17025-1999, section 5.3, Accommodation and environmental conditions, "The laboratory shall ensure that the environmental conditions do not invalidate the results or adversely affect the required quality of any measurement . . . The laboratory shall monitor, control and record environmental conditions . . . where they influence the quality of the results." ANSI/NCSL Z540-1-1994, section 7, Accommodation and environment states, in part, "Laboratory accommodation (facilities), calibration area, energy sources, lighting, temperature, humidity, and ventilation shall be such as to facilitate proper performance of calibrations/verification. The laboratory shall effectively monitor, control and record environmental conditions as appropriate." BSR/ISO/ASQ Q10012:2003(E), section 6.3.1 reads, " . . . Measuring equipment shall be used in an environment that is controlled or known to the extent necessary to ensure valid measurement results. Measuring equipment used to monitor and record the influencing quantities shall be included in the measurement management system." ANSI/ASQC M1-1996, section 4.4, Environmental Controls states, "Environmental controls shall be established and monitored as necessary to assure that calibrations are performed in an environment suitable for the accuracy required." ANSI/ISO/ASQ Q9001-2000, section 6.4, Work environment, states, "The organization shall determine and manage the work environment needed to achieve conformity to product requirements." And finally, NCSL's RP-6, section 5.11, Environmental controls, reads, "The calibration environment need be controlled only to the extent required by the most environmentally sensitive measurement performed in the area. To show compliance with environmental requirements, environmental conditions should be monitored and a record maintained of these conditions."

It's obvious by the inclusion of environmental requirements in each of these standards that the conditions where IM&TE is calibrated and/or used is critical. How important it is can be determined by the IM&TE's tolerances as specified by the manufacturer. Most manufacturers' operating manuals list the operating temperature and humidity limits, as well as any other limiting factors for that particular piece of equipment.

One of the more demanding areas where the environment is critical is a 20°C (68°F) dimensional calibration room. With temperature fluctuations not to exceed a degree or less, the maintenance, monitoring, and use of these areas can be critical. In some cases, an alternative source of heat (a light bulb) is turned on or off whenever a person leaves or enters the room. Temperature and/or humidity recording devices are employed to monitor the area, with their data stored for immediate use and future reference.

In areas that require other types of controls, foresight in laboratory or facility design plans must be accomplished. It is far less expensive to design grounding systems, entry control points to minimize dust or contamination problems, and temperature/humidity controls than it is to upgrade an existing area or lab. Sticky mats, shoe covers, lab coats, smocks, sterile gloves, and so on, are only a start on the list of items possibly required to meet certain regulatory requirements. Some companies can conform to their requirements by monitoring their heating/air conditioning systems or outlets; while others are required to have monitoring devices in every room of their facility. A good guide to selecting laboratory environments can be found in NCSL RP-14.

In most cases, applicable standards or regulations require that the temperature and relative humidity in laboratory are continually monitored and that the current values of those conditions be entered as part of a calibration record. The traditional method of doing this has been to use a circular seven-day chart recorder. While useful, this method does have some problems. It is not possible to analyze the data, the accuracy and resolution are limited, there is more paper to file, and each technician has to look at it to estimate the readings during each calibration. If the laboratory has a computer system, using a high-accuracy temperature/humidity data logger can alleviate all this. The data logger is a small device that can be mounted in a convenient location in the laboratory and connected to a computer. The manufacturers have software that can either monitor it continually or download the accumulated data at intervals. If the data are stored in a database, the random and systematic variations of temperature and humidity can later be statistically evaluated to produce a Type A value for uncertainty analysis. If the laboratory has a network and a calibration management system, another level of automation can be added. The calibration management system software can be programmed to automatically read the data logger at the start of a calibration procedure and put the temperature and humidity into the record. This eliminates a potential error source.

The laboratory should consult standards and recommended practices such as NCSL RP-7 *Laboratory Design*, or NCSL RP-14 *Guide to Selecting Standards—Laboratory Environments*. The laboratory *must also* evaluate the guidance in terms of the measurement standards they actually use. For example, a common guide for the temperature of an electronics calibration laboratory is 23 ± 5 °C. If the laboratory has equipment such as a long-scale digital multimeter (one with a resolution of 1 mV or better on the 10 V range), the real temperature requirement of the equipment is probably 23 ± 1 °C. Those instruments typically have a standardization routine (often called *self-calibration* or *auto-*

**Table 16.1** General-purpose calibration laboratories.

| Measurement Area | Temperature | Stability & Uniformity | Relative Humidity |
|---|---|---|---|
| Dimensional, Optical | 20 ± 1 °C | ± 0.3 °C per hour | 20 to 45% |
| All other areas | 23 ± 5 °C | ± 2.0 °C per hour | 20 to 60% |

**Table 16.2** Standards calibration laboratories or higher-accuracy requirements.

| Measurement Area | Temperature | Stability & Uniformity | Relative Humidity |
|---|---|---|---|
| Dimensional, Optical | 20 ± 0.3 °C | ± 0.1 °C per hour | 20 to 45% |
| Electrical, Electronic | 23 ± 1.0 °C | ± 1.0 °C per hour | 35 to 55% |
| Physical, Mechanical | 23 ± 1.5 °C | ± 1.5 °C per hour | 35 to 55% |

*calibration*) that requires the temperature to be within 1 °C of the temperature the last time that routine was performed, and it must be performed at least once every day if the instrument is being used at its highest accuracy. Close temperature control, with the goal of minimizing variation, is also necessary to reduce thermoelectric effects at connections. Other standards, such as some vector network analyzers, must be restandardized if the temperature changes by 1 °C during the course of the measurements. With these instruments, the dimensions of connectors are critical to quality measurements, and at high microwave frequencies the thermal expansion or contraction from a 1 °C change can be a significant effect.

Some generally recommended temperature and humidity ranges are listed in Tables 16.1 and 16.2.

There are some things to remember when creating or maintaining a laboratory environment. This information is based on more highly detailed sources from ISA, NCSLI and the U.S. Navy, which are listed in the references.

- 45 percent relative humidity is an absolute maximum for dimensional areas to prevent rust and other corrosion.

- 20 percent relative humidity is an absolute minimum for all areas to prevent equipment damage from electrostatic discharge.

- Temperature stability is the maximum variation over time. This is typically measured at the work surface height.

- Temperature uniformity is the maximum variation through the working volume of the laboratory. This is typically measured at several points over the floor area between the average work surface height and one meter higher.

- The air handling system should be set up so that the air pressure inside the laboratory area is higher than the surrounding area. This will reduce dust because air will flow out through doors and other openings.

- Ideally, a calibration lab should not be on exterior walls of a building and should have no windows. This will make temperature control much easier.

- Some measurement areas may have additional limits for vibration, dust particles, or specific ventilation requirements.

- It is important that the working volume of the laboratory is free from excessive drafts. The temperature should be reasonably stable and uniform and any temperature gradients, measured vertically or horizontally, should be small. In order to achieve these conditions, at the standard temperature of 20 °C, good thermal insulation and air-conditioning with automatic temperature control is generally necessary.

- The temperature control necessary depends, to some extent, on the items to be calibrated and the uncertainties required. For general gauge work the temperature of the working volume should be maintained within 20 ± 2 °C. Variations in temperature at any position should not exceed 2 °C per day and 1 °C per hour. These are the minimum expectations for United Kingdom Accreditation Service (UKAS) accreditation.

- For higher grade calibrations demanding smaller uncertainties, such as the calibration of gauge blocks by comparison with standards, the temperature of the working volume should be maintained within 20 ± 1 °C. Variations in temperature at any position should not exceed 1 °C per day and 0.5 °C per hour.

- For the calibration of gauge blocks by interferometry, the temperature within the interferometer should be maintained within 20 °C ± 0.5 °C. Variations in temperature shall not exceed 0.1 °C per hour.

- Within the laboratory, storage space should be provided in which items to be calibrated may be allowed to soak so as to attain the controlled temperature. It is most important that, immediately before calibration, time is allowed for further soaking adjacent to, or preferably on, the measuring equipment. Standards, gauge blocks, and similar items should be laid flat and side by side on a metal plate for a minimum of 30 minutes before being compared. Large items should be set up and left overnight. This is to ensure that temperature differences between equipment, standards, and the item being measured are as small as possible.

# References

ANSI/ASQ M1-1996, *American National Standard for calibration systems*. Milwaukee: ASQC Quality Press.

ANSI/ISO 17025-1999, *American National Standard—General requirements for the competence of testing and calibration laboratories*. Milwaukee: ASQ Quality Press.

ANSI/NCSL Z540-1-1994, *American National Standard for calibration—Calibration laboratories and measuring and test equipment—General requirements*. Boulder, CO: National Conference of Standards Laboratories.

BSR/ISO/ASQ Q10012:2003(E), *Measurement management systems—Requirements for measurement processes and measuring equipment*. Milwaukee: ASQ Quality Press.

The Healthcare Metrology Committee. 1999. *Calibration control systems for the biomedical and pharmaceutical industry*. Recommended Practice RP-6. Boulder, CO: National Conference of Standards Laboratories.

NCSL International. 2000. *Laboratory design*, Recommended Practice, RP-7. Boulder, CO: National Conference of Standards Laboratories.

NCSL International. 1999. *Guide to selecting standards—Laboratory environments*, Recommended Practice. RP-14. Boulder, CO: National Conference of Standards Laboratories.

UKAS Laboratory. Accommodation and Environment in the Measurement of Length, Angle and Form. www.ukas.com/Library/downloads/publications/LAB36.pdf

# Chapter 17
## Industry-Specific Requirements

The phrase "different strokes for different folks" is very applicable for this chapter. One would believe that calibration is calibration is calibration. In most cases, this would be true. But various industries, both in the United States and abroad, have their own particular requirements that must be met to conform to their standards. This chapter will cover the majority of requirements, their specific needs in terms of calibration, and record keeping, and identify where unique emphasis is placed throughout their processes. Each area has unique verbiage, acronyms, and guidelines that the calibration or metrology practitioner must come to know. The bottom line in all of the standards is the same philosophy for any quality system: "say what you do, do what you say, record what you did, check the results, and act on the difference." The slant placed on each of these directives can make the difference between passing an audit, hanging out a shingle that tells the world a company is certified by a particular governing body, or helping to bring new drugs to market faster.

### ISO 17025

ISO/IEC 17025:1999 is the international standard for accreditation of both testing and calibration laboratories. ANSI/ISO/IEC 17025:2000 is the U.S.-adopted version of ISO/IEC 17025. The documents are identical, so this section applies equally to both versions.

ISO/IEC 17025 requirements are documented in two main sections. Section 4 includes the administrative requirements including purchasing, document control, corrective action, internal audits, and management review. Section 5 includes the technical requirements including training, measurement uncertainty, proficiency testing, traceability, and reporting requirements. The main difference between ISO/IEC 17025 and ISO 9000 are the technical requirements in Section 5. Each clause has some differences compared to ISO 9000.

Management review is the main key to the successful operation of any quality management system. ISO/IEC 17025 has very prescriptive requirements for management review. The requirements include a review of audits and proficiency tests, and other factors, such as training and assessments by both internal and external auditors. The prescriptive nature of the management review section is meant to ensure that necessary laboratory operations are reviewed in a systematic nature. The projected volume and type of work is also included in the prescriptive requirements, in order to require the laboratory to examine operations from a business perspective, not just a quality perspective.

Corrective and preventive action, and control of nonconforming calibration each specifically require procedures to be developed and implemented. The organizational structure of the laboratory requires both technical and quality managers to be identified and responsibilities to be defined to prevent conflict of interest and to avoid undue external influence. The two roles can be filled by the same person and usually are in small laboratories.

The laboratory must have a quality policy and specific requirements that all personnel be familiar with the quality system, that the quality system apply both in the laboratory and away from the laboratory, state the objectives of the quality system, and specifically state compliance to ISO/IEC 17025.

Evaluation of subcontractors must be performed by personnel qualified to perform the evaluation. That is, if the laboratory is unable to calibrate a client's gage blocks and sends them to another laboratory, the other laboratory becomes a subcontractor. The evaluation of the subcontracted laboratory must be performed by someone who is familiar with the applicable discipline, in this example gage block calibration.

The technical requirements in Section 5 are extensive. The laboratory must have a plan for personnel training that addresses the current and future needs of the laboratory. Training can be provided either internally or by external sources. Usually, some external training is beneficial. The environmental conditions must be defined and controlled, both in the laboratory and on-site. Control on-site is often difficult or even impossible. There are several ways to address control on-site. The most common method is to establish environmental parameters and cease calibration operations until the client can reestablish environmental control at its location. Other methods may also work for some applications, such as establishing a much wider range for environmental parameters (for example, 50°F to 100°F). If wider parameters are established, then the reported measurement uncertainty must reflect that expanded influence.

There are requirements for equipment, such as requiring a documented calibration program for equipment. Software and procedures that are used for calibration must be validated if developed or modified by the laboratory. This typically requires pre-defined criteria to measure the success of the procedure and a report that describes the observed results of the use of the procedure and specific acceptance by management.

Measurement uncertainty requires a procedure and also must be reported by calibration laboratories. Measurement uncertainty is reported by calibration laboratories using a coverage factor of $k = 2$, which approximates 95% confidence. In reality, the confidence level will fluctuate some. Testing laboratories (also accredited under ISO/IEC 17025) do not report measurement uncertainty unless requested by their clients, but do calculate their measurement uncertainty at 95% confidence and let the coverage factor (the $k$ factor) fluctuate.

Calibration laboratories must be able to demonstrate competence (technical proficiency), measurement capability (the lab must have the personnel, facilities, and equipment to perform the calibrations), and traceability (to SI units through national or international standards). The laboratory also must take actions to ensure the quality of calibration results. The steps that the laboratory takes must be defined and be statistically analyzable. The most common example is proficiency testing. Other methods are included in that section, but proficiency testing is the only method that has international recognition.

The standard requires calibration certificates and reports to include specific information. The information includes the laboratory and client information, traceability path, and uncertainty information. Traceability information should always include the calibration certificate or report number of the last calibration of the standard. That is, the standard used to calibrate an item for a client must be reported, those standards are calibrated, and that calibration is documented on a certificate or report with a unique number. That number is the *traceability path*.

Subcontractor calibration must also be identified. This is often a difficult concept for a laboratory, as there is the obvious concern that the client may bypass the laboratory in favor of the subcontractor for the next calibration. Usually this does not occur because the client then has a need to qualify the subcontractor as a vendor and it is easier to deal with a single vendor.

One special requirement of ISO/IEC 17025 is protection of customer information, otherwise known as confidentiality. This requirement is specifically mentioned in multiple locations, including a specific requirement for a procedure for protecting the client's data and information during electronic storage and transmission of results. Electronic transmission of results includes e-mail and fax transmission. During an accreditation assessment the auditor will typically examine the firewalls and other protections employed by the laboratory to safeguard client information.

# ISO 9000

Before the last half of the twentieth century, the majority of products we encountered were fairly local—they were made in the same country or in a close neighbor. International trade did not have a large economic impact on the average person. Now, in the first decade of the twenty-first century, the majority of products that we encounter are multinational. In order to ensure the quality of products and services all over the world, business needs a way to be assured that they are produced in a quality manner. Standards that are specific to a company, industry, or country are not sufficient in a global economy. There is a need for an internationally recognized standard for the minimum requirements of an effective quality management system. The ISO 9000 series of standards was developed to fill that role. Additional information can be found in Chapter 3.

## Global Trade

While trade has been in evidence well into prehistory (more than 10,000 years), for most businesses it had been a peripheral part of their operations. This has changed over the last 50 years or so, to the point where many products are multinational and even the concept of a country of origin is now questionable. Raw materials, design, parts manufacturing, hardware, assembly, software, machine tools, agricultural products, and more are all part of the global economy. Consider these examples.

- A computer dealer in the United States sells computers under its private brand name. The dealer buys components from various places—China, Taiwan, Korea, Singapore, Japan, and Israel—and the operating system and software from a company in the United States. These items are assembled into a functioning computer system, and the last step in the process is to place a "Made in the USA" label on the back.

- An automobile manufacturer, a joint venture of an American and a Japanese manufacturer, is located in the United States. The design was developed jointly in the USA and Japan. A lot of parts come from other places in the United States, Canada, and Mexico. Some parts come from a United States factory of a German company. The engine is made in Hungary. The transmission is made in another country. A significant number of buyers choose this brand because they are convinced the vehicles are made in America.

- A modern airliner, such as the Boeing 777, includes millions of parts and assemblies from hundreds of suppliers, which may be located in many countries. The aircraft is sold to airline operators in many countries.

- A software company in Atlanta, Georgia, develops code during the day and every evening electronically sends it to a subsidiary in Bangladesh. That organization (during its work day) tests the code for software quality and conformance to technical requirements. The bug reports are sent back to Atlanta in time for the start of the next day's work there.

- A continuous-cast steel mill receives ore and scrap metal from many locations including other countries. The finished product is shipped to locations all over the United States and in other countries.

- A person in a Milwaukee neighborhood grocery store may be buying grapes from Chile, corn from Nebraska, crabs from Japan, mussels from Canada, dates from Palestine, apricots from Turkey, and lamb from New Zealand, all without realizing it. Yet without global commerce, these products would not be available in the customer's local area for most of the year, if at all.

It should be evident that quality standards that are specific to a single company, industry, or country are not sufficient. The ISO 9000:2000 series of quality management system standards, as well as the earlier versions, exist to serve as an aid to international commerce. In line with the mission of ISO, they are a means of reducing technical barriers to trade. The standards provide an internationally recognized set of minimum practices for an effective quality management system. The standards describe what practices should exist and what they should achieve; they do not prescribe how to do it.

### National Versions of International Standards

Many countries have their own national versions of the ISO 9000 series and other international standards. In most cases the national version is a translated edition, sometimes with additional introductory material, that is published by the relevant national standards body. In the United States, for example, the national version of ISO 9001:2000 is ANSI/ISO/ASQ Q9001:2000. While the specified text (usually French and/or English) of the international (ISO) version of a standard is the official authoritative version, many countries designate their national version as the legal equivalent of the international standard. In this book, reference to an international standard by its international designation always includes a national version where applicable; the reverse is also true.

## Important Features of ISO 9000

It is important to understand that a quality management system (QMS) standard only applies to the management system of an organization. It does not have anything to do (directly) with the product. The QMS standard is general because it can apply to any organization in any line of business in any country. The product is the subject of separate specifications and technical requirements. The product specifications are specific because they apply to a particular product or service, and maybe to a specific supplier. The quality standard and the product technical requirements are separate, but complement each other. Both are needed in an agreement between a supplier and a customer.

Another important feature of the ISO 9000 system is the concept of third-party evaluation of an organization's QMS. In a third-party audit, a company is evaluated by a qualified organization that is independent of the customer and supplier, but which is trusted by both. The evidence of conformity to the ISO 9000 requirements is the registration or certification by the assessor. A company may accept that certificate and thereby eliminate the cost of sending people to each supplier to perform quality audits. It is beneficial to suppliers because they are audited to a single set of requirements, and they do not have to host as many audits.

## The ISO 9000 Family

ISO 9000 is a standard, but it is not the only standard. It is not even the one that a third-party registrar will audit an organization against. The term is commonly used as a shorthand reference to the entire family of standards. The complete list changes and is usually available at the ISO Web site. As of June 2003, the ISO 9000 family includes the completed standards listed.

- ISO 9000:2000, Quality Management Systems—Fundamentals and Vocabulary
- ISO 9001:2000, Quality Management Systems—Requirements
- ISO 9004:2000, Quality Management Systems—Guidelines for Performance Improvements.
- ISO 10005:1995, Quality Management—Guidelines for Quality Plans
- ISO 10006:1995, Quality Management—Guidelines to Quality in Project Management
- ISO 10005:1995, Quality Management—Guidelines for Configuration Management
- ISO 10012:2003, Measurement Management Systems—Requirements for Measurement Processes and Measuring Equipment
- ISO 10013:1995, Guidelines for Developing Quality Manuals
- ISO/TR 10014:1998, Guidelines for Managing the Economics of Quality
- ISO 10015:1999, Quality Management—Guidelines for Training
- ISO/TS 16949:2002, Quality Management Systems—Particular Requirements for the Application of ISO 9001:2000 for Automotive Production and Relevant Service Part Organizations

This list uses the international numbers of the standards as published by ISO. Many countries or regions translate and republish national versions of ISO standards. In most cases the content is the same as the international (ISO) version but with changes such as translation to the local language, and making a variation of the standard's number. (ISO standards are published in French and British English.) For example, at the time of this book's publication the United States is in the process of adopting a national version of ISO 10012:2003. During the adoption process, and in this book where the standard is used as a reference, the designation BSR/ISO/ASQ Q10012-2003 is used to identify the draft US verbatim adoption of the ISO 10012:2003, which is available in electronic format from ASQ Quality Press. Upon completion of the adoption balloting process, this standard will be available from Quality Press under the designation ANSI/ISO/ASQ Q10012-2003.

The list of ISO 9000 family standards is available on the ISO Web site and should be checked at intervals, but it is not necessarily fully up-to-date.[1] For example, as of late 2003 a new standard (ISO 19011) was under development. For most industries, ISO 9001:2000 is the only conformance standard and is the only one an organization's quality management system can be audited against. In the automotive industry, ISO/TS 16949:2002 applies. All of the other documents in the ISO 9000 family are guidance to aid in implementation.

## ISO 9000 and Your Business

An organization will be audited against the requirements of one of the conformance standards if it wants its quality management system to be registered (or certified, in many countries). A full discussion of this is outside the scope of this book, but here are a few pointers.

- The ISO 9000 system does not say how to do anything. It describes a set of results to be achieved for an effective QMS, some things that must be done, and provides some guidance. An organization decides on the best way to accomplish them in its own structure. There is no such thing as the ISO way of managing anything.

- A well-run business probably will only have to make a few minor adjustments, if any.

- An organization does not have to reshape its business management system to the standard. All it has to do is describe how its system—whatever it is—meets the requirements of the standard.

- The largest problem areas are documentation, and corrective action and preventive action.

## What the Standard Says About Calibration

ISO 9001:2000 talks about calibration in section 7.6, Control of measuring and monitoring devices. If a device makes a measurement that provides evidence of conformity to requirements, or if it is necessary for the process and if it must be ensured that the measurement results are valid, then certain things must be done.[2]

- The IM&TE must be regularly calibrated against measurement standards traceable to (the SI through) international or national measurement standards, or other accepted standards if there is no such traceability.

- It must be adjusted, if indicated by calibration results.

- It must be identified in a manner that allows the user to determine the calibration status.

- If the item has any controls or adjustments that would invalidate the calibration, then they must be protected from access by the user.

- It must be protected from damage and from environmental conditions that could damage or degrade it whenever it is not in use.

- If a calibration shows that the As Received condition was out of tolerance, the organization must assess the condition and its impact on any product, take appropriate action, and keep appropriate records including results of the calibration.

- If computer software is used to monitor and measure requirements, then the organization must prove that the software operates as intended and produces valid results before placing it in regular use.

As written, ISO 9001:2000 makes reference to ISO 10012-1 and 10012-2 as guidance documents. In May 2003, the publication of ISO 10012:2003 canceled both of those standards and replaced them with the new one. It is discussed later in this chapter.

For most organizations, there is not a lot more that can be said about these requirements. However, calibration laboratories and consultants are very frequently asked one question: "Do I *really* need to have this (whatever it is) calibrated?" The standard says that the answer is "yes" if it (whatever it is) is making a measurement that provides evidence of conformity to requirements. But there are still questions that fall into indeterminate areas as far as the standard (or an auditor) is concerned. In 2000, Philip Stein suggested another test:

> Ask the question: Does it matter whether the answer from this measurement is correct?
>
> – If it does matter, then calibration is needed.
>
> – If it doesn't matter, then why is the measurement being made in the first place?[3]

Stein's test also leads to a risk assessment that can apply to any measurement situation: what can happen if the measurement is wrong, and what can happen if the measurement is not made at all? For example, a number of organizations say that electricians' voltmeters do not need to be calibrated because they are not used for any product realization processes. They are "only used for troubleshooting and repairing plant wiring." Applying a risk assessment reveals the flaw in that line of thinking. If a voltage measurement is wrong, an electrician could believe it is safe to handle a wire that is still live and be injured or killed. As another example, many organizations say that certain meters (voltmeters again, for this example) do not need to be calibrated because they are "only used for troubleshooting and repair of equipment," and calibrated tools are used for final test. Applying a risk assessment here shows that if the voltmeter is reading incorrectly there may be a risk of rework because of units that fail to pass inspection.

So, there are cases where a measuring instrument may not need calibration for a process under the QMS, but calibration may still be required for other reasons. In addition to safety and reduction of rework, other reasons include health and regulatory compliance.

A metrology or calibration organization (a stand-alone company or an in-house department) must be sure to understand the implications of section 7.6. It means something different to a calibration laboratory than it does to a factory turning out 10,000 widgets per hour.

A calibration laboratory (or metrology department or any other variation of the terms) is a service organization. It is providing a service to its customers; the product is the service of calibrating the customer's tool. This means that all of the workload items are customer-owned property that is passing through the laboratory's process (section 7.5.4 of the standard). The monitoring and measuring devices of section 7.6 are the calibration standards—the instruments the laboratory uses when calibrating customer items. This includes their reference and transfer standards. Out on the production floor it may be possible to argue about the *where necessary to ensure valid results* phrase, but in the calibration laboratory no argument is possible. It is clear that the measurement standards are necessary and that their results must be valid. There are two potential problem areas here.

If a calibration shows that the As Received condition was out of tolerance, the organization must assess the condition and its impact on any product, take appropriate action, and keep appropriate records including results of the calibration. If a measurement standard is found to be out of tolerance when calibrated, the calibration laboratory must assess the condition and its effect on the output. It has to determine all of the items that were calibrated by that standard since it was last known to be in calibration. (The ability to do this is called *reverse traceability*.) The laboratory has to compare the out-of-tolerance condition to the performance specifications of the units under test and determine if the error in the standard makes the calibration results invalid. (If there is more than one model number, it has to be done for each model.) In each case where the result would be invalid, the laboratory has to notify the customer about the problem and request return of the item for recalibration. The customer then has to evaluate the impact on his or her own production. As for keeping the results of the calibration, a laboratory should be doing that anyway for all of their measurement standards.

Another potential problem would be if computer software is used to monitor and measure requirements, then the organization must prove that the software operates as intended and produces valid results before placing it in regular use. This applies equally to a calibration laboratory, but may be more difficult to implement. In a lot of cases, computer software is going to be found as part of an automated calibration system. This clause means that the overall operation of the software must be verified. In addition, each separate procedure used for calibration workload items must be validated. If the software collects measurement uncertainty data (and the job will be easier in the long run if it does!), then that function must be validated as well.

A generally useful method of validating calibration software is:

- Calibrate the instrument manually using the same measurement standards and test points that the calibration software will use. (This is to verify that all equipment is working.)

- Calibrate the instrument with the software system and print out full results.

- If there are deficiencies in the software, have them corrected. Repeat the previous step as needed until the software runs to completion with no faults.

- Compare the results to the instrument specifications. Verify that any in- or out-of-tolerance notifications are correct, all required ranges and functions are tested, and that any test accuracy ratios are correct and adequate.

- Perform a software validation (pages 143-44).

## ISO 10012

A note in clause 7.6 of ISO 9001:2000 refers to ISO 10012:2003," *Measurement Management Systems—Requirements for Measurement Processes and Measuring Equipment*. This standard (actually, its predecessors ISO 10012-1:1992 and ISO 10012-2:1997) is suggested as a guidance reference. (The United States is in the process of adopting a national version of this standard, as noted on page 114. During the adoption process, and in this book's references to the content of the standard, the U.S. national version is referred to as BSR/ISO/ASQ Q10012:2003 which is a verbatim copy of the international standard. Once approved and adopted, the U.S. national version will be known as ANSI/ISO/ASQ 10012:2003).

This is not a stand-alone standard. Its intended use is in conjunction with other quality management system standards. It gives generic guidance for two areas: Management of measurement processes as part of a quality or environmental management system, and management of the calibration (metrological confirmation) system for the measuring instruments required by those processes.

Remember that metrology is the science and practice of measurement. With that in mind, there are two recurring phrases in the standard that merit discussion beyond the definitions in that standard.

- Clause 3.6 defines the *metrological function* as the "function with administrative and technical responsibility for defining and implementing the measurement management system." Elsewhere, the standard states that measurement requirements are based on the requirements for the product or, ultimately, the customer requirements. The measurement management system model is similar to the process management model presented in section 0.2 of ISO 9001:2000. It begins with the measurement requirements of the customer. Instead of product realization, the main processes are the measurement process and metrological confirmation. It also includes management system analysis and improvement, management responsibility, and resource management. Effectively, this means that everything that touches product realization and requires or makes measurements is part of the measurement management system. It is not, as many people believe, limited to the calibration activity.

- Clause 3.5 defines *metrological confirmation* as the "set of operations required to ensure that measuring equipment conforms to the requirements for its intended use." This includes calibration of measuring equipment, which is what people in the calibration lab are focused on, but it also includes verification—processes to ensure that the measuring equipment is capable of making the measurements required by the product realization processes and that the measurements being made meet the requirements of the product and the customer. Therefore, it includes even production-level measurement system analysis.

A measurement process may consist of one or many measuring instruments and may occur anywhere in the product realization processes. Measurement processes may exist in design (such as transforming customer requirements to product specifications and tolerances), testing (such as measuring the vibration forces applied to a prototype), production, inspection, and service.

From this, it is clear that the metrological function or measurement management system encompasses all of the product- or process-oriented measurement requirements within an organization. It is not limited to the calibration activity and the calibration recall system. Factors to be considered by the system include, but are not limited to: risk assessment, customer requirements, process requirements, process capability, measuring instrument capability relative to the requirements, resource allocation, training, calibration, measurement process design, environmental requirements of each measurement process and instrument, control of equipment, records, customer satisfaction, and much more.

The metrological function is also responsible for estimating and recording the measurement uncertainty of each measurement process in the measurement management system and ensuring that all measurements are traceable to SI units. This includes traditional gage repeatability and reproducibility studies. It may even require expanding their use, or it may require preparing uncertainty budgets for each measuring system in the product realization processes.

BSR/ISO/ASQ Q10012:2003 specifically states in its scope statement that it is not a substitute for or addition to ISO 17025:1999. This does not mean an accredited laboratory can ignore it, though. An accredited laboratory may be part of an organization that uses this standard's guidance to manage its overall measurement system, or an accredited laboratory may choose to apply parts of the standard to its own system.

## FDA (cGMP)

To paraphrase from the small entity compliance guide (from this point forward it will be referred to as 'the guide') covering current good manufacturing practices (cGMP), "Manufacturers are responsible for ensuring the establishment of routine calibration on their test equipment so it will be suitable for its intended use."[4] Most readers will have heard of the U.S. Food and Drug Administration (FDA), but may not be aware that their requirements fall under those of cGMP. This section gives specific guidance for those companies that operate under cGMP regulations.

Here's a quotation from Ralph E. Bertermann's study guide on *Understanding Current Regulations and International Standards; Calibration Compliance in FDA Regulated Companies*:

> Understanding compliance: . . . Not only is there not just a single document to study, but in addition to GLP and GMP regulations, compliance, which is defined as conformance in fulfilling official requirements, is shaped by a

series of international standards, incorporating industry best practices and a changing and broadening interpretation of regulations as understood by regulatory officials.

The cardinal document that outlines the basic requirements is the FDA regulations that cover cGMP is: TITLE 21—FOOD AND DRUGS, CHAPTER I—FOOD AND DRUG ADMINISTRATION, DEPARTMENT OF HEALTH AND HUMAN SERVICES, PART 820—QUALITY SYSTEM REGULATION—Table of Contents, Subpart A—General Provisions, Sec. 820.1 Scope.

The process starts with defining critical instruments that must be part of a calibration program. A standard instrument, traceable to a national standard, of known and higher accuracy, typically 4 times more accurate . . . following a documented calibration procedure. The result of the calibration, performed by trained technicians, is reported to the user . . . If out of tolerance, the unit is adjusted to an in tolerance condition and an appropriate label attached to the unit, indicating its calibration status. The calibration record is retained . . . The as found historical data is then analyzed periodically to adjust the calibration interval.

In a nutshell, Bertermann covers the requirements for cGMP without going into the detail related in the guide. The following specifics from the guide provide those details with examples for clarity for those new to this area of requirements.

According to the guide, under calibration requirements: the quality system regulation requires in section 820.72(b) that equipment be calibrated according to written procedures that include specific directions and limits for accuracy and precision. For cGMP, calibration requirements are:

- Routine calibration according to written procedures;
- Documentation of the calibration of each piece of equipment requiring calibration;
- Specification of accuracy and precision limits;
- Training of calibration personnel;
- Use of standards traceable to NIST, other recognizable standards, or when necessary, in-house standards; and provisions for remedial action to evaluate whether there was any adverse effect on the device's quality.

The guide goes on to say,

Managers and administrators should understand the scope, significance, and complexity of a metrology program in order to effectively administer it. The selection and training of competent calibration personnel is an important consideration in establishing an effective metrology program. Personnel involved in calibration should ideally possess the following qualities:

- Technical education and experience in the area of job assignment;
- Basic knowledge of metrology and calibration concepts;
- An understanding of basic principles of measurement disciplines, data processing steps, and acceptance requirements;

- Knowledge of the overall calibration program;
- Ability to follow instructions regarding the maintenance and use of measurement equipment and standards; and
- Mental attitude which results in safe, careful, and exacting execution of their duties.

Here's a partial quotation from 21CFR820.72, *Inspection, measuring and test equipment*,

(a) Each manufacturer shall establish and maintain procedures to ensure that equipment is routinely calibrated, inspected, checked, and maintained. These activities shall be documented. (b) Calibration procedures shall include specific directions and limits for accuracy and precision. These activities shall be documented. (1) Calibration standards used for inspection, measuring, and test equipment shall be traceable to national or international standards. (2) The equipment identification, calibration dates, the individual performing each calibration, and the next calibration date shall be documented.

Proper documentation is critical, not only to the process, but also for future statistical analysis, review of calibration intervals, and production trends and root cause analysis when problems are identified. By having the proper documentation in place, you do not have to reinvent the wheel every time a person in a critical or supervisory position changes. Continuity is an important ingredient of any company's production and/or manufacturing processes and procedures.

According to the guide, a typical equipment calibration procedure includes:

- Purpose and scope
- Frequency of calibration
- Equipment and standards required
- Limits for accuracy and precision
- Preliminary examinations and operations
- Calibration process description
- Remedial action for product
- Documentation requirements

Most of these can be found in other standards or regulations (see Chapter 5). By frequency of calibration, the calibration interval must be stated and followed according to the guide. A list of the equipment and standards must also be in the procedure. Within this list, the range and accuracy of both the standards and IM&TE should be listed. Many types of IM&TE must be set-up before any type of calibration can even be started, and that is what they refer to as preliminary examinations and operations. The description of the calibration procedure must be in enough detail as to allow all levels of calibration professionals to be able to follow the instructions as written without additional supervision once they are trained in the use of that procedure and type of IM&TE.

The FDA probably places more emphasis on remedial action for product; and documentation requirements than any other standard or regulation. The safety of the American public is critical during the manufacture of food, drugs, and cosmetics. The actions taken by the manufacturer when IM&TE is found to be out of tolerance during routine calibration is looked at with a very critical eye by the inspectors and auditors. How a company manages this particular process can make the difference between having FDA approval and not being able to produce a particular product. Is recall of product performed? What process is used to make that determination? What effect did the IM&TE have on the process or production? Is there documentation at all levels of decision making? Is that documentation available for inspection?

Some companies use a two-pronged approach to this process and call it *Alert and Action*. Whenever IM&TE is found out of tolerance, Alert procedures go into place. Notification is given to the test equipment owner/user, and repairs and adjustments are recorded and their historical records are archived for future reference. Depending on the criticality of the IM&TE, how it was used in the process, and how much impact it may have had on product quality a second process might be used called Action. In the case of Action Procedures, critical evaluation of product or process is performed to evaluate the quality of product or process, and all levels of supervision up the chain must sign off that quality was not affected by the IM&TE being out of tolerance. This is far more time-consuming and critical to the manufacturing process than an Alert procedure. What system a company puts in place would depend on the criticality of the product produced, the effect on the users of the product, and the possible impact of poor quality for the customer. Of course, both procedures would involve careful documentation and archiving of the records involved with the affected IM&TE.

The documentation requirements are also more stringent when public safety is at risk. Calibration records should meet the following criteria in a cGMP environment:

- Use black or blue ink—no pencil, magic marker, or anything else. Never erase or use Wite Out®.

- Writing over errors is not allowed.

- All entries must be legible.

- All fields are completed or filled in. Place an N/A or None in all unused fields.

- Put a single line out through errors.

- Sign and date all corrections, with explanation of errors.

- Never document or sign another person's work or data.

- Enter data as it is recorded or immediately upon completion.

- Avoid abbreviations, catch phrases, or unprofessional remarks.

- Enter data as received or viewed—never guess or anticipate readings or data.

- Round numbers using accepted procedures (see Chapter 11).

- Never use rubber stamps.

- Never backdate a record or document.

- Additional information added to the record must be signed and dated.

- Identify all measurements with their proper units of measure.

- The record must be reviewed and signed by a competent authority.

- Mark if the IM&TE is in tolerance or out of tolerance.

Calibration of each piece of equipment shall be documented to include: equipment identification, the calibration date, the calibrator, and the date the next calibration is due.

Measuring instruments should be calibrated at periodic intervals established on the basis of stability, purpose, and degree of usage of the equipment. A manufacturer should use a suitable method to remind employees that recalibration is due.

The Quality System regulation requires that standards used to calibrate equipment be traceable to the NIST or other recognized national or international standards. Traceability also can be achieved through a contract calibration laboratory, which in turn uses NIST services.

As appropriate, environmental controls should be established and monitored to assure that measuring instruments are calibrated and used in an environment that will not adversely affect the accuracy required. Consideration should be given to the effects of temperature, humidity, vibration, and cleanliness when purchasing, using, calibrating, and storing instruments. The calibration program shall be included in the quality system audits required by the quality system regulation.

These are the words on paper that dictate the requirements according to the FDA regulations. To some, they are easy to understand and follow. To the vast majority of the public, however, they are complicated and vague. By breaking each section into small pieces, it will be easier to understand what is asked of the common calibration technician or metrology manager.

What does "Manufacturers are responsible for ensuring the establishment of routine calibration" mean in plain English? Simply, an organization is responsible for having a calibration system in place that meets the requirements as outlined in the guide. The easiest way to comply with these requirements is to follow what was written in Chapter 2, "Say what you do, do what you say, record what you did, check the results, and act on the difference." CGMP requirements break these down into very specific demands. The biggest difference between what is required by the FDA for a compliant calibration system compared to other standards is the greater detail in documentation. This includes the record keeping, change control for procedures, security and archiving of records, evaluation of out-of-tolerance IM&TE and how it affected product or processes, and so on. If a company is paperless, it also has to meet the requirements of 21CFR Part 11, Electronic Records and Electronic Signatures.

Usually, one of the first questions asked during an audit or inspection is, "May I see your overdue list?" The answer to this question can either generate more questions or it gives the auditor an overall good impression of the system that is in place. You should have inventory control (all items individually tagged for easy identification, and those items entered into an automated or computerized system); a scheduling system, which includes the calibration intervals setup for your system; and proof that the system has been validated. If there are items on your overdue list, have they been segregated from calibrated IM&TE to prevent use in production and/or manufacturing? If not, what is to keep them from being used? How long have they been overdue calibration? What is the reason for their overdue status? Is there a lack of standards, staff, time, facilities, or a combination of these reasons? Have you documented that you know items are overdue and what you are doing about it, or is this a complete surprise? Being aware of what is going on within your system and having a plan in place to remedy the situation can go a long ways in assuring the auditor that this is a one-time occurrence (if that is the case).

If your standards are overdue or out of tolerance, do you have a recall system for knowing which items those standards were used to calibrate since the last time the standard was certified? Can you readily produce a list of IM&TE to recall? Do you have a system in place for labeling or identifying items that have been recalled or are out of service? Is all of this written down in a procedure for anyone in your department or laboratory to follow?

Have all of your technicians been trained to perform the tasks they are assigned? Do you have documentation to prove who has been trained on what items and when that training was completed?

When it comes to records, can you provide a paper (even if you are paperless) trail showing traceability to a national/international standard? Is it easy to follow, and can your technicians as well as their supervisors accomplish it? Here is an example of a paper trail for a fictitious piece of test equipment.

An auditor observes a scientist using a water bath to produce product. The scientist copies down the identification number for the water bath, the date it was calibrated, when the calibration label shows it is next due calibration, and who performed the calibration. When the auditor is inspecting the calibration system, he or she asks to see the traceability for the calibrated water bath. The calibration record for the water bath, showing the As-Found data, As-Left data (when applicable), dates calibrated and next due, and who performed the calibration is produced for the auditor. There is a statement showing traceability back to a national or international standard (in most cases in the United States, NIST). The record also shows which calibration procedure was used and gives an uncertainty statement (this could be a 4:1 ratio or uncertainty budget showing the actual error). The calibration standards are identified, and their calibration due dates are also shown on the record. The record for the standards used during the calibration of the water bath would then be checked for the same information as stated above, and the record for the standards used for that calibration also inspected. This would continue for each standard used in the traceability chain. Certificates of calibration from outside vendors would also be inspected for the same information, till traceability to a national/international standard can be observed. As each record is produced, the accuracy of the item being calibrated against the standard used must show the test uncertainty ratio (TUR) or uncertainty budget for that particular calibration. Generally, if a TUR of at least four to one (4:1) is observed, then traceability has been accomplished. When TURs of less than 4:1 occur, there should be proof that the user is aware of this and accepts this lower TUR.

The maintenance of this paper trail is valuable for more than just an audit. It provides information for use during recalls of IM&TE and evaluation of calibration intervals, and allows calibration professionals to access the uncertainty of their standards and IM&TE for future calibrations.

## OTHER INDUSTRY REQUIREMENTS

ISO 9001 is supposed to be a *generic* model for a quality management system, applicable to any industry. Because of specific requirements, many industries have developed QMS requirements that are based on ISO 9001 but with added industry-specific requirements. Other industries, particularly those subject to government regulation, may have QMS requirements that are completely separate from the ISO 9000 series. This has created a system where there are two types of industry-specific requirements:

those based on some version of the ISO 9000 series and those that are independently developed.

## Standards Based on ISO 9000

There are four major industry areas, or sectors, that use tailored quality management standards based on a version of the ISO 9000 series. The sector-specific standards contain the requirements of ISO 9001, interpret those requirements in terms of that industry's practice, and include additional sector-specific requirements. Three of the industry areas are aerospace manufacturing (AS9100A), automotive manufacturing (QS9000 and ISO/TS 16949), and telecommunications (TL9000). The fourth area, medical device manufacturing, is discussed in the previous section. Another industry area, computer software, has a standard for applying ISO 9001 to the software development process and a registration program to audit that.

### Aerospace Manufacturing—SAE AS9100A

The aerospace manufacturing industry largely conforms to the AS9100 series of standards. This series of standards is based on the ISO 9001:2000 quality management system standard, with additional requirements specific to aerospace manufacturing. Before adoption of the current revision in August 2001, it had been based on ISO 9001:1994. The AS/EN/JIS 9100 series is now widely accepted in the aerospace industry.[5]

Aerospace quality management standards are coordinated globally by the International Aerospace Quality Group (IAQG), established in 1998. This organization is a cooperative group sponsored by Stanford Applied Engineering (SAE) International, representing North, Central, and South America; the European Association of Aerospace Companies (AECMA), representing Europe and Africa; and the Society of Japanese Aerospace Companies (SJAC), representing the Asia-Pacific region. An important reason for the formation of IAQG is the realization of the members that, where safety and quality are concerned, cooperation has more importance than otherwise normal competition.[6] According to the IAQG charter, the organization was founded to establish and promote cooperation among international aerospace companies with respect to quality improvement and cost reduction. This is achieved by voluntary establishment of common quality standards, specifications, and techniques; continuous improvement processes; sharing results; and other actions.[7] IAQG members are representatives of aircraft and engine manufacturers, and major parts and component suppliers that agree to and sign the charter.

### Automotive Manufacturing

From the early 1980s through 2003, quality management practices in the automotive industry have evolved from supplier requirements set by individual manufacturers through requirements defined by the major manufacturers in a country working together to the current state of having an international technical specification based on ISO 9001:2000. For the benefit of readers who are not familiar with what is meant by the automotive industry, it can be defined as companies eligible to adopt ISO/TS 16949:2002. Eligibility is explained by Graham Hills as the manufacturers of finished automotive products, and:

. . . suppliers that make or fabricate production materials, production or service parts or production part assemblies. It also applies to specific service-oriented suppliers—heat treating, welding, painting, plating or other finishing services. The customers for these types of suppliers must be manufacturers of automobiles, trucks (heavy, medium and light duty), buses or motorcycles. [It] does not apply to manufacturing suppliers for off-highway, agricultural or mining OEMs. It also doesn't apply to service-oriented suppliers offering distribution, warehousing, sorting or non-value-added services. Nor does it apply to aftermarket parts manufacturers.[8]

**QS-9000.** The QS-9000 standard was developed by three large U.S. auto manufacturers (Ford, General Motors, and Chrysler), working with the Automotive Industry Action Group's Supplier Quality Requirements Task Force. The first edition was published in 1994, and revised editions appeared in 1995 and 1998. It incorporated the requirements of ISO 9001:1994 and additional requirements important to the companies. The additional requirements covered business areas such as business planning, product quality planning, customer satisfaction, and continuous improvement.[9] The intent was to make life easier for suppliers by having one quality standard that all agreed on, instead of three separate ones, however, each of the three manufacturers added their own specific additional requirements or interpretations. There are also other interpretations for specific types of suppliers, such as tool manufacturers (the TE Supplement) and calibration laboratories.

QS-9000 is now functionally obsolete. It has been replaced by an international standard, ISO/TS 16949.

**ISO/TS 16949.** After the ISO 9000 series of quality management system standards was first published in 1987, some companies started incorporating it into requirements for their suppliers. This is, of course, an intended application of the standard. By the mid-1990s, automotive manufacturers in particular had extensive requirements for their suppliers based on ISO 9001:1994 by that time. There were four major systems: AVSQ 94 in Italy, EAQF 94 in France, QS-9000 in the United States and VDA 6.1 in Germany. Companies and trade associations started seeing a problem, though. Suppliers often had to become registered to two or more sets of requirements, and they sometimes conflicted. For example, a parts manufacturer in South Carolina sells parts to the three U.S. auto manufacturers, a couple of German manufacturers with plants in the United States and Germany, and a U.S. plant of a Japanese manufacturer. They would have to be registered to the automotive quality system of each country. (And that is better than a few years earlier, when they had to be approved by each company.) By about 1997, the International Automotive Task Force (IATF) started a liaison with ISO technical committee 176 (ISO/TC 176) to establish a single quality management standard for the automotive sector. The first edition of ISO/TS 16949 was published in 1999. It is also a significant item because it marked the first time a sector-specific document was produced using this method.[10]

The current edition is ISO/TS 16949:2002 Quality Management Systems—Particular Requirements for the Application of ISO 9001:2000 for Automotive Production and Relevant Service Part Organizations. This version is aligned with ISO 9001:2000 and was developed by IATF and ISO/TC 176 with input from the Japan Automobile Manufacturers Association (JAMA). Because non-ISO groups developed this document, it is a technical specification, not an international standard. It includes all of the text of ISO 9001:2000, plus additional sector-specific requirements.[11] The intent is for ISO/TS 16949:2000 to be one industry quality management system that is accepted by all manufacturers regardless of location. Each major automotive manufacturer has its own timetable for mandatory conformance by its suppliers, which range from immediate to December 2006.[12] This can be viewed as a two-step process for companies: registration to ISO 9001:2000 since all of the requirements of that standard are included and then certification to the additional ISO/TS 16949:2002 requirements added by the IATF.[13]

Comparing the new version of TS 16949 to the earlier one or to QS-9000 is similar to comparing ISO 9001:2000 to ISO 9001:1994: the differences are too many to list. The major change, of course, is the conceptual shift from conformance to 20 tasks, to the total business and process management concept with management based on data, and the PDSA cycle. All of the information elsewhere in this book about ISO 9001:2000 applies to TS 16949. The added requirements include things such as increased requirements for top management involvement and responsibility, communication with suppliers and customers to determine quality requirements and issues, and employee responsibility for quality.

ISO/TS 16949:2002 has an impact on calibration laboratories, as shown by the requirements of section 7.6.3, laboratory requirements. These requirements are:

- The scope of an organization's internal laboratory is a required part of the quality management system documentation. The scope has to include the capability to perform all of the in-house calibrations with adequate procedures and qualified people. Also, the capability to perform the tests correctly is required, which implies participation in proficiency testing and interlaboratory comparisons. Laboratory accreditation to ISO/IEC 17025 is mentioned, but is not a requirement for an in-house laboratory.

- If an external or independent calibration laboratory is doing work for an automotive manufacturer or supplier, it almost always has to be accredited to ISO/IEC 17025. There are only two exceptions. One is that the automotive customer can have evidence that the laboratory is acceptable, principally by auditing the laboratory to the requirements of ISO/IEC 17025. The other exception is if the calibration activity is part of the manufacturer of the IM&TE being calibrated. In that case the automotive customer has to ensure that the requirements that would apply to its internal laboratory apply.

## Telecommunications—TL 9000

The creation of a telecommunications sector–specific quality management standard was initially conceived in 1996 by a group of four telecommunications service providers in the United States.[14] A separate organization, the QuEST Forum, was created in 1997 to develop and manage the standard in cooperation with ASQ.[15] Membership now includes other service providers, and suppliers of hardware and software. There are also liaison members from other organizations including registrars and training providers.

Like other sector-specific standards, TL 9000 is based on ISO 9001. Portions are also drawn from software quality and engineering standards, preexisting industry requirements and other sources.[16] Requirements of ISO 9001 and other sector-specific requirements are identified as general requirements or ones that apply specifically to hardware, software, or services. Particular emphasis is placed on system and component reliability, with a requirement for at least 99.999 percent availability. For example, this means that a system cannot be out of service for more than approximately five minutes in a year. In addition, there are two specific features that are not yet part of any other sector–specific quality management standard.

The first significant feature of TL9000 is that it defines specific business measurements (metrics) and requires suppliers to report them. Reporting of the specified metrics is required to obtain and maintain registration. The metrics are reported to a secure data repository and used to determine benchmark or best-in-class information for all participants in the standard. All published results are comparative (best, worst, average) and anonymous.[17]

The other significant feature of the TL 9000 system is the QuEST Forum's focus on business excellence and continuous improvement. The Business Excellence Acceleration Model (BEAM) was started in 2000 as an ongoing project to encourage and drive business improvement in the industry.[18] BEAM is largely based on the Malcolm Baldrige National Quality Award model, as well as similar programs in other countries. It is important because it is a move by a major industry sector from a compliance-based system to a system that uses self-assessment based on business excellence.[19]

## Computer Software—ISO 9000-3 and TickIT

As noted recently by ASQ past President Greg Watson, software is increasing in importance. In many cases, capabilities can be added to a product with software changes or additions.[20] Defects in installed software are particularly bad for two reasons. First, the defect repeats every time the product is used, although the effects may differ because of other interactions. In addition, the severity of the defect can range from merely annoying to a level that can shut down a business or cause a major safety problem. Consideration of software issues is important in metrology because of the large and increasing number of measurement systems that use internal software or are controlled by a computer system.

The computer software industry does not have a sector-specific quality management system standard. If computer software is supplied as part of an aerospace, automotive, medical, or telecommunications system, then the software development process is likely to be managed within the scope of the applicable sector-specific quality management standards discussed earlier. The same is true if the software is part of a system in a regulated industry. Otherwise, the industry has two guides for applying ISO 9001 and a separate highly detailed methodology originally based on government requirements. Most of the software-related international standards that exist are focused on the production process and on inspection of the product during development, rather than the overall quality management system.

### ISO 9000-3:1997

ISO 9000-3:1997 is the only international standard that focuses on quality management systems for software development. It is a guide for applying the principles and requirements of ISO 9001:1994 to an organization that produces computer software. The standard explains what the requirements of 9001 mean in the context of a software development organization. As a guidance document, it does not have any requirements, only suggestions for effectively implementing the requirements of the 9001 standard in a software development environment. The wording of ISO 9001:1994 is heavily oriented toward hardware manufacturing. ISO 9000-3 is very useful as a guide for interpreting those requirements and applying them in a way that is more appropriate for a creative enterprise such as software development. As it is a guidance document, it is not auditable by a registrar.

ISO started revision of ISO 9000-3 in 2001. The title of the revised standard will be ISO/IEC 90003 *Software Engineering—Guidelines for the Application of ISO 9001:2000 to Computer Software*.[21] The final draft was approved at the end of September 2003, and at that time it was expected to be approved as an international standard by the end of the year. There are two important administrative changes with this revision. The first is the change in the number, from 9000-3 to 90003. The other change is that the standard is now managed by a subcommittee of the joint ISO/IEC technical committee, instead of ISO TC 176.

### TickIT

TickIT is a program to audit the application of ISO 9001 to software development in organizations that are registered to that standard. TickIT was started in 1991 in Great Britain and Sweden, with a goal of improving quality management systems in the software industry and improving the quality of the resulting products. The name comes from *tick*, the English word for a *checkmark* and *IT*, the acronym for information technology.

The main parts of the TickIT program are guidance in applying ISO 9001:2000 to an organization that develops software, certification (registration) of organizations to ISO 9001 using TickIT procedures, and training and registration of auditors. The source of all TickIT requirements is ISO 9001:2000, with interpretation in the TickIT guidance document. The TickIT guide has an overall introduction, and guidance chapters for customers, suppliers, and auditors. As of July 2003, there were 1157 active registrations in 41 countries, with over 70 percent of the registrations in Great Britain.[22] In that country, TickIT certification is mandatory if an organization is registered to ISO 9001 and any of its processes include software development.

## Standards *Not* Based on ISO 9000

There are a number of other quality management system requirements that are not based on the ISO 9000 standards. These are often sector-specific. They may have been developed within the industry or by a government regulatory agency. A few are discussed here because of their broad impact on society: civil aviation, software development, and civil engineering.

### Civil Aviation

Civil aviation has been subject to government regulation for most of the first 100 years of powered flight. Even the much publicized deregulation of U.S. airlines in 1978 applied mainly to competition in domestic routes and fares; every other aspect of the industry is still highly regulated. Other nations also provide a high degree of regulation of air transportation, primarily in the interest of safety, but, also, in many cases, because of significant government ownership of the companies.

In the United States, government involvement in civil (nonmilitary) aviation started in 1918, when the Post Office Department established an experimental air mail service. Starting in 1925, Congress passed a number of laws that, by 1967, governed every aspect of operating an air transport company.[23] The regulations covered routes, fares, cargo, maintenance, aircraft design, airport operation, navigation, and so on. In many ways the airline industry was similar to a public utility in that it provided services to the public under heavy government regulation.[24] That was changed somewhat by the Airline Deregulation Act of 1978. This eliminated restrictions on domestic routes, virtually eliminated barriers to entering the business, allowed competition based on price, and removed a large number of other restrictions on both competition and cooperative marketing arrangements. Significantly, though, all other areas of regulation were not affected. Of particular interest to people in the calibration industry, aircraft manufacture and maintenance remain heavily regulated, as well as flight control, communication, and navigation systems.

The Federal Aviation Regulations (FAR, or 14CFR) are the U.S. civil aviation regulations.[25] In Europe, the Joint Aviation Authorities (JAA) regulations (JAR) are similar to the FAR in the United States. The FARs as a whole can be viewed as a type of quality management system, even though that phrase (or *quality manual*) does not appear anywhere in them. It also goes well beyond quality management by including detailed requirements for processes, methods, and products. It is a system that is a very rigorously defined and demands full compliance. The FAR has rules on just about any aspect of the aerospace industry. The regulations that apply to areas that affect the largest number of people are those dealing with airlines and their maintenance operations. The necessity to comply with the myriad regulations tends to drive up costs. The effectiveness of the system, however, is shown in the amazingly low fatal accident rate: more than 4000 times lower than the national highway fatal accident rate.[26]

There are two sections of the FAR that most directly apply to airline companies and maintenance operations. They are:

- 14CFR Part 121 has regulations about every person or operation that touches an aircraft operated by an airline, except those affected by deregulation. Examples include training of pilots, flight attendants, mechanics, and flight dispatchers; and requirements if the company maintains or repairs its own aircraft.

- 14CFR Part 145 has regulations for organizations that perform aircraft maintenance or repair for other parties. This includes independent repair stations, and also includes airlines that operate under Part 121 if they provide service for other airlines—and several do.

Calibration is covered in both places. Under Part 121, an airline must have "Procedures, standards, and limits necessary for . . . periodic inspection and calibration of precision tools, measuring devices, and test equipment." (14CFR121.369, *Manual Requirements*.) Under Part 145, a repair station must have a documented system that requires use of calibrated IM&TE (14CFR145.109 *Equipment, Materials and Data Requirements*) and maintaining a documented calibration system (14CFR145.211 *Quality Control System*).

An aviation repair station has additional requirements in an FAA interpretation document, Advisory Circular 145-9 *Guide for Developing and Evaluating Repair Station and Quality Control Manuals*. This document has specific requirements for the two manuals required by the FAA—the Repair Station Manual (RSM) and the Quality Control

Manual (QCM). A significant change for people in this industry is that these no longer have to be two separate documents. The RSM and QCM can be combined, and they can even be combined with an ISO 9000–based quality management system, provided there is a matrix to demonstrate how the manual ensures compliance with each of the regulatory requirements. The FAA has also adjusted its interpretation of calibration traceability to a view that is closer to the real world. There are specific calibration-related requirements and guidance in AC 145-9, but few that are not also required in industry standards such as ANSI/NCSL Z540-1, ISO/IEC 17029, and ISO 10012. Some significant items are:

- If a measurement is traceable to the SI through a national metrology institute other than NIST, then the FAA must approve that other NMI.[27]

- Actual measurement data *should* be collected when equipment is calibrated. If the repair station ever wants to adjust calibration intervals, however, then data *must* be collected because simply noting pass or fail in the record is not sufficient for that purpose.

- The method for adjusting calibration recall intervals has to be documented. Also, there has to be a sufficient calibration history to justify a change. This implies that, of the methods defined in NCSL RP-1, only those that use statistical tests may be used.

- A problem is that the FAA assumes that the manufacturer sets the initial calibration interval of IM&TE. This is becoming less common than it used to be, as current practice is to shift the responsibility for determining calibration intervals to the owner. The implication for a calibration activity (and the equipment owner) is that they must have a documented method for determining an initial calibration interval if one is not recommended.

### Computer Software—SEI Capability Maturity Model

During the late 1980s, the U.S. DOD determined that there was a need to improve the quality of computer software. At that time software was coming into increased use in military systems, but quality problems were driving up costs and development time. In particular, the DOD wanted a method of evaluating a software development organization as part of the selection and procurement process. This was contracted to the Software Engineering Institute (SEI) of Carnegie-Mellon University, and the result became known as the Capability Maturity Model® (CMM®).

The CMM rating system has five levels. An organization's level is determined by an evaluation based on criteria developed by the SEI. The levels have changed slightly over the years, but are currently defined as:

1. **Initial.** The software process is characterized as ad hoc, and occasionally even chaotic. Few processes are defined, and success depends on individual effort and heroics.

2. **Repeatable.** Basic project management processes are established to track cost, schedule, and functionality. The necessary process discipline is in place to repeat earlier successes on projects with similar applications.

3. **Defined.** The software process for both management and engineering activities is documented, standardized, and integrated into a standard software process for the organization. All projects use an approved, tailored version of the organization's standard software process for developing and maintaining software.

4. **Managed.** Detailed measures of the software process and product quality are collected. Both the software process and products are quantitatively understood and controlled.

5. **Optimizing.** Continuous process improvement is enabled by quantitative feedback from the process and from piloting innovative ideas and technologies.[28]

One feature of the software CMM is that the model has included elements of quality management systems and continual improvement from its inception. These are things that were not explicit in ISO 9001 until the 2000 version. There is a considerable amount of overlap between CMM and ISO 9001, and many elements of 9001 are in the CMM. The reverse is not true, though, primarily because the CMM is highly specific to software development.

Early evaluations of 167 software projects in 1989 showed that 86 percent were at level 1 and 13 percent were at level 2.[29] Current evaluations of over 1300 organizations show that about 13 percent are at level 1, and 69 percent are at level 2 or 3.[30] This indicates that software CMM is being used effectively. Another indicator of the usefulness of the capability maturity model is that it is now used in other areas: software/systems engineering, integrated product/process development, and supplier sourcing.

## Civil Engineering

Civil engineering typically does not use the same type of ISO documents that are described in this chapter, although some engineering and construction companies are registered to ISO 9001. Civil engineering is primarily concerned with design and construction of buildings, roads, public works, and related projects. Documented standards that address the areas of concern are typically—but not always—observed in what are known as codes.

Codes may be international, national, or more local. In many countries there is a single organization that develops and administers the codes for that country. In the United States there are many organizations, ranging from organizations such as the International Code Council (ICC) to state, county, and local community levels.

At the state, county, and local level, codes tend to be written to address concerns encountered in that area. For example, a code written in California will often include requirements that are driven by seismic concerns, while a similar code in Michigan will include requirements driven by extremes in heat and cold, and a similar code in Florida will be driven by high wind and water concerns. National and international codes attempt to address the crucial requirements for geographic areas and types of construction that may be covered by the codes.

Calibration laboratories rarely provide direct support to organizations that perform construction or to the oversight organization personnel, often known as *code officials*. Direct support for these organizations and personnel is typically provided by testing laboratories. Code officials and construction organizations usually seek testing laboratories that are accredited to ISO/IEC 17025 and/or have other specific accreditations. Sometimes the requirement for accredited testing laboratories is in the code or in the law or both. An example is a state department of transportation, which may have a requirement for its own personnel to inspect a laboratory for parameters it defines. For this reason, most testing laboratories carry multiple accreditations.

Accredited testing laboratories depend on accredited calibration laboratories for calibration support. Many accredited testing laboratories will perform some limited calibration checks internally also. For example, the testing laboratory will send out items such as gage blocks for calibration and use the gage blocks to check their dial indicators before a test. Accredited calibration and testing laboratories work together to ensure the integrity of constructions performed and of materials used. This cooperation is crucial to ensuring public safety through verified, measurable compliance to codes.

## Endnotes

1. ISO 2003b. www.iso.org/iso/en/iso9000-14000/iso9000/selection_use/iso9000family.html
2. In this usage, *valid* implies that the reading on the instrument objectively represents the true state of the system. It means that someone reading the indicator would normally accept the value as relevant, meaningful, and correct (Merriam-Webster, 2003).
3. Stein, 85.
4. Unless specified otherwise, material in this section is based on U.S. Food and Drug Administration GMP manual, www.fda.gov/cdrh/dsma/gmp_man.htm
5. The AS prefix is used in North and South America. The EN prefix is used in Europe and Africa, and the JIS prefix is used in the Asia/Pacific region.
6. International Aerospace Quality Group (IAQG) 2002.
7. International Aerospace Quality Group (IAQG) 2003.
8. Hills 2003.
9. Okes, Duke and Russell T. Wescott, 379.
10. Bird 2002.
11. Harral, 5–6.
12. Kymal and Watkins, 9–13.
13. Bird 2002.
14. Kempf, 1.
15. Hutchison, E. E. 34–35. (Business operations of the QuEST forum are administered by ASQ).
16. Dandekar, A. V. and E. E. Hutchison, 1–5.
17. Okes, Duke and Russell T. Wescott, 379.; Dandekar, A. V. and E. E. Hutchison, 1–50.
18. Hutchison, E. E., 33–37.
19. Dandekar, A. V. and E. E. Hutchison, 1–5.
20. ASQ 2003.
21. Hailey 2003.
22. TickIT 2003.
23. In 1925 Congress passed the Contract Air Mail Act. This and the Air Commerce Act of 1926 formed the basis of future laws and also helped establish airlines as viable companies. It also had at least six branches of the Federal government involved in regulating or providing services for air transport. The Air Mail Act of 1934, in addition to its principal intent, cut the number of agencies regulating the industry and required aircraft manufacturers to divest their ownership of air transport companies. In 1938, a new law created the Civil Aviation Administration (CAA) and Civil Aeronautics Board (CAB), which became the only agencies with regulatory power over aviation. In 1958 the CAA

became the Federal Aviation Agency, which was reorganized again by Congress in 1967 as the Federal Aviation Administration (FAA).
24. Millbrooke, 9-53; Wells, 58–69.
25. The Federal Aviation Regulations (Title 14, United States Code of Federal Regulations) are available on the Internet at www.access.gpo.gov/nara/cfr/cfrhtml_00/Title_14/14tab_00.html
26. Based on 2001 data from the National Transportation Safety Board and the National Highway Traffic Safety Administration. (Highways: 41,730 deaths, 151 per million miles traveled. Airlines: 266 deaths, 0.038 per million miles flown.) Historically, NTSB does not include aircraft accidents resulting from illegal acts in their statistics, so the victims of the September 11, 2001, terrorist attacks are not included.
27. The requirement that the FAA Commissioner has to approve traceability that goes through a non-U.S. national metrology institute (NMI) is an important improvement for people in the airline industry because the former guidance was that it was flatly forbidden. Measurement traceability had to be to NIST (actually, NBS) only. However, this change also puts the FAA in the curious position of questioning the validity of a large number of international agreements entered into under an international treaty. Most NMIs are members of mutual recognition agreements under CIPM, which is authorized by the Treaty of the Meter. The CIPM also maintains the Key Comparison Index database, so it is relatively easy for any metrologist to determine if the measurement uncertainty from a particular traceability path is adequate.
28. Software Engineering Institute. 2003.
29. Schulmeyers and McManus, 25.
30. Software Engineering Institute. 2003b.

## References

AIAG (Automotive Industry Action Group). 2003. *AIAG history highlights*. Southfield, MI: AIAG. www.aiag.org/about/history.pdf

ANSI/ISO 17025-1999, *American National Standard—General requirements for the competence of testing and calibration laboratories*. Milwaukee: ASQ Quality Press.

ANSI/ISO/ASQ Q9001-2000, *American National Standard—Quality management systems—Requirements*. Milwaukee: ASQ Quality Press.

ANSI/ASQ M1-1996, *American National Standard for calibration systems*. Milwaukee: ASQC Quality Press.

ANSI/NCSL Z540-1-1994, *American National Standard for calibration—Calibration laboratories and measuring and test equipment—General requirements*. Boulder, CO: National Conference of Standards Laboratories.

ASQ. 2003. Greg Watson examines software quality. Online video segment: www.asq.org/news/multimedia/board/watson4.html

Bertermann, Ralph E.; 2002, *Understanding current regulations and international standards; Calibration compliance in FDA regulated companies*; Mt. Prospect, IL: Lighthouse Training Group

Bird, Malcolm. 2002. A few small miracles give birth to an ISO quality management systems standard for the automotive industry. *ISO Bulletin* (August 2002). www.iso.ch/iso/en/commcentre/isobulletin/articles/2002/pdf/automotiveind02-08.pdf

BSR/ISO/ASQ Q10012:2003(E), *Measurement management systems—Requirements for measurement processes and measuring equipment*. Milwaukee: ASQ Quality Press.

Bucher, Jay L. 2000. *When your company needs a metrology program, but can't afford to build a calibration laboratory . . . What can you do?* Boulder, CO: National Conference of Standards Laboratories.

Dandekar, A. V. and E. E. Hutchison. 2002. "The QuEST Forum—The future of telecom quality development." CD-ROM. Milwaukee: ASQ.

Federal Aviation Regulations (Title 14, United States Code of Federal Regulations). Washington, DC: US Government Printing Office.
www.access.gpo.gov/nara/cfr/cfrhtml_00/Title_14/14tab_00.html

FDA Backgrounder, May 3, 1999, Updated August 5, 2002;
www.fda.gov/opacom/backgrounders/miles.html

FDA. Food and Drug Administration—Center for Devices and Radiological Health. *Medical device quality systems manual: A small entity compliance guide, 1st ed.* (Supersedes the *Medical device good manufacturing practices [GMP] manual*). www.fda.gov/cdrh/dsma/gmp_man.html

FDA. Food and Drug Administration. 2001. 21 CFR 820.72, *Inspection, measuring and test equipment*. Washington, DC: FDA:145–146.

Hailey, Victoria A. 2003. Personal correspondence with Graeme C. Payne, October 2003.

Harral, William M. 2003. What is ISO/TS 16949:2002? *Automotive Excellence* (Summer/Fall): 5–6. www.asqauto.org

Hills, Graham. 2003. The effect of ISO/TS 16949:2002. *InsideStandards* (November). www.insidequality.com

Hirano, Hiroyuki. 1995. *5 pillars of the visual workplace: The sourcebook for 5S implementation.* Translated by Bruce Talbot. Shelton, CT: Productivity Press.

Hutchison, E. E. 2001. The Road to TL 9000: From the Bell breakup to today. *Quality Progress* (June): 33–37.

International Aerospace Quality Group (IAQG). 2002. "The IAQG discusses industry initiatives at the 19th FAA/JAA international conference."
www.sae.org/iaqg/about_us/news/june-04-2002.htm

International Aerospace Quality Group (IAQG). 2003. "Excerpts from charter of the International Aerospace Quality Group (IAQG)."
www.iaqg.sae.org/iaqg/about_us/charter.htm

International Organization for Standardization (ISO). 2003a. "Where ISO 9000 came from and who is behind it." www.iso.org/iso/en/iso9000-14000/tour/wherfrom.html

International Organization for Standardization (ISO). 2003b. "The ISO 9000 family."
www.iso.org/iso/en/iso9000-14000/iso9000/selection_use/iso9000family.html

International Organization for Standardization (ISO). 2003c. ISO 10012:2003, *Measurement management systems—Requirements for measurement processes and measuring equipment*. Geneva: ISO.

International Organization for Standardization (ISO). 1997. ISO 9000-3:1997 *Quality management and quality assurance standards—Part 3: Guidelines for the application of ISO 9001:1994 to the development, supply, installation and maintenance of computer software*. Geneva: ISO.

International Organization for Standardization (ISO). 2002. ISO/TS 16949:2002 *Quality management systems—Particular requirements for the application of ISO 9001:2000 for automotive production and relevant service part organizations*. 2nd ed. Geneva: ISO.

Kempf, Mark. 2002. "TL-9000: Auditing the adders." Proceedings of ASQ's 56th Annual Quality Congress. CD-ROM. Milwaukee, WI: ASQ.

Kimothi, S. K. 2002. *The uncertainty of measurements: physical and chemical metrology impact and analysis*. Milwaukee, WI: ASQ Quality Press.

Kymal, Chad and Dave Watkins. 2003. How can you move from QS-9000 to ISO/TS 16949:2002? More key challenges of QS-9000 and ISO/TS 16949:2002 transition. *Automotive Excellence*. (Summer/Fall): 9-13.

Marquardt, Donald W. 1997. "Background and development of ISO 9000 standards." In The ISO 9000 Handbook, 3rd edition. Edited by Robert W. Peach. New York: McGraw-Hill.

Merriam-Webster Online Dictionary. 2003. "valid." www.m-w.com

Millbrooke, Anne Marie. 1999. *Aviation history*. Englewood, CO: Jeppesen Sanderson Inc.

NIST Special Publication 811, 1995 Edition, *Guide for the use of the International System of Units (SI)*. Gaithersburg, MD: National Institute of Standards and Technology.

Okes, Duke and Russell T. Wescott. 2001. *The certified quality manager handbook*. 2nd ed. Milwaukee: ASQ Quality Press.

Pinchard, Corinne. 2001. *Training a calibration technician . . . in a metrology department?* Boulder, CO: National Conference of Standards Laboratories.

Praxiom Research Group Limited, Edmonton, Alberta.

QuEST Forum. 2003. *QuEST Forum member companies*. questforum.asq.org/public/memcomp.shtml

Schulmeyers, G. Gordon and James I. McManus. 1996. *Total quality management for software*. Boston: International Thomson Computer Press.

Software Engineering Institute. 2003a. *Capability Maturity Model® (SW-CMM®) for Software*, July 21, 2003. www.sei.cmu.edu/cmm/cmm.sum.html.

Software Engineering Institute. 2003b. *Process maturity profile—Software CMM® CBA IPI and SPA appraisal results—2003 mid-year update*. September 2003. www.sei.cmu.edu/sema/pdf/SW-CMM/2003sepSwCMM.pdf

Stein, Philip G. 2000. Don't Whine—Calibrate. Measure for Measure column, *Quality Progress* 33. (November): 85.

TickIT. 2001. *The TickIT guide: executive overview*. www.tickit.org/overview.pdf

TickIT. 2003. *TickIT certified organizations*. www.tickit.org/cert-org.htm

Wells, Alexander T. 1998. *Air transportation: A management perspective*. Belmont, CA: Wadsworth Publishing Company.

# Chapter 18
# Computers and Automation

In the twenty-first century, computers and automated calibrations seem to be in use in almost every discipline of metrology and calibration. This was not always the case. Automated calibrations have only been around for the average calibration technician since the early 1990s. Prior to that time, the vast majority of computations, calibrations, and calibration systems were accomplished manually, with written data collection, hard copy records, and if required, calculators or slide rules. An automated calibration system may or may not do the work faster. The real advantages are that the calibration is done the same way every time, the data are (usually) collected automatically, and, if the unit under test is controlled by the system, then the technician is free to calibrate another item. Laboratory information databases allow almost complete elimination of paper in the laboratory, can automatically generate reports, certificates and labels, and provide many more administrative functions. We've come a long way!

This chapter is divided into two sections, with the first covering software and data acquisition, and the second discussing automated calibration and calculation software validation. Even though this is a relatively new area for metrology, the requirements for them are covered in various standards and requirements.

ANSI/ISO 17025-1999, *American National Standard—General Requirements for the Competence of Testing and Calibration Laboratories*, states in section 5.4.7.2, "When computers or automated equipment are used for the acquisition, processing, recording, reporting, storage or retrieval of test or calibration data, the laboratory shall ensure that:

- Computer software developed by the user is documented in sufficient detail and is suitably validated as being adequate for use;

- Procedures are established and implemented for protecting the data; such procedures shall include, but not be limited to, integrity and confidentiality of data entry or collection, data storage, data transmission and data processing;

- Computers and automated equipment are maintained to ensure proper functioning and are provided with the environmental and operating conditions necessary to maintain the integrity of test and calibration data."

BSR/ISO/ASQ Q10012:2003, *Measurement Management Systems—Requirements for Measurement Processes and Measuring Equipment*, section 6.2.2, software, states, "Software used in the measurement processes and calculation of results shall be documented, identified and controlled to ensure suitability for continued use. Software, and any revisions to it, shall be tested and/or validated prior to initial use, approved for use, and archived. Testing shall be to the extent necessary to ensure valid measurement results."

It is up to the software user to make sure it will perform per its intended function(s). Even commercial-off-the-shelf (COTS) packages such as the Microsoft® Excel™ spreadsheet application are subject to computation discrepancies. The following is a news alert that appeared in April 2002 on the Agilent Metrology Forum warning about spreadsheet discrepancies problems: ". . . tests made on Excel have revealed that several functions, such as for standard deviation and regression, are based on defective algorithms or formulae and can produce highly inaccurate results. These are the very functions that feature heavily in metrologists' measurement analyses!" The United Kingdom's National Physical Laboratory (NPL) website on Software Support for Metrology at www.npl.co.uk/ssfm notes the following, "The numerical performance of some of Excel's 'regression' functions can be poor, with results accurate to only a small number of significant figures for certain data sets."

NCSL International's RP-6, *Calibration Control Systems for the Biomedical and Pharmaceutical Industry*, states in section 5.13, Computer software, "The potential for improving calibration-laboratory productivity and quality through the use of a computer-based system is widely recognized. Accordingly, computer software, as an increasingly significant and integral part of measurement systems, must have its development, maintenance, and utilization managed and controlled as carefully as that of the calibration-system hardware. Software validation and certification is as important as hardware calibration in ensuring the quality of measurements that are controlled or assisted by computer. Software changes should be controlled in the same manner as documented processes. Written procedures should be developed, implemented, and maintained detailing such requirements as:

- Software documentation, development, and review: the sequential steps and techniques to be employed in developing the software along with the accompanying internal review and acceptance levels required;

- Software testing and validation: the type of required validation testing to be performed and the manner in which the tests should be performed and recorded;

- Software and hardware configuration control: the steps necessary to ensure that adequate control is maintained over the baseline software materials and identification and the hardware included in the different computer-based systems;

- Software discrepancy corrections and changes: the methods and tracking techniques to be employed in changing or correcting completed computer programs;

- Subcontracted software: control requirements for use of software obtained from outside sources."

FDA regulations and ISO/IEC 17025 have requirements that certain software be validated. FDA applies the requirement to any software that is used in or for the production of medical equipment or to implement associated quality management systems.

ISO/IEC 17025 requires validation of calibration software developed by the user (5.4.7.2) or used with calibration equipment (5.2) and has other requirements for control of data, records, and documents on computer systems. While software verification and validation have been accepted (if not widely used) practices in the software development industry since the 1980s, it is a new concept to most people in the metrology field.

There are a number of resources available on this topic, and the great majority of them are of interest only to professional software developers. In general, guidance for software validation goes beyond simple functional validation. It also includes planning, verification, testing, traceability, configuration management, design control (planning, input, verification, review), documentation of results and other software engineering practices. It is probable that most calibration laboratories do not do enough development to justify a detailed understanding, although many labs probably do more software development than they think. For instance, many laboratories have a calibration automation system, usually with calibration routines provided as part of the system. By itself, that is COTS software that requires minimal validation, however, such systems typically include the capability to modify existing calibration procedures or create new ones that is now "software developed by the user."

## General Guidelines

One of the most general and accessible set of guidelines for software validation is the *General Principles of Software Validation: Final Guidance for Industry and FDA Staff* from the U.S. FDA. This document's guidance is geared toward software life cycle management and risk management, and for application by medical device manufacturers who also develop software for their products. As such, it is largely not applicable for calibration labs. It is, however, a good outline of general software validation principles, including validation of COTS software.

Translating the FDA medical device terminology to metrology applications, this guidance would cover:

- Software that is part of the IM&TE. This includes any built-in firmware.

- Software that is itself the IM&TE. An example would be a virtual meter (on a computer system) that responds to data acquisition system inputs.

- Software used in the calibration of workload, such as automated calibration procedures.

- Software used to implement the quality management system. This can include calibration automation systems and laboratory information management systems.

Software purchased by or developed within a calibration laboratory can fall into any of these categories, although it would be rare for a lab to be developing instrument firmware. Areas of concern include purchased software, obviously, and some other items. Examples of the other things include, but are not limited to:

- New or modified automated calibration procedures (as mentioned earlier).

- Signal acquisition, measurement, control, and presentation applications developed in a graphical development environment.

- Applications built in or using Microsoft Office applications such as Excel or Access. This includes macros developed to perform parts of any of the covered functions.

- Applications built using a scripting language for the Internet or the Microsoft Windows environment.

The FDA also has specific requirements for electronic records and electronic signatures. A laboratory that is not in this regulatory environment may feel they are unnecessary, but actually they are a good practice to follow anyway. This includes requirements for or about user identification (usually meaning password-controlled access), permanent audit trails of records, training and competency, and electronic signatures.[1]

In the software development industry, the words *verification* and *validation*, along with *testing*, are often used as if they are synonyms. The FDA guidance points out the differences. *Verification* is objective evidence that particular phase of the software development life cycle has met all of the phase requirements. Testing is one of many activities used to develop the objective evidence. *Validation* is "confirmation by examination and provision of objective evidence that software specifications conform to user needs and intended uses, and that the particular requirements implemented through software can be consistently fulfilled."[2] Validation is dependent on, among other things, verification and testing through the software development life cycle.

When compared with hardware, software is unique in several ways.

- Software is mostly effort-intensive in the design and development phase. Once a fully functional version is finished, duplication of it is a trivial expense by comparison.

- Software does not wear out and may even improve over time with defect removal. (But defect removal is also opportunity to unintentionally create new ones.)

- Once installed, a clean software version usually requires no maintenance to remain at that level. Some maintenance is occasionally done, typically involving removal of latent defects (bugs) or adding new features or capabilities. Such maintenance, though, requires that the validation be repeated after adding any new or changed requirements to the documentation.

- A lot of software is customizable by the user. For example, a system may allow renaming database fields, creation of custom calculations or event triggers, or changing the visual appearance of the application. If validation is required, then there must be a way to ensure that change is controlled, and changes are validated before large-scale implementation.

Software and hardware can interact in unpredictable ways. Part of this is because computers from different manufacturers that are each compatible with a given industry standard are not necessarily compatible with each other. Also, accessory hardware and other application software may interact differently on different hardware. When a stable operating environment (hardware and software) is determined and validated, that configuration should be placed under change control.

## Software Validation Process

There are certain features that a process for software validation must have.

- **Documented requirements.** A requirements list is the basis for validation of the software.
    - The intended use of the system is documented.
    - The operational and functional requirements of the system are detailed from the user's point of view.
    - If the software will be part of a system that is critical for productivity or the quality management system, a risk assessment including the level of criticality is documented.
    - Any safety requirements are documented. For example, will the software control a high voltage or high current source?
- **Validation tests of the requirements.** The requirements are the stated or implied needs of the user.
    - A test plan should describe how validation of each requirement will be accomplished.
    - A detailed procedure for testing each requirement is documented.
- **Evidence of successful validation is documented and maintained.**

Software and hardware interact as a system, so in any case where software is used at least part of the requirements must be those of the entire system. The two cannot be considered separately. For example, COTS software for office or entertainment applications always has a description of the minimum hardware system listed on the outside of the package.

## Software Validation Tasks

The FDA guidance defines a set of tasks that should be considered in the implementation of software validation. Other sources, Kershaw for example, define somewhat different but similar tasks. The FDA is more detailed, probably because of its regulatory environment. The level of detail and effort may vary depending on the situation and only needs to correspond to the level of risk associated with the software.

- **Quality planning.** Quality planning may include system requirements, risk assessment, configuration management, a software quality assurance plan, and other activities.
- **Requirements definition.** Requirements definition is the most important task, for software purchasing as well as development. Inadequate or incomplete requirement specifications are the principal cause of software project rework, delay, or failure.

    The requirements definition includes all of the software and system requirements for the user's point of view. These should include system inputs and outputs, the functions to be performed, system performance requirements, the user interface, the software to hardware interface when that is a factor, interfaces with other systems, error definition and handling, safety-related requirements, risk assessment, data ranges, limits and defaults, and data validation. This information is usually entered into a matrix format as the starting point for software traceability.

Requirements definition should also include a software validation plan. That plan should include verifying the consistency of all of the requirements, ensure that requirements are expressed in measurable or objectively verifiable terms, and ensure that the tests to be accomplished are traceable to the overall system requirements and risk analysis.

The result of the requirements definition process is a requirements specification that forms a controlled baseline for what the finished product should do.

- **Design.** Software design translates the stated and implied user requirements from the requirements specification to the logical program description. It describes in detail what the software should do and how it should do it (the methods it will use). Software design also includes the acceptance criteria for each module or phase. An important part of the design process is ensuring that all requirement specifications are implemented in the design and all features are traceable to a requirement. This is the core of software traceability. Design review is used to ensure that all elements of the design are correct, consistent, complete, accurate, testable, and traceable to the requirements. Another important task is planning the timeline for the remainder of the development life cycle. Design is usually an iterative process.

- **Construction (coding).** The construction of software is the actual programming and associated activities. It is very important that code be documented with comments to ensure that another person (or the same person!) can understand what is being done if some future maintenance is needed. Construction also includes inspection and testing of the code. Construction is almost always an iterative process and can take most of the time.

- **Testing by the developer.** Testing of the preproduction version is to ensure that all requirements have been met and that all requirements are traceable to tests that are performed. Testing usually runs concurrently with construction as well as after construction is complete. Developer testing is performed on varying levels, from individual code blocks up to functional parts of the system and finally the entire application. At a minimum, testing should include:

    – Tests for normal (expected) inputs

    – Tests to ensure all outputs can be generated

    – Tests with borderline (marginal) and abnormal (unexpected) inputs

    – Tests with combinations of these

- **User testing.** The user implements the documented test plan that has been developed from the requirements. These tests are not necessarily the same as the developer's tests because the finished software application is being tested in its actual operating environment. (That is, after installation on the target computer system and after any customization that has been documented as necessary.) The full range of conditions should be tested. It is important to record correct behavior during the testing, not just failures, because that information is important as an operational baseline. A test report must be produced.

- **Maintenance and changes.** Whenever any maintenance is done that changes the code, or if new features are added, the validation plan must be revised and reaccomplished. There must be provision for corrective action and preventive action.

Most professional software developers implement this type of system in their own businesses. Some will let you obtain a copy of their validation results, although it may be for a nominal fee. It is a good practice for any organization that develops software to implement this type of process, possibly in a compressed form. The principle of keep it simple applies here. It does not make sense to spend a week documenting a day's worth of work; but not having any objective evidence that a defined process was followed is not acceptable either.

## Validating COTS

Under the FDA guidelines, the purchaser of COTS software—the calibration laboratory—is responsible for verifying that the supplier's development process is adequate. Quite often, that is not possible—suppliers often treat their development process and its work products as proprietary, and a typical calibration laboratory does not have the resources to audit the developer. There are other options.

The laboratory may choose to accept the results of a third-party software quality audit. Applicable standards are ISO 9000-3 (soon to be reissued as ISO/IEC 90003), TickIT, and the SEI's Software CMM®. Another choice may be black box testing of the software product to ensure it meets the user's needs and intended uses.

If the software is part of an IM&TE component or system, then it can be argued that passing the calibration (performance verification test) is also functionally validating the software. If a traceable stimulus is applied and the system produces an acceptable result, then everything in between—including the software—must be operating correctly. Naturally, the calibration has to cover all functions and ranges.

When evaluating COTS software, one factor to consider is a risk assessment of the vendor terminating support for the software.

As a general guide, if COTS software has features that are not used, then those features do not necessarily have to be validated. Interactions with parts that are used must be considered and possibly evaluated. Also, there generally is no need to validate computer operating system software—only the application software that is used.

## Validating Automated Calibration Procedures

Validation of automated calibration procedures can be done in at least three different ways. First, though, the laboratory must verify that the software is functional and can run through to completion. This was discussed earlier, in the section on ISO 9000. After that, one or more of the following three methods could be used:

- **Comparison of manual and automated methods.**
    - Calibrate the unit under test (UUT) three to five times using the automated procedure and record all measurement results.
    - Then calibrate the UUT three to five times manually, at the same test points, and record all measurement results.
    - Determine the mean, sample standard deviation and standard deviation of the mean for each test point, in both the automated and manual results.
    - Compare the results using the $t$ test to determine if the mean results at each test point are likely to come from the same population.[3] If the difference between the means is greater than the critical value of $t$ or the standard deviations of the means do not overlap, then the two methods are considered to be different. (Note that this just means different, it does not necessarily mean better or worse; that would be the subject of analyzing the reasons for the difference.)

- **Inspection of procedure.**
    - If the automated procedure can produce a printed output of its commands and computations, compare the method used to a previously used manual procedure. This is obviously a less complete method of validation, but sometimes is the only one possible if the standard or the UUT cannot be operated manually.

- **Interlaboratory comparison.**
    - There may be cases where neither method can be used. Some systems, for example, can be operated only under automated control. One possible way to handle the problem is to send a UUT to another laboratory that uses the same automated method and ask them to calibrate it several times and record all data. When it is returned (or before it is sent), calibrate it on the in-house automated system, recording all data. Compare the results using the $t$ test as just described.

## Endnotes

1. FDA, www.fda.gov/cder/guidance/5667fnl.pdf
2. FDA, www.fda.gov/cdrh/comp/guidance/938.pdf
3. Natrella, 3-22 – 3-30.

## References

ANSI/ISO 17025-1999, *American National Standard—General requirements for the competence of testing and calibration laboratories*. Milwaukee: ASQ Quality Press.

BSR/ISO/ASQ Q10012:2003(E), *Measurement management systems—Requirements for measurement processes and measuring equipment.* Milwaukee: ASQ Quality Press.

FDA (U.S. Food and Drug Administration). 2002. *General principles of software validation: Final guidance for industry and FDA staff.* Rockville, MD: FDA. www.fda.gov/cdrh/comp/guidance/938.pdf

FDA (U.S. Food and Drug Administration). 2003. *Guidance for industry: Part 11 electronic records; Electronic signatures—Scope and application.* Rockville, MD: FDA. www.fda.gov/cder/guidance/5667fnl.pdf

National Conference of Standards Laboratories. 1999. *Calibration control systems for the biomedical and pharmaceutical industry.* Recommended Practice RP-6. Boulder, CO: National Conference of Standards Laboratories.

Natrella, Mary G. 1963. *Experimental statistics* (NBS Handbook 91). Washington, DC: U.S. Government Printing Office. (Reprinted in 1966 with corrections.)

# Part III
# Metrology Concepts

**Chapter 19**  A General Understanding of Metrology
**Chapter 20**  Measurement Methods, Systems, Capabilities, and Data
**Chapter 21**  Specifications
**Chapter 22**  Substituting Calibration Standards
**Chapter 23**  Proficiency Testing, Measurement Assurance Programs, and Laboratory Intercomparisons

# Chapter 19
# A General Understanding of Metrology

The field of metrology spans a multitude of different disciplines. Metrology incorporates an ensemble of knowledge gathered from many diverse fields such as mathematics, statistics, physics, quality, chemistry, and computer science—all applied with a liberal sprinkling of common sense. Essential to the field of metrology is understanding the fundamental methods by which objects and phenomena are measured, as well as the means for assigning values to measurements and the certainty of these assigned values. Encompassing some of these essentials are the establishment and maintenance of units, measurement methods, measurement systems, measurement capability, measurement data, measurement equipment specifications, measurement standards usage, measurement confidence programs, and so on. Remembering the adage "A chain is only as strong as its weakest link," metrology essentials are interdependent because each relies on an assortment of clearly stated definitions and postulations. The ensemble of metrology essentials lays the foundation for realization throughout the world that measurements are accepted and appropriate for their intended purposes. Mathematics is widely considered the universal language; the metrology essentials extend that language into our daily existence as quantifiable attributed information without which our world, as we know it could not exist.

The basic concepts and principles of metrology were formulated from the need to measure and compare a known value or quantity to an unknown in order to define the unknown relative to the known. This may seem like doubletalk but upon further investigation you can see that what is being described is a method for determining the value of an unknown by assigning it a quantity of divisions commonly referred to as units (for example, inches, degrees Celsius, minutes, and so on). Everything we buy, sell, consume, or produce can be compared, measured, and defined in terms of units of a known. Without commonly agreed-upon units, it would not be possible to accurately quantify the passing of time, the length of an object, or the temperature of ones' surroundings. In fact, practically every aspect of our physical world can be related in terms of units. Units allow us to count things in a building-block type fashion so they have meaning beyond a simple descriptive comparison such as smaller than, brighter than, longer than, and so on. Determination of measurement units that are deemed acceptable and repeatable, and maintaining them as measurement standards, lies at the heart of fundamental metrology concepts and principles.

Measurement units must be accepted or recognized and agreed upon in order to conduct most commercial transactions. The VIM defines a unit of measurement as a "particular quantity, defined and adapted by convention, with which other quantities of the same kind are compared in order to express their magnitudes relative to that quantity." The VIM goes on to note that units of measurement have assigned names and symbols such as m for meter and A for ampere. Mutually accepted measurement units for parameters, such as weight and length, provide the means for fair exchange of commodities. For example, the value of one ounce of gold can be equated to its equivalent worth in local currency throughout the world. The same cannot be said when using a nonaccepted unit, as its equivalent worth cannot be easily determined. A rectangle of gold has no defined equivalent worth because rectangle is not an accepted unit for mass.

As well as their importance in commerce, consistent and accepted measurement units are also very important in the sciences and engineering. They serve as a common frame of reference that everyone understands and can relate to. To facilitate the acceptance of units throughout the world, the CGPM established the modern International System of Units (SI) in 1960 as an improvement on earlier measurement units (see Chapter 11). The SI provides a uniform, comprehensive, and coherent system for the establishment and acceptance of units. The SI system is comprised of seven fundamental units. These seven units are used to derive other units as required to quantify our physical world. Congruent with SI units and their use are values that have been measured or determined for fundamental physical constants. The accepted values of these constants, along with the uncertainty associated with them, are published and updated as needed for use in all areas of science, engineering, and technology. A few of the frequently used constants are listed in Table 19.1.

**Table 19.1** Frequently used constants.

| Physical Constant | Value | Standard Uncertainty |
|---|---|---|
| atomic mass constant | $1.660\ 538\ 86 \times 10^{-27}$ kg | $0.000\ 000\ 28 \times 10^{-27}$ kg |
| Avogadro constant | $6.022\ 1415 \times 10^{23}$ mol$^{-1}$ | $0.000\ 0010 \times 10^{23}$ mol$^{-1}$ |
| Boltzmann constant | $1.380\ 6505 \times 10^{-23}$ J K$^{-1}$ | $0.000\ 0024 \times 10^{-23}$ J K$^{-1}$ |
| conductance quantum | $7.748\ 091\ 733 \times 10^{-5}$ S | $0.000\ 000\ 026 \times 10^{-5}$ S |
| electric constant | $8.854\ 187\ 817... \times 10^{-12}$ F m$^{-1}$ | |
| electron mass | $9.109\ 3826 \times 10^{-31}$ kg | $0.000\ 0016 \times 10^{-31}$ kg |
| electron volt | $1.602\ 176\ 53 \times 10^{-19}$ J | $0.000\ 000\ 14 \times 10^{-19}$ J |
| elementary charge | $1.602\ 176\ 53 \times 10^{-19}$ C | $0.000\ 000\ 14 \times 10^{-19}$ C |
| Faraday constant | $96\ 485.3383$ C mol$^{-1}$ | $0.0083$ C mol$^{-1}$ |
| fine-structure constant | $7.297\ 352\ 568 \times 10^{-3}$ | $0.000\ 000\ 024 \times 10^{-3}$ |
| inverse fine-structure constant | $137.035\ 999\ 11$ | $0.000\ 000\ 46$ |
| magnetic constant | $4\pi \times 10^{-7} = 12.566\ 370\ 614... \times 10^{-7}$ N A$^{-2}$ | |
| magnetic flux quantum | $2.067\ 833\ 72 \times 10^{-15}$ Wb | $0.000\ 000\ 18 \times 10^{-15}$ Wb |

*continued*

*continued*

| molar gas constant | 8.314 472 J mol⁻¹ K⁻¹ | 0.000 015 J mol⁻¹ K⁻¹ |
|---|---|---|
| Newtonian constant of gravitation | 6.6742 × 10⁻¹¹ m³ kg⁻¹ s⁻² | 0.0010 × 10⁻¹¹ m³ kg⁻¹ s⁻² |
| Planck constant | 6.626 0693 × 10⁻³⁴ J s | 0.000 0011 × 10⁻³⁴ J s |
| Planck constant over 2 pi | 1.054 571 68 × 10⁻³⁴ J s | 0.000 000 18 × 10⁻³⁴ J s |
| proton mass | 1.672 621 71 × 10⁻²⁷ kg | 0.000 000 29 × 10⁻²⁷ kg |
| proton-electron mass ratio | 1836.152 672 61 | 0.000 000 85 |
| Rydberg constant | 10 973 731.568 525 m⁻¹ | 0.000 073 m⁻¹ |
| speed of light in vacuum | 299 792 458 m s⁻¹ | |
| Stefan-Boltzmann constant | 5.670 400 × 10⁻⁸ W m⁻² K⁻⁴ | 0.000 040 × 10⁻⁸ W m⁻² K⁻⁴ |

SI units, SI-derived units, and fundamental constants form the groundwork for the vast majority of measurement units that are not specialized to a single industry. These measurement units are generated for a variety of measurement parameters in an effort to represent many different measurement technologies employed by a vast array IM&TE and associated calibration standards. Note, the VIM defines a *measurement instrument* as a "device intended to be used to make measurements, alone or in conjunction with supplementary device(s)." See Table 19.2 for some common measurement parameters along with their associated measurement units and typical IM&TE.

**Table 19.2** Common measurement parameters.

| Common Measurement Parameter | Common Units of Measurement | Common Measurement Instruments |
|---|---|---|
| Angular | Radian, Degree, Minute, Second, Gon | Clinomter, Optical Comparator, Radius Gage, Protractor, Precision Square, Cylindrical Square |
| Concentricity (Roundness) | Meter, Inch, Angstrom | Rotary Table with Indicator |
| Current | Ampere | Ampmeter, Current Shunt, Current Probe, Digital Multimeter (DMM) |
| Flatness / Parallelism | Meter, Inch, Angstrom | Optical Flats and Monochromatic Light Source, Profilometer |
| Flow | Liter per Minute (LPM), Standard Cubic Feet per Minute (SCFM), | Flowmeter, Rotameter, Mass Flowmeter (MFC), Anemometer |
| Force (Compression & Tension) | Newton, Dyne, Pound-force | Force Gage, Load Cell, Spring Gage, Proving Ring, Dynamometer |
| Frequency | Hertz (Hz) | Counter, Time Interval Analyzer |
| Hardness | Brinell Hardness Number (BHN), Rockwell Hardness Number | Brinell Hardness Tester, Rockwell Hardness Tester |

*continued*

*continued*

| | | |
|---|---|---|
| Humidity | Dew Point, Relative Humidity | Hydrometer, Psychrometer, Chilled Mirror |
| Impedance | Impedance (Z) | LCR Meter, Impedance Analyzer, Vector Network Analyzer (VNA) |
| Length, Height (Linear Displacement) | Meter, Foot, Inch, Angstrom | Steel Rule, Tape Measure, Caliper, Micrometer, Height / Length Comparator, Laser Interrferometer, Coordinate Measurement Machine (CMM) |
| Luminance | Candela per Square Meter, Lux, Footcandles, Lambert | Lightmeter, Radiometer |
| Mass | Kilogram, Pound, Ounce, Gram, Dram, Grain, Slug | Balance, Weighting Scale |
| Power (RF) | Watt, dBm, dBv | Diode Power Sensor, Thermopile Power Sensor, Thermal Voltage Converter (TVC), Scalar Network Analyzer (SNA), Vector Network Analyzer (VNA) |
| Power (Voltage) | Watt, Joule, Calorie | Wattmeter, Power Analyzer |
| Pressure & Vacuum | Pascal, Pound-Force per Square Inch (psi), Bar, Inches of Water, Inches of Mercury, Atmosphere, Torr | Pressure Gage, Manometer (mercury), Manometer (capacitance), Piranha Gage, Spinning Rotor Gage (SRG) |
| Resistance, Conductance | Ohm, Siemen, Mho, | Ohmmeter, Milli Ohmmeter, Tera Ohmmeter, Digital Multimeter, Current Shunt, LCR Meter, Current Comparator |
| Rotation | Radians per Seconds, Revolutions per Minute (rpm) | Stroboscope, RPM meter |
| Signal Analysis | Hz / dBm (Frequency Domain) | Spectrum Analyzer, Vector Network Analyzer (VNA), Vector Voltmeter, FFT Analyzer |
| Signal Analysis (Time Domain) | Second / Volt | Oscilloscope, Logic Analyzer |
| Temperature | Kelvin, Celsius, Fahrenheit, Pyrometer | Thermocouple, Thermometer (liquid-in-glass), Thermometer (resistance), Infrared Rankine |
| Torque | Newton Meter, Pound-Force Foot, Pound-Force Inch, Ounce-Force Inch | Torque Wrench, Torsion Bar, Torque Cells |
| Vibration / Acceleration | Meter per Second, Meters pk-to-pk | Accelerometer, Velocity Pickup, Displacement Meter, Laser Interferometry |
| Voltage | Volt | Voltmeter, Digital Voltmeter (DVM), Digital Multimeter (DMM), millivoltmeter, nanovoltmeter, HV Probe |

Note: Many of the units of measurement listed, while commonly used in some areas and industries, are not part of the SI and are not generally used in scientific work. The accepted SI unit is listed first in all cases.

# Chapter 19: A General Understanding of Metrology

Not only are measurement units used in defining the quantity of an unknown measurement parameter, they are frequently used when generating known quantities of a parameter. By generating a known quantity of a measurement parameter, equipment used to measure these parameters can be evaluated as to their accuracy (the fundamental principle behind most IM&TE calibrations). See Table 19.3.

The majority of measurement parameters are defined by fundamental principles and concepts. These definitions serve to describe a measurement parameter in terms of the physical world. A simple illustration of this is Force (F), which is defined in terms of mass (m) and acceleration (a). Most measurement parameters can be represented by mathematical formulas that show the relationship between the physical world factors that comprise their makeup. See Table 19.4.

**Table 19.3** Common measurands and equipment used to source them.

| Common Units for Source | Common Source Instruments |
|---|---|
| Angular | Angle Blocks, Sine Plate |
| Current | Current Source, Power Supply, Multifunction Calibrator |
| Flow | Bell Prover, Flow Test Stand, Flow Calibrator |
| Force (Compression & Tension) | Weights, Force Test Stand |
| Frequency | Timebases, Frequency Standards |
| Hardness | Rockwell Hardness Standards |
| Humidity | Environmental Test Chamber, Saturated Salts |
| Impedance | Impedance Test Artifacts (capacitors, inductors, AC resistors) |
| Length, Height (Linear Displacement) | Gage Blocks, Parallels |
| Luminance | Calibration Light Source, Laser |
| Mass | Weights |
| Power (RF) | Power Meter Ref. Output, Generator, Synthesizer |
| Power (Voltage) | Power Supply, Multifunction Calibrator |
| Pressure & Vacuum | Dead Weight Tester, Vacuum Test Stand, Pressure Pimp, Vacuum Pump |
| Resistance | Standard Resistors, Decade Resistors, Multifunction Calibrator |
| Rotation | AC motor |
| Temperature | Oven, Environmental Chamber, Temperature Calibrator, Triple Point of Water, Freezing or Melting point cells |
| Torque | Weights & Torque Arm |
| Vibration / Acceleration | Shaker Table |
| Voltage | Power Supply, Multifunction Calibrator |

**Table 19.4** Common measurands and some of their associated formulas.

| Common Measurement Parameter | Fundamental Formulas | Variable | Variable | Variable |
|---|---|---|---|---|
| Angular | SA = O / H | SA = sine of angle between the hypotenuse and adjacent | O = opposite | H = hypotenuse |
| Current | I = V / R | I = current | V = voltage | R = resistance |
| Flatness / Parallelism | D = (n + 1) / 2 | D = Deviation measured w/ optical flats | n = the number of bands between to high point | |
| Flow | MF = Qv * DT | MF = mass flow | Qv = volume flow rate | DT = weight density of gas or liquid at temperature T |
| Force (Compression & Tension) | F = m * a | F = force | m = mass | a = acceleration |
| Frequency | F = C / S | F = frequency | C = cycle or iterations | S = seconds |
| Humidity | %RH. = (Pv/Ps) * 100 | %RH = percent relative humidity | Pv = pressure of water vapor | Ps = saturation pressure |
| Impedance | Z = Sqrt (L/C) | Z = impedance | L = inductance | C = capacitance |
| Length, Height (Linear Displacement) | Length Change = L * $Lc_{coef}$ * $t_{delta}$ | L = original length | $L_{coef}$ = linear expansion coefficient of material | $T_{delta}$ = change in temperature |
| Luminance | CF = Pref / Pdut | CF = calibration factor at a specific wavelength or wavelength range | Pref = measured reference standard power | Pdut = measured power from device under test |
| Mass | p = m / v | p = absolute density | m = mass | v = volume |
| Power (RF) | $P_{dbm}$ = 10 Log ($P_{meas}$/0.001) | $P_{dbm}$ = power in dbm | $P_{meas}$ = measured power | |
| Power (Voltage) | P = $I^2$ * R | P = power | I = current | R = resistance |
| Pressure & Vacuum | P = F/A | P = pressure | F = force | A = area |
| Resistance | R = V / I | R = resistance | V = voltage | I = current |
| Rotation | w = a / t | w = angular velocity | a = angular displacement | t = elapsed time |
| Temperature | $R_t$ = $R_0$ (1+ At + $Bt^2$) | $R_t$ = resistance at some temperature | $R_0$ = Resistance at 0 Celsius | A & B = constants for a particular element that describes its temperature behavior |
| Torque | T = F x s | T = torque around a point | F = force applied | s = distance through which the force is acting |
| Vibration / Acceleration | As = $E_o$ / Ai | As = accelerometer sensitivity | $E_o$ = electrical signal output | $A_i$ = acceleration input |
| Voltage | V = I * R | V = voltage | I = current | R = resistance |

## References

Anthony, D.M. DATE. *Engineering metrology*. Bristol, UK: Pergamon Press.

Busch, Ted; Roger Harlow, and Richard L. Thompson. 1998. *Fundamentals of dimensional metrology*, 3rd ed. Albany, NY: Delmar Publishers.

de Silva, G.M.S. *Basic metrology for ISO 9000 certification*. Oxford, UK: Reed Educational and Professional Publishing Ltd.

Farago, Francis T. 1982. *Handbook of dimensional measurements*, 2nd ed., New York: Industrial Press, Inc.

Fluke Corporation. 1994. *Calibration: philosophy in practice*. 2nd ed. Everett, WA: Fluke Corporation.

Griffith, Gary. 1986. *Quality technician's handbook*. 3rd ed. Englewood Cliffs, NJ: Prentice Hall.

ISO. 1993. *International vocabulary of basic and general terms of metrology (VIM)*. Geneva: ISO.

Kimothi, S.K. 2002. *The uncertainty of measurements, physical and chemical metrology impact analysis*. Milwaukee: ASQ Quality Press.

NIST. 2001. NIST Special Publication 330 – 2001 Edition, *The international system of units (SI)*. Gaithersburg, MD: U.S. Dept. of Commerce.

NIST. 1995. NIST Special Publication 811 – 1995 Edition, *Guide for the use of the international system of units (SI)*. Gaithersburg, MD: National Institute of Standards and Technology.

U.S.A.F. 1997. HO ESAQR2P031-000-1, *Measurement and calibration handbook*. Keesler AFB, MO : U.S.A.F. Air Education and & Training Command.

U.S. Navy. 1976. NAVAIR 17-35QAL-2 Rev. 3, *Physical measurements*. Pomona, CA: U.S. Navy Metrology and Calibration Program, Metrology Engineering Center.

U.S. Navy. 1978. NAVAIR 17-35QAL-9, Rev. 2, *Dimensional measurements*. Pomona, CA: U.S. Navy Metrology and Calibration Program, Metrology Engineering Center.

White, Robert A. 1993. *Electronic test instruments*. Englewood Cliffs, NJ: Hewlett-Packard Professional Books.

Witte, Robert. 1987. *Electronic test instruments: a user's sourcebook*. Indianapolis, IN: Howard W. Sams & Company.

# Chapter 20

# Measurement Methods, Systems, Capabilities, and Data

## MEASUREMENT METHODS

Measurements are accomplished by employing one or more well-defined measurement methods in an effort to obtain quantifiable information about an object or phenomena. These measurement methods exhibit certain fundamental characteristics, which allow them to be used to categorize different types of measurements. Usually a particular situation or application will obviate the appropriate measurement method(s) needed to achieve the desired measurement data. Experience tells us which specific measurement method will typically yield the best results for a particular situation or application. Factors such as phenomena stability, resolution requirements, environmental influences, timing restraints, and so on. must be considered in order to determine the optimum measurement method(s). Understanding the mechanics and theory behind measurement methods is helpful not only for determining the best method for a particular situation or application but also for understanding their limitations and the measurement data they produce.

Measurement methods employ various metrology related terms that are useful when interpreting the method. The following are definitions from the VIM for some of these frequently used terms:

| Metrology Term | VIM Definition |
|---|---|
| Measurement | Set of operations having the object of determining a value of a quantity |
| Measurand | Particular quantity subject to measurement |
| Method of Measurement | Logical sequence of operations, describe generically, used in the performance of measurement |

Measurement methods are commonly grouped into one or more of seven categories. The seven measurement method categories are: direct, differential, indirect, ratio, reciprocity, substitution, and transfer. The following are generalized definitions for each of the seven categories of measurement methods:

158  Part III: Metrology Concepts

| Measurement Method | Definition |
|---|---|
| Direct | A measurement that is in direct contact with the measurand and provides a value representative of the measurand as read from an indicating device. |
| Differential | A measurement made by comparing an unknown measurand with a known quantity (standard) such that when their values are equal, a difference indication of zero (null) is given. |
| Indirect | A measurement made of a nontargeted measurand that is used to determine the value of the targeted measurand (measurand of interest). |
| Ratio | A measurement made by comparing an unknown measurand with a known quantity (standard) in order to determine how many divisions of the unknown measurand can be comprised within the known quantity. |
| Reciprocity | A measurement that makes use of a transfer function(s) (relationship) in comparing two or more measurement devices subject to the same measurand. |
| Substitution | A measurement using a known measurement device or artifact (standard) to establish a measurand value after which the known measurement device or artifact is removed and an unknown measurement device (unit under test) is inserted in its place so that its response to the measurand can be determined. |
| Transfer | A measurement employing an intermediate device used for conveying (transferring) a known measurand value to an unknown measurement device or artifact. |

Examples of the seven measurement methods are:

| Measurement Method | Examples |
|---|---|
| Direct | A multimeter reading the ACV of a power outlet, using a ruler to measure length, determining temperature by reading a liquid-in-glass thermometer, measuring tire pressure using a pressure gage |
| Differential | Comparison of two voltages using a null meter, measuring the length of a gage block using a gauge block comparator, determining a resistance using a current comparator, measuring a weight using a two-pan balance |
| Indirect | Calculating a current value by measuring the voltage drop across a shunt, determining a temperature by measuring the resistance of a platinum resistive thermometer, determining unknown impedance by measuring reflected voltage |
| Ratio | Creating intermediate voltage values using a Kelvin Varley divider and a fixed voltage source |
| Reciprocity | Determine the sensitivity of a microphone via the response of another microphone |
| Substitution | Measuring weight using a single-pan scale |
| Transfer | Determining ACV using a AC/DC transfer device, determine timing deviations via a portable synchronized quartz clock |

Measurement methods are the schemes by which measurement data are obtained. Determining which type of measurement method is appropriate for a particular situation should be thought-out before engaging in a measurement. Selection of an inappropriate measurement method will result in wasted time and other resources and produce undesirable and /or unreliable measurement data. Looking before leaping is sound advice when it comes to selecting a measurement method.

## MEASUREMENT SYSTEMS

Measurement systems are the means by which measurement data are obtained. A measurement system is an ensemble comprised of various elements including measurement personnel, calibration standards, measurement devices, measurement fixtures, measurement environment and measurement methodology and so on. They are used to obtain quantifiable, attributable data related to an object or phenomena. You can visualize a measurement system as a process of interactive, interrelated activities by which various objects or phenomena are related to measurement data.

Measurement systems are created and used based on needs for specific measurement data. The makeup of a measurement system is determined by an application or particular situation. The adequacy of a measurement system depends on the accuracy and reliability requirements of the measurement data. Less stringent requirements demand less of a measurement system in terms of sophistication, variability, repeatability, and so on. How measurement data will be used will drive the selection, composition and sophistication of a measurement system in order to meet measurement objectives.

Measurement systems are designed to produce measurement data that is assumed to be faithful and representative of the measurand(s) they are intended to measure. The following are VIM definitions applicable to this discussion:

| Term | Definition |
| --- | --- |
| Measurement | Set of operations having the objective of determining a value of a quantity |
| Measurement System | Complete set of measuring instruments and other equipment assembled to carry out specified measurements |
| Measuring Instrument | Device intended to be used to make measurements alone or in conjunction with supplementary device(s) |
| Measurement Signal | Quantity that represents the measurand and which is functionally related to it |
| Measuring Chain | Series of elements of a measuring instrument or system that constitutes the path of the measurement signal from the input to the output |
| Results of a Measurement | Value attributed to a measurand, obtained by measurement |
| Indication (of a Measuring Instrument) | Value of a quantity provided by a measuring instrument |

For a measurement system to be properly constructed, a comprehensive understanding of applicable measurement application(s) is required. This understanding will direct the development and definition of the measurement system. Also, a thorough understanding of measurement data requirements will provide guidance as to the confidence level and reliability traits required of the measurement system. NASA Reference Publication 1342, *Metrology—Calibration and Measurement Processes Guidelines*, notes the following 10-stage sequence in defining measurement requirements:

| Stage | Description |
|---|---|
| Mission Profile | What is to be accomplished—what is the measurement system purpose? |
| System Performance Profile | Define the needed capability of the measurement system that meet the mission profile. |
| System Performance Attributes | Define the functions and features of the measurement system—performance profile. |
| Component Performance Attributes | Define the functions and features of each component of the measurement system that combines to describe the system's performance attributes. |
| Measurement Parameters | Define the measurable characteristics that describe component and/or performance attributes. |
| Measurement Process Requirements | Define the measurement parameter values, ranges and tolerances, uncertainty limits, confidence levels, etc., associated with measurement parameter requirements. |
| Measurement System Design | Define the physical measurement system that meets the measurement process requirements. |
| Calibration Process Requirement | Define calibration requirements of the measurement system. |
| Calibration System Design | Define the functions and features of the calibration system that meet the calibration process requirement. |
| Measurement Traceability Requirements | Define the progressive chain of calibration process requirements that provide traceability. |

S. K. Kimothi in his book *The Uncertainty of Measurements: Physical and Chemical Impact and Analysis*, lists entities that can be considered part of the measurement system as follows:

- Measurement/test method
- Measurement equipment and measurement setup
- Personnel involved in the maintenance and operation of the measuring equipment and the measurement process
- Organizational structure and responsibilities
- Quality control tools and techniques
- Preventive and corrective mechanisms to eliminate defects in operations

The Measurement Systems Analysis (MSA) Reference Manual specifies the following as fundamental properties that define a good measurement system:

- Adequate discrimination and sensitivity.
- Be in a state of statistical process control.
- For product control, exhibit small variability compared to specification limits.
- For process control, exhibit small variability compared to manufacturing process variations and demonstrate effective resolution.

Measurement systems produce data within a window normally associated with a probability or likelihood that the data obtained faithfully represent their intended measurand(s). This likelihood is, as a rule, described in terms of standard deviation in regards to a normal frequency distribution curve (see Chapter 27) such that a percentage of probability of occurrence equates to an area under the normal distribution curve. See Figure 20.1.

The following are percentages of probability of occurrence for various multiples of standard deviations in regards to a normal frequency distribution curve:

| Standard Deviation | Area Under the Normal Frequency Distribution Curve |
|---|---|
| 0.6745 | 50% |
| 1 | 68.30% |
| 1.036 | 70.00% |
| 1.282 | 80% |
| 1.645 | 90% |
| 1.96 | 95% |
| 2 | 95.50% |
| 2.58 | 99% |
| 3 | 99.75% |
| 3.291 | 99.90% |
| 4 | 99.9940% |
| 5 | 99.99994% |

Information such as this may be used to derive reliability targets in terms of risk assessment associated with the probability that IM&TE will drift out of specification during its calibration interval (see Chapter 10) or in the estimation that measurements may be expected to lie within a probable range (see Chapter 29).

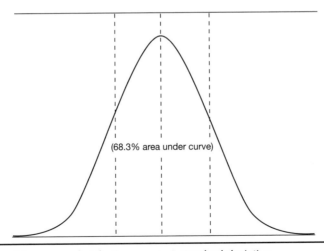

**Figure 20.1** Normal frequency distribution curve–1 standard deviation.

Calibration considerations in regards to measurement systems are of particular interest to metrologists. NASA Reference Publication 1342 notes that "Measurement processes are accompanied by errors and uncertainties that cannot be eliminated. However, they can be quantified and limited or controlled to 'acceptable' levels. Calibration is done for this purpose." Fluke's *Calibration: Philosophy in Practice* notes that calibration is, "A set of operations, performed in accordance with a definite, documented procedure, that compares the measurements performed by an instrument to those made by a more accurate instrument or standard, for the purpose of detecting and reporting, or eliminating by adjustment, errors in the instrument tested." Calibration can thus be considered as defining attributes about a measurement system in terms of how well they can correlate an unknown (object or phenomena being measured) to a known (calibration standard). Calibration relates measurement systems to performance indices thereby providing the means to estimate expected performance.

Measurement systems, like manufacturing processes, transform inputs to desired outputs and are composed of various interactive, interdependent elements that determine the quality and makeup of these outputs. Careful consideration must be given to the development or selection of a measurement system in order for it to correspond with measurement application requirements thus ensuring the validity and usability of derived measurement data. Measurements are metrological tools used in performing measurement tasks. As the age-old adage goes, 'The right tool for the right job,' cannot be overemphasized.

## MEASUREMENT CAPABILITIES

Measurement systems embody a variety of measurement capabilities inherent in their design for their intended purpose(s). Measurement capabilities are basically attributes of a measurement system that determine the extent to which measurements may be made within some qualifying restraints such as measurement range, ambient conditions, required input amplitude, and so on. Measurement system capabilities should be congruent with the requirements of the measurement application it is intended for. Determining whether a measurement system has the required capabilities to meet a measurement application is not always readily apparent and must often be established through user-assessment activities. It is ultimately the responsibility of the measurement system user to ascertain whether a measurement system is capable of meeting the requirements of a particular measurement application. Intentional/unintentional use of a measurement system in terms of operation beyond its established capabilities normally results in measurement data with unknown uncertainties at best or totally erroneous measurement data in worst case scenarios.

The adequacy of a measurement system to fulfill the requirements of a measurement application is addressed within the ISO 9001:2000 quality standard. The section Control of Monitoring and Measuring Devices states that an organization shall (1) determine what monitoring and measurement needs to be performed and (2) use monitoring and measuring systems that provide evidence that processes and product conform within established limits. Similarly, the section Monitoring & Measurement of Processes states that an organization shall apply suitable methods for the monitoring and measurements of processes and that these methods demonstrate the ability of the processes to achieve desired results. It is implicit that monitoring and measurement systems be capable of meeting the needs of product/process conformity assessment.

A measurement system's capabilities are often characterized in terms of bias, linearity, repeatability, reproducibility, and stability. VIM defines *bias* (of a measuring instrument) as "systematic error of the indication of a measurement instrument" noting that bias is normally established by averaging the error of indication over an appropriate number of repeated measurements. These measurements are assumed to be of the same measurand using the same measurement system. Bias is frequently referred to as systematic offset. Some possible causes for bias are:

- Measurement system needs calibration or has been improperly calibrated
- Measurement system is defective, worn, contaminated
- Measurement system is inadequate or inappropriate for the measurement application
- Environmental conditions are excessive
- Compensation not applied
- Operator error
- Computational error

VIM defines a *linear scale* as, "scale in which each scale spacing is related to the corresponding scale interval by a coefficient of proportionality that is a constant throughout the scale." VIM subsequently notes that a nonlinear scale coefficient of proportionality is non-constant such as in the case of a logarithmic scale or square-law scale. Relating this definition in terms of bias, linearity can be thought of as the proportional difference in bias throughout a scale or range. This bias can be constant (linear) or nonconstant (nonlinear). Some possible causes for linearity errors are:

- Measurement system needs calibration or is improperly calibrated.
- Measurement system is defective, worn, or contaminated.
- Measurement system environment is excessive and/or unstable.
- Measurement system is inadequately maintained.

Continuing with VIM definitions, *repeatability* (of results of measurements) is defined as, "ability of a measuring instrument to provide closely similar indications for repeated applications for the same measurand under the same conditions of measurement." The same conditions of measurement assumes:

- The same operator makes the measurements.
- The same measurement procedure is used.
- The same measurement equipment is used.
- The same calibration standard used.
- It's the same location.
- Environmental conditions are the same.
- Measurements are performed over a short period of time.

Repeatability is commonly referred to as within-system variation or equipment variation.

VIM defines *reproducibility* (of results of measurements) as, "closeness of the agreement between the results of measurements of the same measurand carried out under changed conditions of measurements." Some possible changed conditions include:

- Different measurement technique
- Different measurement system
- Different environmental conditions
- Different operator makes the measurements
- Different calibration standard used
- Different time and/or location

Reproducibility is often referred to as the average variations between measurement systems or the average variations between changing conditions of a measurement.

VIM defines *stability* as, "ability of a measurement instrument to maintain constant its metrological characteristics with time." Stated a little differently, stability is a measure of a measurement systems total variation in regards to a specific measurand over some time interval. Quite simply, stability is the change in a measurement systems bias over time. Some possible causes for a measurement system bias changes over time are:

- The measurement system needs calibration.
- The measurement system is defective, worn, or contaminated.
- The measurement system is aging.
- The measurement system environment is changing.
- The measurement system is inadequately maintained.

Nonstability is closely related to *drift*, which the VIM defines as, "slow change in metrological characteristic of a measurement instrument."

## Statistical Process Control and Control Charts

The variability of a measurement system may be determined via the use of control charts. Shewhart, the creator of statistical process control (SPC), pioneered the use of control charts, frequently referred to as Shewhart charts, during the 1920s while working for Bell Telephone Laboratories. Juran's, *Quality Control Handbook*, defines *SPC* as "the application of statistical techniques for measuring and analyzing the variation in processes." Shewhart analyzed many different processes and identified two variation components common to all; a steady component inherent to the process and an intermittent component. The steady variation components he referred to as random variations attributable to chance and undiscovered causes such that when averaged its variance is about the same as the parameter being measured. The intermittent components, he concluded, could be attributed to assignable sources (systematic) and as such be removed from a process. He went on to say that a process with only random components could be said to be in a state of statistical control. (Note: A process can be in a state of statistical control and not meet specifications, as statistical control merely means that only random variations are present.) Shewhart envisioned using control charts as a means of applying statistical principles to identify and monitor process variation using intuitive graphics. Control chart limits, based on statistical variations of a process in terms of multiples of standard deviation (one, two, or three standard deviations), are normally included in control charts as a ready means of determining whether a process is in a state of statistical control. The two most popular control charts deal with measurement averages (X-bar charts) and measurement ranges (R-bar charts). The GOAL/QPC *Memory Jogger II* recommends the following steps in constructing a control chart:

- Select process to be charted.
- Determine sampling method and plan.
- Initiate data collection.
- Calculate the appropriate statistics.

Table 20.1 is an example of time interval measurements (in seconds) made of four thermocouples in an environmental chamber. Measurements are made once a day for 25 consecutive days. Table 20.2 list values used for calculating control limits. Figure 20.2 X-bar chart and Figure 20.3 R-bar chart are derived from Table 20.1 calculations.

**Table 20.1** X-bar and R-bar control chart calculations.

| No. | TC 1 | TC 2 | TC 3 | TC 4 | Meas. Means | Meas. Range |
|---|---|---|---|---|---|---|
| 1 | 27.347 | 27.501 | 29.944 | 28.212 | 28.251 | 2.597 |
| 2 | 27.797 | 26.150 | 31.213 | 31.333 | 29.123 | 5.183 |
| 3 | 33.533 | 29.330 | 29.705 | 31.053 | 30.905 | 4.203 |
| 4 | 37.984 | 32.269 | 31.917 | 29.443 | 32.903 | 8.541 |
| 5 | 33.827 | 30.325 | 28.381 | 33.701 | 31.559 | 5.44 |
| 6 | 29.684 | 29.567 | 27.231 | 34.004 | 30.121 | 6.773 |
| 7 | 32.626 | 26.320 | 32.079 | 36.172 | 31.799 | 9.852 |
| 8 | 30.296 | 30.529 | 24.433 | 26.852 | 28.027 | 6.096 |
| 9 | 33.533 | 29.330 | 29.705 | 31.053 | 30.905 | 4.203 |
| 10 | 37.984 | 32.269 | 31.917 | 29.443 | 32.903 | 8.541 |
| 11 | 33.827 | 30.325 | 28.381 | 33.701 | 31.559 | 5.446 |
| 12 | 29.684 | 29.567 | 27.231 | 34.004 | 30.121 | 6.773 |
| 13 | 26.919 | 27.661 | 31.469 | 29.669 | 28.930 | 4.551 |
| 14 | 28.465 | 28.299 | 28.994 | 31.145 | 29.226 | 2.846 |
| 15 | 32.427 | 26.104 | 29.477 | 37.201 | 31.302 | 11.097 |
| 16 | 28.843 | 30.518 | 32.236 | 30.471 | 30.517 | 3.393 |

*continued*

continued

| | | | | | | |
|---|---|---|---|---|---|---|
| 17 | 30.751 | 32.999 | 28.085 | 26.200 | 29.509 | 6.799 |
| 18 | 31.258 | 24.295 | 35.465 | 28.411 | 29.857 | 11.170 |
| 19 | 28.278 | 33.949 | 30.474 | 28.874 | 30.394 | 5.671 |
| 20 | 26.919 | 27.661 | 31.469 | 29.669 | 28.930 | 4.551 |
| 21 | 28.465 | 28.299 | 28.994 | 31.145 | 29.226 | 2.846 |
| 22 | 32.427 | 26.104 | 29.477 | 37.201 | 31.302 | 11.097 |
| 23 | 28.843 | 30.518 | 32.236 | 30.471 | 30.517 | 3.393 |
| 24 | 30.751 | 32.999 | 28.085 | 26.200 | 29.509 | 6.799 |
| 25 | 31.258 | 24.295 | 35.465 | 28.411 | 29.857 | 11.170 |

| XBar Control Chart | |
|---|---|
| UCL = xbar + A2 × Rbar | 34.928 |
| LCL = xbar − A2 × Rbar | 25.653 |
| CL = xbar | 30.290 |
| RBar Control Chart | |
| UCL = D4 × Rbar | 14.517 |
| LCL = D3 × Rbar | 0.000 |
| CL = Rbar | 6.361 |

| | | |
|---|---|---|
| UCL | Upper Control Limit | |
| LCL | Lower Control Limit | |
| CL | Center Line | |
| n | Sample Size | 4 |
| Rbar | Mean of Ranges | 6.361 |
| Xbar | Mean of Meas. Means | 30.290 |

**Table 20.2** Table for calculating the control limits.

| n | A2 | D3 | D4 | n | A2 | D3 | D4 |
|---|---|---|---|---|---|---|---|
| 2 | 1.88 | 0 | 3.27 | 14 | 0.235 | 0.328 | 1.672 |
| 3 | 1.023 | 0 | 2.57 | 15 | 0.223 | 0.47 | 1.653 |
| 4 | 0.729 | 0 | 2.28 | 16 | 0.212 | 0.363 | 1.637 |
| 5 | 0.577 | 0 | 2.12 | 17 | 0.203 | 0.378 | 1.622 |
| 6 | 0.483 | 0 | 2 | 18 | 0.194 | 0.391 | 1.608 |
| 7 | 0.419 | 0.076 | 1.92 | 19 | 0.187 | 0.403 | 1.597 |
| 8 | 0.373 | 0.136 | 1.86 | 20 | 0.18 | 0.415 | 1.585 |
| 9 | 0.337 | 0.184 | 1.82 | 21 | 0.173 | 0.425 | 1.575 |
| 10 | 0.308 | 0.223 | 1.78 | 22 | 0.167 | 0.434 | 1.566 |
| 11 | 0.285 | 0.256 | 1.74 | 23 | 0.162 | 0.443 | 1.557 |
| 12 | 0.266 | 0.283 | 1.72 | 24 | 0.157 | 0.451 | 1.548 |
| 13 | 0.249 | 0.307 | 1.69 | 25 | 0.153 | 0.459 | 1.541 |

Note: To avoid errors associated with small sample sizes (< 25) the above control chart table values are used for A2, D3, D4 as condensed from Juran's *Quality Control Handbook, 4th edition*, Table A, Factors for Computing Control Charts.

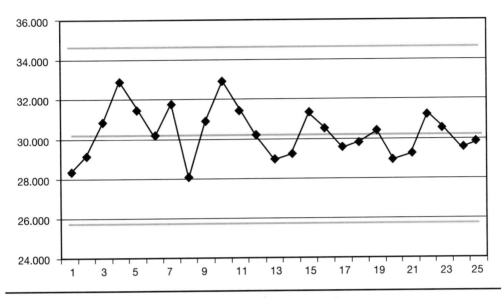

**Figure 20.2** Xbar chart for example of time interval measurements.

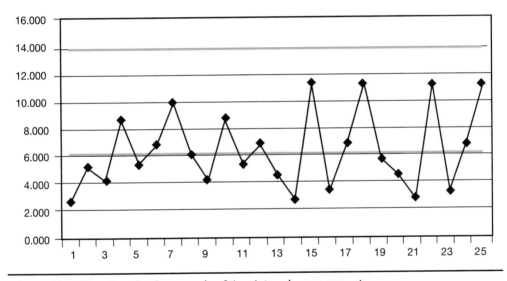

**Figure 20.3** Rbar chart for the example of time interval measurement.

The GOAL/QPC, Memory Jogger II gives the following criteria for determining if your process is out of control by dividing a control chart into different zones:

**Control Chart Zones**

| Upper Control Limit (UCL) |
|---|
| Zone A |
| Zone B |
| Zone C |
| **Average** |
| Zone C |
| Zone B |
| Zone A |
| **Lower Control Limit (LCL)** |

- One or more points fall outside of the control limits.
- Two points, out of three consecutive points, are on the same side of average, in Zone A or beyond.
- Four points, out of five consecutive points, are on the same side of average, in Zone B or beyond.
- Nine consecutive points on one side of average.
- There are six consecutive points increasing or deceasing.
- There are 14 consecutive points that alternate up and down.
- There are 15 consecutive points within Zone C (above and below the average).

Gary Griffith, author of *The Quality Technician's Handbook,* recommends you always annotate the control chart for any of the following:

- Out-of-control conditions and the cause
- Reason the chart was stopped such as during equipment down time
- Reason the chart was started again such as a new setup
- When control limits were recalculated
- Any adjustments made to the process
- Any other pertinent information about the process

# GAGE R&R

A method for assessing the capability of a measurement system is known as a *gage repeatability and reproducibility assessment* or more commonly called a *gage R&R study.* The gage R&R study is designed to measure both the repeatability and reproducibility of a measurement process. *The Quality Technician's Handbook* recommends the following elements of planning prior to performing a gage R&R study:

- Select the proper instrument.
- Make sure the measurement method is appropriate.
- Follow the 10 percent rule of discrimination.
- Look for obvious training/skill problems with observers.
- Make sure all observers use the same gage (or same type of gage).
- Make sure all the measuring equipment is calibrated.

The range method is the most common method for performing a gage R&R study. The range method for gage R&R study is similar to setting up a X-bar and R-bar control charts in terms of data acquisition and computing the means of observations and the ranges of these observations. Normally a gage R&R study involves more than one observer (know as an appraiser) performing measurements. The following parameters are typically calculated in a range method gage R&R study: equipment variation (EV), equipment variation percentage (% EV), appraiser variation (AV), appraiser variation percentage (% AV), repeatability & reproducibility (R&R), R&R percentage (% R&R), part variation (PV), part variation percentage (% PV), and total variation (TV).

The following is an example of a range R&R study involving three appraisers measuring ten parts. See Table 20.3.

**Table 20.3** Range R&R study.

| Measurements | | | | | | | | | | | | |
|---|---|---|---|---|---|---|---|---|---|---|---|---|
| Part # | 1 | 2 | 3 | 4 | 5 | 6 | 7 | 8 | 9 | 10 | Average | |
| Appraiser 1 Trial # 1 | 0.39 | −0.56 | 1.34 | 0.47 | −0.70 | 0.03 | 0.59 | −0.31 | 3.36 | −1.36 | 0.33 | |
| Appraiser 1 Trial # 2 | 0.41 | −0.67 | 1.17 | 0.50 | −0.93 | −0.11 | 0.75 | −0.30 | 1.99 | −1.35 | 0.15 | |
| Appraiser 1 Trial # 3 | 0.64 | −0.57 | 1.37 | 0.64 | −0.74 | −0.31 | 0.66 | −0.17 | 3.01 | −1.31 | 0.32 | |
| Average | 0.480 | −0.600 | 1.293 | 0.537 | −0.790 | −0.130 | 0.667 | −0.260 | 2.787 | −1.340 | 0.26 | Xbar₁ |
| Range | 0.25 | 0.11 | 0.2 | 0.17 | 0.23 | 0.34 | 0.16 | 0.14 | 1.37 | 0.05 | 0.302 | Rbar₁ |
| Part # | 1 | 2 | 3 | 4 | 5 | 6 | 7 | 8 | 9 | 10 | Average | |
| Appraiser 2 Trial # 1 | 0.07 | −0.47 | 1.19 | 0.01 | −0.56 | −0.3 | 0.47 | −0.63 | 1.7 | −1.67 | −0.02 | |
| Appraiser 2 Trial # 2 | 0.35 | −1.33 | 0.94 | 1.03 | −1.3 | 0.33 | 0.55 | 0.07 | 3.13 | −1.63 | 0.21 | |
| Appraiser 2 Trial # 3 | 0.07 | −0.67 | 1.34 | 0.3 | −1.37 | 0.06 | 0.73 | −0.34 | 3.19 | −1.5 | 0.18 | |
| Average | 0.163 | −0.823 | 1.157 | 0.447 | −1.077 | 0.030 | 0.583 | −0.300 | 2.673 | −1.600 | 0.125 | Xbar₂ |
| Range | 0.28 | 0.86 | 0.40 | 1.02 | 0.81 | 0.63 | 0.26 | 0.70 | 1.49 | 0.17 | 0.662 | Rbar₂ |
| Part # | 1 | 2 | 3 | 4 | 5 | 6 | 7 | 8 | 9 | 10 | Average | |

*continued*

*continued*

| | | | | | | | | | | | |
|---|---|---|---|---|---|---|---|---|---|---|---|
| Appraiser 3 Trial # 1 | 0.04 | −1.37 | 0.77 | 0.14 | −1.46 | −0.39 | 0.03 | −0.46 | 1.77 | −1.49 | −0.24 | |
| Appraiser 3 Trial # 2 | −0.11 | −1.13 | 1.09 | 0.3 | −1.07 | −0.67 | 0.01 | −0.46 | 1.44 | −1.77 | v0.24 | |
| Appraiser 3 Trial # 3 | −0.14 | −0.96 | 0.67 | 0.11 | −1.44 | −0.49 | 0.31 | −0.49 | 1.77 | −3.16 | −0.38 | |
| Average | −0.070 | −1.153 | 0.843 | 0.183 | −1.323 | −0.517 | 0.117 | −0.470 | 1.660 | −2.140 | −0.287 | $Xbar_3$ |
| Range | 0.18 | 0.41 | 0.42 | 0.19 | 0.39 | 0.28 | 0.30 | 0.03 | 0.33 | 1.67 | 0.420 | $Rbar_3$ |
| Part # | 1 | 2 | 3 | 4 | 5 | 6 | 7 | 8 | 9 | 10 | | |
| Average of Trial Averages | 0.191 | −0.859 | 1.098 | 0.389 | −1.063 | −0.206 | 0.456 | −0.343 | 2.373 | −1.693 | 0.034 | A_Xbar |

| Computations | | | | | |
|---|---|---|---|---|---|
| Average of ($Rbar_1$, $Rbar_2$, $Rbar_3$) | A_Rbar | = | 0.4613 | | |
| Range of ($Xbar_1$, $Xbar_2$, $Xbar_3$) | R_Xbar | = | 0.5513 | | |
| Range of Averages of Trial Averages | RA_Xbar | = | 4.0667 | | |
| Number of Parts | n | = | 10 | | |
| Number of Trial | t | = | 3 | | |
| Equipment Variation | EV | = | A_Rbar × $K_1$ | = | 0.27256 |
| Appraiser Variation | AV | = | Sqrt ((R_Xbar × $K_2$)^2−(EV^2 / (n × t))) | = | 0.28408 |
| Repeatability & Reproducibility | RR | = | Sqrt (EV^2 + AV^2) | = | 0.39368 |
| Part Variation | PV | = | RA_Xbar × $K_3$ | = | 1.27937 |
| Total Variation | TV | = | Sqrt (RR^2 + PV^2) | = | 1.33857 |
| % Equipment Variation | % EV | = | 100 × (EV/TV) | = | 20.36% |
| % Appraiser Variation | % AV | = | 100 × (AV/TV) | = | 21.22% |
| % Repeatability & Reproducibility | % RR | = | 100 × (RR/TV) | = | 29.41% |
| % Part Variation | % PV | = | 100 * (PV/TV) | = | 95.58% |

| Constants Values | Trial # | 2 | 3 | 4 | 5 | 6 | 7 | 8 | 9 | 10 |
|---|---|---|---|---|---|---|---|---|---|---|
| | $D_4$ | 3.27 | 2.57 | 2.28 | 2.11 | 2.00 | 1.92 | 1.86 | 1.82 | 1.78 |
| | $K_1$ | 0.8862 | 0.5908 | | | | | | | |
| | $K_2$ | 0.7071 | 0.5231 | | | | | | | |
| | # of Parts | 2 | 3 | 4 | 5 | 6 | 7 | 8 | 9 | 10 |
| | $K_3$ | 0.7071 | 0.5231 | 0.4467 | 0.4030 | 0.3742 | 0.3534 | 0.3375 | 0.3249 | 0.3146 |

CHAPTER 20: MEASUREMENT METHODS, SYSTEMS, CAPABILITIES, AND DATA   171

Figure 20.4 is another way to look at the R&R study involving three appraisers.

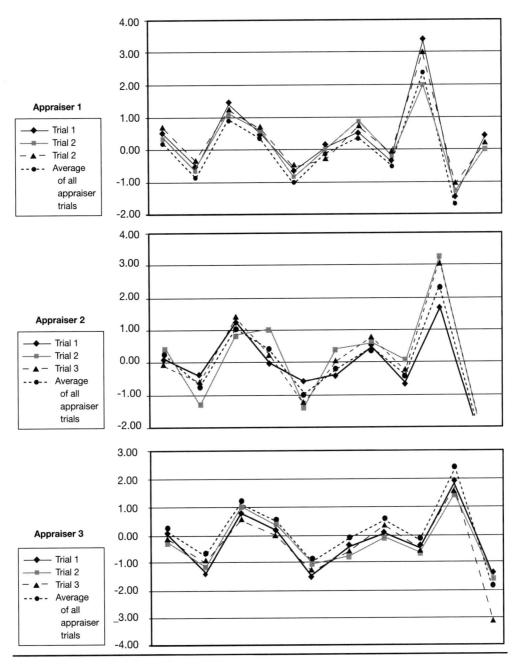

**Figure 20.4** Example of appraiser variations.

## MEASUREMENT DATA

A measurement systems principal purpose is to generate measurement data. Measurement data come in many varieties, such as alphanumeric characters, plots and graphs, increasing or decreasing audiovisual displays, limit indicators, and so on. Measurement data type should be compatible with the requirements of the intended measurement application. Measurement data are often the only basis for making decisions as to whether a process is in statistical control or a product is in conformance with published specifications. In this context, measurement data need to be of a type and quality sufficient to provide adequate information about a measurement application in order to make informed decisions about it. Inappropriate measurement data type, format or quality can be misleading, resulting in erroneous assumptions about a measurement application. An example of inappropriate measurement data type would be trying to use the graphic display of an oscilloscope to derive numeric data about a process having both very large and very small amplitude changes. In this case the digital readout of a digital multimeter, having both sufficient resolution and acquisition speed, would probably be a better choice. To avoid masking or distorting relevant information, measurement data type, format and quality should always be considered when evaluating measurements made by measurement systems.

Measurement data, to be useful, must be faithful to the represented measurement application. Measurement data considerations should be addressed in order to insure that the data are accurate, credible, and usable for its intended purpose(s). The following are some key measurement data considerations:

- **Format.** This refers to the way measurement data are oriented (layout), type of graphic (bar, pie, and so on), font type and font size, numerical convention, date convention, and so on.

- **Resolution.** This is the smallest or least significant digit (LSD) distinguishable within measurement data. VIM defines *resolution* (of a displaying instrument) as the "smallest difference between indications of a displaying device that can be meaningfully distinguished." VIM notes that for a digital measuring device this is the change in the indication when the LSD changes by one step.

- **Readability.** This refers to the ergonomic way measurement data are presented, in terms of how easy it can be read by observers. Note this does not imply comprehension of the data, but rather that the data are presented in a way that its intent can be readily determined.

- **Suitability.** This refers to how measurement data are presented in regards to both the application it is derived from and the intent of how the data will be used. An example of this would be in presenting seldom occurring, small changes in a large quantity of measurement data. Presenting the measurement data in table format would not readily identify these small, seldom occurring changes, whereas a log-linear bar chart would allow for easier identification.

- **Confidentiality.** This refers to protection and control issues focusing on both measurement data and the source(s) from which it was obtained. Often measurement data are used in benchmarking and/or proficiency evaluations and as such has the potential for:

- Competitors to use it to their advantage
- Publicize uncomplimentary performance
- Be interpreted outside of its intended context
- Disclosing capabilities or limitations
- Giving insight into programs or products in development

Measurement data confidentiality should be explicitly addressed and proper safeguards incorporated to prevent unauthorized disclosure.

Measurement data considerations can, if not satisfactorily addressed, render good data unusable for its intended purpose(s). Without a satisfactory understanding of a measurement systems output data in relation to the requirements of a measurement application, said data consideration can result in wasted time and effort, and, ultimately, bad decisions being made. Measurement practitioners and those interpreting measurement data would do well to consider how measurement data are to be used before selecting a measurement system for a measurement application to avoid many of the aforementioned pitfalls.

## CALIBRATION METHODS AND TECHNIQUES

IM&TE requires the use of various methods and techniques for determining whether a unit is operating to its published specifications. These methods and techniques are used to establish a relationship between an applied signal and the corresponding IM&TE measurement display; cancel out residuals that can offset the measurand of interest; provide sufficient sensitivity to determine small differences between a unit and a calibration standard; and so on. Often a calibration method/technique is recommended by the OEM as a necessary step to be performed prior to using a unit.

The selection of a particular calibration method/technique is dictated by various factors. These factors include the measurand of interest: the inherent functionality and/or limitations of the IM&TE, the measurement scenario and its associated environment, the operator knowledge and skill, OEM recommendations, and so on. Selecting an inappropriate calibration method/technique can often mask an IM&TE measurement response, rendering it inaccurate. Not performing an OEM recommended calibration method/technique can also degrade a units performance. It is essential that calibration practitioners be aware of the calibration methods/techniques appropriate for a particular application in order to insure their evaluations are based on reliable measurement data and to avoid misadjusting a unit as a result of misleading measurement data.

Calibration methods and techniques can be adapted to a wide variety of IM&TE types. Some of the most common calibration methods and techniques follow.

- **Linearization.** This is a method by which IM&TE is corrected for a linear response such that a step change in an applied signal will result in a corresponding step change in the IM&TE indication. This method may also be used to correct a nonlinear output using a linear responding measuring device. Linearization is typically used to correct for nonlinear measurement sensors such as those used to make high temperature measurement.

- **Nulling.** This is a method by which two applied signals are algebraically summed whereby the common portion of each signal cancels out leaving only the difference between the signals. This allows for very small differences in signal to be measured that would otherwise be very difficult to detect due to the size of the applied signals relative to these differences. Nulling may also be used to establish a known quantity from an unknown quantity by increasing or decreasing the unknown quantity until the difference between it and the known quantity is sufficiently small. Nulling is frequently used in intercomparing 10VDC calibration standards allowing for difference measurements at the parts-per-million (ppm) level.

- **Spanning.** This is a method by which an IM&TE specific range is defined. Spanning typically involves bringing an IM&TE measurement readout scale in agreement with intended range of the unit via adjustments, changing component values, or firmware correction. Spanning helps ensure that a unit's high, low, and mid-range measurement responses correspond to its high, low, and mid-scale measurement readout values. This method is commonly used to set up the range of a pressure gauge.

- **Spot Frequency.** This is a method by which specific outputs or measurement ranges, commonly referred to as sweet spots, are enhanced via correction factors. Outputs and measurement ranges between sweet spots normally include additional interpretation uncertainties as a result of not being directly compared to calibration standards. This method is commonly used to correct IM&TE AC voltage and current response at standardized amplitudes and frequencies as provided directly by AC voltage and current calibration standards.

- **Zeroing.** This is a method by which an IM&TE measurement readout offset present in the absence of an applied signal is excluded from measurements via hardware adjustment or algebraic cancellation. IM&TE numeric indication in the absence of an applied signal after this exclusion is nominally zero. Zeroing is commonly used to establish a datum, or starting or reference point, such as when zeroing a height gauge on a surface plate before measuring the height of gauge blocks placed on the plate.

## References

ANSI/ISO/ASQ Q9001-2000, *American National Standard—Quality management systems—Requirements*. Milwaukee: ASQ Quality Press.

Brassard, Michael and Dianne Ritter. 1994. *The memory jogger II*. Salem, NH: GOAL/QPC.

Daimler-Chrysler Corporation, Ford Motor Company, General Motors Corporation. *Measurement system analysis (MSA)*, 3rd ed. Southfield, MI: Automotive Industry Action Group (AIAG)

Doebelin, Ernest O. 1989. *Measurement systems, application and design*. 4th ed. New York: McGraw-Hill.

Fluke. 1994. *Calibration: philosophy in practice*. 2nd ed. Everett, WA: Fluke Corporation.

Griffith, Gary K. 1986 *Quality technician's handbook.* 3rd ed., Englewood Cliffs, NJ: Prentice-Hall, Inc.

ISO. 1993. *International vocabulary of basic and general terms of metrology (VIM).* Geneva: ISO.

Juran. 1974. *Juran's quality control handbook,* 4th ed. Wilton, CT: McGraw-Hill.

Kimothi, S.K. 2002. *The uncertainty of measurements, physical and chemical metrology impact analysis.* Milwaukee: ASQ Quality Press.

NASA. 1994. NASA Reference Publication 1342, *Metrology-calibration and measurement processes guidelines.* Pasadena, CA: NASA.

# Chapter 21
# Specifications

Understanding specifications (or tolerances) is important when applying measurement theory to the real world. Specifications define the limits within which an instrument is able to be useful. Properly interpreted and used, specifications are useful to evaluate the performance of IM&TE during calibration and for evaluating the capability of a measurement standard to perform a calibration.

The terms *specification* and *tolerance* are closely related and often confused. While both terms define quantitative limits in relation to some nominal value, they are used in different situations. In general, a tolerance applies when something is being manufactured and a specification applies when something is being used or evaluated.[1]

- A *tolerance* is a design feature that defines limits within which a quality characteristic is supposed to be on individual parts. A tolerance has to balance the perfection desired by the designer with what is economically achievable from the reality of the manufacturing process.[2] A part that is outside the tolerance limits is rejected for later rework or scrapping.

- *Specifications* define the expected performance limits of a large group of substantially identical finished products—all units of a specific model of digital thermometer, for example. Customers use specifications to determine the suitability of a product for their own applications. A product that performs outside the limits when tested (calibrated) is rejected for later adjustment, repair, or scrapping. When in use, the specifications of an instrument can be used to estimate the uncertainty of the measurement being made.

The essential result of a calibration (ignoring the data for now) is a pass/fail decision about the item. The decision is based on the results of one or more measurements. Historically, the terminology has been that the result of a measurement is in tolerance or out of tolerance. This is probably from the sense that calibration is a process that acts on one item at a time, so the result applies to only one item. It should be understood, however, that the performance specifications of an instrument or tool are the basis of a calibration procedure, as they are what the performance is evaluated against. It is more correct to say that the result of a measurement is within specification or out of specification.

## TYPES OF SPECIFICATION LIMITS

There are two general types of specification limits. Which is used is dependent on the equipment, application, and what the information is used for.

1. A *one-way specification* allows variation only in one direction from the nominal value. The result of a measurement may be either more or less than one of the limits, but not both.[3] The nominal value is either a lower limit or an upper limit of the specification. This type of limit is often found in instruments used for work related to safety. Some companies may refer to a one-way specification as unilateral or single-sided.[4] Two examples of one-way specification follow.

   - An insulation tester may have a specification that the applied test voltage is 500 V, −0% +10% when the test voltage selector is in the 500 V position. The purpose of this is to ensure that at least the specified test voltage is applied to the tested part to ensure that it meets safety requirements. For the calibration technician, any measured value from 500 V to 550 V would pass, but 499.9 and 550.1 would both fail.

   - An appliance safety tester may specify that an alarm is to sound if a leakage current is 1 mA, −25% +0%. The purpose of this test is to ensure that the user of the appliance cannot receive a dangerous shock. In this case, the instrument would fail if the applied current exceeds 1 mA and the alarm still has not sounded.

2. A *two-way specification* allows variation in either direction from the nominal value. The nominal value is a target, and the specification defines lower and upper limits of acceptable values.[5] The nominal value is not necessarily centered between the limits. Some companies may refer to a two-way specification as bilateral, or as double-sided.[6] Three examples of a two-way specification follow.

   - The output of a voltage source may have a specification of ± (0.005% of setting + 10 µV). If the calibration technician is testing it at 500 mV, the specification limits would be 499.965 to 500.035 mV.

   - A signal generator may have an output flatness specified as ±2 dB with respect to a 1.000 mW reference level. If a meter scaled in dB is used to make the measurement, the limits are −2 to +2 dB. Sometimes it may happen that the only available meter is scaled in watts. The limits need to be recalculated using the relationship

     $$\frac{P_2}{P_{ref}} = 10^{(dB/10)}$$

     This results in a nominal value of 1.000 mW, and the specification limits are 0.631 to 1.585 mW. Note that the nominal value is not centered between the limits.

   - A Grade 3 gage block with a nominal size of 2 inches has a specification of +8 µinch, −4 µinch.[7] The nominal dimension is not centered between the specification limits.

## CHARACTERISTICS OF A SPECIFICATION

For an item of IM&TE, a specification is a condensed way of giving information about the probable uncertainty of a measurement made with the instrument. As noted earlier, a specification applies to all instruments of that make and model. As defined by Fluke,[8] there are three groups of uncertainty terms that make up a specification:

- *Baseline* specifications describe the basic performance of the instrument. Terms for the output, scale and floor are included, although not all three are included every time. Baseline specifications that include one or two of these terms are more common that those with all three.

- *Modifier* specifications apply changes to the baseline, usually to express variation caused by environmental factors.

- *Qualifier* specifications are other important factors that may need to be considered in the application of the instrument.

Specifications are defined in relation to a nominal value. For fixed-value devices such as gage blocks, the nominal value is the specified size or other value. For any variable source or measuring instrument, the specification tables usually state the nominal value in terms of a range. The specification applies to any specific value in that range.

> Specification tables for instruments may also contain values that are essentially qualitative specifications. Typical examples include:
>
> - Design parameters and other specifications that cannot be verified by a performance test
>
> - Typical (and other) specifications that are not guaranteed by the manufacturer
>
> - Specifications that are not critical to the quality of the measurements performed by the instrument (determined by the user)
>
> In most cases, calibration procedures do not attempt to verify such qualitative specifications, even if they are shown in quantitative terms.

### Baseline

A baseline specification that contains all three of the terms defined by Fluke could be shown as

Nominal value ± (output + scale + floor)

where

Output = a percentage or parts per million (ppm) of the nominal value.

Scale = a percentage or ppm of the range or full-scale value.

Floor = a specific value expressed in the applicable SI units.

Note that the nominal value may be an operating range rather than a specific value. The metrologist may have to apply the specification to each specific value actually used.

## Baseline Specification: Output Term

The *output* term is a fixed percentage or ppm ratio of the output (for a source) or the input (for a measuring instrument). Its value is always the input or output value, multiplied by the percentage or ppm that applies to the range. For a physical device such as a gage block (or any fixed device), this part of the specification is normally omitted.

## Baseline Specification: Scale Term

The *scale* term can be expressed in different ways and can be the most confusing to interpret. On instruments with analog meters, it is usually a percentage of the full-scale value of the meter. For example, a profilometer has an analog meter with a full-scale value of 100 µin and a specification of ±2% of full scale. This means that the possible error of the reading is ± (100 × 2%) or ±2 µin at any point on the scale. It is apparent that this value becomes more significant as the measured value is smaller, which is why it is common practice to select ranges that allow analog meter readings to be in the upper ⅔ of their scale.

On many digital instruments, the scale term is expressed as a number of digits or counts. This always refers to the least significant digits—the ones at the right of the display. The actual value they represent depends on the meter scale being used. For example: a 4½ digit meter is measuring current on the 20 A range, and the specification is ± (0.1% reading + 5 digits). The maximum possible displayed value on this range is 19.999, so the least significant digit represents steps of 0.001 A. In SI units, the scale term of this specification is (5 × 0.001) or 0.005 A.

---

### About Display Size: Digits and Counts

Instruments with a digital display (numbers instead of a moving-pointer meter) are often described with having a display of a certain number of digits (such as 4½) or counts (such as 20,000.) These are two different ways of talking about the same thing.

In general, any given position of a digital display can show any of the counting digits, 0 through 9. Historically, the left-most digit position is often designed to show a 1 only or be blank. So, for example, a typical display could show any value from 0000 through 19999 (ignoring any decimal points or polarity signs.) This would be a 4½ digit display—four positions that display 0 through 9, and one position at the left that is either blank or 1. The number of digits of the display is the maximum number of nines that can be displayed, plus ½ (or some other fraction) to represent the incomplete left-most digit.

In this example, it can also be seen that the range from 0000 through 19999 covers 20,000 distinct values. (Remember that zero is a counting figure.) Therefore, this can be referred to as a 20,000-count display. The number of counts of the display is the total number of discrete values included by the range zero through the maximum displayed value.

---

*continued*

*continued*

> Many newer digital instruments do not have the restriction of having the most significant digit being 1 or blank. A popular handheld digital multimeter, for example, has a maximum displayed value of 3999 (ignoring the decimal point location) on most of its ranges. For this reason, it is becoming more practical to refer to a digital meter display in terms of the number of counts instead of terms such as 4½ digits. This meter could be described as having a 4000-count display (0000 through 3999).
> 
> Digital displays usually have some of their specifications expressed in terms of digits or counts. Again, these terms are two different ways of talking about the same thing. They always refer to the least significant digits—the ones at the right end of the display. A single count is the smallest interval that can be displayed, the least significant digit, which also defines the resolution of the display. The number of digits or counts in a specification is multiplied by the resolution to get the actual value. For example, assume that a gauge specification is
> 
> $$\pm(0.05\% \text{ indicated value} + 12 \text{ counts})$$
> 
> and the gauge display is 100.000 kPa.
> 
> Then,
> 
> > The digit/count/resolution value is 0.001 kPa.
> > 
> > The value of the output term of the specification is $100 \times 0.0005 = 0.05$ kPa.
> > 
> > The value of the scale term of the specification is $0.001 \times 12 = 0.012$ kPa.
> > 
> > The sum—which represents the uncertainty of the gauge at this value—is ±0.062 kPa.

On some digital meters, especially higher-accuracy models, the scale term is expressed as a percent of range. Use caution with this type of specification, and read the instrument manual closely! This usually—but not always—means the range as labeled on the front panel of the meter or as listed in the specification table.

Many digital multimeters have an over range display; that is, the 1 V range may display values higher than that. The over range may go up to a value such as 1.200. There are some models with 100% over range, which means that the 1.000 V range will read up to 1.999 V. In most cases, the percent of range would apply to the name of the range (1 V), but in a few cases the manufacturer intends it to apply to the maximum displayed value. Check the manual to be sure. If the scale term is listed as percent of scale, then it usually means the maximum possible reading on that range (name) of the meter. Check the manual to be sure. For example, a long-scale digital meter has an accuracy specification of ± (0.0009% reading + 0.0010% scale) on the 1.000000 V range. The maximum reading on that range is 1.200000 V. The scale term becomes (1.2 × 0.0010%) or 0.000012 V.

For an artifact such as a gage block, the scale term is not as obvious. Also, the factors that equate to the scale term are often not stated in a specification sheet because they are contained in other documents that are included by reference and considered to be common knowledge among users of the devices. One part is the dimension specification, which varies by nominal length. This normally does not apply to an individual block, but rather to a full set. For example, a Grade 1 1-inch block has a specification of ±2 µinch. A 20-inch block has a specification of ±20 µinch.[9] The other part is the thermal coefficient of expansion. This is a constant for the particular material, affects every block, and must be considered whenever the measurement is not being done at 20 °C. In the inch/pound system, the expansion coefficient is given in terms of inches per degree per inch of length.

## Baseline Specification: Floor Term

The *floor* value is expressed in SI units and is a fixed amount that is added to the output at any value on the applicable range. At low levels, the floor term can be a significant part of the uncertainty. Here's an example. On the 1.000000 V DC range, a thermocouple simulator voltage source specification is ±(0.0005% output + 5 µV). At an output setting of 0.450000 V, the floor value is only 0.001% of the output. At 0.080000 V output, the floor value is up to 0.006% of the output. At 0.001174 V output, the floor is more than 0.4% of the voltage output and can be very significant. Since this instrument is a thermocouple simulator, the floor on the last value is even more significant. The voltage represents a thermocouple output at 23.0 °C, and, in those terms, the temperature uncertainty due to the floor specification is approximately 0.1°.

The floor term is usually not explicitly found in dimensional measurements, but the effect is there. Using a Grade 0.5 gage block set specification as an example, all blocks 1 inch and smaller in nominal dimension have a specification of ±1 µinch.[10] This effectively places that as a limit on the best available measurement uncertainty and is therefore a floor.

The most commonly seen forms of these three terms are scale only (usually on analog meters and fixed devices), output plus scale, or output plus floor.

## Forms of Writing Specifications

There are several ways that equipment manufacturers use to print their specifications, with three being the most common. In a strict mathematical sense, the expressions mean different things and can result in different uncertainty intervals for the instrument. The three most common forms are:

- (± output ± scale) This format implies that both terms are random and therefore could be combined using the rss method. Using example values of 0.005 and 0.01, the result would be a range of ±0.011.

- (± output + scale) This format implies that the first term is random and the second term always acts as a bias in a specific direction. Using example values of 0.005 and 0.01, the result would be a range of 0.005 to 0.015.

- ± (output + scale) This format implies that the specification as a whole is random. Using example values of 0.005 and 0.01, the result would be a range of ±0.015. *This is the most conservative method of showing specifications and is the one we recommend.*

When comparing instruments, determining performance verification limits, writing calibration procedures, or any other work involving specifications, it is much easier and fewer errors are made if all specifications are expressed in the same format first.

## Modifiers

As noted by Fluke, modifiers change the baseline specification. If present, a modifier may indicate differences in performance relative to time, ambient conditions, load, or line power. Other conditions may be mentioned occasionally, but these are the most common.

**Modifiers: Time Term.** Most specifications of electronic equipment include a time during which the specification is valid. Many bench or system digital multimeters, for example, have performance specifications for 24 hours since calibration, for 30 and 90 days, and for one year. (Other time periods are also common.) Mechanical measuring instruments are not often specified this way because they do not drift in the same way electronic ones do and because their physical wear rate is highly dependent on the type and frequency of use.

Some users have tried to set specifications for their instruments by doing a linear interpolation between the published time intervals. Experiments performed by Fluke indicate that this is not a good practice.[11] While the long-term drift of a standard may be nearly linear, it is subject to short-term random variation due to noise and other factors. A more conservative approach is to use the specification for the time period that has not yet passed. The most conservative approach is to always use the specification that corresponds to the recalibration interval that has been set. For example, if the instrument is calibrated every 12 months, then use the one-year performance specification all the time.

What if there is no time term and no suggested calibration interval? Then the laboratory and the customer have to work together to select an appropriate calibration interval. See Chapter 10 for more information on this.

If the calibration activity uses data to adjust the calibration interval of instruments, note that changing the interval does not affect which (time-related) specifications are used for calibration. For example, consider a group of instruments with 12-month specifications that have been calibrated every 12 months for a few years. Analysis of the data may indicate that the interval can be extended to 14 months and maintain the same reliability (probability of being in-tolerance at the end of that time.) Or, the analysis may indicate that the interval must be shortened to 10 months to achieve the desired reliability. In either case, when the instruments are calibrated the 12-month specifications will still be used. Those specification limits are part of the analysis process in that they are the basis for pass/fail decisions. (See Chapter 10 for more information on calibration intervals.)

Time is also a factor in other performance specifications that may be given. Two important ones that may be seen are drift and stability.

- *Drift* is a long-term characteristic of electronic circuits. Drift is normally specified over a time interval that may range from a month to a year or more.

- *Stability* (sometimes called *jitter*) is a short-term characteristic of electronic circuits. Stability is normally specified over a time interval that may range from seconds to days, but rarely more than a week.

Drift and stability are measures of the long-term and short-term change in the IM&TE performance when all other factors are accounted for. These effects are present to some degree in all electronic devices and are commonly specified in oscillators. In other instruments, they may be included as part of the overall performance specification.

Some instruments, such as counters or digital oscilloscopes, have time-related specifications on an even shorter scale. A frequency counter always has a half-count (minimum) uncertainty because the master timebase in the counter is not synchronous with the signal being counted. Other timing and pulse characteristics may cause other effects as well.

**Modifiers: Temperature Term.** Temperature affects all measuring instruments to some degree. A reference temperature is an important part of a performance specification. Performance of dimensional/ mechanical/physical measuring instruments is normally specified at 20 °C even if that is not explicitly stated on the specification sheet. If the actual temperature is different at the time of measurement, thermal effects on the measuring instrument and the item being measured must be considered and possibly accounted for. Performance of electronic instruments is normally (but not always) specified at 23 °C. There will often be a band around that temperature where the performance is expected to meet the specification. If the actual calibration temperature is outside that band, a correction may be needed.

Some instruments do not have a temperature band specified, so the temperature correction must be made any time the actual temperature is different from the specified reference temperature.

If the item is mounted in an equipment rack when it is calibrated, the temperature inside the rack should be measured and appropriate corrections made. This may be the case, for example, when performing an in-place calibration of an automated test system.

**Modifiers: Line Power Term.** In some cases, the output or measuring capability of an electronic instrument may vary with changes in the applied line power. The effect is usually slight, but should be evaluated if the IM&TE is used in an environment where the line power fluctuates or where stability of the output is a critical parameter.

**Modifiers: Load Term.** In some cases, the output or measuring capability of an electronic instrument may vary with changes in the load applied to the front panel inputs (for a measuring instrument) or outputs (for a source). A current source, for example, may be suitable for use when supplying direct current, but have limitations on the output when supplying alternating current. That is due largely to the compliance voltage generated across reactive components in the load.

## Qualifiers

Qualifier specifications are other factors that may affect the usability of the IM&TE for a particular location or for a particular application, however, they rarely affect the general operating specifications and are rarely checked during the performance verification accomplished by a calibration procedure. Many qualifiers fall into the categories of conformance to other standards such as safety or electromagnetic interference. Others are operating or storage environmental conditions; these do not normally affect calibration because the laboratory environment is controlled. There is one type of qualifier, however, that may affect any calibration laboratory (relative versus absolute specifications), and one environmental qualifier that may affect a few (operating altitude).

A manufacturer may describe performance specifications in two ways: in reference to the standards used to calibrate the instruments on the production line (relative specifications) or in reference to the SI values of the particular parameter (absolute or total specifications.) The absolute specification is, of course, the most useful to a calibration laboratory. Some manufacturers even list the specifications that way, at least for their standards-grade instruments. Other manufacturers state that the specifications are relative to their calibration standards. The laboratory may have to add uncertainty to them to account for their measurement traceability. Most manufacturers don't explicitly state what type of specification it is. In those cases, if the type cannot be determined, the most conservative path is to assume that the specifications are relative. Always check the IM&TE manual to be sure.

Many electronic devices (IM&TE and other things) are specified for operation up to a maximum altitude (or elevation) of about 10,000 feet (about 3050 m). For most of us, this is not a concern. If the work site happens to be in a place such as Leadville, Colorado, (about 3110 m or 10,200 ft) or La Paz, Bolivia, (about 3625 m or 11,900 ft) then it is a definite problem. Remember, though, the specifications must be read with care. As an example, a very popular 6½-digit digital multimeter has a specified maximum operating altitude of 2000 m (about 6562 ft). If this meter is used in Denver, Colorado, (elevation 1600 m or 5260 ft) it is within its altitude specifications. If it is taken 42 km (26 miles) southwest to the town of Evergreen (elevation 2145 m or 7040 ft), it will be over the specification limit. If an organization works in a mountain area, it may have to perform some type of designed experiment to determine how elevation affects the measuring instruments. There are several reasons for altitude effects.

- As altitude increases, there is less mass of air to absorb heat generated in the instrument, so cooling efficiency is reduced.[12]

- Altitude is also a problem where high-voltage circuits are concerned, because the dielectric strength of air reduces with altitude.[13]

- Mechanical devices may also have problems. For example, all computers and many other electronic devices now include a hard disk drive (HDD) for mass storage of operating programs and/or data. Most HDDs are limited to a maximum altitude of 3050 m (about 10,000 ft.)[14] As the air density becomes lower with altitude, it will reach a point where there is not enough air to provide the aerodynamic lift needed to keep the read-write heads flying a few micrometers above the spinning platters. The results of a read-write head touching the spinning platter surface at a relative velocity of 24 m/s or more are always fatal to the drive and the data stored on it!

- A minor consideration is that sealed electronic components undergo more physical stress as the surrounding air pressure is lowered and may possibly change value or rupture.

Note that a user cannot rely on weather-related barometric pressure reports to determine elevation. In order to maintain worldwide consistency, and as an important aviation safety element, those reports are always corrected to what the pressure would be at mean sea level at that location and instant of time.

## Specification Tables

Following are some examples of specification tables for various kinds of electronic IM&TE. The format and values are abstracted from actual data sheets, but are not complete. Not all functions or ranges are shown. They are not intended to represent any specific instrument or to be examples of anything other than what may be seen in practice.

## Example 1. In this example, the source instrument is a calibrator.

Direct voltage source:[15]

| Voltage output | 12-month accuracy ±(% of output + floor) | Max. current | Resolution |
|---|---|---|---|
| 0.000 mV to 320.000 mV | 0.006% + 4.16 µV | 20 mA | 1 µV |
| 3.2001 V to 32.0000 V | 0.0065% + 416 µV | 20 mA | 100 µV |
| 320.01 V to 1050.00 V | 0.006% + 19.95 mV | 6 mA | 10 mV |

Note: Specifications apply:

> To positive and negative polarities
>
> Within ±5 °C of the temperature at last calibration
>
> With added load regulation error if load is < 1 MΩ

Alternating voltage source:

| Voltage output | Frequency (Hz) | 12-month accuracy ±(% of output + floor) | Max. current | Total harmonic distortion (% of output) | Resolution |
|---|---|---|---|---|---|
| 0.32001 V to 3.20000 V | 10 – 3 k | 0.04% + 192 µV | 20 mA | 0.06% | 10 µV |
| | 3 k – 10 k | 0.04% + 256 µV | 20 mA | 0.10% | 10 µV |
| | 10 k – 30 k | 0.06% + 480 µV | 20 mA | 0.13% | 10 µV |
| | 30 k – 50 k | 0.09% + 960 µV | 10 mA | 0.20% | 10 µV |
| | 50 k – 100 k | 0.20% + 2.56 mV | 10 mA | 0.32% | 10 µV |

Note: Specifications apply:

> To sine wave only
>
> Within ±5 °C of the temperature at last calibration
>
> With added load regulation error if load is < 1 MΩ

Direct current source:

| Current output | 12-month accuracy ±(% of output + floor) | Compliance voltage (at lead end) | Resolution |
|---|---|---|---|
| 3.2001 mA to 32.0000 mA | 0.014% + 900 nA | 4 V | 100 nA |
| 0.32001 A to 3.20000 A | 0.060% + 118 µA | 2.2 V | 10 µA |
| 10.5001 A to 20.0000 A | 0.055% + 4.50 mA | 2.2 V | 100 µA |

Note: Specifications apply:

> To positive and negative polarities.
>
> Within ±5 °C of the temperature at last calibration.

When using special test cable (instrument accessory)

With maximum duty cycle of 1:4 on 20 A output range

*Example 2. Here the measuring instrument is a digital multimeter.*
Direct voltage measurement ± (% of reading + % of range):[16]

| Range | 24 hour<br>23 ± 1 °C | 90 days<br>23 ± 5 °C | 12 months<br>23 ± 5 °C | Temp. coeff.<br>0-18 & 28-55 °C |
|---|---|---|---|---|
| 100.0000 mV | 0.0030 + 0.0030 | 0.0040 + 0.0035 | 0.0050 + 0.0035 | 0.0005 + 0.0005 |
| 10.00000 V | 0.0015 + 0.0004 | 0.0020 + 0.0005 | 0.0035 + 0.0005 | 0.0005 + 0.0001 |
| 1000.000 V | 0.0020 + 0.0006 | 0.0035 + 0.0010 | 0.0045 + 0.0010 | 0.0005 + 0.0001 |

Notes:

Specifications apply after one hour warm-up, 6½ digits, AC filter set to slow.

Specifications are relative to calibration standards.

There is a 20% over-range capability on all ranges except 1000 V DC.

Observe these points:

- The specifications are larger both below and above the 10 V range and as the time interval increases.

- When calculating the values in SI units, the percent of range applies to the value in the range column. The over-range amount is not included.

- The temperatures listed in the 24-hour, 90-day and 12-month columns can be taken as the required temperature limits in the calibration lab.

- The temperature coefficient is a per degree term. For example, assume a meter is being used to measure 5.0 V and is mounted in an automated test system rack with an internal temperature of 32 °C.
  - Using the 12-month specifications, the basic uncertainty is
    ±(5 × 0.0035% + 10 × 0.0005%) or ± 0.00022 V.
  - The temperature coefficient is ±(5 × 0.0005% + 10 × 0.0001%) per degree. The number of degrees to use is (32 − 28) = 4. The result is 0.000035 × 4, or 0.00014.
  - The uncertainty of the 5 V measurement is the sum of those two values, or ±0.00036 V.

*Example 3. In this example, the measuring instrument is a digital oscilloscope.*
Sensitivity is 5 mV/division to 50 V/division in 1, 3, 5 sequence.
Vertical resolution is 8 bits.

An oscilloscope display is like a graph, almost always with eight major divisions up the vertical scale and 10 major divisions along the horizontal scale.[17] An important specification is the amplitude per division on the vertical scale. The specification above indicates the sensitivity range and indicates that it can be set to 5 mV, 10 mV, 30 mV, 50 mV, 100 mV, 300 mV, 500 mV, 1 V, 3 V, 5 V, 10 V, 30 V, or 50 V per division.

The vertical resolution indicates the available resolution of the vertical scale at any particular sensitivity setting. This indicates an eight-bit analog to digital converter is used, which means that any input voltage will be one of $2^8$ or 256 possible output positions. This means that the vertical scale has a resolution of (sensitivity × 8) ÷ 256.

- If the sensitivity is set to 5 mV/division, the resolution is (5 × 8) ÷ 256 or 0.156 mV.
- If the sensitivity is set to 10 V/division, the resolution is (10 × 8) ÷ 256 or 0.312 V.

**Unusual Terminology**

There are occasions where some work, or contact with the manufacturer, is required to interpret a specification. For example, a laboratory in southeastern United States received a new handheld insulation resistance tester from a customer. The customer did have the operating manual. The tester is manufactured in Germany. The specification data table in the manual was fairly straightforward, with the expected columns for measuring range, test voltage and so on. For the performance specifications, however, there were two columns instead of the usual one; they were labeled Intrinsic Error and Measuring Error.

| Range | Test voltage | Intrinsic error | Measuring error |
|---|---|---|---|
| 200 kΩ to 10 GΩ | 500 V, +15% – 0% | ±(5% reading + 3 digits) | ±(7% reading + 3 digits) |

It took several e-mail exchanges with the manufacturer to determine that the Intrinsic Error is the performance specification for calibration and the Measuring Error is what the customer could expect when using the tester in field conditions.

## Comparing Specifications

Sometimes a technician will have to compare specifications of instruments. He or she may need to compare specifications of two instruments to see if they are equivalent. Other times, the need may be to compare the specifications of a measuring instrument to the requirements of the measurement to be made to determine if the instrument has the required capability. There are well-documented methods for doing these comparisons, and this section is a quick review of them.

**Comparing the Specifications of Two Instruments.** There are occasions when a calibration laboratory may need to compare specifications of one instrument to another, either to determine if they are equivalent or to determine if one can be used to calibrate the other. A common reason for doing this is when using a calibration procedure that may have been written many years ago and lists equipment models that the lab no longer has. The lab may need to see if its current measurement standards are equal to or better than the ones specified (equivalency) or if the measurement standards have the capability to calibrate the item in question.

In order to compare specifications, several things must be done.[18] All of these are necessary to ensure that equal values are being compared. The process must:

- Identify the specifications to be compared.
- Convert the specifications to equal units of measure, of the same order of magnitude.
- Apply all modifications required by the specifications, for each value.
- Adjust the uncertainties according to the stated confidence intervals.

For example, say the measurement requirement is to monitor a voltage that is to be maintained at 28.0 V AC, 400 Hz, ± 300 mV. The measurement will be made in an environment where the air temperature is 32 °C (about 90 °F). The meter specified in the procedure (Meter A) is no longer available, but specifications are published in an old manufacturer's catalog. The engineer wants to see if the meter currently available (Meter B) is equivalent—that is, it can be used to make the measurement with equal or better accuracy. Table 21.1 lists the specifications of each meter as they apply to making this measurement. Note these features as they apply to the description of specifications.

**Table 21.1** Specifications of meters A and B.

| Specification | Meter A | Meter B |
|---|---|---|
| Range (V) | 30.0000 | 100.0000 |
| Full scale (V) | 30.1000 | 120.0000 |
| Resolution (V) | 0.0001 | 0.0001 |
| Accuracy (% reading) | 0.26% | 0.06% |
| Accuracy (digits) | 102 | |
| Accuracy (% range) | | 0.03% |
| Specification period (months) | 12 | 12 |
| Temperature | $T_{cal} \pm 5$ °C | 23 ±5 °C |
| $T_{cal}$ Range | 20 to 30 °C | |
| Operating temperature | 0 to 55 °C | 0 to 55 °C |
| Temperature coefficient | Included | |
| TC (% reading) | | 0.005% |
| TC (% range) | | 0.003% |
| Confidence level | unknown | unknown |
| Relative/Absolute | unknown | Relative |
| Conditions | Auto zero ON<br>Sine wave<br>> 10% full scale | Slow AC filter<br>Sine wave<br>> 5% range<br>Altitude < 2000 m |

- Accuracy in percent of reading is the output term. (Since this is a measuring instrument it is really the input, but the term is the same for consistency.)
- Accuracy in number of digits or in percent of range is the scale term.
- These meters do not have a floor term in their specifications.
- For the purposes of this example, both meters are calibrated at 23 °C.
- Meter A does not have a specification for temperature coefficient on AC voltage measurements, so it is assumed to be included in the performance specification.
- For Meter B, the temperature coefficient terms are per °C away from the lower or upper limits of the temperature specification. For example, if the meter was being used at 15 °C (or 31 °C)—three degrees outside the 23 ±5 °C band—then the two temperature coefficient values would be multiplied by three.
- Meter A gives no indication of whether the performance specifications are absolute or relative to the calibration standards. Therefore, the conservative assumption for this type of meter is that the specifications are relative. Since Meter B states that, they can both be treated the same.
- Since the confidence level for both meters is unknown, the most conservative assumption to make is that they both represent a uniform distribution at about 99% level. Since this applies to both, there is no reason to adjust the values. If one of the meters specified the confidence interval or coverage factor, however, that adjustment would have to be made. Methods for this are detailed in the *ISO Guide to the Expression of Uncertainty in Measurement* (GUM) and equivalent documents.
- The conditions listed are examples of specification qualifier terms.

Table 21.2 shows the conversion of the specifications to units of measure (volts, in this case). At the end, each column is added to get a final result. The uncertainty of Meter B is lower, so it is suitable for this measurement.

An additional result of this comparison is that it reveals one of the ugly little secrets of performance specifications. The natural tendency of most users is to accept

**Table 21.2** Measurement unit values at 28 V AC.

| Specification | Meter A | Meter B |
| --- | --- | --- |
| Accuracy (% reading) Volts | 0.0728 | 0.0168 |
| Accuracy (digits) Volts | 0.0102 | |
| Accuracy (% range) Volts | | 0.0300 |
| Operating temperature (°C) | 32.0 | 32.0 |
| Temperature coefficient | included | |
| TC (% reading) | | 0.0056 |
| TC (% range) | | 0.0120 |
| TOTAL ± (Volts) | 0.0830 | 0.0644 |

the meter's reading at face value—particularly if it is a digital display. ("If it has more digits, that means it must be better!") Those of us working in metrology, however, must be more skeptical. The uncertainty numbers calculated in this example take up three of the four digits to the right of the decimal point. So, as a practical matter, the usable resolution of either of these meters is only 0.1 V (or possibly 0.01 V) when making this measurement, not the 0.0001 V that the naive user might believe. If a measurement result was recorded to the full resolution of the meter's display, the effect would be assigning two or three more significant figures to the result than can be justified by the uncertainty.

Fluke has a detailed comparison using a pair of calibrators.[19]

### Case Study: Comparing the Ratio Function of Two Digital Multimeters

A calibration laboratory's customer was trying to determine an equivalent meter to replace an obsolete one. The measurement function being evaluated was DC Ratio. The metrology engineer was asked to assist, but an opportunity to observe the work was not available.

The meter originally specified for the measurement is a Data Precision 3500. The customer has one that is no longer economical to support. There is a manual, dated 1978. The customer also has the instructions for the measurement process being performed.

The observed display is described as .0001 and the reference input is 5 V. This implies that the Data Precision 3500 is in DC Ratio mode and the 10 range is selected.

- The measurement input and the reference appear to be measured on the same range.

- The 10 range has a full scale display of 11.9999.

- The displayed value is 10 times the actual ratio, so the true maximum ratio is 1.19999:1.

- The actual ratio is 0.00001, which implies that the test input voltage is 0.00005 V, or 50 µV.

The ratio performance specifications for the Data Precision 3500 on that range are:

± (0.008% reading + 0.001% full scale + 1 digit) × (10 ÷ Reference Voltage)

The specification period is within six months of calibration with the temperature at 23 ±5 °C.

| | | |
|---|---|---|
| 0.008% reading = | (0.00008 * 0.0001) = | 0.000000008 |
| 0.001% full scale = | (0.00001 * 11.9999) = | 0.000119999 |
| 1 least digit = | | 0.0001 |
| **Sum =** | | 0.000220007 |
| 10 ÷ Reference Voltage = | | 2 |

*continued*

# References

Agilent Technologies. 2003. *Agilent 34401A product overview*, publication number 5968-0162EN. Palo Alto, CA: Agilent Technologies.

ASQC Statistics Division. 1996. *Glossary and tables for statistical quality control*, 3rd ed. Milwaukee: ASQC Quality Press.

Blair, Thomas. 2003. Letter to author (Payne), 28 September 2003.

Fluke Corporation. 1984. *Calibration: Philosophy in practice*. 2nd ed. Everett, WA: Fluke Corporation.

FS GGG-G-15C. 1975. *Federal specification: Gage blocks and accessories (inch and metric)*. U.S. General Services Administration. Washington DC: U.S. Government Printing Office.

Griffith, Gary. 1986. *Quality technician's handbook*. Englewood Cliffs, NJ: Prentice Hall.

Kimothi, S. K. 2002. *The uncertainty of measurements: physical and chemical metrology impact and analysis*. Milwaukee: ASQ Quality Press.

Shewhart, Walter A. 1939. *Statistical method from the viewpoint of quality control*. Washington DC: Graduate School of the Department of Agriculture. New York: Dover Publications, 1986.

Tektronix. 2002. *Handheld battery operated Oscilloscope/DMM/Power Analyzers, THS700 series: Characteristics*. Beaverton, OR: Tektronix Inc.

Wavetek Corp. 1998. *User's handbook for the model 9100 universal calibration system: Volume 2—Performance*, Issue 8.0.

# Chapter 22
# Substituting Calibration Standards

In an ideal world, a technician would be able to select a calibration procedure, always select the exact measurement standards specified, and always follow the procedure exactly. We must deal with the world as it really is, however, not as we wish it to be. In the real world, there will be cases where the laboratory does not have—and cannot easily obtain—the specified calibration standards. In those cases, a way must be found to make the required measurements using equivalents or substitutes for the originally specified equipment.

## MEASUREMENT STANDARD SUBSTITUTION

### Why Might a Substitute for a Measurement Standard Be Needed?

Industrial and commercial systems are designed to have a useful life cycle on the order of decades. For example, the hydroelectric power plant at Niagara Falls (Ontario, Canada) was put in service in 1905 and retired at the end of 1999.[1] Many commercial airliners are up to 30 or more years old,[2] and new ones are expected to last at least that long. When system maintenance requirements are developed during the prime system development phase, IM&TE is selected from what is available at that time. Since that IM&TE must be calibrated, this inevitably means that there will always be equipment to be calibrated that is much older than current measurement standards. In itself, that is not a problem. The problem is that the documentation is often of the same vintage, if it still exists. Quality management systems require that a documented procedure has to exist and be followed, but that can be hard to achieve when the specified equipment is not available, and the companies that made it may not exist. Therefore, the calibration laboratory needs a method to identify suitable equivalents or substitutes for the unavailable equipment.

### How Is a Suitable Substitute Selected?

An instrument that will be used as a substitute for another one must have certain technical characteristics and one practical consideration. When compared to the instrument originally specified:

- It must be capable of measuring the same parameter at the same level.
- The resolution must be equal or better.
- The accuracy and precision (or measurement uncertainty, if stated directly) must be equal or better.
- It should be something that is readily available.

In an ideal case, one would be able to compare the specifications of the old and new to make a determination. That is not always possible and is often not practical. In a lot of cases, the original manufacturers of the test equipment have been absorbed into another company or have disappeared completely. Even if the company still exists, it often does not keep information on equipment that is beyond its support life (or the information may have been lost by fire, flood, effects of aging, or other events.) Also, it is not unknown to find that the equipment originally specified is not capable of making the measurement adequately when evaluated with modern criteria. In a lot of cases the documentation of the unit under test will have to be studied to determine what measurements are being made anyway. So that is probably the best way to examine the problem from the start.

*A fundamental requirement is that in all cases, the calibration standards have to be capable of making the required measurement with a specified minimum TUR.* This must be the basis from which any substitution decision is made. Even if the laboratory can determine that the substitute instrument is equivalent to the original based on specifications, it must still be examined against the measurement requirements to see if it is adequate. Once the measurement requirements have been determined, it is usually easy to see what standards are adequate and available.

### Example: Determining Substitutes for Obsolete Calibration Standards

In 2003, a maintenance department of a large company sent a new test set to the corporate electronics calibration laboratory. The lab had never calibrated the test set model before, so the manuals were requested and received. The test set can best be described as new/old—it is a physically new unit, manufactured in 2002 by ABC Company, but the design is old because it was originally built by XYZ Company for about 20 years starting in the mid-1960s. The manual provided by ABC Company is a photocopy of the old XYZ Company manual with the most recent modifications (1989) and revisions (1993). Due to a series of corporate acquisitions in the past 15 years, XYZ Company no longer exists.

The test set manual includes a calibration procedure, but in spite of the chapter title it is actually an adjustment/alignment procedure. There is enough information for the metrology engineer to produce an in-house calibration (performance verification) procedure, but the original calibration standards specified are a problem. The recommended items are listed in Table 22.1.

**Table 22.1** Measurement standards in original procedures.

| Description | Example Type |
|---|---|
| Differential AC-DC Voltmeter[3] | John Fluke 803[4] |
| Audio Oscillator | Hewlett-Packard[5] 200AB |
| Wheatstone Bridge | Leeds & Northrup 5300 |

*continued*

| | continued |
|---|---|
| DC power source, 0 to 27 V @ 1 A | (None) |
| Resolver Bridge | Julie Research PRBAB-5 |
| Synchro Bridge | Julie Research PSBAB-5 |
| Phase Angle Voltmeter | North Atlantic VM-204 |

Preliminary research on the calibration standards produced this information:

- The Fluke 803 was discontinued before 1980. It was replaced by the Fluke 893A, which was discontinued after 1991. Performance specifications of the 893A were found on the Internet.[6]

- The Hewlett-Packard 200AB was discontinued after 1975, but performance specifications were found on the Internet. The *1994 Hewlett-Packard Logistics Manual* suggests its 3325B as a suitable replacement.[7] The calibration lab does have one of these.

- The Leeds & Northrup 5300 was discontinued and the company no longer exists. Performance specifications were available from a company that specializes in calibration and service of Leeds & Northrup equipment as well as other legacy equipment.

- The Julie Research bridges were discontinued (probably more than 30 years ago) and the company no longer exists, but performance specifications were available from the successor company.[8]

- The North Atlantic VM 204 was discontinued before (probably well before 1980) and no data are available from the manufacturer.

Since the calibration lab does not have any of the listed standards, and a replacement for only one was readily identifiable, the lab had to attack the problem from the other direction. The actual measurement requirements were collected from the test set manual and analyzed. The identified measurement requirements are listed in Table 22.2.

In Table 22.2, the first four columns identify the parameter, value, and tolerance as stated in the test set manual. The last column (Minimum Required Accuracy) lists the minimum performance requirement of a calibration standard based on a 4:1 test accuracy ratio. The requirement is stated as a percentage of the measured value and in the applicable SI units. That can be compared to the performance specifications of the laboratory's available calibration standards. The results are shown in Table 22.3.

No lower-accuracy digital multimeter in the laboratory's inventory can meet the performance requirement for the first two AC voltage measurements in Table 22.2. Since the HP 3458A was specified for AC voltage, the equipment list and the calibration process can be simplified by using the same meter for the DC volt and resistance measurements as well. Its performance is one or two orders of magnitude better than the minimum requirement for those measurements, but that is not a reason to specify an additional meter.

The phase angle tests in the original XYZ Company manual are based first on the assumption that the synchros need to be rezeroed (which is not a valid assumption for performance verification) and also on the capabilities of test equipment available in 1965. The purpose of the test is to verify the bearing accuracy of the synchro and

Table 22.2 Measurement requirements.

| Parameter | Value | Frequency | Tolerance | Min. required accuracy | |
|---|---|---|---|---|---|
| AC voltage measure | 0.220 V | 400 Hz | ± 1% | ± 0.25% | (550 µV) |
| AC voltage measure | 0.440 V | 400 Hz | ± 1% | ± 0.25% | (1.10 mV) |
| AC voltage source & measure | 1.5 V | 400 Hz | ± 2% | ± 0.5% | (7.50 mV) |
| AC voltage source & measure | 3.0 V | 400 Hz | ± 2% | ± 0.5% | (15.0 mV) |
| AC voltage source & measure | 7.07 V | 1000 Hz | ± 0.5% | ± 0.125% | (8.84 mV) |
| AC voltage source & measure | 26.0 V | 400 Hz | ± 1% | ± 0.25% | (65.0 mV) |
| AC voltage source & measure | 36.0 V | 400 Hz | ± 2% | ± 0.5% | (180 mV) |
| AC voltage/phase measure | 26.0 V | 400 Hz, 90° | None | Set & used as phase reference | |
| DC voltage source & measure | 0.080 V | | ± 1 mV | ± 0.31% | (0.25 mV) |
| DC voltage source & measure | 0.150 V | | ± 1% | ± 0.25% | (0.375 mV) |
| DC voltage source & measure | 0.150 V | | ± 3 mV | ± 0.5% | (0.75 mV) |
| DC voltage source & measure | 0.45 V | | ± 1% | ± 0.25% | (1.125 mV) |
| DC voltage source & measure | 16.0 V | | ± 1% | ± 0.25% | (40.0 mV) |
| DC voltage source & measure | 27.5 V | | ± 1% | ± 0.25% | (68.75 mV) |
| Phase Angle measure | 0 ... 360 | every 10° | ± 0.10° | ± 0.025° | |
| Resistance measure | 110 | | ± 2% | ± 0.5% | (0.55 Ω) |
| Resistance measure | 500 | | ± 1% | ± 0.25% | (1.25 Ω) |

resolver, which the angle position indicator is well suited for. (If the synchro or the resolver actually need adjustment, then a phase angle voltmeter would be needed, but that is adjustment or repair, not calibration.)

Analysis of the requirements determined that AC frequency is not a critical parameter. The signal generator originally specified had a frequency accuracy of ±3%. The 3325B frequency accuracy is ±5 ppm or better, and this would be similar for any modern signal generator selected.

At this point, it was possible for the metrology engineer to rewrite the procedure so that it is actually a performance verification using the substituted calibration standards.

## EQUIPMENT SUBSTITUTION AND CALIBRATION PROCEDURES

When an equivalent or substitute measurement standard has been identified, the calibration procedure may need to be examined as well. If the procedure is focused on

**Table 22.3** Equivalent standards to substitute.

| Original | Equivalent Replacement |
|---|---|
| Fluke 803 | Hewlett-Packard 3458A digital multimeter, based on requirement for measuring 220.000 mV AC ± 0.550 mV and 440.000 mV AC ± 1.100 mV, both at 400 Hz |
| Hewlett-Packard 200AB | Hewlett Packard 3325B Option 002 function generator, based on requirement for output of 36.0 V at 400 Hz |
| Leeds & Northrup 5300 | Hewlett-Packard 3458A digital multimeter because is it is already required for other measurements |
| Julie Research PRBAB-5<br>Julie Research PSBAB-5<br>North Atlantic VM-204 | North Atlantic 8810 angle position indicator, based on the phase angle measurement requirement of ±0.025° |
| DC Volt power supply | Above 10.0 V:<br>  Vector-Vid WP-707B DC power supply, based on the requirement for 27.5 V output and the implied requirement for regulation and PARD[9] below the measurement resolution<br>Up to 10.0 V:<br>  General Resistance DAS-46A precision voltage source, based on the requirement for voltage source uncertainties of 375 mV or less at 150 mV output |

Note: The replacement standards shown above may not be the best possible ones to use. They are merely the most suitable ones available to that particular calibration lab at that specific time, which is the point of the example.

adjustment instead of performance verification it will have to be revised as discussed in Chapter 5. But other things may have to be examined as well, depending on its age:

- If the original procedure was written before 1990, measurements involving voltage or resistance should be examined. The conventional values of the SI volt and ohm were adjusted at the beginning of 1990.

- If the original procedure was written before 1990, any temperature measurements will have to be examined. The ITS-90 thermodynamic scale went into effect then, replacing IPTS-68. While the differences are relatively minor in commonly used ranges, they still need to be evaluated. Also, some items may have been written using the old IPTS-48 (from 1948) temperature scale, and there are significant changes from that.

- If the original procedure was written before 1968 (or even after) and contains values expressed in terms of the old centimeter, gram, second (CGS) or gravitational systems, or old meter, kilogram, second (MKS) derived units, they may need to be converted to SI units.

- Finally, any instructions that are written for using or operating a specific measurement standard must be revised to accommodate the new one or to a more generic format.

## Measurement Systems

Since the early 1800s there have been at least four major measurement systems used in science and engineering. The distinguishing features are the definitions of units of mass, length, time, and force.

- The gravitational system is built into the old English system, the basis of the common measurement system used in the United States. In this system, the defined units are the *foot* (length), *pound* (force), *second* (time); and a derived unit is the *slug* (mass).
- The CGS system was published in 1874, with defined units of the *centimeter* (length), *gram* (mass), and *second* (time); and a derived unit of the *dyne* (force).
- The MKS or metric system was published in 1889, with defined units of the *meter* (length), *kilogram* (mass) and *second* (time); and a derived unit of the *newton* (force).
- The SI or modern metric system uses the same defined units as the MKS system and was essentially created in 1954 when the *ampere*, *kelvin* and *candela* were added to the MKS system as defined units. The name was changed in 1960, and the *mole* was added as a seventh defined unit in 1971.

The history of the English system and other historical customary measurement systems is covered in Chapter 1. This system is no longer of any major importance for scientific, engineering, or technical measurements, and is important for common measures only in the United States.

The CGS system was the principal measurement system used by the researchers who built the foundation of modern physics in the nineteenth century. This system was expanded several times and, as a result, many CGS derived units appear in technical literature. These units are all replaced by SI units and should not be used in metrology or technical writing: erg, dyne, poise, stoke, gauss, oersted, maxwell, stilb, and phot.

The MKS system was published by CIPM and its defined (or base) units persist in the SI. Derived units of the MKS system were often inconsistent with those of the CGS system, however. Work on unifying the two systems started as early as 1901.

The SI is built on the MKS system and, with the four additional base units, forms an internally consistent system of measurement for all disciplines. Also, note that the SI now defines the second based on atomic physics, but the three earlier systems all used a second derived from astronomical observations.

## Endnotes

1. Ontario Power Generation. 2003.
2. According to the Boeing Internet site (www.boeing.com/commercial/flash.html), their 727 was in production from 1963 to 1984. Production of the 737 started in 1963 and is ongoing. The older Boeing 707 made its first flight in 1954. The Boeing 727 is still in service with many air carriers. The 737 is used by a large number of air carriers; at least one company has it as the only aircraft type in its fleet. Although the 707 is no longer widely used as a commercial airliner, some cargo carriers use it and it is the airframe of military aircraft such as the KC-135 air refueling tanker and E-3 airborne warning and control system.
3. Some readers may not be familiar with differential voltmeters. Before the appearance of low-cost high-accuracy digital meters in the 1980s, differential voltmeters were among the highest accuracy voltage measurement instruments available. A typical differential voltmeter consisted of a high-accuracy high voltage source, a Kelvin-Varley divider, and a null meter. Measurement accuracy in differential mode was determined by the reference voltage accuracy, and the accuracy and linearity of the divider. A major advantage of the differential voltmeter was that the effective input impedance is nearly infinite when at the null point. That eliminated any loading of the circuit being measured and was not matched by other common test equipment of the time or even by most current digital multimeters.
4. Now called Fluke Corporation.
5. Hewlett-Packard Company split in 1999, and all of the non-computer–related product and service lines (including test and measurement equipment) became a new company, Agilent Technologies.
6. Fluke Corporation. Various catalogs archived by Testmart. Internet location www.testmart.com/advice/advicetmp.cfm
7. Hewlett-Packard Company. Various catalogs archived by Testmart. Internet location www.testmart.com/advice/advicetmp.cfm
8. Process Instruments 2003.
9. Periodic and random deviation, PARD, is a term that includes the sum of all variation from the mean output of a power supply. The variation influences include ripple, noise, temperature coefficient, line and load regulation, and stability. PARD may be expressed as a RMS or peak value, and usually has a frequency bandwidth.

## References

Agilent Technologies. 2003. *Agilent Technologies fact book*. Palo Alto, CA: Agilent Technologies Corporate Media Relations. www.agilent.com/about/newsroom/facts/agilentfactbook.pdf

Ontario Power Generation. *2003 hydroelectric generating stations: Niagara plant group*. www.opg.com/ops/Stations/OntarioPower.asp

Process Instruments. 2003. *History of Julie Research Laboratories*.www.procinst.com/jrlhistory.htm

Taylor, Barry N. 2001. *NIST Special Publication 330, 2001 edition: The international system of units (SI)*. Gaithersburg, MD: National Institute of Standards and Technology. physics.nist.gov/Pubs/SP330/sp330.pdf

Taylor, Barry N. 1995. *NIST Special Publication 811, 1995 edition: Guide for the use of the International System of Units (SI)*. Gaithersburg, MD: National Institute of Standards and Technology. physics.nist.gov/Document/sp811.pdf

# Chapter 23

# Proficiency Testing, Measurement Assurance Programs, and Laboratory Intercomparisons

## PROFICIENCY TESTING

The term *proficiency testing* is misunderstood when laboratories seeking accreditation hear about it the first time. Many think that its purpose is to test employees' skills. While indirectly it is a result of the employees' skills and other factors, directly it is a means of determining the laboratory's competence compared to other laboratories performing the same kind of work.

ISO IEC Guide 2 defines *proficiency testing* as "determination of laboratory testing performance by means of Interlaboratory comparisons."

ISO Guide 43-1 defines *interlaboratory comparisons* as "organization, performance and evaluation of tests on the same or similar test items by two or more laboratories in accordance with predetermined conditions."

Laboratory proficiency testing is an essential element of laboratory quality assurance. In addition to laboratory accreditation and the use of validated methods, it is an important requirement of the accreditation process. More customers demand independent proof of competence from laboratories. It is also a requirement if the laboratory is involved in any regulated industry such as nuclear food, drug, and pharmaceuticals. Laboratories can determine how well they perform against other laboratories and how their measurements compare with others.

For participating laboratories proficiency testing is a comparison of its individual performance against industry performance. It identifies areas for improvement in measurement and techniques. It may identify best practices by comparing other laboratories in the same field.

For the customers of laboratory services, proficiency testing:

- Establishes confidence and is a demonstration of accreditation.

- Helps the customer decide if the laboratory meets its measurement, calibration and testing requirements.

- Acts as a measure for ensuring that the laboratory will continuously meet its quality requirements.

Consistent participation in proficiency testing is also a requirement by some accrediting bodies for a laboratory to maintain accreditation. An example is the ISO 17025 standard, which requires laboratories to participate in proficiency testing schemes. ISO 17025 section 5.9, Assuring the quality of test and calibration results, states that: "The laboratory shall have quality control procedures for monitoring the validity of tests and calibrations undertaken. The resulting data shall be recorded in such a way that trends are detectable and, where practicable, statistical techniques shall be applied to the reviewing of the results. This monitoring shall be planned and reviewed and may include, but not be limited to, the following:

   a) regular use of certified reference materials and/or internal quality control using secondary reference materials;

   b) participation in interlaboratory comparison or proficiency-testing programmes;

   c) replicate tests or calibrations using the same or different methods;

   d) retesting or recalibration of retained items;

   e) correlation of results for different characteristics of an item.

   Note: The selected methods should be appropriate for the type and volume of the work undertaken."

Various schemes for organizing proficiency testing, coordination and analysis of data exist and are referenced in the standards. This chapter covers the more popular schemes and its associated statistics.

## Measurement Comparison Scheme

In this scheme, an artifact with an assigned value is circulated between participating laboratories. Data are collected and published with the appropriate statistics. A national laboratory usually provides the artifact's reference value. Laboratories follow a predetermined process (procedure) when testing the artifact and providing the measurement data. Data are collected and published with the appropriate statistics

In the following example, Figure 23.1, an artifact with a true value of 10.0000 units is measured by the reference laboratory and assigned a reference value of 9.99998 units.

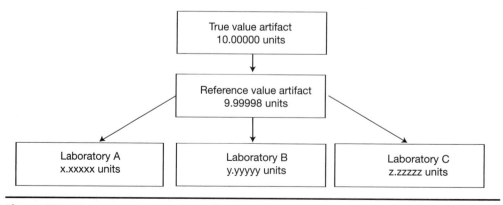

**Figure 23.1** Measurement comparison scheme.

**Table 23.1** Assigning a reference value.

| Specification | ± 0.00001 |
|---|---|
| Uncertainty of the calibrator (k = 1) | 5.7735E-06 |
| Laboratory | Reference |
| Measurement Observation 1 | 9.99999 |
| Measurement Observation 2 | 10.00000 |
| Measurement Observation 3 | 9.99999 |
| Measurement Observation 4 | 10.00000 |
| Measurement Observation 5 | 10.00000 |
| Measurement Observation 6 | 10.00000 |
| Measurement Observation 7 | 9.99999 |
| Measurement Observation 8 | 9.99999 |
| Measurement Observation 9 | 10.00000 |
| Measurement Observation 10 | 10.00000 |
| Sum | 99.99998 |
| Mean | 10.00000 |
| Maximum Value | 10.00000 |
| Minimum Value | 9.99999 |
| Range | 0.00001 |
| Standard Deviation | 0.00001 |
| Median | 10.00000 |
| Uncertainty (Type A) | 0.00001 |
| Combined Uncertainty (k=2) | 0.00002 |
| Reference Value | 10.00000 |

The assigned value is provided by the reference or pivot laboratory making 10 measurements on the artifact and calculating the mean of the 10 measurements as shown in Table 23.1. The uncertainty of the measurement is also determined.

The artifact with the assigned reference value is then circulated to 10 other laboratories participating in the proficiency-testing program, labeled 1 through 10. The laboratories are coded to ensure the confidentiality of the participants. The participating laboratory's data are summarized in Table 23.2.

It is easy to visualize the data if they are graphed. The data are graphed in Figure 23.2 with the mean of each laboratory and the ± 3 standard deviations.

From the data in Table 23.2 and Figure 23.2, it should be noted that laboratory 2 and 6 have a higher variability in their data. This is based on the specification (or capability of the laboratory) as shown in Table 23.2. It is also important to consider the mean value of the data generated by the laboratories and ensure that they do fall within ± 3 standard deviations of the reference laboratory data. Another way to look at individual laboratory's data is to look at the individual laboratory mean ± U (k=2) data as shown in

**Table 23.2** Measurement comparison scheme: raw data and calculations.

| Specification of Equipment | ± 0.00001 | ± 0.005 | ± 0.03 | ± 0.001 | ± 0.0005 | ± 0.005 | ± 0.015 | ± 0.005 | ± 0.005 | ± 0.005 | ± 0.001 |
|---|---|---|---|---|---|---|---|---|---|---|---|
| Uncertainty of Calibrator (k = 1) | 5.7735E-06 | 0.0029 | 0.0173 | 0.0006 | 0.0003 | 0.0029 | 0.0087 | 0.0029 | 0.0029 | 0.0029 | 0.0006 |
| Laboratory | Reference | 1 | 2 | 3 | 4 | 5 | 6 | 7 | 8 | 9 | 10 |
| Measurement Observation 1 | 9.99999 | 10.0011 | 9.983 | 9.9968 | 10.00042 | 9.9984 | 10.0054 | 10.0046 | 10.0021 | 10.0045 | 9.9992 |
| Measurement Observation 2 | 10.00000 | 10.0002 | 9.970 | 9.9966 | 10.00002 | 9.9958 | 10.0148 | 10.0032 | 10.0014 | 10.0027 | 10.0001 |
| Measurement Observation 3 | 9.99999 | 9.9958 | 10.030 | 9.9953 | 10.00020 | 9.9991 | 9.9929 | 10.0042 | 10.0023 | 10.0048 | 10.0002 |
| Measurement Observation 4 | 10.00000 | 10.0004 | 10.009 | 9.9961 | 10.00008 | 10.0022 | 9.9929 | 9.9992 | 9.9969 | 9.9982 | 10.0001 |
| Measurement Observation 5 | 10.00000 | 9.9999 | 10.008 | 9.9955 | 10.00006 | 9.9971 | 10.0081 | 10.0029 | 9.9987 | 10.0031 | 9.9995 |
| Measurement Observation 6 | 10.00000 | 10.0041 | 10.007 | 9.9966 | 9.99987 | 10.0017 | 9.9979 | 10.0040 | 9.9985 | 9.9961 | 10.0006 |
| Measurement Observation 7 | 9.99999 | 9.9951 | 10.027 | 9.9965 | 10.00036 | 10.0014 | 9.9961 | 9.9956 | 10.0007 | 9.9976 | 10.0009 |
| Measurement Observation 8 | 9.99999 | 10.0031 | 10.016 | 9.9957 | 9.99971 | 9.9987 | 10.0121 | 10.0018 | 9.9965 | 10.0024 | 9.9993 |
| Measurement Observation 9 | 10.00000 | 9.9998 | 10.000 | 9.9963 | 9.99969 | 10.0002 | 10.0090 | 10.0047 | 10.0036 | 10.0009 | 9.9994 |
| Measurement Observation 10 | 10.00000 | 9.9964 | 9.989 | 9.9966 | 10.00013 | 9.9951 | 9.9890 | 9.9951 | 10.0004 | 9.9990 | 10.0002 |
| Sum | 99.999977 | 99.99571 | 100.0398 | 99.96209 | 100.000548 | 99.98958 | 100.01815 | 100.01534 | 100.00112 | 100.00927 | 99.99956 |
| Mean | 9.999998 | 9.99957 | 10.0040 | 9.99621 | 10.000055 | 9.99896 | 10.00181 | 10.00153 | 10.00011 | 10.00093 | 9.99996 |
| Maximum Value | 10.000005 | 10.00405 | 10.0300 | 9.99684 | 10.000415 | 10.00216 | 10.01477 | 10.00470 | 10.00361 | 10.00477 | 10.00085 |
| Minimum Value | 9.999990 | 9.99507 | 9.9704 | 9.99530 | 9.999694 | 9.99509 | 9.98899 | 9.99509 | 9.99646 | 9.99608 | 9.99919 |
| Range | 0.000015 | 0.00898 | 0.0596 | 0.00154 | 0.000721 | 0.00706 | 0.02578 | 0.00961 | 0.00714 | 0.00868 | 0.00167 |
| Standard Deviation | 0.000006 | 0.00299 | 0.0189 | 0.00054 | 0.000243 | 0.00246 | 0.00912 | 0.00365 | 0.00239 | 0.00304 | 0.00057 |
| + 3 Sigma | 10.000016 | 10.00854 | 10.0608 | 9.99783 | 10.000783 | 10.00634 | 10.02919 | 10.01247 | 10.00728 | 10.01006 | 10.00165 |
| - 3 Sigma | 9.999980 | 9.99060 | 9.9472 | 9.99458 | 9.999326 | 9.99157 | 9.97444 | 9.99060 | 9.99295 | 9.99179 | 9.99826 |
| Median | 9.999999 | 10.00004 | 10.0078 | 9.99641 | 10.000074 | 9.99890 | 10.00164 | 10.00303 | 10.00056 | 10.00162 | 10.00013 |
| Uncertainty (Type A) | 0.000006 | 0.00299 | 0.0189 | 0.00054 | 0.000243 | 0.00246 | 0.00912 | 0.00365 | 0.00239 | 0.00304 | 0.00057 |
| Expanded Uncertainty (k=2) | 0.000017 | 0.008314 | 0.051321 | 0.001583 | 0.000754 | 0.007588 | 0.025159 | 0.009301 | 0.007493 | 0.008391 | 0.001616 |
| $E_n$ |  | 0.051 | 0.078 | 2.393 | 0.076 | 0.137 | 0.072 | 0.165 | 0.015 | 0.111 | 0.026 |
| Difference |  | -0.0004 | 0.0040 | -0.0038 | 0.0001 | -0.0010 | 0.0018 | 0.0015 | 0.0001 | 0.0009 | 0.0000 |
| Percent Difference |  | -0.0043% | 0.0398% | -0.0379% | 0.0006% | -0.0104% | 0.0182% | 0.0154% | 0.0011% | 0.0093% | -0.0004% |
| Z-Score |  | -0.107 | 0.530 | -0.593 | -0.037 | -0.196 | 0.217 | 0.177 | -0.029 | 0.089 | -0.051 |

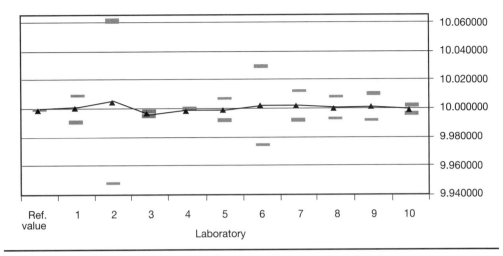

**Figure 23.2** Measurement comparison scheme: Mean ± 3 standard deviations.

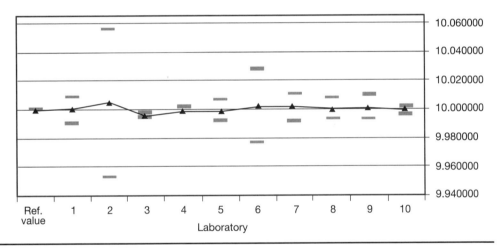

**Figure 23.3** Measurement comparison scheme: Mean ± U (k = 2).

Figure 23.3. Other statistical tests can also be performed using the data generated to test for statistical significance. Most computer spreadsheet software enables this to be done easily. Discussion of the acceptability of data using En numbers is covered later in this chapter.

It is also important to note the individual laboratory method used, operator training, environment tested, calibration of the measurement system (including measurement uncertainty analysis), and any other important parameter of the process.

Another popular graphic technique that gives a good visual presentation of a participant's test data and associated uncertainty compared to a test artifact's assigned value and its associated uncertainty does so by displaying the boundaries created by participant's and test artifact's uncertainties and evaluating whether there is an overlap

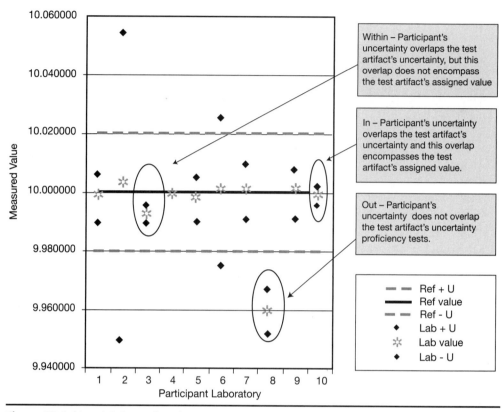

**Figure 23.4** Uncertainty overlap plot.

between them and to what degree (see Figure 23.4). The following is commonly used to evaluate performance levels for this type of uncertainty overlap graph:

- **In.** Participant's uncertainty overlaps the test artifact's uncertainty and this overlap encompasses the test artifact's assigned value.

- **Within.** Participant's uncertainty overlaps the test artifact's uncertainty, but this overlap does not encompass the test artifact's assigned value.

- **Out.** Participant's uncertainty does not overlap test artifact's uncertainty proficiency tests.

## Interlaboratory Testing Scheme

In this scheme, homogeneous material such as a batch of compounded rubber is split among the laboratories and tested simultaneously under agreed-upon conditions. The test data are sent to the proficiency testing coordinator. The material sometimes has an assigned value. Also, the assigned value for the material can be derived from the results of the tests.

In this example, the assigned value is derived from five laboratories' test data. See Table 23.3. It is always easy to visualize the data better if it is presented in graphical form. The laboratory's data are presented in Figure 23.5.

## Chapter 23: Testing, Assurance Programs, and Laboratory Intercomparisons

**Table 23.3** Interlaboratory testing comparison data.

| Laboratory | A | B | C | D | E |
|---|---|---|---|---|---|
| 1 | 99.983 | 99.9997 | 99.9851 | 99.9994 | 100.00033 |
| 2 | 99.993 | 100.0013 | 100.0043 | 99.9997 | 100.00043 |
| 3 | 100.027 | 100.0023 | 100.0057 | 100.0007 | 100.00002 |
| 4 | 99.986 | 100.0039 | 99.9850 | 100.0006 | 99.99952 |
| 5 | 100.009 | 99.9965 | 100.0112 | 100.0008 | 99.99985 |
| 6 | 99.982 | 99.9956 | 100.0011 | 100.0000 | 100.00008 |
| 7 | 100.013 | 99.9955 | 100.0049 | 100.0008 | 99.99957 |
| 8 | 99.977 | 100.0021 | 99.9968 | 100.0009 | 100.00010 |
| 9 | 100.024 | 99.9968 | 100.0038 | 99.9990 | 99.99995 |
| 10 | 99.982 | 100.0034 | 100.0144 | 99.9997 | 100.00030 |
| Sum | 999.97501 | 999.99705 | 1000.01230 | 1000.00166 | 1000.00015 |
| Mean | 99.99750 | 99.99970 | 100.00123 | 100.00017 | 100.00001 |
| Maximum Value | 100.02661 | 100.00387 | 100.01438 | 100.00087 | 100.00043 |
| Minimum Value | 99.97742 | 99.99552 | 99.98504 | 99.99905 | 99.99952 |
| Range | 0.04919 | 0.00836 | 0.02934 | 0.00182 | 0.00091 |
| Standard Deviation | 0.01886 | 0.00333 | 0.00979 | 0.00067 | 0.00031 |
| Median | 99.98946 | 100.00051 | 100.00402 | 100.00031 | 100.00005 |
| Established Ref. Value | 99.99972 | 99.99972 | 99.99972 | 99.99972 | 99.99972 |
| + 3 Std. Dev. | 100.02717 | 100.02717 | 100.02717 | 100.02717 | 100.02717 |
| − 3 Std. Dev. | 99.97228 | 99.97228 | 99.97228 | 99.97228 | 99.97228 |

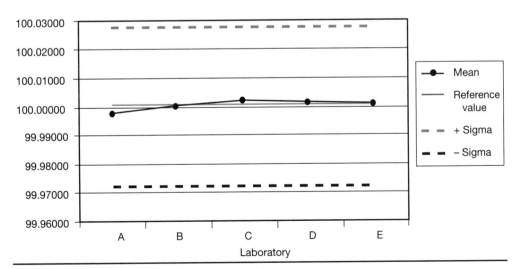

**Figure 23.5** Interlaboratory testing comparison data.

**Table 23.4** Derivation of consensus or reference value.

| Sum | 4999.98616 |
|---|---|
| Mean | 99.99972 |
| Maximum Value | 100.02661 |
| Minimum Value | 99.97742 |
| Range | 0.04919 |
| Standard Deviation | 0.00930 |
| Median | 100.00009 |
| + 3 Std. Dev. | 100.02763 |
| − 3 Std. Dev. | 99.971812 |

Results from the data are used to calculate summary statistics to assign a reference value to the material being tested or measured. See Table 23.4.

## Split-Sample Testing Scheme

When customers of laboratory services wish to compare the performance of a particular laboratory, they may split the homogeneous sample material or circulate a known-value artifact (value unknown to the laboratories) among the laboratories. Data sent by the laboratories are analyzed by the customer to select a particular laboratory's services.

In the example shown on Table 23.5, a homogeneous material with a known true value of 100.00 units is distributed to the two laboratories competing for the customer's business. Twenty measurements are taken at each laboratory and the data are sent to the customer for analysis.

Analysis of variance (ANOVA) techniques can be used in the selection of the laboratory by subjecting the split-sample data from Table 23.5 to ANOVA analysis.

Looking at just the repeatability data (standard deviation) of the two laboratories by itself (Table 23.5), it would be hard to judge if there were much difference. A case can be made to select Laboratory A because its repeatability is better than Laboratory B's and the mean value of the data are closer to the true value of 100.00 units. The ANOVA test (Figure 23.6) also shows that there is not a significant statistical difference between the data provided by the two laboratories (between groups and within groups). A brief explanation on how the ANOVA standard output can be interpreted is given in Figure 23.7.

Based on the data, the F value of 3.09971 is less than the F crit value of 4.098169 from the F distribution table (see CD ROM Appendix IV-a). Therefore there is no statistical difference in mean value of data from laboratory A or B, and either can be chosen based on the one criterion.

**Table 23.5** Split-sample data analysis.

| Laboratory | A | B |
|---|---|---|
| 1 | 100.0025 | 99.00752 |
| 2 | 99.9977 | 99.00601 |
| 3 | 99.9992 | 99.00902 |
| 4 | 100.0047 | 99.00282 |
| 5 | 100.0020 | 99.00581 |
| 6 | 100.0043 | 99.00422 |
| 7 | 99.9957 | 99.00001 |
| 8 | 99.9955 | 99.00198 |
| 9 | 100.0003 | 99.00571 |
| 10 | 100.0036 | 98.99759 |
| 11 | 99.9952 | 99.00170 |
| 12 | 100.0036 | 99.01047 |
| 13 | 100.0034 | 99.01045 |
| 14 | 99.9953 | 99.01258 |
| 15 | 100.0030 | 99.00031 |
| 16 | 99.9996 | 98.99871 |
| 17 | 100.0018 | 99.00280 |
| 18 | 100.0031 | 99.00805 |
| 19 | 100.0028 | 99.01215 |
| 20 | 99.9982 | 99.00293 |
| Mean | 100.00058 | 99.005042 |
| Standard Deviation | 0.00327 | 0.004461 |

Anova: Single Factor

SUMMARY

| Groups | Count | Sum | Average | Variance |
|---|---|---|---|---|
| A | 20 | 2000.177 | 100.0088 | 7.28E-06 |
| B | 20 | 2000.145 | 100.0072 | 8.83E-06 |

ANOVA

| Source of Variation | SS | df | MS | F | P-value | F crit |
|---|---|---|---|---|---|---|
| Between Groups | 2.5E-05 | 1 | 2.5E-05 | 3.099271 | 0.086375 | 4.098169 |
| Within Groups | 0.000306 | 38 | 8.06E-06 | | | |
| | | | | | | |
| Total | 0.000331 | 39 | | | | |

**Figure 23.6** Analysis of variance.

Anova: Single Factor

SUMMARY

| Groups | Count | Sum | Average | Variance |
|---|---|---|---|---|
| A | 5 | 49 | 9.8 | 11.2 |
| B | 5 | 77 | 15.4 | 9.8 |
| C | 5 | 88 | 17.6 | 4.3 |
| D | 5 | 108 | 21.6 | 6.8 |
| E | 5 | 54 | 10.8 | 8.2 |

ANOVA

| Source of Variation | SS | df* | MS** | F*** | P-value**** | F crit***** |
|---|---|---|---|---|---|---|
| Between Groups | 475.76 | 4 | 118.94 | 14.75682 | 9.13E-06 | 2.866081 |
| Within Groups | 161.2 | 20 | 8.06 | | | |
| | | | | | | |
| Total | 636.96 | 24 | | | | |

\*      df = Number of groups – 1  
\*\*     MS = SS/df (mean square = sample variance)  
\*\*\*    Ratio of Between groups MS/Within Groups MS  
\*\*\*\*   P-value: Probability of Significance  
\*\*\*\*\* From F-Table: Degree of Freedom  
       (Numerator) / Degree of Freedom (Denominator)

If $F$ is greater than $F_{crit}$, then there is a significant difference between the means of data sets.

**Figure 23.7** Interpretation of one-way ANOVA data.

## Acceptability of Data

Acceptance of proficiency testing data are based on several factors. Statistical tests are one way to determine compliance and to form the basis for such things as data outliers. ASTM E1301 and ISO Guide 43 documents provide a limited discussion on statistical methods. It is important that the proficiency testing coordinator have good statistical support base to ensure that the correct, unbiased assumption about data is made and reported.

At a minimum, the following statistical parameters should be considered when making assumptions about the proficiency testing data:

Mean

Standard deviation

Range (range can be a good estimator of variability)

Statistical significance using $z$, $t$, or $F$ tests (see Part IV)

The following information is normally analyzed to compare the performance of the laboratories from the data. There are rules to determine the Satisfactory versus Unsatisfactory performance of the lab.

# Chapter 23: Testing, Assurance Programs, and Laboratory Intercomparisons

**Difference.** This is an arithmetic difference between the reference value and that of the participant laboratory.

$$x_{\text{Lab value}} - X_{\text{Assigned value (ref)}}$$

x: participant's result
X: Assigned value

**Percent Difference.** This is an arithmetic difference expressed as a percentage between the reference value and that of the participant laboratory.

$$\frac{(x_{\text{Lab value}} - X_{\text{Assigned value (ref)}})}{X_{\text{Assigned value}}} \times 100$$

x: participant's result
X: Assigned value

**z-Score.** Here the participant laboratory's result is converted to standardized z-Score and the result compared.

$$z = \frac{(x_{\text{Lab value}} - X_{\text{Assigned value (ref)}})}{s}$$

$$|z| \leq 2 = \text{Satisfactory}$$
$$2 < |z| < 3 = \text{Questionable}$$
$$|z| \geq 3 = \text{Unsatisfactory}$$

x: participant's result
X: Assigned value
s: standard deviation of the participant's data unless assigned.

**$E_n$ Number.** The $E_n$ number takes into consideration the expanded measurement uncertainty (at usually k = 2) of the measured artifact when comparing the performance of the laboratory. Laboratories participating in formalized proficiency testing programs will see this number reported.

Care should be taken to ensure that laboratories do not give importance to just one pass/fail criteria in the proficiency testing program.

$$E_n = \frac{(x_{\text{Lab value}} - X_{\text{Assigned value (ref)}})}{\sqrt{U^2_{\text{LAB}} + U^2_{\text{ref}}}}$$

$$|E_n| \leq 1 = \text{Satisfactory}$$
$$|E_n| > 1 = \text{Unsatisfactory}$$

x: participant's result
X: Assigned value
$U^2_{\text{LAB}}$: Uncertainty of participant's result
$U^2_{\text{ref}}$: Uncertainty of ref laboratory's assigned value

**Other Considerations.** It is critical that the confidentiality of the laboratories be maintained when the data are reported publicly. The testing coordinator also should ensure and maintain neutrality and report data in an unbiased manner.

Laboratories should ensure that their processes are in statistical control before participating in the proficiency testing program. Use of Shewhart X-bar, R control charts (see Part IV) in the laboratory calibration and maintenance program is one way to do this in a preventive manner.

Process control should cover operator training, controlled procedures, and measuring equipment repeatability and reproducibility studies.

Development of measurement uncertainty budgets and measurement uncertainty analysis of the measurement process is another important consideration when reporting the measurement data for proficiency testing. See Chapter 29.

## MEASUREMENT ASSURANCE PROGRAMS

It is one thing to compare a laboratory's proficiency against another laboratory. But it is equally important in a laboratory's day to day operations to ensure that there is measure of confidence in testing and calibration work performed.

Some good elements of a measurement assurance program (MAP) include:

- Scheduled traceable calibration of laboratory's standards
- Use of check standards
- Comparison of reference standards against check standards and working standards
- Continuous evaluation of the measurement uncertainties associated with the standards and test and calibration processes
- Implementation of a quality system in accordance with recognized standards such as ISO 17025 and ANSI Z540-1
- Controlled, documented test and calibration procedures established under a quality system
- Trained calibration and test technicians (metrologists)
- A formal internal audit program to track quality of calibration and test performed
- A MAP for control charting and analyzing the results in a timely, proactive manner

Before implementing a MAP, it is important to understand what a laboratory is trying to accomplish. The objectives of a MAP should be defined. The following should be considered as a minimum:

- Select the process to monitor.
- Select the check standard.
- Establish a value for the check standard through traceable calibration or other means (for reference materials).
- Ensure that the check standard is controlled so that its established value is not averse to drift to external factors that can be controlled.

- Monitor and record the process data at a regular interval determined.
- Chart the data. Control charting is a powerful technique.
- Analyze data using statistical techniques. Observe any trends.
- Make an objective decision based on the data and statistical evidence.

## Establishing a MAP

The rest of this chapter is an example of how a laboratory can establish a MAP.

The laboratory has a 10.0 gram weight that it is using as a check standard. It also has another 10.0 gram weight that is designated as a site reference standard. The check standard is compared against the reference standard on a scheduled basis and the data recorded and charted. The reference standard is sent to an accredited primary standards laboratory to obtain traceable calibration per its calibration interval.

The laboratory wants to monitor a digital scale that it uses to calibrate customers' weights. The check standard has an established reference value of 10.001 grams. For simplifying this example, all other factors are assumed constant. In practical issues more issues should considered when making a measurement (environment, operator, training, method, and so on).

The laboratory measures the check standard with digital scale once every week and records the data on the control chart. Since this measurement does not take long, the laboratory makes five measurements. For more than one measurement, the laboratory will utilize an X-bar $R$ chart.

If the measurement took a longer time, the laboratory may choose to make a single measurement every week. For a single measurement, the laboratory will utilize an individual $X$ $R$ chart.

Table 23.6 is used to calculate control limits.

**Table 23.6** Control chart constants.

| | Variables Data | | | |
|---|---|---|---|---|
| | Control limit constants | | | |
| n | A2 | D3 | D4 | d2 |
| 2 | 1.88 | 0 | 3.267 | 1.128 |
| 3 | 1.023 | 0 | 2.574 | 1.693 |
| 4 | 0.729 | 0 | 2.282 | 2.059 |
| 5 | 0.577 | 0 | 2.115 | 2.326 |
| 6 | 0.483 | 0 | 2.004 | 2.534 |
| 7 | 0.419 | 0.076 | 1.924 | 2.704 |
| 8 | 0.373 | 0.136 | 1.864 | 2.847 |
| 9 | 0.337 | 0.184 | 1.816 | 2.97 |
| 10 | 0.308 | 0.223 | 1.777 | 3.078 |

The formula for calculating control limits for individual moving range chart is:

$$UCL_{IX} = \overline{IX} + A_2 \overline{MR}$$
$$LCL_{IX} = \overline{IX} - A_2 \overline{MR}$$
$$UCL_{MR} = D_4 \overline{MR}$$
$$LCL_{MR} = 0$$

The formula for calculating control limits for Xbar, Range chart should be:

$$UCL_X = \overline{\overline{x}} + A_2 \overline{R}$$
$$LCL_X = \overline{\overline{x}} - A_2 \overline{R}$$
$$CL_X = \overline{\overline{x}}$$
$$UCL_R = \overline{R} D_4$$
$$LCL_R = \overline{R} D_3$$
$$CL_R = \overline{R}$$

UCL = Xbar Upper Control Limit
LCL = Xbar Lower Control Limit
$UCL_R$ = Range Upper Control Limit
$LCL_R$ = Range Lower Control Limit

The data for individual measurements are shown on Table 23.7.

**Table 23.7** Data for individual measurements.

| Date | Scale | Range | Mean | Range mean | Xbar UCL | Xbar LCL | Range UCL |
|---|---|---|---|---|---|---|---|
| 3-Jan-03 | 10.000 |  | 10.001 | 0.0007 | 10.0026 | 9.9991 | 0.0022 |
| 10-Jan-03 | 10.000 | 0.000 | 10.001 | 0.0007 | 10.0026 | 9.9991 | 0.0022 |
| 17-Jan-03 | 10.000 | 0.000 | 10.001 | 0.0007 | 10.0026 | 9.9991 | 0.0022 |
| 24-Jan-03 | 10.001 | 0.001 | 10.001 | 0.0007 | 10.0026 | 9.9991 | 0.0022 |
| 31-Jan-03 | 10.001 | 0.000 | 10.001 | 0.0007 | 10.0026 | 9.9991 | 0.0022 |
| 7-Feb-03 | 10.000 | 0.001 | 10.001 | 0.0007 | 10.0026 | 9.9991 | 0.0022 |
| 14-Feb-03 | 10.000 | 0.000 | 10.001 | 0.0007 | 10.0026 | 9.9991 | 0.0022 |
| 21-Feb-03 | 10.001 | 0.001 | 10.001 | 0.0007 | 10.0026 | 9.9991 | 0.0022 |
| 28-Feb-03 | 10.000 | 0.001 | 10.001 | 0.0007 | 10.0026 | 9.9991 | 0.0022 |
| 7-Mar-03 | 10.001 | 0.001 | 10.001 | 0.0007 | 10.0026 | 9.9991 | 0.0022 |
| 14-Mar-03 | 10.001 | 0.000 | 10.001 | 0.0007 | 10.0026 | 9.9991 | 0.0022 |
| 21-Mar-03 | 10.000 | 0.001 | 10.001 | 0.0007 | 10.0026 | 9.9991 | 0.0022 |
| 28-Mar-03 | 10.001 | 0.001 | 10.001 | 0.0007 | 10.0026 | 9.9991 | 0.0022 |

*continued*

| | | | | | | | continued |
|---|---|---|---|---|---|---|---|
| 4-Apr-03 | 10.002 | 0.000 | 10.001 | 0.0007 | 10.0026 | 9.9991 | 0.0022 |
| 11-Apr-03 | 10.000 | 0.001 | 10.001 | 0.0007 | 10.0026 | 9.9991 | 0.0022 |
| 18-Apr-03 | 10.000 | 0.000 | 10.001 | 0.0007 | 10.0026 | 9.9991 | 0.0022 |
| 25-Apr-03 | 10.001 | 0.000 | 10.001 | 0.0007 | 10.0026 | 9.9991 | 0.0022 |
| 2-May-03 | 10.001 | 0.001 | 10.001 | 0.0007 | 10.0026 | 9.9991 | 0.0022 |
| 9-May-03 | 10.001 | 0.000 | 10.001 | 0.0007 | 10.0026 | 9.9991 | 0.0022 |
| 16-May-03 | 10.000 | 0.000 | 10.001 | 0.0007 | 10.0026 | 9.9991 | 0.0022 |
| 23-May-03 | 10.001 | 0.001 | 10.001 | 0.0007 | 10.0026 | 9.9991 | 0.0022 |
| 30-May-03 | 10.000 | 0.001 | 10.001 | 0.0007 | 10.0026 | 9.9991 | 0.0022 |
| 6-Jun-03 | 10.001 | 0.000 | 10.001 | 0.0007 | 10.0026 | 9.9991 | 0.0022 |
| 13-Jun-03 | 10.002 | 0.001 | 10.001 | 0.0007 | 10.0026 | 9.9991 | 0.0022 |
| 20-Jun-03 | 10.001 | 0.001 | 10.001 | 0.0007 | 10.0026 | 9.9991 | 0.0022 |
| 27-Jun-03 | 10.002 | 0.001 | 10.001 | 0.0007 | 10.0026 | 9.9991 | 0.0022 |
| 4-Jul-03 | 10.000 | 0.001 | 10.001 | 0.0007 | 10.0026 | 9.9991 | 0.0022 |
| 11-Jul-03 | 10.001 | 0.000 | 10.001 | 0.0007 | 10.0026 | 9.9991 | 0.0022 |
| 8-Jul-03 | 10.001 | 0.001 | 10.001 | 0.0007 | 10.0026 | 9.9991 | 0.0022 |
| 25-Jul-03 | 10.001 | 0.001 | 10.001 | 0.0007 | 10.0026 | 9.9991 | 0.0022 |
| 1-Aug-03 | 10.000 | 0.000 | 10.001 | 0.0007 | 10.0026 | 9.9991 | 0.0022 |
| 8-Aug-03 | 10.000 | 0.000 | 10.001 | 0.0007 | 10.0026 | 9.9991 | 0.0022 |
| 15-Aug-03 | 10.002 | 0.002 | 10.001 | 0.0007 | 10.0026 | 9.9991 | 0.0022 |
| 22-Aug-03 | 10.001 | 0.000 | 10.001 | 0.0007 | 10.0026 | 9.9991 | 0.0022 |
| 29-Aug-03 | 10.001 | 0.000 | 10.001 | 0.0007 | 10.0026 | 9.9991 | 0.0022 |
| 5-Sep-03 | 10.001 | 0.000 | 10.001 | 0.0007 | 10.0026 | 9.9991 | 0.0022 |
| 12-Sep-03 | 10.000 | 0.001 | 10.001 | 0.0007 | 10.0026 | 9.9991 | 0.0022 |
| 19-Sep-03 | 10.000 | 0.000 | 10.001 | 0.0007 | 10.0026 | 9.9991 | 0.0022 |
| 26-Sep-03 | 10.002 | 0.001 | 10.001 | 0.0007 | 10.0026 | 9.9991 | 0.0022 |
| 3-Oct-03 | 10.002 | 0.000 | 10.001 | 0.0007 | 10.0026 | 9.9991 | 0.0022 |
| 10-Oct-03 | 10.002 | 0.000 | 10.001 | 0.0007 | 10.0026 | 9.9991 | 0.0022 |
| 17-Oct-03 | 10.000 | 0.002 | 10.001 | 0.0007 | 10.0026 | 9.9991 | 0.0022 |

Control charts to monitor the measurements are shown in Figures 23.8 and 23.9.

**218** Part III: Metrology Concepts

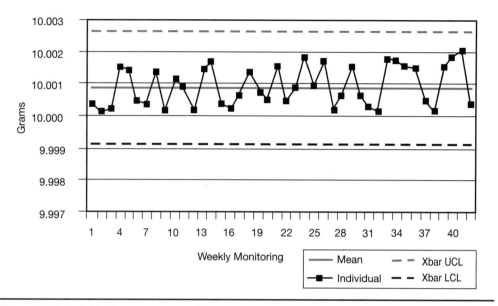

**Figure 23.8** Individual measurements chart.

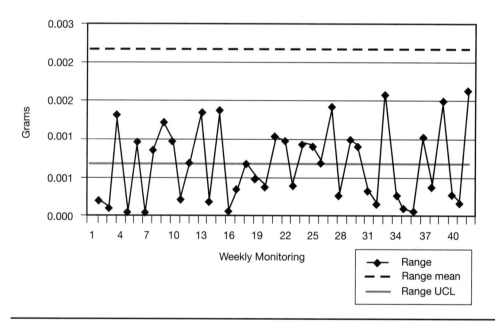

**Figure 23.9** Moving range chart.

The data for multiple measurements are shown on Table 23.8.

**Table 23.8** Multiple measurement data.

| Date | 1 | 2 | 3 | 4 | 5 | Xbar | Range | Mean | Range Mean | Xbar UCL | Xbar LCL | Range UCL |
|---|---|---|---|---|---|---|---|---|---|---|---|---|
| 3-Jan-03 | 10.001 | 10.001 | 10.000 | 10.001 | 10.001 | 10.0008 | 0.001 | 10.0010 | 0.001 | 10.0017 | 10.0003 | 0.0025 |
| 10-Jan-03 | 10.000 | 10.001 | 10.001 | 10.001 | 10.000 | 10.0009 | 0.001 | 10.0010 | 0.001 | 10.0017 | 10.0003 | 0.0025 |
| 17-Jan-03 | 10.001 | 10.000 | 10.002 | 10.001 | 10.002 | 10.0011 | 0.002 | 10.0010 | 0.001 | 10.0017 | 10.0003 | 0.0025 |
| 24-Jan-03 | 10.001 | 10.001 | 10.002 | 10.001 | 10.002 | 10.0014 | 0.001 | 10.0010 | 0.001 | 10.0017 | 10.0003 | 0.0025 |
| 31-Jan-03 | 10.002 | 10.000 | 10.000 | 10.001 | 10.002 | 10.0011 | 0.002 | 10.0010 | 0.001 | 10.0017 | 10.0003 | 0.0025 |
| 7-Feb-03 | 10.002 | 10.002 | 10.001 | 10.000 | 10.002 | 10.0010 | 0.002 | 10.0010 | 0.001 | 10.0017 | 10.0003 | 0.0025 |
| 14-Feb-03 | 10.000 | 10.001 | 10.001 | 10.001 | 10.002 | 10.0009 | 0.001 | 10.0010 | 0.001 | 10.0017 | 10.0003 | 0.0025 |
| 21-Feb-03 | 10.001 | 10.000 | 10.001 | 10.001 | 10.000 | 10.0008 | 0.001 | 10.0010 | 0.001 | 10.0017 | 10.0003 | 0.0025 |
| 28-Feb-03 | 10.001 | 10.000 | 10.002 | 10.001 | 10.001 | 10.0011 | 0.001 | 10.0010 | 0.001 | 10.0017 | 10.0003 | 0.0025 |
| 7-Mar-03 | 10.001 | 10.000 | 10.001 | 10.001 | 10.001 | 10.0009 | 0.001 | 10.0010 | 0.001 | 10.0017 | 10.0003 | 0.0025 |
| 14-Mar-03 | 10.000 | 10.000 | 10.002 | 10.000 | 10.001 | 10.0007 | 0.001 | 10.0010 | 0.001 | 10.0017 | 10.0003 | 0.0025 |
| 21-Mar-03 | 10.000 | 10.001 | 10.001 | 10.002 | 10.001 | 10.0010 | 0.002 | 10.0010 | 0.001 | 10.0017 | 10.0003 | 0.0025 |
| 28-Mar-03 | 10.000 | 10.000 | 10.001 | 10.001 | 10.002 | 10.0009 | 0.002 | 10.0010 | 0.001 | 10.0017 | 10.0003 | 0.0025 |
| 4-Apr-03 | 10.001 | 10.002 | 10.001 | 10.001 | 10.002 | 10.0013 | 0.001 | 10.0010 | 0.001 | 10.0017 | 10.0003 | 0.0025 |
| 11-Apr-03 | 10.000 | 10.000 | 10.000 | 10.000 | 10.001 | 10.0005 | 0.001 | 10.0010 | 0.001 | 10.0017 | 10.0003 | 0.0025 |
| 18-Apr-03 | 10.002 | 10.001 | 10.002 | 10.001 | 10.002 | 10.0016 | 0.000 | 10.0010 | 0.001 | 10.0017 | 10.0003 | 0.0025 |
| 25-Apr-03 | 10.000 | 10.000 | 10.001 | 10.000 | 10.001 | 10.0004 | 0.001 | 10.0010 | 0.001 | 10.0017 | 10.0003 | 0.0025 |
| 2-May-03 | 10.001 | 10.002 | 10.001 | 10.001 | 10.002 | 10.0014 | 0.001 | 10.0010 | 0.001 | 10.0017 | 10.0003 | 0.0025 |
| 9-May-03 | 10.001 | 10.001 | 10.000 | 10.001 | 10.000 | 10.0005 | 0.000 | 10.0010 | 0.001 | 10.0017 | 10.0003 | 0.0025 |
| 16-May-03 | 10.001 | 10.001 | 10.000 | 10.001 | 10.001 | 10.0007 | 0.000 | 10.0010 | 0.001 | 10.0017 | 10.0003 | 0.0025 |
| 23-May-03 | 10.001 | 10.001 | 10.001 | 10.001 | 10.001 | 10.0010 | 0.000 | 10.0010 | 0.001 | 10.0017 | 10.0003 | 0.0025 |

*continued*

| | | | | | | | | | | | | *continued* |
|---|---|---|---|---|---|---|---|---|---|---|---|---|
| 30-May-03 | 10.001 | 10.002 | 10.001 | 10.001 | 10.001 | 10.0013 | 0.001 | 10.0010 | 0.001 | 10.0017 | 10.0003 | 0.0025 |
| 6-Jun-03 | 10.000 | 10.001 | 10.000 | 10.001 | 10.001 | 10.0005 | 0.001 | 10.0010 | 0.001 | 10.0017 | 10.0003 | 0.0025 |
| 13-Jun-03 | 10.001 | 10.001 | 10.001 | 10.002 | 10.002 | 10.0013 | 0.001 | 10.0010 | 0.001 | 10.0017 | 10.0003 | 0.0025 |
| 20-Jun-03 | 10.000 | 10.001 | 10.001 | 10.002 | 10.001 | 10.0009 | 0.002 | 10.0010 | 0.001 | 10.0017 | 10.0003 | 0.0025 |
| 27-Jun-03 | 10.002 | 10.001 | 10.001 | 10.001 | 10.002 | 10.0012 | 0.001 | 10.0010 | 0.001 | 10.0017 | 10.0003 | 0.0025 |
| 4-Jul-03 | 10.001 | 10.000 | 10.001 | 10.000 | 10.000 | 10.0007 | 0.001 | 10.0010 | 0.001 | 10.0017 | 10.0003 | 0.0025 |
| 11-Jul-03 | 10.002 | 10.001 | 10.000 | 10.000 | 10.002 | 10.0010 | 0.002 | 10.0010 | 0.001 | 10.0017 | 10.0003 | 0.0025 |
| 18-Jul-03 | 10.001 | 10.000 | 10.001 | 10.001 | 10.000 | 10.0009 | 0.001 | 10.0010 | 0.001 | 10.0017 | 10.0003 | 0.0025 |
| 25-Jul-03 | 10.000 | 10.000 | 10.000 | 10.000 | 10.001 | 10.0004 | 0.001 | 10.0010 | 0.001 | 10.0017 | 10.0003 | 0.0025 |
| 1-Aug-03 | 10.001 | 10.002 | 10.000 | 10.001 | 10.001 | 10.0011 | 0.001 | 10.0010 | 0.001 | 10.0017 | 10.0003 | 0.0025 |
| 8-Aug-03 | 10.002 | 10.001 | 10.001 | 10.002 | 10.002 | 10.0013 | 0.001 | 10.0010 | 0.001 | 10.0017 | 10.0003 | 0.0025 |
| 15-Aug-03 | 10.000 | 10.002 | 10.001 | 10.000 | 10.000 | 10.0009 | 0.002 | 10.0010 | 0.001 | 10.0017 | 10.0003 | 0.0025 |
| 22-Aug-03 | 10.001 | 10.001 | 10.002 | 10.001 | 10.000 | 10.0011 | 0.002 | 10.0010 | 0.001 | 10.0017 | 10.0003 | 0.0025 |
| 29-Aug-03 | 10.000 | 10.001 | 10.001 | 10.002 | 10.002 | 10.0010 | 0.001 | 10.0010 | 0.001 | 10.0017 | 10.0003 | 0.0025 |
| 5-Sep-03 | 10.000 | 10.001 | 10.000 | 10.001 | 10.001 | 10.0008 | 0.001 | 10.0010 | 0.001 | 10.0017 | 10.0003 | 0.0025 |
| 12-Sep-03 | 10.002 | 10.002 | 10.001 | 10.001 | 10.001 | 10.0014 | 0.001 | 10.0010 | 0.001 | 10.0017 | 10.0003 | 0.0025 |
| 19-Sep-03 | 10.001 | 10.001 | 10.001 | 10.001 | 10.000 | 10.0009 | 0.001 | 10.0010 | 0.001 | 10.0017 | 10.0003 | 0.0025 |
| 26-Sep-03 | 10.002 | 10.001 | 10.001 | 10.000 | 10.002 | 10.0012 | 0.002 | 10.0010 | 0.001 | 10.0017 | 10.0003 | 0.0025 |
| 3-Oct-03 | 10.000 | 10.001 | 10.001 | 10.001 | 10.002 | 10.0008 | 0.001 | 10.0010 | 0.001 | 10.0017 | 10.0003 | 0.0025 |
| 10-Oct-03 | 10.002 | 10.001 | 10.001 | 10.001 | 10.001 | 10.0011 | 0.001 | 10.0010 | 0.001 | 10.0017 | 10.0003 | 0.0025 |
| 17-Oct-03 | 10.002 | 10.002 | 10.002 | 10.002 | 10.002 | 10.0016 | 0.000 | 10.0010 | 0.001 | 10.0017 | 10.0003 | 0.0025 |

Control charts to monitor the measurements are shown in Figures 23.10 and 23.11.

If the measurements are normally within the control limits, there is no need for concern. By graphically monitoring data, one can observe rising or declining trends, erratic up and down fluctuations and other trends. SPC books describe how to interpret Control Charts in detail.

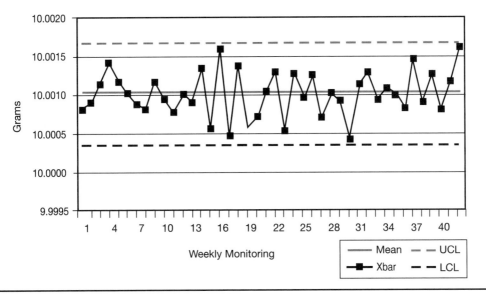

**Figure 23.10** Xbar chart.

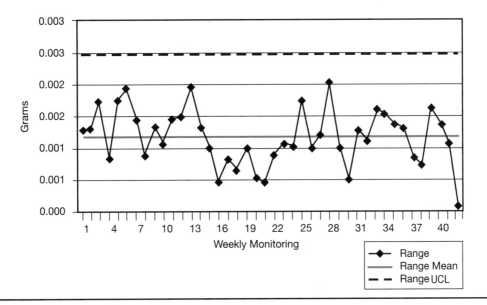

**Figure 23.11** Range chart.

Monitoring the performance of equipment with control charts helps the laboratory catch conformance related problems relating to the measurement in a proactive way before the problem really becomes a problem. It provides assistance in quantifying drift when used with other statistical tools. It provides guidelines on determining calibration intervals.

MAPs provide confidence that the accuracy of standards or equipment is maintained when used in a proper manner.

## References

ANSI/ISO 17025-1999, *American National Standard—General requirements for the competence of testing and calibration laboratories.* Milwaukee: ASQ Quality Press.

ASTM. 1999. ASTM E 691 – 1999, *Standard practice for conducting an interlaboratory study to determine the precision of a test method.* West Conshohocken, PA: ASTM.

ASTM. 1995. ASTM E 1301 – 1995, *Standard guide for proficiency testing by interlaboratory comparisons.* West Conshohocken, PA: ASTM.

ISO. 1997. ISO Guide 43-1 and 2, *Proficiency testing by interlaboratory comparisons.* 2nd ed. Geneva: ISO.

ISO. 1994. ISO 5725 Parts 1-4. Geneva: ISO.

NCSL International. 1999. *Guide for interlaboratory comparisons* RP-15. Boulder: NCSL International.

# Part IV
## Mathematics and Statistics: Their Use in Measurement

**Chapter 24**  Number Formatting
**Chapter 25**  Unit Conversions
**Chapter 26**  Ratios
**Chapter 27**  Statistics
**Chapter 28**  Mensuration, Volume, and Surface Areas

# Chapter 24
# Number Formatting

## SIGNIFICANT DIGITS

The definition of a *significant digit* is, "Any digit in a number that is necessary to define a numerical value of a quantity is said to be significant." Wherever possible the number of significant digits must be determined and stated in the formatting of a measurement result or in the accompanying text. The following are examples of numbers with a statement of resolution or precision in the accompanying text (followed in parentheses by the resulting number of significant digits):

| | |
|---|---|
| 130,000,000 | reported in millions (three significant digits) |
| 307,450,000 | measured in thousands (six significant digits) |
| 805,006,700 | resolved in units (nine significant digits) |
| 2.030, 0.0140, 0.68240 | precise to thousandths (four, three, and three significant digits, respectively) |

Where clarification of intended meaning is not present by accompanying text or other means, the assumptions about the number of significant digits could change. For instance, for the same data items we could conclude the following:

| | |
|---|---|
| 130,000,000, | (two significant digits) |
| 307,450,000, | (five significant digits) |
| 805,006,700, | (seven significant digits) |
| 2.030, 0.0140, 0.68240, | (three, two, and four significant digits, respectively) |

The number of significant digits could change because the determination of significant digits is based either on assumption or knowledge about how each datum (reference, measurement) was taken and the degree of precision intended to be asserted, as presented.

When a value is stated in scientific notation with trailing zeroes on the right, the implication is that these digits are significant. Examples of this are:

2.8600 × 10⁻⁴ A,    here five significant digits are implied

3.780 × 10⁻⁸ A,    here four significant digits are implied

1.24400 × 10⁹ Hz.    here six significant digits are implied

This implication of resolution/significance is a unique aspect of the scientific notation data format.

As leading zeros are automatically omitted, and magnitude is contained in the power of 10 exponent (for example, $10^{-4}$, E – 4, and so on), the number of significant digits may be clearly stated and understood by the value in the coefficient. Whichever number format is used, a conservative practice is if a number is stated to a given number of digits, without further available basis information, assume all the stated digits are significant.

In math operations exact numbers are considered to have an infinite number of significant digits and, therefore, are not considered in determining the number of significant digits in a result. Actual counts, not estimates, are exact numbers. So are defined numbers (1 foot = 12 inches).

Certain irrational number constants (for example, $\pi$, $\varepsilon$, and so on), commonly used in mathematical operations, occupy a kind of middle zone of significant digits. These constants may be determined to any number of significant digits—more than sufficient to not be a limitation on the accuracy of the math operation they are used in. When as a shorthand they are used with a low number of significant digits in the calculation (the most common of course is $\pi$, where it often is stated to only three significant digits, 3.14), however, that must be considered in determining the significant digits stated in the calculation result.

There are two approximate rules for determining significant digits in the results of math operations on data. Note that these rules may yield results that are too small by one or two digits in some cases:

1. **Addition and subtraction significant digits rule.** The answer should contain no more significant digits, relative to the decimal marker, further to the right than is present in the least precise number. For example:

    $$134{,}000{,}000 \text{ (in millions)} + 1{,}843{,}991 - 4.600 \times 10^5 = 135383991$$

    This number is adjusted to the precision of the least precise number, 134,000,000 (stated in millions), giving 135,000,000, or $1.35 \times 10^8$.

2. **Multiplication and division significant digits rule.** The product or quotient should contain no more significant digits than the number with the fewest significant digits. For example:

    $$134{,}000{,}000 \text{ (stated in millions)} / (1{,}843{,}991 \times 4.600 \times 10^{-1}) = 157.9749293$$

    This number is adjusted to three significant digits, giving a result of 150, or $1.50 \times 10^2$.

Including digits that are not significant would imply resolution not appropriate for the measurement statement. This burdens further uses of the stated measurement result with transcription and computation complexity that has no value. It would be misconstrued as result resolution, or even worse, precision and accuracy that is not warranted.

It is important to note that there is some further degradation in resulting significant digits in exponentiation, transcendental, and other complex math operations. Use of more advanced methods, such as error calculus, are required to adjust for these effects.

## NOTATION METHODS

When measurements are made and their values recorded, a normal requirement is to communicate those values to others. Most often the purpose for communicating these results is to enable either subsequent information gathering or decision making. Nearly always the recorded or communicated values are expressed as numbers. There is a need to express these numerical values in a consistent, comparative, and understandable format.

There have been many ways in which numbers have been represented over the years. The simplest is to express decimal numbers as significant digits related to zero by the location of the decimal marker (sometimes called *decimal point* or *decimal place*).

There are various notation and formatting methods in use for representing numerical data. Some of these are:

Standard notation

Scientific notation

SI prefix system (See Chapter 25 on conversions.)

Coding (These will not be discussed.)

### Standard Notation

The term *standard notation* is used to describe a numerical value that is expressed simply as a number whose magnitude is conveyed by the placement of the decimal marker. Hence, the magnitude of 1000.1 is 10 times the magnitude of 100.01, and the magnitude of 0.0034 is one-hundredth the magnitude of 0.34.

From this we can see that moving the decimal marker one place to the right has the same effect as multiplying the number by 10, and moving it one place to the left has the same effect as dividing it by 10. Moving two places to the right increases the value by a factor of one hundred, and to the left decreases the value by a factor of one hundredth. Moving the decimal marker by three places changes the value by a factor of 1000, and so on.

The resulting numbers in standard notation from measurements can vary from quantities that are comparatively small in magnitude to ones that are quite large.

In the following, u = standard uncertainty.

For the definition of standard uncertainty, see Chapter 29 and Appendix D.

**Avogadro's number,**
602214199000000000000000 $mol^{-1}$,
u = 47000000000000000 $mol^{-1}$,
which specifies the number of atoms, molecules, and so on in a gram mole of any chemical substance.

**Loschmidt constant** (given at 273.15 K, 101.325 kPa),
26867775000000000000000000 m$^{-3}$,
u = 470000000000000000000 m$^{-3}$,

which specifies the number of molecules in a cubic centimeter of a gas under standard conditions (often confused with Avogadro's Number).

The **Andromeda galaxy** (closest to our Milky Way galaxy) is estimated to contain at least 200,000,000,000 stars.

The neutron **magnetic moment**,
− 0.00000000000000000000000009662364 J-T$^{-1}$,
u = 0.000000000000000000000000000000023 J-T$^{-1}$.

The **neutron mass**,
0.00000000000000000000000000167492716 kg,
u = 0.0000000000000000000000000000000013 kg.

One thing that's immediately obvious in these measurement numbers is the number of zeroes involved in making the standard notation numerical measurement statement. Standard notation clearly is a poor way of representing very large and small values.

## Scientific Notation

Fortunately these values can be expressed much more efficiently in another numerical format called *scientific notation*. The scientific notation method of formatting numbers is based on the observation that the value of a number is unchanged if you multiply and divide it by the same constant. For instance, for the value 0.00512, we can see that 0.00512 × 1000 / 1000 = 0.00512. Doing this in parts, 0.00512 × 1000 = 5.12, and then 5.12 /1000 = 0.00512.

We also know that 1/1000 can be written as $10^{-3}$ (where 10 is the base and –3 is the exponent of the expression $10^{-3}$). Saying $10^{-3}$ is the exponential form of representing divide by 1000, or multiply by 1/1000.

This understanding, and these relationships, allow us to change 0.00512 (in standard notation) to be transformed to scientific notation as $5.12 \times 10^{-3}$.

In scientific notation (E–format) we express the above standard notation values as:

**Avogadro's number**
6.02214199E+23 mol$^{-1}$,        u = 4.70E + 16 mol$^{-1}$

**Loschmidt constant** (273.15 K, 101.325 kPa)
2.6867775E+25 m$^{-3}$,          u = 4.7E + 19 m$^{-3}$

The **Andromeda galaxy** contains at least 2.0E+11 stars.

**Magnetic moment**
− 9.662364E-27 J-T$^{-1}$,        u = 2.3E-33 J-T$^{-1}$

**Neutron mass**
1.67492716E-27 kg,              u = 1.3E-34 kg

By these comparisons you can immediately see how much more compact scientific notation is. There are several benefits of using scientific notation. It is compact and efficient both for the purposes of reading and comparing values, since the exponent concisely gives magnitude as a number (rather than by counting zeroes). It reduces typographical errors that could come from erroneously adding or omitting zeros. Using values in subsequent math operations is greatly simplified, as will be seen.

There are several frequently used formats for stating a number in scientific notation. In scientific notation numbers are always expressed in relation to an exponent of the base 10.

In the following, $N$ is the number we wish to express, with the decimal marker placed to the right of the first digit, $M$ is the exponent of the base 10, and $E$ is sometimes used to mean $10^{exponent}$.

The three formats in most frequent use are:

- **Superscript format**. This is used in scientific publications.

$$\pm N.N \times 10^{\pm M}. \text{ For example, } 5.12 \times 10^{-3}.$$

- **E-format**. This is used in spreadsheet displays and in most programming languages.

(MS Excel, FORTRAN).

$$\pm N.N \: E \pm M. \text{ For example, } 5.12 \: E - 03.$$

- **Carat format**. This is used in some texts and in spreadsheet formulas entered at the formula bar.

(MS Excel).

$$\pm N.N * 10^\wedge \pm M. \text{ For example, } 5.12 * 10^\wedge - 3.$$

The notation of a value expressed in scientific notation has three parts:

± N.N     A *numerical value*, sometimes called the *coefficient*, having an absolute magnitude greater than or equal to 1, and less than 10. Here, the decimal marker is placed immediately to the right of the left-most nonzero digit. This value may be positive or negative signed (±).

× 10     A *base*—scientific notation is always base 10.

± M     An *exponent* that, with the base, reflects the magnitude of the numerical value. This value may be positive or negative signed (±).

A number is transformed from standard to scientific notation by recognizing a relationship of powers of 10 as shown here:

$0.000002 = 2.0 \times 10^{-6}$     =     $2.0 \times 10^{-1} \times 10^{-1} \times 10^{-1} \times 10^{-1} \times 10^{-1} \times 10^{-1}$

$0.00002 = 2.0 \times 10^{-5}$     =     $2.0 \times 10^{-1} \times 10^{-1} \times 10^{-1} \times 10^{-1} \times 10^{-1}$

$0.0002 = 2.0 \times 10^{-4}$     =     $2.0 \times 10^{-1} \times 10^{-1} \times 10^{-1} \times 10^{-1}$

| | | |
|---|---|---|
| $0.002 = 2.0 \times 10^{-3}$ | = | $2.0 \times 10^{-1} \times 10^{-1} \times 10^{-1}$ |
| $0.02 = 2.0 \times 10^{-2}$ | = | $2.0 \times 10^{-1} \times 10^{-1}$ |
| $0.2 = 2.0 \; 10^{-1}$ | = | $2.0 \times 10^{-1}$ |
| $2.0$ | = | $2.0 \times 10^{0}$ |
| $20.0 = 2.0 \times 10^{1}$ | = | $2.0 \times 10$ |
| $200.0 = 2.0 \times 10^{2}$ | = | $2.0 \times 10 \times 10$ |
| $2000.0 = 2.0 \; v \; 10^{3}$ | = | $2.0 \times 10 \times 10 \times 10$ |
| $20000.0 = 2.0 \times 10^{4}$ | = | $2.0 \times 10 \times 10 \times 10 \times 10$ |
| $200000.0 = 2.0 \times 10^{5}$ | = | $2.0 \times 10 \times 10 \times 10 \times 10 \times 10$ |
| $2000000.0 = 2.0 \times 10^{6}$ | = | $2.0 \times 10 \times 10 \times 10 \times 10 \times 10 \times 10$ |

This obviously can, beneficially, be extended to any magnitude, large or small!

Any standard notation value may be easily converted to scientific notation. This is done by recognizing that a value is unchanged when the decimal marker is moved to behind the first nonzero digit when one adds a $\times 10^{\pm M}$ multiplier that adjusts for this movement, where $M$ is the number of digits that the decimal marker was moved (+) to the left or (–) to the right, respectively.

As just mentioned, one more, very important advantage of using scientific notation is in the efficiency of doing math (addition, subtraction, multiplication, division, and raising to whole or fractional powers) on numerical results.

## Math Operations Using Scientific Notation

When doing math operations on measurement numbers near the same order of magnitude as each other, ordinary math methods can be employed. When the order of magnitude of any of the numbers involved in a math operation or operations is substantially different from other numbers, however, all the numbers should be converted to scientific notation before performing the required operations.

The rules for doing math operations on numbers formatted in scientific notation relate to the operation being performed, as will now be explained.

### Addition/Subtraction

The steps to take to add or subtract using scientific notation are:

1. Convert the exponents to the same magnitude before performing the operation. This will often require changing the location of the decimal marker for the numerical value or coefficient.

2. Perform the addition / subtraction.

3. Adjust for required significant digits in the result, as required.

4. Restore to scientific notation format, as required.

   For example, add $4.875324 \times 10^{6}$ and $5.02 \times 10^{2}$. Follow the steps.

1. Change the decimal marker of $4.875324 \times 10^{6}$ to result in $48753.24 \times 10^{2}$.

2. $48753.24 \times 10^2 + 5.02 \times 10^2 = 48758.26 \times 10^2$.

3. Restore to scientific notation format, or $4.875826 \times 10^6$, and there is no adjustment for significant digits.

A more direct method would have been to change $5.02 \times 10^2$ to $0.000502 \times 10^6$ and add to $4.875324 \times 10^6$. For addition, converting to the algebraically more positive exponent (here, 6) often avoids final adjustment for scientific notation formatting.

In another example, subtract $5.02 \times 10^2$ from $4.87532448 \times 10^6$. Follow the steps.

1. Change the decimal marker of $5.02 \times 10^2$ to result in $0.000502 \times 10^6$.

2. $4.87532448 \times 10^6 - 0.000502 \times 10^6 = 4.87482248 \times 10^6$, which is already in scientific notation format.

3. Adjust for the significant digits in $0.000502 \times 10^6$, giving $4.874822 \times 10^6$.

## *Multiplication*

The steps to multiply using scientific notation are:

1. Algebraically multiply the coefficients and add the values of the exponents.

2. Round to the number of significant digits of the lower significant digit number.

3. Adjust for required significant digits in the result, as required.

4. Restore to scientific notation format, if required.

For example, multiply $4.875324 \times 10^6$ and $5.02 \times 10^2$. Follow the steps.

1. $(4.875324 \times 10^6)(5.02 \times 10^2) = (4.875324)(5.02) \times (10^6)(10^2)(24.474126) \times (10^{6+2}) = 24.5 \times 10^8$.

2. Note, this number was rounded to the number of significant digits (three) of the lower resolution number, $5.12 \times 10^2$.

3. Restore to scientific notation format, giving $2.45 \times 10^9$.

## *Division*

The steps to divide using scientific notation are:

1. Algebraically divide the coefficients.

2. Subtract the value of the denominator exponent (below the division line) from the numerator exponent. Or, change the signs of the exponents below the division line (– to +, + to –) and add these to the sum of those above the division line.

3. Round to the number of significant digits of the lower significant digit number.

4. Adjust for required significant digits in the result, as required.

5. Restore to scientific notation format, if required.

For example, divide $4.875324 \times 10^6$ by $5.02 \times 10^2$. Follow the steps.

1. $(4.875324 \times 10^6) / (5.02 \times 10^2) = ((4.875324) / (5.02)) \times (10^6)(10^{-2}) = (0.97118008) \times (10^{6-2}) = 0.971 \times 10^3$.

2. Note, this number was rounded to the number of significant digits of the lower resolution number, $5.02 \times 10^2$.

3. Restore to scientific notation format, giving $9.71 \times 10^3$.

### Raising to Powers (Integral and Some Fractional)

Raising to an integral (that is, whole number) power is like multiplying the number by itself the number of times given by the power expressed by the exponent. So, use the multiplication methods. (Though not as intuitive, the same is also true for fractional exponentiation, involving a kind of repeated division and multiplication, not discussed here.)

For example, square $5.02 \times 10^2$. Follow the steps for multiplication.

1. Squaring, $(5.02 \times 10^2)^2$, is like multiplying $5.02 \times 10^2$ and $5.02 \times 10^2$.

2. $(5.02 \times 10^2)(5.02 \times 10^2) = (5.02)(5.02) \times (10^2)(10^2) = (5.02)^2 \times (10^2)^2 = 25.2 \times 10^4 = 2.52 \times 10^5$.

Obviously, we can extend this to any integral power.

Raising to a fractional power is somewhat like raising to integral powers, except the fractional power must be resolved to bring the result into scientific notation format.

The steps to raise to a fractional power are:

1. Algebraically raise the coefficients to the given power (integral or fractional).

2. Raise the exponents to the given power by multiplying (not adding) the power of 10 exponent by the separate power exponent. If the resulting power of (base) 10 exponent is not a whole number exponent, factor the fractional part to a coefficient and resulting whole number exponent.

3. Round to the number of significant digits of the lower significant digit number.

4. Adjust for required significant digits in the result, as required.

5. Restore to scientific notation format, if required.

For example, raise $5.02 \times 10^2$ to the one-half power (take the square root of $5.02 \times 10^2$).

$(5.02 \times 10^2)^{1/2} = (5.02)^{1/2} \times (10^2)^{1/2} = 2.24053565 \times 10^1 = 2.24 \times 10^1$.

In another example, raise $5.02 \times 10^2$ to the one-third power (take the cube root of $5.02 \times 10^2$).

1. $(5.02 \times 10^2)^{1/3} = (5.02)^{1/3} \times (10^2)^{1/3} = 1.712252881 \times 10^{2/3}$.

2. Note that $10^{2/3}$ cannot be reduced to a whole number exponent. Using a calculator, we see $10^{2/3}$ has a value of 4.641588834 and make this a factor of the result, giving

3. $1.712252881 \times 4.641588834 = 7.95 \times 10^0$, or 7.95.

Notice, in all cases we do not adjust for the number of significant digits until after all the math operations are performed and we have the result. This is so that we do not lose any available resolution of the result due to the math operations performed. Fortunately, we have computers and spreadsheet programs that make the math portion of this easy. The assignment of significant digits, however, currently requires our intervention in the formatting of the result.

### *Estimation*

Another benefit of using the scientific notation format is the ease and efficiency of estimating approximate results. Using the previous example, estimate the result of 134,000,000 (stated in millions) divided by (1,843,991 multiplied by $4.600 \times 10^{-1}$).

Restating these in scientific notation gives $1.34 \times 10^8$ / ($1.843991 \times 10^6 \times 4.600 \times 10^{-1}$). Changing these to estimated values yields $1.50 \times 10^8$ / ($2.000000 \times 10^6 \times 4.600 \times 10^{-1}$), giving a result of 150, or $1.50 \times 10^2$.

In this case the estimated result is the same as that obtained using the full precision of the originating data. Obviously, care must be taken in the judgments made by estimation. In this case, had we estimated the first number to be $1.00 \times 10^8$ instead of $1.50 \times 10^8$, the result would have been $1.00 \times 10^2$ instead of $1.50 \times 10^2$, an error of 50%.

One method for minimizing accumulated estimation errors is to make the estimate values close to the actual data. Another is to alternate the directions of the estimates relative to the data: one higher, the next lower, and so on. Yet another is to alternate the directions of the estimates above and below a divisor line when present.

## SI PREFIX SYSTEM

The SI unit prefix system has the advantage of aggregating measurements into easily-recognized groups of magnitudes. See Chapter 25 for these units and how they are used. These SI unit prefixes permit compactly expressing large and small orders of magnitude and easily comparing units of similar magnitude. On the other hand, it will be seen that it is not as easy to compare relative orders of magnitude, numerically, as compared to the scientific notation method.

Here's a comparison of standard, scientific, and SI prefix notation. One set of measurements is reported in standard and scientific notation (in superscript format) for electrical current measurements in amperes:

| **Standard notation** | **Scientific notation** | **SI prefix notation** |
| --- | --- | --- |
| 0.000286 A | $2.86 \times 10^{-4}$ A | 0.286 mA |
| 126.78 A | $1.2678 \times 10^{-2}$ A | 0.12678 kA |
| 0.45965 A | $4.5965 \times 10^{-1}$ A | 0.45965 A |
| 0.0000000378 A | $3.78 \times 10^{-8}$ A | 37.8 nA, or 0.0378 pA |

and for frequency measurements in Hertz:

| **Standard notation** | **Scientific notation** | **SI prefix notation** |
| --- | --- | --- |
| 1,244,000,000 Hz. | $1.244 \times 10^9$ Hz | 1.244 GHz |
| 3.43 Hz | 3.43 Hz | 3.43 Hz |

| | | |
|---|---|---|
| 60.25 Hz | $6.025 \times 10^1$ Hz | 60.25 Hz, or 6.025 daHz |
| 56,734,200 Hz | $5.67342 \times 10^7$ Hz | 56.7342 MHz |

## ISO PREFERRED NUMBERS

We are all familiar with products that are manufactured in a series of standard sizes. It is often more efficient for the manufacturer to produce the ideal spacing of sizes so that a range is covered with the fewest number of intermediate sizes and with ones that are evenly spread across the range. In the fluid volume series "tablespoon, fluid ounce, quarter-cup, gill, cup, pint, quart, pottle, gallon," each successive unit increases by a factor of two. For the SI system there is a preference for decimal units, and, additionally, a method is given for distributing evenly spaced values between successive powers of 10, when needed.

In 1877 a French military engineer, Col. Charles Reynard was charged with standardizing the size of mooring cables used for balloons that did surveillance during wartime. He reduced the sizes used from 425 to 17 by developing a geometric series basis that resulted in every fifth step increasing by a factor of 10. We now call such series the *Reynard series*. ISO has defined four such basic (Reynard) series of preferred numbers, with the designator $R$ in honor of Reynard. Listed here are values from 10 to 100 for these series:

- R5: 10, 16, 25, 40, 63, 100.
- R10: 10, 12.5, 16, 20, 25, 31.5, 40, 50, 63, 80, 100.
- R20: 10, 11.2, 12.5, 14, 16, 18, 20, 22.4, 25, 28, 31.5, 35.5, 40, 45, 50, 56, 63, 71, 80, 90, 100.
- R40: 10, 10.6, 11.2, 11.8, 12.5, 13.2, 14, 15, 16, 17, 18, 19, 20, 21.2, 22.4, 23.6, 23.6, 25, 26.5, 28, 30, 31.5, 33.5, 35.5, 37.5, 40, 42.5, 45, 47.5, 50, 53, 56, 60, 63, 67, 71, 75, 80, 85, 90, 95, 100.

By moving the decimal marker to the left or right, these same series may be extended to any magnitude number. For instance, the R5 series could be used to define a preferred number sequence between 0.1 and 1.0, as 0.10, 0.16, 0.25, 0.40, 0.63, or 1.00. One would use this method where it is desired to have a range of produced results be restated as values conforming to the spacing of a preferred number series.

As an example, say a series of measurements has been made with the results being 0.4812, 0.0125, 0.9823, 0.5479, 0.2258, 0.7601, 0.4271, 0.15812, and 0.7013. Report these in a series conforming to the ISO R10 preferred number series.

The adjusted R10 equivalent series would be:
R10': 0.01, 0.125, 0.16, 0.20, 0.25, 0.315, 0.40, 0.50, 0.63, 0.80, 1.00.

The R10 preferred number transformation of the results would then be:
0.50, 0.01, 1.00, 0.50, 0.20, 0.80, 0.40, 0.16, 0.63.

Note: If there are paired results, adjusting one set of the paired values to a preferred number spacing will require interpolating the results of the other set in the pair for each value in the series. (See Chapter 27.)

# NUMBER ROUNDING METHODS

The purpose for rounding numbers is to present data in more concise format and appropriate to the purpose at hand. For instance, a bar of steel has a diameter of 9/16 in (0.5625 in). If the purpose at hand is to compare it to other choices of ½ in (0.50 in) and ¾ in (0.75 in), it is more convenient (and may be more appropriate) to report the 9/16 inch measurement decimally as 0.56 in.

Caution needs to be exercised in employing rounding methods to ensure that the necessary precision in numbers or measurements is not sacrificed by rounding merely for the sake of brevity.

There are some uses of data where reporting data to a greater number of significant digits is advantageous. Some of those applications include, but are not limited to, comparisons involving small differences in measurements, and some statistical calculations (paired t-tests, ANOVA, design of experiments, correlations, autocorrelations, non-parametric tests).

There are several number rounding methods in common use. Prominent among these are:

1. **ISO rounding (preferred method).** This method is similar to the "round up for 5 or over" rule, but it balances what is done when the discarded digit is 5 followed by all zeros. In the special case where the discarded digit is 5 followed by all zeros, the right-most retained digit, before rounding, determines if it is kept unchanged or increased by one. If before rounding this digit is even, it is left unchanged. If before rounding this digit is odd, it is increased by one to make it even. This method is preferred for most scientific measurements. It is required for most weights and measures reporting.

2. **Round up for 5 or over.** Here the right-most retained digit is unchanged if the discarded digit is less than 5 and increased by one if the discarded digit is 5 or over. This is the most-often taught method. Be aware that this is the method implemented in the displayed result as well as the ROUND() function for MS Excel and some other spreadsheet programs.

3. **Truncation.** Here deleted digits on the right side of a number are simply omitted with no accommodation to remaining digits. The obvious problem with this method is the right-most retained digit may not reflect the precision implied from data in the deleted digits: $0.343 \times 10^3$ could be the truncated result of either $0.343999 \times 10^3$ or $0.343001 \times 10^3$ actual measurements.

4. **Round to closest ISO preferred number.** Use one of the four ISO basic series of preferred numbers (ISO R5, R10, R20, R40). This method ensures that reported values aggregate into preferred values. This method is recommended for use in design and both eases comparison as well as increases standardization in reporting of results.

5. **Round to nearest stated multiple of significance.** In spreadsheet programs, such as MS Excel, other methods are available:
   - Floor(), rounds toward zero, to the nearest stated multiple of significance
   - Ceiling(), rounds away from zero, to the nearest stated multiple of significance

Such methods are useful for cases where a governing standard or practice specifies a multiple of significance that a result must be rounded to (for example, ASTM standards that specify rounding a measurement to an stated increment, such as to the nearest 1%, 100 kPa, and so on).

As previously emphasized, do not round intermediate results of math operations. Round only the final result.

Here are examples of rounding by various methods. The measurement is 1⅞ inches (1.875 in or 47.625 mm, exactly). Referring to the five most prominent methods:

1. ISO rounding to

    3 significant digits: 1.88 in, 47.6 mm
    2 decimal places:     1.88 in, 47.62 mm

2. Round up for 5 or over to

    3 significant digits: 1.88 in, 47.6 mm
    2 decimal places:     1.88 in, 47.63 mm
    1 decimal place:      1.9 in, 47.6 mm

3. Truncate to

    3 significant digits: 1.87 in, 47.6 mm
    1 decimal place:      1.8 in, or 47.6

4. Round to closest ISO preferred number:

    R5:   1.6 in, 40 mm
    R10:  2.0 in, 50 mm
    R20:  1.8 in, 50 mm
    R40:  1.9 in, 47.5 mm

5. Round the value, 126.34, to nearest stated multiple of significance (for example, with MS Excel spreadsheet or similar built-in functions):

    — Floor (126.34,25) = 125, Floor (0.623,0.25) = 1.0

    — Ceiling (126.34,25) = 150, Ceiling (0.123,0.25) = 0.25

## OTHER NUMBER FORMATTING ISSUES

The symbol for:

Degrees (often written as the abbreviation, deg) is °, (small superscript circle)

Minutes (often written as the abbreviation, min) is ', (single quotation mark)

seconds (often written as the abbreviation, sec) is ", (double quotation mark)

Angles with a magnitude less than one degree are written beginning with 0°, as follows:

0° 43'

0° 0.35', or 0° 0' 30"(preferred)

## Leading Zeroes

"Leading zeroes" for values are stated in standard format on drawings, with magnitudes less than unity:

Customary (U.S.) unit values are formatted with no zero preceding the decimal marker. SI units are to have a zero preceding the decimal marker. SI units whose magnitude is less than unity are to have a leading zero to the left of the decimal marker. Examples of these are .3085 inches and 0.43 mm.

## Customary Number Formatting

In the United States customary unit numbers with four or more digits to the left of the decimal marker have frequently been represented with commas separating each group of three digits, and a decimal point as a decimal marker. In most other countries the comma has been widely used as a decimal marker.

To minimize confusion it is presently recommended that numbers have a decimal point for a decimal marker, and that numbers to the left and right of the decimal point be separated into groups of threes, with a space between the groupings. Groups of four numbers on either side of the decimal marker optionally may be grouped together without spacing.

Examples of these representations are:

| SI recommended | Not recommended |
|---|---|
| 0.284 5 or 0.2845 | |
| 4 683 or 4683 | 4,683 |
| 2 352 491 | 2,352,491 |
| 44 231.112 34 | 44,231.11234 |
| 1403.2597 or 1 403.259 7 | 1,403.2597 |

## References

American National Standards Institute. 1973. ANSI Z17.1-1973. *American National Standard for preferred numbers.* New York: ANSI.

American Society for Testing Materials. 1990. ASTM E 29-90. *Standard practice for using significant digits in test data to determine conformance with specifications.* West Conshohocken, PA: IEEE Press.

Institute of Electrical and Electronic Engineers, and American National Standards Institute. 2002. IEEE. ASTM SI 10-2002. *American National Standard for use of the international system of units(SI): The modern metric system.* West Conshohocken, PA: ASTM.

ISO/ANSI. 1973. *Preferred number standards.* Geneva: ISO.

___ ISO 3-1973 *Preferred numbers—Series of preferred numbers.* Geneva: ISO.

___ ISO 17-1973 *Guide to use preferred numbers and of series of preferred numbers.* Geneva: ISO.

___ ISO 497-1973, *Guide to the choice of series of preferred numbers and of series containing more rounded values of preferred numbers.* Geneva: ISO.

Taylor, Barry N. 1995. *NIST Special Publication 811, 1995 edition: Guide for the use of the International System of Units (SI).* Gaithersburg, MD: National Institute of Standards and Technology. physics.nist.gov/Document/sp811.pdf

Wilson, Bruce A. 1996. *Dimensioning and tolerancing.* ASME Y 14.5. Chicago: Goodheart Willcox.

# Chapter 25

# Unit Conversions

SI is in use throughout the world and is the most widely recognized and accepted system of measurement units. The benefits of use of the SI in industry, trade, and the sciences are numerous. SI units enable far-improved comparability of measurements in both similar or widely varying magnitudes. Products and components dimensioned in SI units afford more direct engineering and scientific analysis, as the units are often more directly associated with underlying physical quantities.

Another widely used system is often called the English, customary, conventional, or inch-pound system of units of measure. These units, based on inch / pound / horsepower / and so on units of measure, are still strongly in use today even though SI is the legal basis for all measurements. Their use is found mainly in the United States, but also to some degree worldwide in countries that have had significant trade with and/or influence from both British and U.S. commerce and industry.

Because of this continuing divergence from the practical use of a single system of units in the world, it continues to be important to accurately and easily convert from measurements made in one system, say SI units, to those of the other, customary units. To be able to compare measurements it is necessary to convert to and from each unit with confidence.

This chapter will provide the tools for correct and adequate conversion of units between SI and customary units.

## SI UNITS

SI has currently adopted seven base units. The names of SI units are not capitalized; and the symbol is a capital letter only if the name is derived from a person's proper name. These are:

| Symbol | Unit | Quantity |
|--------|------|----------|
| m | meter | length |
| kg | kilogram | mass |
| s | second | time |
| A | ampere | electric current |

| | | |
|---|---|---|
| K | kelvin | thermodynamic temperature |
| mol | mole | amount of substance |
| cd | candela | luminous intensity |

Each of these units is based either on primary artifact standards, on defined relationships to physical quantities that may be reproduced to very high precision, or on exact mathematical relationships defined by their relationship to each other. Let's discuss each of these and their bases[1]:

- A *meter* is the length of the path traveled by light in vacuum during a time interval of 1/299 792 458 of a second. (Defined at the 17th CGPM, 1983).

- A *kilogram* is the unit of mass equal to the mass of the international prototype of the kilogram. (Defined at the 3rd CGPM, 1901).

- A *second* is the duration of 9 192 631 770 periods of the radiation corresponding to the transition between the two hyperfine levels of the ground state of the cesium-133 atom. (Defined at the 13th CGPM, 1967).

- An *ampere* is the constant current, which, if maintained in two straight parallel conductors of infinite length, of negligible circular cross-section, and placed 1 meter apart in a vacuum, would produce between these conductors a force equal to $2 \times 10^{-7}$ newton per meter of length. (defined at the 9th CGPM, 1948).

- A *kelvin* is the unit of thermodynamic temperature as 1/273.16 of the thermodynamic temperature of the triple point of water. (Defined at the 13th CGPM, 1967).

- A *mole* is the amount of substance of a system that contains as many elementary entities as there are atoms in 0.012 kilogram of carbon 12. NOTE: When the mole is used, the elementary entities must be specified and may be atoms, molecules, ions, electrons, other particles, or specified groups of such particles. (Defined at the 14th CGPM, 1971).

- A *candela* is the luminous intensity, in a given direction, of a source that emits monochromatic radiation of $540 \times 10^{12}$ hertz and has a radiant intensity in that direction of $1/683$ watt per steradian. (Defined at the 16th CGPM, 1979).

Additionally defined are the following nondimensional units:

- A *radian*, symbol rad, is the plane angle between two radii of a circular arc that has a length equal to the radius.

- A *steradian*, symbol sr, is the solid angle such that where its vertex is at the center of a sphere, cuts off a surface area on the spherical surface equal to that of a square with sides of arc length equal to the radius of the sphere.

In discussions of the SI base and derived units, forms of the words *realize* and *represent* are often used (for example, *realized, realization, representation,* and so on.) While similar, the meanings are different in practice. A *unit* can always be represented (with some level of uncertainty) by a laboratory with sufficient expertise and the proper equipment, but it is often impossible to directly realize the definition of a unit with the current state of the art. For example, to realize (to bring into concrete existence [2]) the definition of the ampere requires a pair of infinitely long conductors and the ability to measure the force between them; likewise, to realize the definition of the mole requires an exact count of the number of atoms in 0.012 kg of pure carbon 12. Both are clearly impossible with today's technology. We can, however, produce an ampere of current, and we can determine a molar quantity by mass ratios. As explained by Dr. Barry Taylor,

> To realize the definition of the ampere and to realize the unit "ampere" are not necessarily the same. To realize the definition of the ampere requires one to measure the force between current carrying conductors since that is how it is defined; to realize the unit "ampere" one can use any means at one's disposal. For example, by combining a Josephson effect voltage standard and a quantum Hall effect resistance standard one can realize the unit "ampere" with a much smaller uncertainty than one can realize the definition of the ampere. The bottom line is use "representation" for the embodiment of a unit. . . .[3]

In a metrology laboratory, therefore, the measurement standards are representations of the SI units. This is true whether you are using an intrinsic standard such as a Josephson junction array or a triple point of water cell, or if you are using a standard such as a multifunction calibrator or gage block. In almost all cases, the use of *realize* or *realization* should be avoided. The principal exception is when referring to the prototype kilogram at BIPM. Since mass is the only unit still defined by an artifact, the prototype kilogram is both a realization and a representation.

## COHERENCE OF SI UNITS

The SI system of units is called *coherent* because all other units are derived from these *base units* by the rules of multiplication and division, with no numerical factor other than unity. Hence, all SI-recognized units are direct (by multiplication and/or division) combinations of base units. Power of 10 SI prefixes are added for representing larger and smaller magnitudes, making the entire SI system decimally based. One exception is time, which still has recognized derived units from the second of minute, hour, day, and so on.

(Note that chronological time, which still has recognized derived units from the second of minute, hour, day, and year, is a set of non-SI units that are acceptable for use along with the SI.[4] The second itself is a unit of time interval and is the interval that the International Atomic Time (TAI) scale is based on. TAI is not practical for most chronological uses, so Universal Coordinated Time (UTC) has been developed. UTC is derived from TAI but is kept in step with the rotation of the Earth by adding leap seconds as needed and is the reference for chronological time.)

This decimal basis is a quite distinctive feature of the SI system of units. For comparison, customary units often are products of doubles (for example, pint vs. quart), triples (for example, foot vs. yard), combinations of these (as in inch vs. foot), sexagesimal (degree, minute, second—with origins as early as ancient Babylonian usage), and other integral and nonintegral multiples. In fact, the decimal system's attractiveness is based on the great simplification that it provides over competing unit systems, providing for this coherence that, with SI prefixes and scientific notation (see Chapter 24) yields expression of wide ranges of magnitudes in a way that is easily understood and compared.

## SI DERIVED UNITS

A significant number of units have been developed to satisfy specific application needs that are based on these base SI units. Some of these derived units have names expressed in terms of the base units from which they are formed.

Table 25.1 shows some of these units and how they are derived from base units. Table 25.2 shows other derived units with specialized names and symbols. Other derived units are themselves expressed in terms of derived units, some with special names, as the examples listed in Table 25.3.

**Table 25.1** SI units derived from base.

| Quantity | SI-derived unit | |
| --- | --- | --- |
| | Name | Symbol |
| area | square meter | $m^2$ |
| volume | cubic meter | $m^3$ |
| speed, velocity | meter per second | m / s |
| acceleration | meter per second squared | $m / s^2$ |
| wave number | reciprocal meter | $m^{-1}$ |
| mass density (density) | kilogram per cubic meter | $kg / m^3$ |
| specific volume | cubic meter per kilogram | $m^3 / kg$ |
| current density | ampere per square meter | $A / m^2$ |
| magnetic field strength | ampere per meter | A / m |
| amount-of-substance concentration (concentration) | mole per cubic meter | $mol / m^3$ |
| luminance | candela per square meter | $cd / m^2$ |

**Table 25.2** Derived units with specialized names and symbols.

| Derived quantity | SI-derived unit | | Expression in terms of | |
| --- | --- | --- | --- | --- |
| | Name | Symbol | other SI units | base SI units |
| absorbed dose, specific energy imparted, kerma | gray | Gy | J / kg | $m^2\ s^{-2}$ |
| activity (of a radionuclide) | becquerel | Bq | | $s^{-1}$ |
| angle, plane | radian rad | | | $m\ m^{-1} = 1$ |
| angle, solid | steradian | sr | | $m^2\ m^{-2} = 1$ |
| capacitance | farad | F | C / V | $m^{-2}\ kg^{-1}\ s^4\ A^2$ |
| catalytic activity | katal | kat | | $mol\ s^{-1}$ |
| Celsius temperature | deg Celsius | C° | | K |

*continued*

*continued*

| | | | | |
|---|---|---|---|---|
| dose equivalent; ambient, directional, or equivalent dose | sievert | Sv | J / kg | $m^2\ s^{-2}$ |
| electric charge, quantity of electricity | coulomb | C | | s A |
| electric conductance | siemens | S | A / V | $m^{-2}\ kg^{-1}\ s^3\ A^2$ |
| electric potential, potential difference, electromotive force | volt | V | W / A | $m^2\ kg\ s^{-3}\ A^{-1}$ |
| electric resistance | ohm | Ω | V / A | $m^2\ kg\ s^{-3}\ A^{-2}$ |
| energy, work, quantity of heat | joule | J | N m | $m^2\ kg\ s^{-2}$ |
| force | newton | N | | $m\ kg\ s^{-2}$ |
| frequency (of a periodic phenomenon) | hertz | Hz | | $s^{-1}$ |
| illuminance | lux | lx | lm / $m^2$ | $m^2\ m^{-4}\ cd = m^{-2}\ cd$ |
| inductance | henry | H | Wb / A | $m^{-2}\ kg\ s^{-2}\ A^{-2}$ |
| luminous flux | lumen | lm | cd sr | $m^2 m^{-2} cd = cd$ |
| magnetic flux | weber | Wb | V s | $m^2\ kg\ s^{-2}\ \Omega\ A^{-1}$ |
| magnetic flux density | tesla | T | Wb / $m^2$, N / (AΩm) | $kg\ s^{-2}\ A^{-1}$ |
| power, radiant flux | watt | W | J / s | $m\ kg\ s^{-3}$ |
| pressure, stress | pascal | Pa | N / $m^2$ | $kg\ m^{-1} s^{-2}$ |

**Table 25.3** Other derived units.

| Derived Quantity | SI-derived unit | | Expression in terms of SI base units |
|---|---|---|---|
| | Name | Symbol | |
| absorbed dose rate | gray per second | Gy / s | $m^2\ s^{-3}$ |
| angular acceleration | radian per second squared | rad / $s^2$ | $m\ m^{-1}\ s^{-2} = s^{-2}$ |
| angular velocity | radian per second | rad / s | $m\ m^{-1}\ s^{-1} = s^{-1}$ |
| electric charge density | coulombs per cubic meter | C / $m^3$ | $m^{-3}\ s\ A$ |
| electric field strength | volt per meter | V / m | $m\ kg\ s^{-3}\ A^{-1}$ |
| electric field strength | newtons per coulomb | N / C | $m\ kg\ s^{-3}\ A^{-1}$ |
| electric flux density | coulomb per square meter | C / $m^2$ | $m^{-2}\ s\ A$ |
| energy density | joule per cubic meter | J / $m^3$ | $m^{-1}\ kg\ s^{-2}$ |

*continued*

*continued*

| | | | |
|---|---|---|---|
| entropy | joule per kelvin | J / K | $m^2$ kg $s^{-2}$ $K^{-1}$ |
| exposure (x and g rays) | coulomb per kilogram | C / kg | $kg^{-1}s$ A |
| heat capacity | joule per kelvin | J / K | $m^2$ kg $s^{-2}$ $K^{-1}$ |
| heat flux density, irradiance | watt per square meter | W / m2 | kg $s^{-3}$ |
| molar energy | joule per mole | J / mol | m kg $s^{-2}$ $mol^{-1}$ |
| molar entropy molar heat capacity | joule per mole kelvin | J / (mol K) | m kg $s^{-2}$ $K^{-1}$ $mol^{-1}$ |
| moment of force | newton meter | N m | $m^2$ kg $s^{-2}$ |
| permeability (magnetic) | henry per meter | H / m | m kg $s^{-2}$ $A^{-2}$ |
| permittivity | farad per meter | F / m | $m^{-3}$ $kg^{-1}s^4$ $A^{-2}$ |
| power density | watt per square meter | W / m2 | kg $s^{-3}$ |
| radiance | watt per square meter steradian | W / (m2 sr) | kg $s^{-3}$ |
| radiant intensity | watt per steradian | W / sr | $m^2$ kg $s^{-3}$ |
| specific energy | joule per kilogram | J / kg | $m^2$ $s^{-2}$ |
| specific entropy | joule per kilogram kelvin | J / (kg K) | $m^2$ $s^{-2}$ $K^{-1}$ |
| specific heat capacity, specific entropy | joule per kilogram kelvin | J / (kg K) | $m^2$ $s^{-2}$ $K^{-1}$ |
| surface tension | newton per meter | N / m | kg $s^{-2}$ |
| surface tension | joule per square meter | J / $m^2$ | kg $s^{-2}$ |
| thermal conductivity | watt per meter kelvin | W / (m K) | m kg $s^{-3}$ $K^{-1}$ |
| viscosity, dynamic | pascal second | Pa s | $m^{-1}$ kg $s^{-1}$ |
| viscosity, kinematic | square meter per second | $m^2$ / s | $m^2$ $s^{-1}$ |

## SI Prefixes

Table 25.4 lists currently recognized SI unit prefixes. Some common examples of use of these SI prefixes are the following:

      A ruler that is graduated in 0.5 centimeter units, or 0.5 cm graduations

      A microprocessor operating at 4 gigahertz, or 4 GHz

      A voltmeter that has a resolution to 10 nanovolts direct current, or 10 nV DC

      A power meter that measures in 5 kilowatt units, or 5 kW units

      A current meter that is sensitive to 2.0 femtoamperes, or 2.0 fA

      A pressure gage that is accurate to 5 kilopascal, or 5 kPa

**Table 25.4** Currently recognized SI unit prefixes.

| Multiplication factor | Prefix | Symbol |
|---|---|---|
| $10^{24}$ | yotta | Y |
| $10^{21}$ | zetta | Z |
| $10^{18}$ | exa | E |
| $10^{15}$ | peta | P |
| $10^{12}$ | tera | T |
| $10^{9}$ | giga | G |
| $10^{6}$ | mega | M |
| $10^{3}$ | kilo | k |
| $10^{2}$ | hecto | h |
| $10^{1}$ | deka | da |
| $10^{-1}$ | deci | d |
| $10^{-2}$ | centi | c |
| $10^{-3}$ | milli | m |
| $10^{-6}$ | micro | μ |
| $10^{-9}$ | nano | n |
| $10^{-12}$ | pico | p |
| $10^{-15}$ | femto | f |
| $10^{-18}$ | atto | a |
| $10^{-21}$ | zepto | z |
| $10^{-24}$ | yocto | y |

Note that there is a major difference between the meaning of an uppercase and a lowercase letter. For example, M is $10^6$, but m is $10^{-3}$. The correct case must be used or the magnitude of the number will be incorrect.

These SI prefixes may be used singly as prefixes to base or named derived units. In other words, compound prefixes are not permitted. 1.45 nm should be used instead of 1.45 mμm, and 4.65 pF instead of 4.65 μμF.

## SI-Named Units That Are Multiples of SI Base or Derived Units

There are a number of widely used units that are multiples or submultiples of either SI base units or SI-derived units. Many of these have special names as well. Table 25.5 is a partial list of these.

## Units Not to Be Used Within the SI System of Units

SI has designated certain units as not to be used. These units, many of which are often found in literature, are discouraged because equivalent SI units are available and are now preferred. Some of these prior units to be avoided are cgs-based (centimeter-gram-second) special units, such as: lambert, emu, esu, gilbert, biot, franklin, and names prefixed with *ab-* or *stat-*. Table 25.6 shows all of the following units are to be avoided within SI usage.

**Table 25.5** SI-named units that are multiples of SI base or derived units.

| Quantity | SI derived unit | | Value in SI Units |
|---|---|---|---|
| | Name | Symbol | |
| angle, plane | degree | ° | $\pi / 180$ rad |
| angle, plane | minute | ' | $(1/60)° = \pi / 10\,800$ rad |
| angle, plane | second | " | $(1/60)' = \pi / 648\,000$ rad |
| angle, plane | revolution, turn | r | $2\pi$ rad |
| area | hectare | ha | $hm^2 = 10^4\,m^2$ |
| mass | metric ton | t | $Mg = 10^3$ kg |
| time | minute | min | 60 s |
| time | hour | h | 60 min = 3 600 s |
| time | day | d | 24 h = 86 400 s |
| volume | liter | L | $dm^3 = 10^{-3}\,m^3$ |

**Table 25.6** Units not to be used within SI system of units.

| Symbol | Unit Name | Value in SI units |
|---|---|---|
| A° | angstrom | $10^{-10}$ m |
| a | are | 100 m² |
| atm | atmosphere, standard* | 101.325 kPa |
| at | atmosphere, technical | 98.0665 kPa |
| bar | bar | 100 kPa |
| barn | b | $10^{-28}\,m^2$ |
| cal | calorie, physics | 4.184 J |
| Cal | calorie, nutrition | 4.184 kJ |
| | candle | cd |
| cp | candlepower | cd |
| dyn | dyne | $10^{-5}$ N |
| erg | erg | $10^{-7}$ J |

*continued*

*continued*

| | | |
|---|---|---|
| fermi | fermi | $10^{-15}$ m |
| $g_n$ (G, g) | gravity, standard acceleration due to* | 9.80665 m / s$^2$ |
| Gal | gal | $10^{-2}$ m/s$^2$ |
| γ | gamma | nT = $10^{-9}$ T |
| G | gauss | $10^{-4}$ T |
| gon | gon, grad, grade | ($\pi$ / 200) rad |
| kcal | kilocalorie | 4.184 kJ |
| kgf | kilogram force * | 9.80665 N |
| 1000 L | kiloliter | m$^3$ |
| cal / cm2 | langley | 4.184 x 10$^4$ J/m$^3$ |
| Mx | maxwell | $10^{-8}$ Wb |
| | metric carat | 2 x $10^{-4}$ kg |
| | metric horsepower | 735.4988 W |
| μ | micron | $10^{-6}$ m |
| mmHg | millimeter of mercury | 133.3 Pa |
| mμ | millimicron | $10^{-9}$ m |
| mho | mho | S |
| Oe | oersted | (1000 / 4$\pi$) A/m |
| ph | phot | $10^4$ lx |
| P | poise | 0.1 Pa s |
| st | stere | m$^3$ |
| sb | stilb | $10^4$ cd / m2 |
| St | stokes | cm$^2$ / s |
| Torr | torr | (101325 / 760) Pa |
| xu (Cu Ka1) | Cu x unit | 1.00207703 x $10^{-13}$ m |

## SI Units Arranged by Unit Category

SI units in particular, relate to the physical world we live in. They represent quantities that may be categorized by the general kind of unit, or the branch of science, that they are related to. Table 25.7 shows the current base and derived units arranged by unit category.

**Table 25.7** SI units arranged by unit category.

| Unit Category | Quantity | Symbol | Name |
|---|---|---|---|
| **electricity & magnetism** | | | |
| | charge, electric, electrostatic | A-hr | ampere-hour |
| | charge, electric, electrostatic, quantity of electricity | C | coulomb |
| | current density | A / m$^2$ | ampere per square meter |
| | electric capacitance | F | farad |
| | electric charge density | C / m$^3$ | coulomb per cubic meter |
| | electric current | A | ampere |
| | electric dipole moment | C - m | coulomb meter |
| | electric field strength | N / C | newton per coulomb |
| | electric field strength | V / m | volt per meter |
| | electric flux density | C / m$^2$ | coulomb per square meter |
| | electric inductance | H | henry |
| | electric potential difference, electromotive force | V | volt |
| | electric resistance | W | ohm |
| | electrical conductance | S | siemens |
| | inductance, electrical | H | henry |
| | magnetic constant | N / A$^2$ | newton per square ampere |
| | magnetic field strength | A / m | ampere per meter |
| | magnetic flux | Wb | weber |
| | magnetic flux density, induction | T | tesla |
| | magnetic flux density, induction | Wb / m$^2$ | weber per square meter |
| | magnetomotive force | A | ampere |
| | magnetomotive force | A | ampere-turn |
| | magnetomotive force | Oe-cm | oersted-centimeter |
| | permeability (magnetic) | H / m | henry / meter |
| | permittivity | F / m | F / m |
| | resistance length | W-m | ohm meter |
| | resistance, electrical | W | ohm |
| **light** | | | |
| | illuminance | lm / m$^2$ | lumen per square meter |
| | illuminance | lx | lux |
| | irradiance, heat flux density, heat flow rate / area | W / m$^2$ | watt per square meter |

*continued*

| | | | |
|---|---|---|---|
| **light** | luminance | cd / m² | candela per square meter |
| | luminous flux | lm | lumen |
| | luminous intensity | cd | candela |
| | radiance | W / (m² - sr) | watt per square meter steradian |
| | radiant intensity | W / sr | watt per steradian |
| **mechanics** | | | |
| | angular momentum | kg - m² / s | kilogram meter squared per second |
| | catalytic activity | kat | katal |
| | coefficient of heat transfer | W / (m² K) | watt per square meter-kelvin |
| | concentration (of amt. of substance) | mol / m³ | mole per cubic meter |
| | density, mass / volume | kg / m³ | kilogram per cubic meter |
| | density, mass / volume | kg / L | kilogram per liter |
| | density, specific gravity | kg / m³ | kilogram per cubic meter |
| | energy | eV | electron volt |
| | energy | J | joule |
| | energy | W s | watt second |
| | energy density | J / m3 | joule per cubic meter |
| | energy per area | kJ / m² | kilojoule per square meter |
| | energy per mass | J / kg | joule per kilogram |
| | energy per mass, specific energy | J / kg | joule per kilogram |
| | energy per mole | J / mol | joule per mole |
| | energy, work | N – m | newton-meter |
| | force | N | newton |
| | force per length | N / m | newton per meter |
| | frequency (periodic phenomena) | Hz | cycles per second |
| | fuel economy, efficiency | L / (100 km) | liter per hundred kilometer |
| | heat capacity, enthalpy | J / K | joule per kelvin |
| | heat capacity, entropy | J / K | joule per kelvin |
| | heat flow rate | W | watt |
| | insulance, thermal | K m² / W | kelvin square meter per watt |
| | linear momentum | kg – m / s | kilogram meter per second |
| | mass | kg | kilogram |
| | mass | t | ton (metric), tonne |
| | mass | u | unified atomic mass unit |
| | mass / area | kg / m² | kilogram per square meter |
| | mass / energy | kg / J | kilogram per joule |

*continued*

| | | | |
|---|---|---|---|
| **mechanics** | mass / length | kg / m | kilogram per meter |
| | mass / time | kg / s | kilogram per second |
| | mass / volume | kg / m³ | kilogram per cubic meter |
| | mass per mole | kg / mol | kilogram per mole |
| | molar entropy, molar heat capacity | J / (mol – K) | joule per mole kelvin |
| | molar gas constant | R | J mol$^{-1}$ K$^{-1}$ |
| | molar heat capacity | J / (mol-K) | joule per mole kelvin |
| | moment of force, torque, bending moment | N-m | newton meter |
| | moment of inertia | kg – m² | kilogram meter squared |
| | moment of section | m⁴ | meter to the fourth power |
| | per viscosity, dynamic | 1 / (Pa-s) | per pascal second |
| | permeability | m² | square meter |
| | power | W | watt |
| | power density, power / area | W / m² | watt per square meter |
| | power density, power / area | W / sr | watt per steradian |
| | pressure, stress | kg / m³ | kilogram per cubic meter |
| | pressure, stress | Pa | pascal |
| | quantity of heat | J | joule |
| | section modulus | m³ | meter cubed |
| | specific heat capacity, specific entropy | J / (kg-K) | joule per kilogram kelvin |
| | specific volume | m³ / kg | cubic meter per kilogram |
| | surface tension | J / m² | joule per square meter |
| | temperature | K | kelvin |
| | thermal conductance | W / (m²-K) | watt per square meter-kelvin |
| | thermal conductivity | W / (m-K) | watt per meter-kelvin |
| | thermal diffusivity | m² / s | square meter per second |
| | thermal insulance | K m² / W | kelvin square meter per watt |
| | thermal resistance | K / W | kelvin per watt |
| | thermal resistivity | K m / W | kelvin meter per watt |
| | thrust / mass | N / kg | newton per kilogram |
| | viscosity, dynamic | mk | kilogram per meter-second |
| | viscosity, dynamic | mN | newton-second per meter squared |
| | viscosity, dynamic | Pa-s | pascal second |
| | viscosity, kinematic | m² / s | meter squared per second |
| | volume / energy | m³ / J | cubic meter per joule |

*continued*

*continued*

| | | | |
|---|---|---|---|
| **other** | | | |
| | count | n/a | n/a |
| **radiology** | | | |
| | absorbed dose | Gy | gray |
| | absorbed dose | J / kg | joule per kilogram |
| | absorbed dose rate | Gy / s | gray per second |
| | activity | Bq | becquerel |
| | activity | s$^{-1}$ | per second (disintegration) |
| | dose equivalent | Sv | sievert |
| | exposure (x and gamma rays) | C / kg | coulomb per kilogram |
| **space & time** | | | |
| | acceleration, angular | rad / s$^2$ | radian per second squared |
| | acceleration, linear | m / s$^2$ | meter per second squared |
| | angle, plane | ° | degree |
| | angle, plane | ' | minute (arc) |
| | angle, plane | rad | radian |
| | angle, plane | r | revolution |
| | angle, plane | " | second |
| | angle, solid | sr | steradian |
| | area, plane | ha | hectare |
| | area, plane | hm$^2$ | square hectometer |
| | area, plane | m$^2$ | square meter |
| | length | m | meter |
| | time | d | day (24h) |
| | time | h | hour |
| | time | min | minute |
| | time | s | second |
| | velocity / speed | m / s | meter per second |
| | velocity, angular | rad / s | radian per second |
| | velocity, angular | r / s | revolution per second |
| | volume / time, (flow rate) | m$^3$ / s | cubic meter per second |
| | volume / time, (flow rate) | L / s | liter per second |
| | volume / time, (leakage) | slpm | standard liter per minute |
| | volume, capacity | m$^3$ | cubic meter |
| | volume, capacity | L | liter |
| | wavenumber | m$^{-1}$ | per meter |

## CONVERSION FACTORS AND THEIR USES

Equivalent quantities that are stated in different units are related to each other by a conversion factor that is the ratio of the differing units. Because of the need to state quantities in units required by a user or specifying organization, the conversion of a quantity from one unit to another occurs often.

A conversion factor is based on the ratio of the units between which a conversion is needed. For instance, the often-required conversion from inches to millimeters involves multiplying a number in inches by the conversion factor 25.4 to obtain the equivalent number in millimeters.

Appendix E lists a number of frequently required unit conversions between various customary units, SI units, and also between many customary and SI units. The conversion factors given are accurate to the number of decimal places shown or, where boldfaced, are exact conversion factors with zero.

The format for conversion factor listings most often follows one of two formats: the from/to list and from/to matrix table. In Appendix E, the from/to list is given for a wide range of common required unit conversions.

### From / To Lists

The following example will show how these lists are arranged:

| To convert from | to | multiply by: |
|---|---|---|
| inch | millimeter | **2.54 E+01** |
| circular mil | square meter (m2) | 5.067 075 E-10 |

**Table 25.8** Conversion matrix: plane angle units.

| From | | | | To Output Value | | | | | | |
|---|---|---|---|---|---|---|---|---|---|---|
| Symbol | Name | deg | grad grade gon | mil | min | rad | rev | quad | sec |
| deg | degree | 1 | 1.1111 | 17.7778 | 60 | 0.0175 | 0.0028 | 0.0111 | 3600 |
| grad, grade, gon | grad, grade, gon | 0.9000 | 1 | 16 | 0.015 | 0.0157 | 0.0025 | 0.01 | 0.0003 |
| mil | mil | 0.0563 | 0.0625 | 1 | 0.0009 | 0.001 | 0.0002 | 0.0006 | 2E-05 |
| min | minute | 0.0167 | 66.667 | 1066.67 | 1 | 0.0003 | 5E-05 | 0.0002 | 60 |
| rad | radian | 57.2958 | 63.662 | 1018.59 | 3437.7 | 1 | 0.1592 | 0.6366 | 206265 |
| rev | revolution | 360 | 400 | 6400 | 21600 | 6.2832 | 1 | 4 | 1E+06 |
| quad | quadrant | 90 | 100 | 1600 | 5400 | 1.5708 | 0.25 | 1 | 324000 |
| sec | second | 0.0003 | 4000 | 64000 | 0.0167 | 5E-06 | 8E-07 | 3E-06 | 1 |

This example shows **2.54 E+01** in boldface. Remember, in these lists boldface values are exact conversions, with zero uncertainty or loss in resolution for any number of significant digits.

## From / To Matrix Table

In these tables, a matrix of from and to values are listed in column and row headings, with the conversion factors for each combination given at the intersection of the column and row. Table 25.8 is an example of how these tables are arranged.

## UNCERTAINTIES FOR SOME FUNDAMENTAL UNITS (STANDARD, AND RELATIVE)

Table 25.9 is compiled from NIST physics.nist.gov/cuu/Constants/index.html and is a table of frequently used constants.

## NUMBER BASES

A number base is the exponent base that defines the magnitude of each digit in a number, where the position of each digit determines its value by the relationship:

$$\text{Digit Value} = \text{Number} \times \text{base}^N,$$

where N is the number of positions to the left or right of the decimal marker, with those on the left being positive N and those on the right being negative N.

For example, for the decimal (base 10) number 453.0, the 5 has a value of 50 because

$$5 \times 10^2 = 50$$

Without thinking, we interpret the value of numbers by using this almost unconscious rule for decimal numbers. Some, to clarify the base that a number is related to, will list that base as a trailing subscript, as in $453.0_{10}$. 453.0 is interpreted as four hundred, fifty, and 3 units.

Other number bases that are used quite often in computer programming are binary, octal, and hexadecimal. Whereas in decimal numbers each digit can take on values from 0 to 9 (10 values), binary digits can take on only 0 and 1 (two values), octal 0 to 7 (eight values) and hexadecimal 0 to F (zero to nine plus A, B, C, D, E, F, or 16 values).

All SI units and the vast majority of customary unit measurements are made in decimal (base 10) units and, therefore, have a base of 10. Digital computers and calculators use binary numbers internally, however, and many computer programming languages use hexadecimal numbers.

## CONVERSION FACTORS FOR CONVERTING BETWEEN EQUIVALENT MEASUREMENT UNITS

Unit conversion factors, otherwise known as conversion factors, are made widely available in various forms and extents in texts, publications, and standards, and are often

**Table 25.9** Frequently used constants.

| NIST Category | Name | Symbol | Value | Units | Standard uncertainty | Relative standard uncertainty | Equivalent Equation |
|---|---|---|---|---|---|---|---|
| PC | Avogadro constant | $N_A$, "L" | 6.02214199E+23 | $mol^{-1}$ | 4.70E+16 | 7.80E-08 | |
| PC | Boltzmann constant | $k$ | 1.3806503E-23 | $J\,K^{-1}$ | 2.40E-29 | 1.74E-06 | $k = R/N_A$ |
| EM | conductance quantum | $G_0$ | 7.748091696E-05 | S | 2.80E-13 | 3.61E-09 | $G_0 = 2e^2/h$ |
| UN | electric constant | $e_0$ | 8.854187817E-12 | $F\,m^{-1}$ | 0 | 0 | |
| AN | electron mass | $\mu_e$ | 9.10938188E-31 | kg | 7.20E-38 | 7.90E-08 | |
| NS | electron volt | $eV$ | 1.60217646E-19 | J | 6.30E-27 | 3.93E-08 | $1\,eV = (e/C)\,J$ |
| EM | elementary charge | $e$ | 1.602176462E-19 | C | 6.30E-27 | 3.93E-08 | |
| PC | Faraday constant | $F$ | 96485.3415 | $C\,mol^{-1}$ | 0.0039 | 4.04E-08 | |
| AN | fine-structure constant | $\alpha$ | 7.297352533E-03 | | 2.70E-11 | 3.70E-09 | $\alpha = e^2/4\pi e_0 \hbar c$ |
| UN | magnetic constant (permeability in a vacuum) | $m_0$ | 1.2566370614E-06 | $N\,A^{-2}$ | 0 | 0 | |
| EM | magnetic flux quantum | $\Phi_0$ | 2.067833636E-15 | Wb | 8.1E-23 | 3.92E-08 | $\Phi_0 = h/2e$ |
| PC | molar gas constant | $R$ | 8.314472 | $J\,mol^{-1}\,K^{-1}$ | 0.000015 | 1.80E-06 | |
| UN | Newtonian constant of gravitation | $G$ | 6.673E-11 | $m^{-3}\,kg^{-1}\,s^{-2}$ | 1E-13 | 1.50E-03 | |
| UN | Planck constant | $h$ | 6.62606876E-34 | J s | 5.2E-41 | 7.85E-08 | |
| UN | Planck constant over 2 pi | $\hbar$ | 1.054571596E-34 | J s | 8.2E-42 | 7.78E-08 | |
| AN | proton mass | $m_p$ | 1.67262158E-27 | kg | 1.3E-34 | 7.77E-08 | |
| AN | proton-electron mass ratio | $m_p/m_e$ | 1836.1526675 | | 3.9E-06 | 2.12E-09 | |
| AN | Rydberg constant | $R_x$ | 10973731.568549 | $m^{-1}$ | 8.3E-05 | 7.56E-12 | $R_x = a^2 m_e c / 2h$ |
| NS | speed of light in vacuum | $c$, $c_0$ | 299792458 | $m\,s^{-1}$ | 0 | 0 | |
| PC | Stefan-Boltzmann constant | $\sigma$ | 5.6704E-08 | $W\,m^{-2}\,K^{-4}$ | 4E-13 | 7.05E-06 | $\sigma = (\pi^2/60)\,k^4/\hbar^3 c^2$ |
| NS | unified atomic mass unit | 1 u | 1.66053873E-27 | kg | 1.3E-34 | 7.83E-08 | $1\,u = m_u =$ $(1/12)m(12C) = (10^{-3}\,kg\,Mol^{-1})$ |

supplied as appendices in catalogs of manufactured products, as well as being provided in specific measurement instrumentation catalogs.

Conversion factors typically follow the format:

To convert from (first) given unit to (second) desired unit, multiply (first unit) measurement by conversion factor given. Generally such conversion factors consist of dividing the relationship of second unit to the first unit.

Appendix E, contains conversion factors for many common sets of units. Those listed are all those found in both SI-10, and in NIST Special Publication 811, over 500 factors in all. Two lists are provided. The first is conversion factors are sorted by category and then by alphabetical unit. This table is useful if you know the category of physical unit and are looking for options within that category. The other list, where conversion factors sorted by alphabetical unit only, is useful if you know the unit you want to convert from and to, irrespective of the physical unit's category. An example of the second table is as follows:

| To Convert | From | | To | | Multiply by |
|---|---|---|---|---|---|
| Category | Unit A | | Unit B | | Factor = Unit B / Unit A |
| | Symbol | Name | Symbol | Name | |
| electric current | abA | abampere | A | ampere | 10 |
| electric capacitance | abF | abfarad | F | farad | 1000000000 |
| electric inductance | abH | abhenry | H | henry | 1000000000 |

As an example, using this table: To convert a 25 abampere value to amperes, multiply by the conversion factor shown, 10, giving 250 amperes. These tabled values are computed from fundamental units and are accurate to the number of digits shown in the table. Generally these factors are accurate to at least 10 decimal places, with many to 15 decimal places. They are computed using the base factor relationships found in the CD spreadsheet tool. The vast majority of factors have zero uncertainty, so the percent uncertainty of the given value determines the related percent uncertainty of the converted value. The few conversion factors that are empirically, rather than rule-based, are shown with their relative uncertainties. References to the source documents are provided as well.

## THE FOLLOWING ARE ADDITIONAL NOTES FOR USING APPENDIX IV AND V ON CD-ROM

The CD included with this book contains a Microsoft Excel workbook file labeled Appendix IVa–IVc. It contains the following worksheets:

Unit Conversion

Base Factors

References

Properties

NIST

CODATA

All worksheets as provided are password protected, with the initial password given in the cover worksheet. Those worksheets that should not be sorted have cautionary statements at the top of each worksheet.

## Unit Conversion

This worksheet contains all conversion factors listed in Appendix V and are automated such that one may easily enter a given unit to be converted and directly obtain the desired conversion unit value or convert back from an output unit to the original input unit value. When no input value is entered, the output cell is blanked.

Units shown in blue font are SI units, those in boldfaced blue font are preferred SI units. For example, meter is a preferred SI unit of length, whereas centimeter is only a SI unit (not preferred).

Conversion factors in green font are based on direct-stated base unit conversions (see workbook tab "Base Factors." Conversion factors in black font are computed using two or more base factors.

Since the factors shown are based on base factors that are generally exact (the uncertainty, $U = u / y$ column shows the uncertainty value: most are zero), the conversion resulting from use of this workbook is highly accurate. It even contains a number of error corrections for values found in the most prominent current listed conversion factor tables.

## Base Factors

This worksheet contains the defining rules for converting from one unit to another for those unit conversions listed in SI 10 and NIST SP 811. Most listed units will be seen to have zero uncertainty. This basically means that the originating or built-up conversion factor is exact. Standard ($u$) and relative ($U = u / y$) uncertainties are given for those few units that are not exact conversions.

The source for each of the base factors listed is shown by a reference number, found on another worksheet.

*Do not sort this worksheet* (as is cautioned at the top), as these values are linked to the base factors and unit conversion worksheets, and all links would be lost.

## References

This worksheet contains the source references for both the base factors and unit conversions listed in the CD workbook, arranged by reference number.

## Properties

This worksheet contains some commonly occurring empirical values of densities of water and mercury, used in pressure unit conversions.

*Do not sort this worksheet* (as is cautioned at the top), as these values are linked to the base factors and unit conversion worksheets, and all links would be lost.

**NIST**

This worksheet contains the physical constants published by NIST that are also found on the NIST Website.

*Do not sort this worksheet* (as is cautioned at the top), as these values are linked to the base factors and unit conversion worksheets, and all links would be lost.

This worksheet contains a complete listing of the 2000 CODATA fundamental unit constants. Most of these are empirical, with currently known Standard (u) and relative (U = u / y) uncertainties shown. Also shown are the base units on which these fundamental constants are based and their symbols.

*Do not sort this worksheet* (as is cautioned at the top), as these values are linked to the base factors and unit conversion worksheets, and all links would be lost if sorted.

## Endnotes

1. IEEE. Institute of Electrical and Electronic Engineers. IEEE 1451.2. *The representation of physical units.* West Conshohocken, PA: IEEE Press.
2. Gove, Philip Babcock, editor. 2002. *Webster's third new international dictionary, unabridged.* 3rd ed. New York: Merriam-Webster. On-line edition, www.m-w.com
3. Taylor, Barry N. 1995. *NIST Special Publication 811, 1995 edition: Guide for the use of the International System of Units (SI).* Gaithersburg, MD: National Institute of Standards and Technology. physics.nist.gov/Pubs/SP811/contents.html. (section 7.9 on SI prefixes).
4. Wildi, Theodore. 2001. *Metric units and conversion charts.* Weinheim, Germany: John Wiley & Sons.

# Chapter 26

# Ratios

## DECIBEL MEASURES

All measurements given in decibel are statements of a ratio between either a measurement (a) and a reference value (b) or between two measurements (a / b).

Decibels are measures stated using logarithms. A few simple rules of logarithms need to be reviewed first to enable better understanding calculations involving decibels.

A *logarithm* is an exponential function of an input variable (the number we supply) and an exponential base. The logarithm of a variable, x, is defined as the following function:

$$f(x) = \log_c(x), \text{ or conversely, } x = c^{f(x)}$$

Said another way, let y = f (x). Then

$$y = \log_c(x), \text{ or conversely, } x = c^y$$

Here it is required that x > 0, c > 0 and not = 1

Some of the nice things about using logarithms is they make multiplication and division transform into simple addition and subtraction. Anyone that has used a slide rule knows this, as the scales on a slide rule were proportioned logarithmically. So,

$$\log_c(f) + \log_c(g) = \log_c(fg), \text{ or conversely, for } f = c^{y_1} \text{ and } g = c^{y_2},$$

we get $fg = c^{y_1 + y_2}$

and

$$\log_c(f) - \log_c(g) = \log_c(f/g), \text{ or conversely, for } f = c^{y_1} \text{ and } g = c^{y_2},$$

we get $f / g = c^{y_1 - y_2}$.

Similarly, we get

$$\log_c(f^n) = n \log_c(f), \text{ or conversely, for } f^n = (c^y)^n, \text{ we get } f^n = c^{n(y)} = c^{ny}.$$

One thing we note is that the argument for the log() function cannot be negative or zero.

The bel is named after Alexander Graham Bell, hence the capital $B$ used as the symbol. To express a value in bels, we use the logarithm function and the base of $c = 10$, giving:

$$B = \log_{10}(x), \text{ or conversely, } x = 10^B.$$

The general form of the equation for determining a value in decibels is

$$dB = 10 \log_{10}(a/b), \text{ or conversely, } a/b = 10^{dB/10}.$$

Where (often) the ratio being examined is a squared ratio (as in inputs to power calculations), this equation takes a slightly different form:

$$dB = 10 \log_{10}(a^2/b^2) = 20 \log_{10}(a/b), \text{ or conversely, } a/b = 10^{dB/20}.$$

With these two expressions we have the basis for calculating all decibel results. Where one of the input values (b) is a reference value, we merely supply that in the denominator.

Logarithms are useful because they are a shorthand for obtaining products and quotients of numbers. The following are useful decibel relationships:

| An increase / decrease * in dB | is equivalent to an increase / reduction by a factor (ratio) of |
|---|---|
| 3 | 2 |
| 6 | 4 |
| 10 | 10 |
| 20 | 100 |
| 30 | 1000 |
| n × 10 | $10^n$ |

* increasing dBs are positive, decreasing dBs are negative.

From this we immediately conclude that:

A 56 dB measurement describes a value of a parameter that is roughly twice the magnitude of a 53 dB measurement.

A 46 dB measurement describes a value of a parameter that is roughly on tenth the magnitude of a 56 dB measurement.

A 46 dB measurement describes a value of a parameter that is roughly 1000 times the magnitude of a 16 dB measurement.

## An Example of Decibel Use

The power in a sound wave, generally, goes with the square of the peak pressure of the sound wave. Sound pressure level measurements are generally based on a reference sound pressure level of 20 micropascals, 20 µPa, 0.02 mPa, or $2.0^{-5}$ Pa.

Note: Alternatively, 1 microbar has also gained wide acceptance for calibration of certain transducers and sound measurements in liquids. Also, unless otherwise specifically given, sound pressure is taken to be the effective (RMS) pressure.

Sound power levels, distinct from sound pressure levels, are expressed in relation to a reference power level of one picowatt ($1.0 \times 10^{-12}$ W) exactly.

Unless otherwise explicitly stated, it is to be understood that the sound pressure value used is the effective (root-mean-square) of the measured sound pressure. Also, in many specific sound fields the sound pressure ratio is not the square root of the corresponding power ratio.

If a sound has a pressure level of $p_2$ = 1.00 Pa, to express this sound pressure level in decibels the result is obtained by use of the following formula:

$$\text{Sound pressure level, dBspl} = 20 \log (p_2 / p_1).$$

or, sound pressure level, dBspl) = 20 log (1.0 / 0.00002). = 94.0 dB

## Various Decibel Scales in Use

Decibel scales have been specialized for use in the fields of acoustics, electricity and electromagnetics, differing from each other by chosen reference value and units employed.

### dB (acoustics, sound power level)

Here, the reference level is 1 picowatt (1.0 pW). The corresponding level of sound intensity, $1.0 \times 10^{-12}$ W / m², corresponds to a sound power level of 0 dB, which is the lower limit threshold of normal hearing.

### dBA and dBC scales (acoustics)

The most widely used sound level filter is the A scale, which roughly corresponds to the inverse of the 40 dB (at 1 kHz) equal-loudness curve. Measurements are expressed in dBA, and sound meter readings on this scale are less sensitive to very high and low frequencies.

The C scale is nearly linear from 80 to 2.5 kHz, becoming less sensitive below and above this range. Another scale, B, is rarely used, and is midway between the A and C scales.

### dB (general, electrical signals)

When impedances are equal:

$$dB = 10 \log (P_2 / P_1) = 20 \log (E_2 / E_1) = 20 \log (I_2 / I_1).$$

When impedances are unequal:

$$dB = 10 \log (P_2 / P_1) = 20 \log [(E_2 \sqrt{Z_1}) / (E_1 \sqrt{Z_2})]$$

$$= 20 \log [(I_2 \sqrt{Z_2}) / (I_1 \sqrt{Z_1})]$$

Here, P refers to power in watts, rms; E refers to voltage, rms; I refers to current, rms; and Z refers to impedance in the general form, including inductance and capacitance.

### dB/bit (electrical signals)

$$dB/bit = 20 \log(2)/bit = \text{approximately } 6.02 \text{ dB} / \text{bit}$$

This is commonly used for specifying the dynamic range or resolution for pulse coded modulation (PCM) systems.

### dB/Hz (electrical signals)

This refers to dB measurements of relative noise power in a 1 Hz bandwidth and is used in determining a laser's relative intensity noise (RIN).

### dBi (electrical signals)

This refers to dB isotropic, with reference to defining antenna gain.

### dBm scale (electric signals, also dBmW)

$$dBm = 10 \log (W_2 / 1.0 \text{ mW}_{rms})$$

Here 1 milliwatt (1 mW) across a specified impedance is the reference level. For example, a 0 dBm signal in a circuit with an impedance of 600 $\Omega$ corresponds to approximately 0.7746 V rms.

In most cases, the specified reference impedance is assumed from the nature of the circuit.

- Audio and communications circuits and IM&TE typically use 600 $\Omega$. In audio circuits, this is also the same as indications on a VU (volume unit) meter.

- RF circuits and IM&TE typically use 50 $\Omega$.

- Cable television systems (and some other systems) and IM&TE typically use 75 $\Omega$.

If the impedance is different from one of these customary values, it must be explicitly stated.

### dBm scale (electric signals)

$$dBm = 10 \log (W_2 / 1.0 \text{ µW}_{rms})$$

Here 1 microwatt (1 µW) is the reference level.

### dBr (electrical signals)

$$dBr = 20 \log (V_2 / V_1).$$

Here the reference level is specified in the immediate context of the value.

### dBu (preferred) or dBv (electrical signals)

$$dBu = 20 \log (V_2 / 0.775 \text{ V}_{rms})$$

Here the reference level is defined as 0.775 volt, rms (0.775 V, rms), across any impedance. Compare to dBm.

### dBuV (electrical signals)

$$dBuV = 20 \log (V_2 / mV_{rms})$$

This is commonly used for specifying RF levels to a communications receiver. Here the reference level is defined as 1.0 microvolt, rms (1.0 µV, rms), across any impedance. Compare to dBm.

### dBV (electrical signals)

$$dBV = 20 \log (V_2 / V_1)$$

Here the reference level is 1 volt, rms (1.0 V, rms), across any impedance.

Also, the following is a useful table of dBm Equivalents

| | | |
|---|---|---|
| dBV | = | dBm − 13.0 |
| dBmV | = | dBm + 47.0 |
| dBuV | = | dBm + 107.0 |
| V | = | $10^{(dBm - 13.0)/20}$ |
| mV | = | $10^{(dBm + 47.0)/20}$ |
| uV | = | $10^{(dBm + 107.0)/20}$ |

### dBW scale (electric signals)

$$dBW = 10 \log (W_2 / 1.0 W_1)$$

Here the reference level is 1 watt (1.0 W). Usually the impedance is 50 Ω.

### DBW/K-Hz scale (electric signals)

$$DBW/K\text{-}Hz = 10 \log (k) / K\text{-}Hz = 10 \log (1.38065^{-23})$$
$$= -228.5991631 \, DBW/K\text{-}Hz.$$

Here the reference level is defined in relation to Boltzmann's constant. This is commonly used when analyzing carrier-to-noise (C/N) in communication links.

## LOGARITHMS IN MICROSOFT EXCEL

In Microsoft Excel there are three logarithmic functions available: LOG(number,base), LOG10(number), and LN(number).

In the LOG(number,base) function, number is the value you provide, and base is the base of the logarithm. If the base is omitted, then Excel assumes it to be 10.

In the LOG10(number) function, the base is 10. Actually, if you omit the base argument in LOG(), that is provide only LOG(number), the calculation assumes the base is 10 and computes identically to LOG10(number).

In the LN(number) function the base is the naperian constant, e, an irrational number that to nine decimal places is 2.71828183. We can obtain this constant within Excel, where needed, by using the Microsoft Excel formula "=EXP(1)." We will not consider the uses of this function.

## TYPES OF LINEARIZING TRANSFORMATIONS

In general, linearizing transforms convert one or more data variables in a way that results in the plot of a variable against another resulting in a graphical straight line relationship between pairs of the data sets. Exponential transforms raise or lower one or more data variables by an exponential power or fractional power such that the transformed data sets plot as a straight line when graphed against one another. In log transformations one or more data variables are transformed by taking the logarithm of the data variable, and when the log of the data is plotted against other variables the result is a linear (straight line) relationship.

## GRAPHS

Graphs are means of visualizing the shape of data sets when compared by plotting one data variable against another. Examples of kinds of plots are:

**Scatter plots.** These graphs plot one or more variable against a common reference variable. Examples of these are X – Y charts.

**Histograms.** These graphs show the aggregation of data in bins, which indicate the distribution of data items over a range of a given variable.

**Bar (or column) graphs.** These graphs show the relative magnitude of categories of variables, compared side-by-side. Bar graphs have the magnitudes compared by lengths of horizontal bars, one below the other. Column graphs, similar to bar graphs, show the same side-by-side comparison of categories, but the magnitudes are in a vertical direction.

**Pie (or doughnut) charts.** These graphs compare relative magnitudes of categories by their portion of a circle (360° being equal to the sum, 100 percent, of the magnitudes).

**Radar (or spider) graphs.** These graphs show comparative magnitudes of categories of data by the radial distance from a center point, arranged around a circle, with the adjacent endpoints connected.

**Bubble graphs.** These graphs compare three separate data variables: two variable ones and one categorical one. Categories of data are circles, whose location in an X – Y plane, and the size of the circle depict the three magnitudes.

**Three-D (or surface) graphs.** These graphs plot three continuous variables as a surface in two-D space.

**Line (or run) charts.** These graphs show the running relationship of one variable vs. another, by data item number or sequence (one form of which is by time).

Microsoft Excel has a rich variety of graphing wizards and tools for achieving a wide range of data graphing results. All of the above graph types can be easily constructed from tabular data in a few easy steps in Excel.

# Chapter 27
# Statistics

## FUNDAMENTAL MEASUREMENT ASSUMPTIONS

The four basic assumptions upon which the validity of all measurements depend are random distribution, fixed model, fixed variation, and fixed distribution. The majority of statistical tests depend very heavily on the assumption that the data were taken in a random manner. This is an important assumption because of the significant probability during measurement of the occurrence of aggregations, changes in condition, assignable causes, trends, cyclical effects, and progressive learning, and drift in natural, industrial, and human processes. The random distribution assumption is most particularly important when calculating statistics that are the basis of an inference or comparison. For example, taking a sample that will be the basis of making an inference as to the population it is drawn from. Another example of random distribution assumption is in experiments, where the results of one treatment are compared to that of another factor setting.

A commonly used definition for normality of data distributions is IIDN $(0, \sigma^2)$ (identically and independently distributed in a normal distribution with mean zero and variance $\sigma^2$).[1] This assumption is often made of data that have been randomly selected for use in statistical calculations.

## DEGREES OF FREEDOM

Degrees of freedom (often represented by $n$ or the Greek letter equivalent $v$, nu), related to the count of data items on which a statistic is based, refers to the number of ways a statistic's value can vary with a variation in each of the source data items. It is one of the first items determined in the calculation of many statistics. When the number of data items is very large or includes the total population of items being on which an inference is being made, the symbol N is often used. Conversely, when the degrees of freedom (DOF) relates to a statistic based on a sample from a larger population that is being described, the symbols $n$ or $v$ are most often used. This is to help identify the meaning to the user.

Because determining DOF generally involves summation of the number of data items being evaluated (minus a constant in some cases), the resulting DOF, being a count, is dimensionless (not associated with a unit). The determination of DOF depends on the net ways that a statistic may vary with variation in the data elements going into it. For instance, in the calculation of the mean statistic based on 20 values, the resulting statistic can vary 20 different ways with the change of any of the source data values. Hence the DOF for mean is merely the count, N, of the data items.

On the other hand, as mentioned, some statistics have a DOF that is the count of data items used in the statistic calculation minus a constant. For instance, in the calculation of the sample standard deviation, part of the calculation involves first calculating the mean. Because the mean statistic is part of the sample standard deviation calculation, one degree of freedom is lost due to the use of the mean and the resulting DOF is the data count minus one.

Since this is not a book on statistical theory, the formulas for various statistics will be given and their DOF basis merely listed without discussion as to the mathematical basis for it.

## RESIDUALS

A *residual* is the difference between the value of a data item and a statistic describing it. For instance, if the average of a set of measurements is x_bar = 4.583, and the first two values are $x_1 = 4.282$ and $x_2 = 4.632$, the residuals of these values is the difference between the measurement value and the average, or $r_1 = -0.101$, and $r_2 = 0.049$.

Because the operation involved in calculating residuals is subtraction, the resulting statistic has the same units as the data from which it is calculated. For instance, if measurements of voltage are used to calculate the mean and residuals for some or all of the data, the units for the residuals are volts.

## CENTRAL TENDENCY

This obviously is the case because the objective quite often is to understand where the central value is. In mathematical terms such statistics are said to determine central tendency. The primary statistics for describing central tendency are the mean, median and mode.

### Mean

One of the most common and often used or referred to statistics is the mean or average. Because the operations involved in calculating the mean are addition and division, the resulting statistic has the same units as the data from which it is calculated. The mean or average is given by the following equation:

$$\bar{x} = \frac{\sum_{i=1}^{n} x_i}{n}$$

For example, in the data series of amperage measurements,

13, 14, 23, 23, 32, 33, 45, 99, and 105 A

the mean (or average) value is 44.375 A.

Use Excel's AVERAGE() function. For example, for the data named range DATA, use the Excel function expression "=AVERAGE(DATA)."

## Median

The median statistic is the middle value in a set of data items, arranged in order of increasing value. That is, the median is the value where half the data items are less in value and half are greater value. More often we are hearing this statistic being used for economic and demographic purposes, as in median income, median house value, median age.

To determine the median of a set of measurements, arrange the values in order of increasing value. If the number of values in the data set is odd, the median is the middle value in this ordered set. If the number of values in the data set is even, the median is the average of the middle two values.

For example, in the data series of amperage measurements,

13, 14, 23, 23, 32, 33, 45, 99, and 105 A

the median (or middle) value is 28 A (midway between 23 and 32 A).

Use Excel's MEDIAN() function. For example, for the data named range DATA, use the Excel function expression "=MEDIAN(DATA)."

## Mode

The mode statistic describes the most often (in statistics, described as most frequently) occurring value.

For example, in the data series,

13, 14, 23, 23, 32, 33, 45, 99, and 105 A

the mode (or most often occurring) value is 23 A.

Use Excel's MODE() function. For example, for the data named range DATA, use the Excel function expression "=MODE(DATA)."

# BIMODAL DISTRIBUTIONS

Some distributions have either more than one mode or have frequencies of occurrence of values where the histograms indicate more than one peak.

Such distributions are not normal distributions, that is, distributions that likely fit a normal distribution model and where statistics based on the assumption of a normal distribution may not be valid. Figure 27.1 is a sample graph of bimodal distribution.

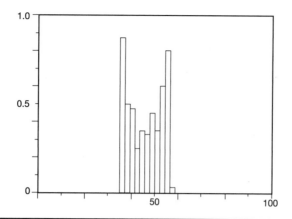

**Figure 27.1** Bimodal distribution.

## CENTRAL LIMIT THEOREM

The central limit theorem consists of three statements.[2]

1. The mean of the sampling distribution of means is equal to the mean of the population from which the samples were drawn.

2. The variance of the sampling distribution of means is equal to the variance of the population from which the samples were drawn divided by the size of the samples.

3. If the original population is distributed normally (in other words, bell shaped), the sampling distribution of means will also be normal. If the original population is not normally distributed, the sampling distribution of means will increasingly approximate a normal distribution as sample size increases (in other words, when increasingly large samples are drawn).

## ROOT MEAN SQUARE

The root mean square (RMS) statistic is most often associated with continuous variables, such as time. There are occasions where RMS is associated with discrete variables as well, such as specific measurements of a variable, for example, weight. Basically, as the name implies, RMS involves squaring the values of data, finding the mean (average) of these values, followed by taking a (square) root of this mean. As such, the unit of the calculation is the same as that of the data. For instance, if the RMS value of a set of data in feet is taken, the RMS result also has the unit, feet.

The unit of RMS is the same as the measurement unit. Root mean square (RMS) is given by the following equation:

$$RMS_{xx} = \sqrt{\frac{\sum_{i=1}^{n} Y^2}{n}}$$

For instance, in the data series of amperage measurements,

13, 14, 23, 23, 32, 33, 45, 99, and 105 A

the RMS value is 54.15 A, rounded to two decimal places.

## SUM OF SQUARES

Sum of Squares (SS) is often used as one step in ANOVA calculations. The unit of SS is the measurement unit squared. The sum of squares (SS) is given by the following equation(s):

$$SS_{xx} = \sum_{i=1}^{n} Y^2$$

or... $SS = X_1^2 + X_2^2 + X_3^2 + \ldots + X_n^2$

For instance, in the data series of amperage measurements,

13, 14, 23, 23, 32, 33, 45, 99, and 105 A

the SS value is 26,387 A², rounded to two decimal places.

## ROOT SUM OF SQUARES

The unit of RSS is the same as the measurement unit. The root sum of squares (RSS) is given by the following equation(s):

$$RSS_{XX} = \sqrt{\sum_{i=1}^{n} Y^2}$$

$$\text{or } RSS = \sqrt{X_1^2 + X_2^2 + X_3^2 + \ldots + X_n^2}$$

For instance, in the data series of amperage measurements,

13, 14, 23, 23, 32, 33, 45, 99, and 105 A

the RSS value is 162.44 A, rounded to two decimal places.

## VARIANCE

Unlike standard deviation, variances may be combined by addition. The population variance $V$ is given by the following equation:

$$V = \sigma^2 = \frac{\sum_{i=1}^{n}(X_i - \overline{X})^2}{n}$$

For example, in the data series of amperage measurements,

13, 14, 23, 23, 32, 33, 45, 99, and 105 A

the variance $V$ value is 1201.23 A, rounded to two decimal places.

Use Excel's VARP() function. For example, for the data named range DATA, use the Excel function expression "=VARP(DATA)."

$$\sigma = \sqrt{\frac{\sum_{i=1}^{n}(X_i - \overline{X})^2}{n}} = \sqrt{\frac{\sum_{i=1}^{n} X_i^2}{n} - \overline{X}}$$

For example, in the data series of amperage measurements,

13, 14, 23, 23, 32, 33, 45, 99, and 105 A

the population standard deviation $\sigma$ value is 34.66 A, rounded to two decimal places.

Use Excel's STDEVP() function. For example, for the data named range DATA, use the Excel function expression "=STDEVP(DATA)."

## SAMPLE VARIANCE

As noted, variances may be combined by addition. The sample variance $v$ is given by the following equation:

$$v = s^2 = \frac{\sum_{i=1}^{n}(X_i - \overline{X})^2}{n-1}$$

For example, in the data series of amperage measurements,

13, 14, 23, 23, 32, 33, 45, 99, and 105 A

the sample variance $v$ value is 1372.84 A, rounded to two decimal places.

Use Excel's VAR() function. For example, for the data named range DATA, use the Excel function expression "=VAR(DATA)."

## SAMPLE STANDARD DEVIATION

The sample standard deviation is used in the determination of several other statistics, for example, in determining a confidence interval and for hypothesis testing. The sample standard deviation $s$ is given by any of the following equations:

$$s = \sqrt{\frac{\sum_{i=1}^{n}(X_i - \overline{X})^2}{n-1}} \qquad s = \sqrt{\frac{\sum_{i=1}^{n}\left(X_i - \frac{\sum X_i}{n}\right)^2}{n-1}} \qquad s = \sqrt{\frac{n\sum_{i=1}^{n} X_i^2 - \left(\sum_{i=1}^{n} X_i\right)^2}{n(n-1)}}$$

For example, in the data series of amperage measurements,

13, 14, 23, 23, 32, 33, 45, 99, and 105 A

the sample standard deviation $s$ value is 37.05 A, rounded to two decimal places.

Use Excel's STDEV() function. For example, for the data named range DATA, use the Excel function expression "=STDEV(DATA)."

## STANDARD ERROR OF THE MEAN

Standard error of the mean (SEM) does not assume a normal distribution. Many applications of SEM do assume a normal distribution. For SEM, the larger the sample size the smaller the standard error of the mean. In other words, the size of SEM is inversely proportional to the square root of the sample size.

SEM, or s_y_bar, is given by the following formula:

$$s_{\bar{y}} = \frac{s}{\sqrt{n}}$$

where s = sample standard deviation and n = number of items in the sample.

For example, in this data series of amperage measurements, the sample standard deviation $s$ was 37.05 and the number of items $n$ was 8, giving the SEM to be 13.10, rounded to two decimal places.

## SKEWNESS

Data sets generally are not completely normally distributed. There may be more low values than high values, and so on, resulting in histograms of the data that are higher on one side or the other of the mean.

There are data sets where more values are below the mean value, and the graphed shape of the distribution appears to lean toward higher, more positive values. Such distributions have a positive skewness (the value of the distribution's skew is positive, or greater than zero).

There are other data sets where more values are above (at a more positive value than) the mean, and the distribution appears to lean toward the more negative values. Such distributions have a negative skewness (the value of the distribution's skew is negative, or less than zero).

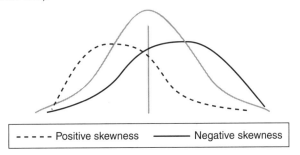

----- Positive skewness      ——— Negative skewness

Distribution skewness is determined by the following formula:

$$skew = \frac{n}{(n-1)(n-2)} \sum_{i=1}^{n} \left( \frac{x_i - \bar{x}}{s} \right)^3$$

For example, in the data series of amperage measurements,

13, 14, 23, 23, 32, 33, 45, 99, and 105 A

the skewness value is 1.15, dimensionless, rounded to two decimal places.

Use Excel's SKEW() function. For example, for the data named range DATA, use the Excel function expression "=SKEW(DATA)."

## KURTOSIS

Other data sets, not completely normally distributed, may have more values toward the tails than toward the mean, and so on, resulting in histograms of the data that are more rectangular-shaped than bell-curve–(normal distribution) shaped.

In the data sets where more values are toward the tails, and the distribution appears to be more rectangular or flat, the distribution is said to have negative kurtosis. Alternately, if more data is located closer to the mean, the distribution appears to be more triangular or peaked, and the distribution is said to have positive kurtosis.

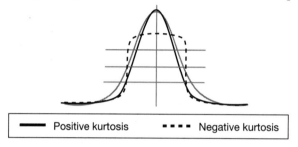

Distribution kurtosis is determined by the following formula:

$$kurtosis = \frac{n(n+1)}{(n-1)(n-2)(n-3)} \sum_{i=1}^{n} \left(\frac{x_i - \bar{x}}{s}\right)^4 - \frac{3(n-1)^4}{(n-2)(n-3)}$$

For example, in the data series of amperage measurements,

13, 14, 23, 23, 32, 33, 45, 99, and 105 A

the kurtosis value is –0.41, dimensionless, rounded to two decimal places.

Use Excel's KURT() function. For example, for the data named range DATA, use the Excel function expression "=KURT(DATA)."

## CORRELATION

When a plot of data is made between one variable and another, with the presence of some amount of random variation in one of the variables, the resulting graph can look like one of the following ones.

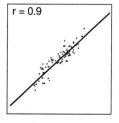
Positively correlated

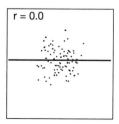
Negatively correlated

When the general shape of the data is upward sloping to the right, the data is said to be positively correlated, and when downward sloping to the right, it is negatively correlated.

Mathematically, data are described by the statistic $r$, or Pearson's correlation coefficient, defined by the following formula:

$$r = \frac{n\sum_{i=1}^{n} x_i y_i - \left(\sum_{i=1}^{n} x_i\right)\left(\sum_{i=1}^{n} y_i\right)}{\sqrt{\left[n\sum_{i=1}^{n} x_i^2 - \left(\sum_{i=1}^{n} x_i\right)^2\right]\left[n\sum_{i=1}^{n} y_i^2 - \left(\sum_{i=1}^{n} y_i\right)^2\right]}}$$

$$= \frac{\sum_{i=1}^{n}(x_i - \bar{x})(y_i - \bar{y})}{\sqrt{\sum_{i=1}^{n}(x_i - \bar{x})^2(y_i - \bar{y})^2}} = \frac{\sum_{i=1}^{n} x_i y_i - \overline{xy}}{(n-1)s(x)s(y)}$$

where $s()$ = sample standard deviation.

Positive correlated data is where there is a trend in the data where, generally, an increase in one variable relates to increasing values in the other variable.

For example, in the pair of data series of amperage measurements,

13, 14, 23, 23, 32, 33, 45, 99, and 105 A,

and induced amperage measurements,

34.60, 36.43, 59.23, 59.56, 82.03, 111.91, 240.50, and 255.19, respectively

the Pearson coefficient of variation $r$ value is 0.99998, dimensionless.

Use Excel's PEARSON() function. For example, for the data named ranges DATA1 and DATA2, use the Excel function expression "=PEARSON(DATA1, DATA2)."

## LINEAR RELATIONSHIPS

Many physical relationships are known to be generally linear relationships to one another. As an example, many elastic relationships are assumed to follow this rule. The relationship between the deflected length of a helical spring and the resulting force it produces is considered linear over a portion of its available deflection range. The current flowing in a purely resistive circuit is linearly related to voltage drop over a range of current values.

Departures from a linear relationship occur at the limits of such linear ranges. For example, compression springs are known to be nonlinear near the end limits of their available stroke (near their free length and also near the solid height). The relationship of current to voltage in an electrical circuit can become nonlinear when the value of resistance (or impedance) varies with the magnitude of current through it (for example, when heating effects change the magnitude of resistance).

A linear relationship is one where, if two or more pairs of data values (for example, two or more points in an x – y relationship) are plotted against one another, the resulting graph produces a straight line.

## Two-Point Slope-Intercept Relationship in Linear Data Sets

When there are only two points, the line connecting them is simply a straight line. Such a linear graphical relationship can be described by a formula with two constants: one for the slope of the line and another for the intercept of the line with the y-axis. This equation is the slope-intercept formula:

$$y = mx + b$$

where the formula for the slope is:

$$m = \frac{y_2 - y_1}{x_2 - x_1}$$

The units of slope are a ratio: unit of y divided by the unit of x. And the formula for the intercept is:

$$b = y_1 - mx_1$$

The units of intercept are the same unit as y.

For example, in the x, y data pairs, x1 = 12.5 mm, y1 = 4.2 N, x2 = 22.1 mm, y2 = 8.4 N, the slope is m = 0.4 N / mm, and the intercept is b = −1.3 N, rounded to one decimal place.

Use Excel's INTERCEPT() function. For example, for the data named ranges DATA1 and DATA2, use the Excel function expression "=INTERCEPT(DATA1, DATA2)."

## Linear Regression, Best-Fit Line Through a Data Set

Often data that are generally well-correlated can be described by a best-fit line through the set of measurements. The method by which the best-fit line is determined is through use of the least-squares method.

Data sets of linear data containing some random variation result in deviations from a straight line relationship. For such data, the best fit line through the data may need to be determined.

The general equation for the best-fit line through a set of input-output data values is as follows:

$$b = y - mx$$

where for the slope:

$$m = \frac{n \sum_{i=1}^{n} x_i y_i - \left( \sum_{i=1}^{n} x_i \right)\left( \sum_{i=1}^{n} y_i \right)}{n \sum_{i=1}^{n} x_i^2 - \left( \sum_{i=1}^{n} x_i \right)^2}$$

and for the intercept:

$$m = \overline{Y} - b\overline{X} = \sum_{i=1}^{n} Y_i - b \sum_{i=1}^{n} X_i$$

For example, in the following two pairs of data series of amperage measurements,

13, 14, 23, 23, 32, 33, 45, 99, and 105 A,

and induced amperage measurements,

34.60, 36.43, 59.23, 59.56, 82.03, 111.91, 240.50 and 255.19 A, respectively

the slope of the best-fit line through the data pairs is 2.39, dimensionless, and the intercept is 3.66 A, rounded to two decimal places.

Use Excel's SLOPE() function. For example, for the data named ranges DATA1 and DATA2, use the Excel function expression "=SLOPE(DATA1, DATA2)."

Also, the intercept of the best-fit line through the data pairs is 3.66 A, rounded to two decimal places.

Use Excel's INTERCEPT() function. For example, for the data named ranges DATA1 and DATA2, use the Excel function expression "=INTERCEPT(DATA1, DATA2)."

## ZERO AND SPAN RELATIONSHIPS

In many measuring and process control instruments it is desired to have the displayed value, or value sent to a control final element (valve, motor, and so on) be scaled such that a given lower output is used as a zero value, and the difference between an upper output and the lower zero value termed as the *span*.

See CD spreadsheet tool, worksheet "Calibration," workbook "Appendix IV-b, Metrology, math, statistics and engineering" for an adjustment protocol and Excel tool to adjust for required zero and span.

## INTERPOLATION

Many numerical relationships are given in table (tabular) form. Often acceptance criteria, standard values, and empirical data are transmitted in such format. The user of these data formats often needs to obtain intermediate values, obtained by interpolation.

A related term, *extrapolation* refers to estimating a value that is outside the range of given data, at a higher or lower value. Generally, extrapolation is to be discouraged unless specifically allowed by written guidance from the governing source of the data.

Interpolation is determining an intermediate output value in relation to one or more intermediate input values.

## FORMATS OF TABULAR DATA

**One-way tabulations.** Tabular data are most often provided in input–output format. Each input data item value is related in a one-to-one relationship to an output value.

**Two-way tabulations.** Some data, such as in published steam tables, relate more than one input variable to an output variable. For example, in two-way tabulations, the first input is listed down a side column and the second across a row (usually at the top of the table), with the output variable listed as an array on a page. An example of this would be relating gas volume (output) to pressure and temperature (input variables).

**Three-way tabulations.** This is where three input variables are associated with an output variable. Extending the two-way thought, such tabulations have separate pages or tables for each value of the third input variable. An example of this is the F-distribution table.

## LINEAR INTERPOLATION METHODS

Fortunately, much available data are either strictly linearly related or may be approximated satisfactorily with linear interpolation methods, over small known increments of the available tabular data. Depending on the type of data and the degree of accuracy required the following interpolation methods are used:

### One-way, Two-point Interpolation, Linear

One-way interpolation is used for graphical or tabular data, where the objective is to determine an intermediate output value in relation to a given intermediate input value.

A special case of one-way interpolation is where the input–output relationship is either known or assumed to be linear. This form of interpolation is that which is most-often used in practice, where the desired intermediate values are taken as linearly related.

The method for one-way, two-point linear interpolation involves locating known higher and lower input values, their related higher and lower output value, and use of the intermediate input value to compute the desired intermediate output value by the following formula:

$$y(x) = y_1 + \frac{x - x_1}{x_2 - x_1}(y_2 - y_1)$$

Here,

$y(x)$ is the desired interpolated output value.

$x_1, x_2, y_1,$ and $y_2$ are the tabulated high and low input and output values, respectively.

For example, for the following source data voltage values, for points x and y,

| x | y |
|---|---|
| 1 | 8 |
| 2 | 1 |

the interpolated value of y(x) at x = 1.6 V is y(x) = 3.8 V.

Use the CD spreadsheet tool in Appendix IV-c, Interpolation worksheet "two-point interpolation, linear, one-way."

### Two-way, Two-point Interpolation, Linear

For this requirement the preceeding formula is used incrementally, as follows. Using the formula and the intermediate value of the first input variable, determine two intermediate output values for the second input variable at both the high and low value of the first variable. Then repeat this process using the intermediate value of the second input variable, now having the values of the output variable at the intermediate value of the first input variable, compute the intermediate value of the output variable (which is the interpolated output value related to both intermediate input values).

For example, we have the following table of source data voltage values, for voltages at points w and x, related to a temperature, y °C, the interpolated value of y(x) at $w_i = 77$ V, and $x_i = 86$ V is $y(w, x) = 0.507$ V.

| | | Source data table for output temperature, y °C | | |
|---|---|---|---|---|
| | | $x_1$ | $x_2$ | |
| | | 80 | 90 | |
| $w_1$ | 70 | 0.483 | 0.534 | y |
| $w_2$ | 80 | 0.474 | 0.524 | y |

Use the CD spreadsheet tool in Appendix IV-c, "two-way linear" worksheet, "two-point interpolation, linear, two-way."

In the CD this calculation is done automatically by providing the desired intermediate and tabulated input and output values from known tabular data.

## Three-way, Interpolation, Linear

Three-way interpolation is merely an extension of two-way, following the same logic. Obviously, the process becomes more and more repetitive and time consuming.

In the CD the calculation is done automatically by providing the desired intermediate and tabulated input and output values from known tabular data.

For example, we have the following table of source data voltage values, for voltages at points w and x, related to a temperature, y °C, the interpolated value of z ($y_i$) at $w_i = 77$ V, $x_i = 86$ V and $y_i = 0.50$, is y (w, x, y) = 0.517 V.

| | | Source data tables for output temperature, y °C | | | | | | | |
|---|---|---|---|---|---|---|---|---|---|
| $Y_1 = 0.25$ | $x_1$ | $x_2$ | | | $Y_2 = 0.75$ | | $x_1$ | $x_2$ | |
| | 80 | 90 | | | | | 80 | 90 | |
| $w_1$ 70 | 0.483 | 0.534 | z (y1) | | $w_1$ | 70 | 0.503 | 0.554 | z (y2) |
| $w_2$ 80 | 0.474 | 0.524 | z (y1) | | $w_2$ | 80 | 0.494 | 0.544 | z (y2) |

Use the CD spreadsheet tool in Appendix IV-c, "three-way linear" worksheet, "two-point interpolation, linear, three-way."

# INTERPOLATION METHODS FOR NONLINEAR DATA

Data that are slightly nonlinear, where higher accuracy interpolated values are required, or for substantially nonlinear data where known values between input and output values are more widely separated, requires use of nonlinear interpolation methods. Generally, interpolation methods for such data are restricted to one-way (one input variable associated with each output variable) data, as the complexity and uncertainty can increase in nonlinear relationships.

**One-way, Three-point Interpolation—Quadratic**
When the data relationship is slightly nonlinear, a better and more accurate interpolation method is to use three-point, quadratic interpolation, with the following formula:

$$y(x) = \frac{(x-x_2)(x-x_3)}{(x_1-x_2)(x_1-x_3)} y_1$$
$$+ \frac{(x-x_1)(x-x_3)}{(x_2-x_1)(x_2-x_3)} y_2$$
$$+ \frac{(x-x_1)(x-x_2)}{(x_3-x_1)(x_3-x_2)} y_3$$

Here,

$y(x)$ is the desired interpolated output value.

$x_1, x_2, x_3, y_1, y_2$, and $y_3$ are the tabulated high, intermediate and low input and output values, respectively.

For example, for the following source data voltage values, for points **x** and **y**,

| x | y |
|---|---|
| 1 | 8 |
| 2 | 1 |
| 4 | 5 |

the interpolated value of **y(x)** at **x** = 1.6 V is **y(x)** = 3.08 V.

Use the CD spreadsheet tool in Appendix IV-c, Interpolation worksheet, "Three-point Interpolation, Quadratic, Second Order Exponential."

$$y(x) = \sum_{k=1}^{n} L_k(x) y(x_k)$$

where

$$L_k = \prod_{i=1}^{n} \left[ \frac{(x-x_i)}{(x_k-x_i)} \bigg|_{i \neq k}, 1 \bigg|_{i=k} \right]$$

**One-way, n-Point Interpolation—Nonlinear (Lagrangian)**
Both the linear and quadratic interpolation equations just mentioned are special cases of the general Lagrangian interpolation equation. This equation may be extended to any number of input/output pairs for consideration in producing even more accurate interpolation results.

For example, for the following source data voltage values, for points **x** and **y**,

| x | y |
|---|---|
| 1 | 12 |
| 2 | 4 |
| 4 | 16 |
| 5 | 25 |

the 4-point Interpolation, Cubic, 3rd Order Exponential interpolated value of **y(x)** at **x = 1.5 V** is **y(x) = 6.3 V**, rounded to one decimal place.

Use the CD spreadsheet tool in Appendix IV-c, Interpolation worksheet, "4-point Interpolation, Cubic, 3rd Order Exponential."

## TYPES OF DISTRIBUTIONS AND THEIR PROPERTIES

### Normal (Gaussian) Cumulative Distribution

The normal distribution, or gaussian distribution, describes randomly occurring events and their frequency relative to the magnitude of their comparative measurement values.

The normal (gaussian) cumulative distribution is given by the formula:

$$f(x, \mu, \sigma) = \frac{1}{\sqrt{2\pi}\sigma} e^{-\frac{(x-\mu)^2}{2\sigma^2}}$$

### Standard Normal (Gaussian) Cumulative Distribution

The standard normal (gaussian) cumulative distribution, or z-distribution is normalized to produce an area under the normal curve of 1.0, and place the peak of the curve at z = 0. This permits the convenient use of the z-statistic to locate the point where the area under the curve, at greater or lesser values than the z-statistic, to be associated with the probability of occurrence of events having any of those z-statistic values.

The standard normal (gaussian) cumulative (z-) distribution is given by the formula:

$$f(z, 0, 1) = \frac{1}{\sqrt{2\pi}} e^{-\frac{x^2}{2}}$$

*standard normal deviate*

### t-Distribution

The t-distribution relates the distribution of a sample to the z-distribution's z-statistic and probabilities. As the sample size is increased to larger and larger values, the t-distribution approaches, and in the limit of infinite sample size, becomes identical to the z-distribution.

### F-Distribution

The F-distribution is used to compare the variances of two sample or population statistics. The F-statistic is the ratio of the two variances.

It is used in ANOVA and other variance-based statistic calculations.

### $\chi^2$ (Chi-squared)-Distribution

The Chi-squared is used to compare two distributions to test if they may be assumed to represent the same population, or for samples if the samples may be considered to be drawn from a population having a common distribution.

## Weibull Distribution

Weibull distributions are used in reliability statistics, for estimating expected failure rates over the life of an item under consideration. We will not consider this distribution in this text.

## Hypergeometric Distribution

The hypergeometric distribution is used to determine the number of successes or failures in a population of events, based on a given sample from that population. It is often the basis for determine an acceptance sampling plan and its performance at correctly identifying the presence of a condition.

The hypergeometric distribution gives the probability of the number of successes, given the number of population successes, sample size, and population size.

The hypergeometric distribution is given by the formula:

$$P(X=x) = p(x,n,M,N) = \frac{\binom{M}{x}\binom{N-M}{n-x}}{\binom{N}{n}} = \frac{\left(\frac{M!}{x!(M-x)!}\right)\left(\frac{(N-M)!}{(n-x)!((N-M)-(n-x))!}\right)}{\left(\frac{N!}{n!(N-n)!}\right)}$$

## Binomial Distribution

The binomial distribution is often used in situations where there is a fixed number of trials or tests, when the outcomes are pass/fail, trials are statistically independent on one another, and the probability of success is constant.

When the probability $p$ is small, and the number of samples $n$ is large, for a fixed np in the limit the binomial distribution approaches the poisson distribution.

The binomial distribution is given by the formula:

$$P(m) = \frac{n!}{m!(n-m)!} p^m (1-p)^{n-m}$$

The cumulative binomial distribution is given by the formula:

$$P\{m \leq c\} = \sum_{m=0}^{c} \frac{n!}{m!(n-m)!} p^m (1-p)^{n-m}$$

## Poisson Distribution

A common application of the Poisson distribution is predicting the number of events that will occur over a specified period of time. The Poisson distribution has often been called the small probability distribution because it is usually applied in situations where there is a small, per unit, probability $p$ of an occurrence of an event in a given standard sample unit size. Multiples of the stated sample unit on which the probability

$p$ is based are labeled $n$ for the number or size of sample being considered. The probability for the sample size $n$ therefore is the product, or np.

The Poisson distribution is given by the formula:

$$P(k) = \frac{p^{-\lambda}\lambda^k}{k!}$$

The cumulative Poisson distribution is given by the formula:

$$P\{k \leq c\} = \sum_{k=0}^{c} \frac{p^{-\lambda}\lambda^k}{k!}$$

## Uncertainty, Normal Distribution

Many measurements are made under the assumption that the underlying uncertainty is normally distributed.

For an assumed normal distribution, the Type B standard uncertainty $u_j$ where 2a is the range including approximately 67%(2σ) probability that the value lies in the interval –a to +a, is given by the following formula,

$$u_j = a$$

For example, if an ammeter is assumed to have a approximately 100% probability that the reading lies in the interval of ± 0.02 mA, with a normal distribution, the standard uncertainty $u_j$ then is 0.02 mA.

## Uncertainty, Rectangular Distribution

Rectangular (or uniform) distributions, as their name implies, have distributions that are rectangular–shaped, with a constant frequency of occurrence over the range of the distribution's measurement values.

For an assumed rectangular distribution, the Type B standard uncertainty $u_j$ where 2a is the range including approximately 100% probability that the value lies in the interval –a to +a, is given by the following formula,

$$u_j = \frac{a}{\sqrt{3}}$$

For example, if a volumetric meter is assumed to have a approximately 100% probability that the reading lies in the interval of ± 0.15 m³, with a rectangular distribution, the standard uncertainty $u_j$ then is 0.087 m³.

## Uncertainty, Triangular Distribution

Triangular distributions, as their name implies, have distributions that are isosceles triangular–shaped, with frequencies and probabilities linearly decreasing in magnitude for values above or below the distribution mean (which is at the peak of the triangle).

For an assumed triangular distribution, the Type B standard uncertainty, $u_j$, where 2a is the range including approximately 100% probability that the value lies in the interval −a to +a, is given by the following formula,

$$u_j = \frac{a}{\sqrt{6}}$$

For example, if a weigh scale is assumed to have a approximately 100% probability that the reading lies in the interval of ± 0.015 N, with a triangular distribution, the standard uncertainty $u_j$ then is 0.0061 N.

## Uncertainty, U-Shaped Distribution

U-shaped distributions generally apply to distributions where the frequency of occurrence of the measurement is lowest at the mean of the distribution, increasing in some manner for values above or below the distribution mean.

For an assumed u-shaped distribution, the Type B standard uncertainty $u_j$ where 2a is the range including approximately 100% probability that the value lies in the interval −a to +a, is given by the following formula,

$$u_j = \frac{a}{\sqrt{2}}$$

For example, if a thermometer is assumed to have a approximately 100% probability that the reading lies in the interval of ± 0.1 °K, with a U-shaped distribution, the standard uncertainty, $u_j$, then is 0.07 °K.

## Relative Uncertainty

Relative uncertainty, $U_r$, is a dimensionless value that expresses a unit's uncertainty independent of the unit's size. It is the measurement uncertainty, U, divided by the magnitude of the unit, y, given by the following formula:

$$U_r = \frac{u}{y}$$

For instance, if an ammeter measures (y = ) 0.75 A, and the uncertainty, u, of the measurement is known to be 0.005 A, the relative uncertainty $U_r$ then is 0.00667, dimensionless.

## AUTO-CORRELATION

One important test of the assumption that source data is normally distributed is to determine the auto-correlation coefficient for the data set, a value between 0 and 1. If the auto-correlation coefficient is near zero, the data set may be more confidently assumed to be normally distributed. If it is larger, this indicates the presence of cyclic variation in the data, and the presence of a sequence- or time-based factor in the data.

$$\text{autocorr }(k) = \frac{\sum_{i=0}^{N-1}(x(i)-\bar{x})(x(i+k)-\bar{x})}{\sum_{i=0}^{N-1}(x(i)-\bar{x})^2}$$

for series, $x(i)$ where $i = 0...N-1$

If the autocorrelation coefficient is calculated for all lags, k = 0...N–1, the resulting series is called the autocorrelation series, or correlogram.

## References

Box, George E.P., William G. Hunter, and J. Stuart Hunter. 1978. *Statistics for experimenters, an introduction to the design, data analysis and model building.* New York: Wiley & Sons.

Central Limit theorum, www.animatedsoftware.com/statglos/sgcltheo.htm

# Chapter 28

# Mensuration, Volume, and Surface Areas

In this chapter, we cover the basic formulas for common calculations used in measurement of quantities dependent on geometry and known dimensions. This section enumerates often-encountered methods for calculating lengths, angles, arcs, and the more common plane areas, volumes, and surface areas. For other formulas and methods, consult general references containing formulas for specific areas of interest.

All equations in this section are also automated in a spreadsheet on the CD, in Appendix IV-b, Formulas: Metrology, math, statistics, engineering.

## MENSURATION, LENGTHS, AND ANGLES

*Mensuration* is defined as the act or process of measuring. More often the term *mensuration* is associated with dimensional measurement and is often associated with the determination, by computation from known characteristics, the lengths of lines, arcs, angles, and distances.

The following are formulas that are often encountered in determining both length and graphical relationships.

### Length

#### y-Intercept of a Line

This equation is used to determine the point where a line crosses the y-axis at x = 0.

Formula:

$$b = y_i - mx_i$$

where:  $b$, y-intercept of line (intercept at $x = 0$)

$m$, slope of line, dimensionless

$x_i$, value of individual measurement

$y_i$, value of individual measurement

An example of use of this formula is:

| y₁ | x₁ | m | b |
|---|---|---|---|
| m | m | dmls | m |
| 6.1 | 3.8 | 1.2 | 1.54 |

## Slope of a Line

Formula:

$$m = \frac{y_2 - y_1}{x_2 - x_1}$$

where: $m$, slope of line, dimensionless

$x_1, x_2$, values of individual measurements

$y_1, y_2$, values of individual measurements

An example of use of this formula is:

| y₁ | y₂ | x₁ | x₂ | m |
|---|---|---|---|---|
| m | m | m | dmls | m |
| 6.1 | 10.6 | 3.8 | 6.8 | 1.5 |

## Slope-Intercept Equation of a Line

Formula:

$$y = mx + b$$

where: $m$, slope of line, dimensionless

$x$, value of measurement independent variable

$y$, value of measurement dependent variable

$b$, $y$-intercept of line (intercept at $x = 0$)

An example of use of this formula is:

| x₁ | m | b | y₁ |
|---|---|---|---|
| m | m | dmls | m |
| 6.8 | 1.5 | 2.1 | 12.3 |

## Point-Slope Equation of a Line
Formula:
$$(y - y_i) = m(x - x_i)$$

where: $m$, slope of line

$x_i$, value of individual measurement

$y_i$, value of individual measurement

## Distance Between Two x - y Points
Formula:
$$d = \pm\sqrt{(x_2 - x_1)^2 + (y_2 - y_1)^2}$$

where: $d$, distance between two points in a plane

$x_i$, value of individual measurements

$y_i$, value of individual measurements

An example of use of this formula is:

| $y_1$ | $y_2$ | $x_1$ | $x_2$ | d |
|---|---|---|---|---|
| m | m | m | dmls | m |
| 6.1 | 10.6 | 3.8 | 6.8 | 5.40833 |

## Distance in a General Plane Between Two Points
Formula:
$$d = \pm\sqrt{(x_2 - x_1)^2 + (y_2 - y_1)^2 + (z_2 - z_1)^2}$$

where: $d$, distance between two general points

$x_i$, value of individual measurements

$y_i$, value of individual measurements

An example of use of this formula is:

| $y_1$ | $y_2$ | $x_1$ | $x_2$ | $z_1$ | $z_2$ | d |
|---|---|---|---|---|---|---|
| m | m | m | m | m | m | m |
| 6.1 | 10.6 | 3.8 | 10.6 | 3.8 | 6.8 | 8.6885 |

## Perpendicular Lines, Slope Relationship
Formula:

$$m_1 = -\frac{1}{m_2}$$

where: $m$, slope of line, dimensionless

An example of use of this formula is:

| m₁ | m₂ |
|---|---|
| dmls | dmls |
| 3.4 | -0.2941 |

# Circle

## Circle, General
Formula:

$$r = \sqrt{(x-h)^2 + (y-k)^2}$$

where: $r$, radius of circle

$x$, value of x-coordinate

$h$, x-offset

$y$, value of y-coordinate

$k$, y-offset

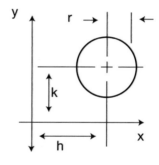

An example of use of this formula is:

| x | h | y | k | r |
|---|---|---|---|---|
| m | m | m | m | m |
| 4.3 | 2.1 | 5.1 | 7.3 | 3.11127 |

# Ellipse

## Ellipse, Major Axis Horizontal
Formula:

$$y = k \pm b \sqrt{1 - \frac{(x-h)^2}{a^2}} \qquad 1 = \frac{(x-h)^2}{a^2} + \frac{(y-k)^2}{b^2}$$

where:  major axis length: 2a

minor axis length: 2b

x, value of x-coordinate

h, x-offset, center of ellipse from x = 0

y, value of y-coordinate

k, y-offset, center of ellipse from y = 0

An example of use of this formula is:

| x | h | a | b | k | y₁ | y₂ |
|---|---|---|---|---|---|---|
| m | m | m | m | m | m | m |
| 4.1 | 2.1 | 4.3 | 2.4 | 2.1 | -0.0246 | 4.2246 |

## Ellipse, Major Axis Vertical

Formula:

$$y = k \pm a\sqrt{1 - \frac{(x-h)^2}{b^2}} \qquad 1 = \frac{(x-h)^2}{b^2} + \frac{(y-k)^2}{a^2}$$

where:  major axis length: 2a

minor axis length: 2b

x, value of x-coordinate

h, x-offset, center of ellipse from x = 0

y, value of y-coordinate

k, y-offset, center of ellipse from y = 0

An example of use of this formula is:

| x | h | a | b | k | y 1 | y 2 |
|---|---|---|---|---|---|---|
| m | m | m | m | m | m | m |
| 4.1 | 2.1 | 4.3 | 2.4 | 2.1 | -0.2769 | 4.4769 |

# Angle

## Degree to Radian

Formula:

$$\deg = \frac{\pi}{180} \text{rad}$$

An example of use of this formula is:

| angle | | angle | |
|---|---|---|---|
| degree | radian | radian | degree |
| 3.4 | 0.05934 | 2.4 | 137.51 |

## General Plane Triangle Relationships—Right Triangle

$\sin \theta = \dfrac{\text{side opposite}}{\text{hypotenuse}}$

$\cos \theta = \dfrac{\text{side adjacent}}{\text{hypotenuse}}$

$\tan \theta = \dfrac{\text{side opposite}}{\text{side adjacent}}$

$\csc \theta = \dfrac{\text{hypotenuse}}{\text{side opposite}} = \dfrac{1}{\sin \theta}$

$\sec \theta = \dfrac{\text{hypotenuse}}{\text{side adjacent}} = \dfrac{1}{\cos \theta}$

$\cot \theta = \dfrac{\text{side adjacent}}{\text{side opposite}} = \dfrac{1}{\tan \theta}$

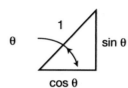

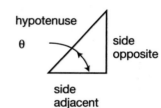

## Pythagorean (Right Angle Triangle) Theorem

Formula:

$$c = \pm\sqrt{a^2 + b^2}$$

where:  $a, b$, length of sides

$c$, length of hypotenuse

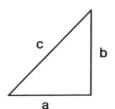

An example of use of this formula is:

| a | b | c |
|---|---|---|
| m | m | m |
| 6.1 | 3.8 | 7.18679 |

## Law of Sines

Formula:

$$\frac{a}{\sin\alpha} = \frac{b}{\sin\beta} = \frac{c}{\sin\gamma}$$

where: $a$, side opposite angle $\alpha$

$b$, side opposite angle $\beta$

$c$, side opposite angle $\gamma$

An example of use of this formula is:

| β | α | β | sin α | sin β | a |
|---|---|---|---|---|---|
| m | deg | deg | dmls | dmls | m |
| 6.1 | 60 | 30 | 0.86603 | 0.5 | 10.5655 |

## Law of Cosines

Formula:

$$c = \sqrt{a^2 + b^2 - 2ab\,\cos(\gamma)}$$

where: $a$, side opposite angle $\alpha$

$b$, side opposite angle $\beta$

$\gamma$, angle between sides $a$ and $b$

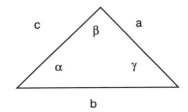

An example of use of this formula is:

| a | b | g | cos b | c |
|---|---|---|---|---|
| m | deg | deg | dmls | m |
| 1 | 1.41421 | 45 | 0.70711 | 1 |

## Plane Area

### Rectangle

Formula:

$$A = bh$$

where: $A$, plane area

$b$, length of base

$h$, length of height

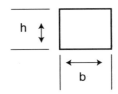

An example of use of this formula is:

| b | h | A |
|---|---|---|
| m | m | m ^2 |
| 1 | 2 | 2 |

## Parallelogram

Formula:

$$A = bh$$

where: $A$, plane area

$b$, length of base

$h$, perpendicular distance from base to top

An example of use of this formula is:

| b | h | A |
|---|---|---|
| m | m | m ^2 |
| 1 | 2 | 2 |

## Trapezoid

Formula:

$$A = \frac{h(b_1 + b_2)}{2}$$

where: $A$, plane area

$b_2$, length of base, top

$b_1$, length of base, bottom

$h$, perpendicular distance from base to top

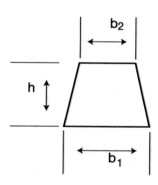

An example of use of this formula is:

| $b_1$ | $b_2$ | h | A |
|---|---|---|---|
| m | m | m | m ^2 |
| 1 | 2 | 3 | 4.5 |

## Right Triangle

Formula:

$$A = b\frac{h}{2}$$

where:  A, plane area

    b, length of base

    h, perpendicular distance from base to apex

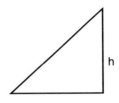

An example of use of this formula is:

| b | h | A |
|---|---|---|
| m | m | m ^2 |
| 3.45 | 2.2 | 3.795 |

## Oblique Triangle

Formula:

$$A = b\,\frac{c\sin\alpha}{2}$$

where:  A, plane area

    b, length of base

    c, length of side adjacent to b

    a, angle between sides b and c

    h, perpendicular distance from base to apex

$h = c\sin\alpha$

An example of use of this formula is:

| b | c | γ | sin β | c |
|---|---|---|-------|---|
| m | deg | deg | dmls | m |
| 1 | 1.41421 | 45 | 0.70711 | 0.5 |

## Circle

Formula:

$$A = \pi r^2 = \pi\,\frac{d^2}{4}$$

where:  A, plane area

    d, diameter

    r, radius

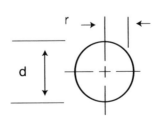

An example of use of this formula is:

| r | A |
|---|---|
| m | m ^2 |
| 0.5 | 0.7854 |

| d | A |
|---|---|
| m | m ^2 |
| 1 | 0.7854 |

## Ellipse

Formula:

$$A = \pi ab = \pi \frac{CD}{4}$$

where:  $A$, plane area

$a$, major axis radius length

$b$, minor axis radius length

$C$, major axis length: $2a$

$D$, minor axis length: $2b$

An example of use of this formula is:

| a | b | A |
|---|---|---|
| m | m | m ^2 |
| 0.5 | 0.5 | 0.7854 |

| C | D | A |
|---|---|---|
| m | m | m ^2 |
| 1 | 1 | 0.7854 |

## Perimeter

### Rectangle

Formula:

$$P = 2(b + h)$$

where:  $P$, length of perimeter

$b$, length of base

$h$, length of height

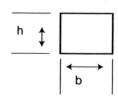

An example of use of this formula is:

| b | h | P |
|---|---|---|
| m | m | m |
| 0.78 | 1.32 | 4.2 |

## Right Triangle

Formula:

$$P = a + b + \sqrt{a^2 + b^2}$$

where: $P$, length of perimeter

a, length of side

b, length of base

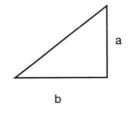

An example of use of this formula is:

| a | b | P |
|---|---|---|
| m | m | m |
| 1.8 | 3.4 | 9.04708 |

## Circle

Formula:

$$P = 2\pi r = \pi d$$

where: $P$, length of perimeter

d, diameter

r, radius

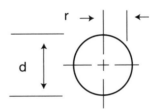

An example of use of this formula is:

| r | P |
|---|---|
| m | m |
| 0.5 | 3.14159 |

| d | P |
|---|---|
| m | m |
| 1 | 3.14159 |

## Ellipse

Formula:

$$P = \pi(a + b) = \pi\left(\frac{C + D}{2}\right)$$

where: $P$, length of perimeter

a, major axis radius length

b, minor axis radius length

C, major axis length: $2a$

D, minor axis length: $2b$

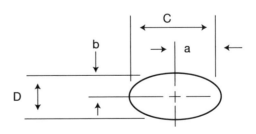

An example of use of this formula is:

| a | b | P |
|---|---|---|
| m | m | m |
| 0.5 | 0.5 | 3.14159 |

| C | D | P |
|---|---|---|
| m | m | m |
| 1 | 1 | 3.14159 |

## Volume

### Rectangular Prism

Formula:

$$V = lwh$$

where: $V$, volume

$l$, length

$w$, width

$h$, height

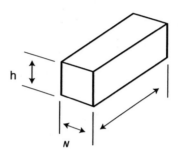

An example of use of this formula is:

| l | w | h | V |
|---|---|---|---|
| m | m | m | m^3 |
| 1.24 | 3.8 | 2.76 | 13.005 |

### Sphere

Formula:

$$V = \frac{4}{3}\pi r^3 = \frac{1}{6}\pi d^3$$

where: $V$, volume

$d$, length, spherical diameter

$r$, length, spherical radius

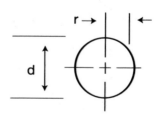

An example of use of this formula is:

| r | V |
|---|---|
| m | m^3 |
| 0.5 | 0.5236 |

| d | V |
|---|---|
| m | m^3 |
| 1 | 0.5236 |

## Ellipsoid

Formula:

$$V = \frac{4}{3}\pi abc = \frac{1}{6}\pi ABC$$

where:  $V$, volume

$a$, length, major axis radius

$b$, length, first minor axis radius

$c$, length, second minor axis radius

$A$, lengths, first thru third axes

$B$, lengths, first thru third axes

$C$, lengths, first thru third axes

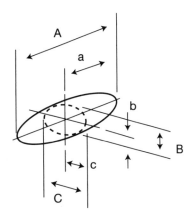

An example of use of this formula is:

| a | b | c | V |
|---|---|---|---|
| m | m | m | m ^3 |
| 0.5 | 0.5 | 0.5 | 0.5236 |

| A | B | C | V |
|---|---|---|---|
| m | m | m | m ^3 |
| 1 | 1 | 1 | 0.5236 |

## Pyramid

Formula:

$$V = \frac{1}{3} A_{base}\, h$$

where:  $V$, volume

$A_{base}$, plane area of base

$h$, height from base

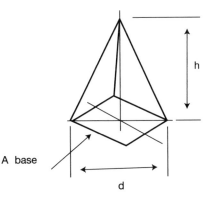

An example of use of this formula is:

| $A_{base}$ | h | V |
|---|---|---|
| m ^2 | m | m ^3 |
| 0.5 | 1 | 0.16667 |

## Truncated Pyramid

Formula:

$$V = \frac{1}{3} h \left( A_1 + \sqrt{A_1 A_2} + A_2 \right)$$

where: $V$, volume

$A_1$, plane area of base 1

$A_2$, plane area of base 2

$h$, height from base

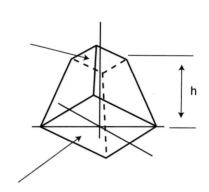

An example of use of this formula is:

| $A_1$ base | $A_2$ base | h | V |
|---|---|---|---|
| m ^2 | m | m | m ^3 |
| 0.5 | 1 | 1 | 0.7357 |

## Cone

Formula:

$$V = \frac{1}{3} \pi r^2 h = \frac{1}{12} \pi d^2 h$$

where: $V$, volume

$r$, base radius

$d$, base diameter

$h$, height from base

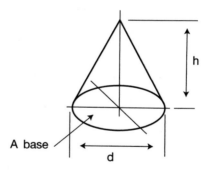

An example of use of this formula is:

| r | h | V |
|---|---|---|
| m | m | m ^3 |
| 0.5 | 1 | 0.2618 |

| d | h | V |
|---|---|---|
| m | m | m ^3 |
| 1 | 1 | 0.2618 |

## Truncated Cone

Formula:

$$V = \frac{\pi}{12} h \left(d_1^2 + d_1 d_2 + d_2^2\right) = \frac{1}{3} h \left(A_1 + \sqrt{A_1 A_2} + A_2\right)$$

where: $V$, volume

$A_1$, plane area of base 1

$A_2$, plane area of base 2

$d_1$, diameter of base 1

$d_2$, diameter of base 2

$h$, height from base

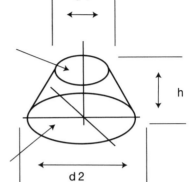

An example of use of this formula is:

| $d_1$ | $d_2$ | h | V |
|---|---|---|---|
| m | m | m | m ^3 |
| 0.25 | 0.25 | 0.5 | 0.02454 |

| $A_1$ | $A_2$ | h | V |
|---|---|---|---|
| m ^2 | m ^2 | m | m ^3 |
| 0.04909 | 0.04909 | 0.5 | 0.02454 |

## Surface Area

### Rectangular Prism

Formula:

$$A_S = 2(lw + lh + wh)$$

where: $A_S$, surface area

$l$, length

$w$, width

$h$, height

An example of use of this formula is:

| l | w | h | $A_S$ |
|---|---|---|---|
| m | m | m | m ^2 |
| 1 | 2 | 3 | 22 |

## Sphere

Formula:

$$A_S = 4\pi r^2 = \pi d^2$$

where: $A_S$, surface area

d, length, spherical diameter

r, length, spherical radius

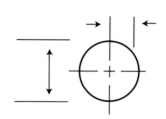

An example of use of this formula is:

| r | V |
|---|---|
| m | m ^2 |
| 0.5 | 3.14159 |

| d | V |
|---|---|
| m | m ^2 |
| 1 | 3.14159 |

## Prolate Ellipsoid of Revolution

Physical examples of this shape are eggs.
Formula:

$$A_S = 2\pi \left[ b^2 + \frac{a^2 b}{\sqrt{a^2 - b^2}} \sin^{-1}\left( \sqrt{\frac{a^2 - b^2}{a^2}} \right) \right]$$

where: $A_S$, surface area

a, length, major axis radius

b, length, minor axis radius

Note: a not equal to b

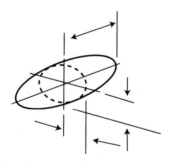

An example of use of this formula is:

| a | b | V |
|---|---|---|
| m | m | Unit ^2 |
| 0.503 | 0.497 | 3.12902 |

## Oblate Ellipsoid of Revolution

Physical examples of this shape are large rising bubbles.
Formula:

$$A_S = 2\pi \left[ a^2 + \frac{ab^2}{\sqrt{a^2-b^2}} \ln\left(\frac{a+\sqrt{a^2-b^2}}{b}\right) \right]$$

where: $A_S$, surface area

$a$, length, major axis radius

$b$, length, minor axis radius

Note: $a$ not equal to $b$

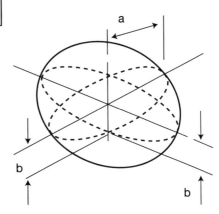

An example of use of this formula is:

| a | b | V |
|---|---|---|
| m | m | Unit ^2 |
| 0.503 | 0.497 | 3.15415 |

# Part V
## Uncertainty in Measurement

**Chapter 29**  Uncertainty in Measurement

# Chapter 29
# Uncertainty in Measurement

When one makes a measurement, the measurement in some way is assumed to be wrong. The difference between the value of the measurement and the value of the parameter being measured (the measurand) is known as the *error*. The total value of this error is made up of a number of error sources. The indication of the wrongness of the measurement is the *uncertainty of the measurement*.

The types of errors that affect a measurement are classified as random and systematic. An error is classified as *random* if its effect is varied with each repeat of a measurement and systematic if it does not. While GUM and ANSI/NCSLI Z540-2-1997 discourage using the words *random* and *systematic errors*, it is useful to discuss them briefly as they are used elsewhere in quality assurance and quality control applications.

When 10 measurements are made, their standard deviation can be a measure of the random error associated with the measurement.

The error associated with a repeat measurement on a mercury level of a barometer made by the same observer is considered systematic. The measurement on the mercury level of the barometer by different observers is considered random.

Systematic errors are difficult to quantify. Sometimes, one has to rely on the experience of the technician to estimate its effect.

Error sources have various distributions associated with them. The most common distribution associated with errors is the Gaussian or normal distribution. There are other distributions that one should familiarize oneself with. The three other distributions encountered in estimating measurement uncertainty are: rectangular, triangular, and U-shaped.

## DETERMINING MEASUREMENT UNCERTAINTY

When one makes a measurement, there is an uncertainty associated with it. In an ideal world the measurement made is absolute and has a true value associated with it; we would not have to worry about it. The author describes this as the one measurement bliss or one measurement quandary depending on whether one is an optimist or a pessimist. Unfortunately, we do not live in an ideal world and there are many factors (error sources) that contribute to the uncertainty of measurement.

There are situations where one measurement is all that is possible to make even though many factors contribute to the uncertainty of measurement. That is why careful consideration should be given to estimating the uncertainty of measurement.

There are many reasons why it is important to determine measurement uncertainty:

- It estimates the error associated with the measurement in a numerical value.
- It provides a level of confidence in one's measurement.
- It is a good practice.
- It is required for laboratory accreditation.

This section describes the process of determining measurement uncertainty in a test and calibration environment. Practical examples using various parameters are used to illustrate the process of determining measurement uncertainty.

## Measurement Uncertainty Determination

The process of determining measurement uncertainty can be broken down in seven basic steps:

1. Identify the uncertainties in the measurement process.
2. Evaluate and classify type of uncertainty (Type A or B).
3. Quantify (evaluate and calculate) individual uncertainty by various methods.
4. Document in an uncertainty budget.
5. Combine uncertainty using the RSS method.
6. Assign appropriate k factor multiplier to combined uncertainty to report expanded uncertainty.
7. Document in an uncertainty report with the appropriate information (add notes and comments for future reference).

It is important to understand that before any measurement uncertainty determination is made, the process (calibration or test) must be in a state of statistical control. Many practitioners ignore this fact and find that the uncertainty budgets for their measurement parameters cannot be validated or verified at a later date. For more information on statistical process control and control charts, refer to Part III–Chapter 20.

A detailed process for determining measurement uncertainty using the seven steps outlined follows:

### Identify the Uncertainties in the Measurement Process

This is a brainstorming exercise where technicians and engineers familiar with the process (test or calibration) determine the factors affecting the measurement. Typical examples of some of the factors affecting the measurement are:

- Environment (temperature, humidity, vibration)
- Accuracy of measurement equipment
- Stability

- Instrument resolution
- Instrument calibration
- Repeatability
- Reproducibility
- Operator
- Measurement setup
- Method (procedure)
- Software

**Evaluate and Classify Type of Uncertainty (Type A or B)**

Once the factors affecting the uncertainty are identified, it is important to classify the type of uncertainty. The GUM classification of types of uncertainty follow.

*Type A evaluation method.* The method of evaluation of uncertainty of measurement by the statistical analysis of series of observations. Two examples are: Standard deviation of a series of measurements, and other statistical evaluation methods such as ANOVA and design of experiments (DOE).

A series of measurements are taken to determine the uncertainty of measurement. Ten readings are taken. See Table 29.1.

The standard uncertainty of measurement is the standard deviation of the 10 readings. This is derived using a statistical method (standard deviation) and is considered Type A uncertainty.

Table 29.1 Individual data.

| Type A Uncertainty Example | |
|---|---|
| 1 | 10.05 |
| 2 | 9.98 |
| 3 | 9.97 |
| 4 | 9.98 |
| 5 | 10.01 |
| 6 | 10.02 |
| 7 | 10.03 |
| 8 | 10.01 |
| 9 | 10.05 |
| 10 | 10.00 |
| Sum | 100.11 |
| Mean | 10.01 |
| Standard Deviation | 0.029 |

It is important to note that there are three standard deviation calculations:

1. Population standard deviation

$$\sigma = \sqrt{\frac{\sum_{i=1}^{i=n}(\bar{x}-x_i)^2}{n}}$$

2. Sample standard deviation

$$s = \sqrt{\frac{\sum_{i=1}^{i=n}(\bar{x}-x_i)^2}{n-1}}$$

3. Standard deviation of the mean

$$S_{\bar{x}} = \frac{s}{\sqrt{n}}$$

where: $\sigma$ = population standard deviation
$s$ = sample standard deviation
$S_{\bar{x}}$ = standard deviation of the mean
$n$ = number of measurements
$\bar{x}$ = average of data
$i$ = index

In most measurement uncertainty calculations, the sample standard deviation and standard deviation of the mean are normally encountered.

If the data in Table 29.1 were an average of five individual measurements as shown in Table 29.2, then calculating the standard deviation of the mean to determine measurement uncertainty would be more appropriate.

*Type B evaluation method.* The method of evaluation of uncertainty of measurement by means other than the statistical analysis of series of observations. Some examples are the history of parameter, other knowledge of the process parameter, based on specification, and reference data, for example, a physics handbook.

### Distributions Associated with Measurement Uncertainty

There is usually another piece of information that is required to determine standard uncertainty. One has to determine the type of distribution that the Type B uncertainty falls under.

Usually, there are four distributions that one can classify individual uncertainty information under. They are:

- Normal distribution (also known as bell curve or Gaussian distribution)
- Rectangular distribution
- Triangular distribution
- U- shaped distribution

The *normal distribution* is usually associated with Type A uncertainty. Examples of a normal distribution are the Type A uncertainty data where a series of measurements are recorded and the uncertainty calculated using the standard deviation.

**Table 29.2** Standard deviation of the mean calculation.

| | Type A Uncertainty Example—Means | | | | | |
|---|---|---|---|---|---|---|
| | Mean | 1 | 2 | 3 | 4 | 5 |
| 1 | 10.054 | 10.02 | 10.02 | 10.09 | 10.06 | 10.07 |
| 2 | 9.981 | 10.02 | 9.95 | 9.95 | 10.04 | 9.95 |
| 3 | 9.970 | 9.95 | 9.93 | 9.98 | 9.92 | 10.07 |
| 4 | 9.983 | 9.90 | 10.07 | 10.04 | 9.93 | 9.97 |
| 5 | 10.015 | 9.97 | 10.02 | 9.98 | 10.07 | 10.03 |
| 6 | 10.025 | 10.02 | 10.05 | 9.99 | 10.03 | 10.03 |
| 7 | 10.029 | 10.08 | 9.97 | 9.97 | 10.06 | 10.07 |
| 8 | 10.009 | 9.98 | 10.08 | 9.98 | 9.92 | 10.09 |
| 9 | 10.050 | 10.05 | 10.10 | 9.94 | 10.06 | 10.10 |
| 10 | 9.998 | 9.92 | 9.96 | 10.03 | 10.02 | 10.05 |
| Sum | 100.11 | | | | | |
| Mean | 10.01 | | | | | |
| Standard deviation | 0.029 | | | | | |
| Standard deviation of the mean | 0.012884 | | | | | |

The *rectangular distribution* is one where there is an equal probability of a measurement occurring within the bound limits. An example of a rectangular measurement is the specification data normally supplied by a manufacturer of an instrument.

Note: When the frequency distribution of a particular component of uncertainty cannot be determined, the GUM suggests assuming the rectangular distribution and thereby erring on the conservative side.

Here's an example of determining measurement uncertainty. Say the accuracy specification for a voltmeter at 20 volt scale is ± 0.02 volts.

The measurement uncertainty associated with this statement for the voltmeter at this scale is determined in the following manner.

It is classified Type B uncertainty because the information provided does not state how the accuracy specification is derived. Information like this usually found in the manufacturer's manual or data sheets. Therefore, it is classified as Type B.

The *distribution* that this specification falls under is *rectangular*. This is because the specification states that the measurement made has an equal chance of being anywhere within ± 0.02 volts.

In another example, the accuracy specification for the voltmeter at 20 volt scale is ± 0.02 volts.

This information would be considered as a rectangular distribution. To convert it to standard uncertainty, divide 0.02 volts by the square root of 3 as shown here.

$$\frac{0.020}{\sqrt{3}} = 0.01155 \text{ (Standard uncertainty attributed to the voltage specification).}$$

The fourth item in a correction factor table is not necessarily a distribution, but is mentioned as a special case. It is attributed to the rectangular distribution and shown best by the next example.

Say the resolution of a digital multimeter (DMM) is 0.001 volts. This is referred to as a 3½ digit meter. The standard uncertainty contribution due to the resolution of the multimeter is determined this way:

The last digit of the DMM will either read 0 or 1 depending on the fourth invisible digit resolving the meter third decimal digit.

Taking the rectangular distribution correction approach, the uncertainty associated with the resolution of the DMM is:

$$\frac{0.0005}{\sqrt{3}} = 0.000288675$$

Or divide the minimum resolution by square root of 12.

$$\frac{0.001}{\sqrt{12}} = \frac{0.001}{2\sqrt{3}} = 0.000288675$$

Particular care must be exercised when using this approach. If one does not know how the digits are rounded in the instrumentation, it is better to take the conservative approach and divide the minimum resolution by the square root of 3 as shown here:

$$\frac{0.001}{\sqrt{3}} = 0.00057735$$

In another example, a manufacturer specifies that the *class zyx* gage block has a specification of ± 0.001 inches.

The variance associated with this rectangular distribution is:

$$u^2 = \frac{(0.001)^2}{3} = 333.33\,E - 09$$

The standard uncertainty for this rectangular distribution is:

$$u = \sqrt{\frac{(0.001)^2}{3}} = 577.35\,E - 06$$

The *triangular distribution* is one where there is a central tendency for a measurement to occur with a few dispersing values of a measurement. An example of a triangular distribution is that of a frequency measurement where there is a fixed frequency value with its associated harmonics.

An example of a triangular distribution is: say a series of measurements taken indicate that most of the measurements fall at the center with a few spreading equally (±) 0.5 units away from the mean.

The variance associated with this triangular distribution is:

$$u^2 = \frac{(0.5)^2}{6} = 41.667\,E - 03$$

The standard uncertainty for this triangular distribution is:

$$u = \sqrt{\frac{(0.5)^2}{6}} = 204.124 \; E-03$$

The *U-shaped distribution* is one where there is less chance of a measurement occurring at the mean value and more chance of a value occurring at the outer bound limits. A cyclical or sinusoidal measurement usually falls under a U-shaped distribution.

An example of a U-shaped distribution is: say the temperature of the oil bath stated by the manufacturer is 100.00° ± 0.2° Celsius.

The variance associated with this U-shaped (trough) distribution is:

$$u^2 = \frac{(0.2)^2}{2} = 20.0 \; E-03$$

The standard uncertainty for this U-shaped (trough) distribution is:

$$u = \sqrt{\frac{(0.2)^2}{2}} = 141.421 \; E-03$$

Quantify (evaluate and calculate) individual uncertainty by various methods.

This process sometimes works in conjunction with the evaluation process. Sometimes it is a separate process where calculations are made depending on the evaluation selected.

## Document in an Uncertainty Budget

The uncertainty budget lists the uncertainty contributors of the measurement process in a list with its individual uncertainties. See Figure 29.1.

| | Type A Uncertainty | | | | | |
|---|---|---|---|---|---|---|
| | Uncertainty Description | Uncertainty | Distribution | Divisor | Standard Uncertainty | Variance |
| 1 | Repeatability | 6.46E-07 | normal | 1 | 6.46E-07 | 4.17E-13 |
| 2 | | | | | | |
| | | | Combined Type A Uncertainty | | 6.46E-07 | 4.17E-13 |
| | Type B Uncertainty | | | | | |
| | Uncertainty Description | Uncertainty | Distribution | Divisor | Standard Uncertainty | Variance |
| 1 | Resolution (1E–7) | 1.00E-07 | Rectangular | 3.4641016 | 2.89E-08 | 8.33E-16 |
| 2 | 10 Volt standard (± 1E–10) | 1.00E-10 | Rectangular | 1.7320508 | 5.77E-11 | 3.33E-21 |
| 3 | Thermal Stability (0.01 ppm/V/°C) | 2.00E-07 | U-shaped | 1.4142136 | 1.41E-07 | 2.00E-14 |
| 4 | Calibration (1.5E–8) k=2 | 1.50E-08 | 2U | 2 | 7.50E-09 | 5.63E-17 |
| | | | Combined Type B Uncertainty | | 1.4453E-07 | 2.09E-14 |

**Figure 29.1** An example of an uncertainty budget.

Note that before the uncertainty contributors are combined, they should be normalized to standard uncertainty. One cannot combine rectangular distribution with triangular or normal distributions. The GUM provides the following correction factors for the other three nonnormal distributions.

| Distribution | Divide by | Divisor | 1/Divisor |
|---|---|---|---|
| Rectangular | Square-root 3 | 1.7321 | 0.5774 |
| Triangular | Square-root 6 | 2.4495 | 0.4082 |
| U-shaped | Square-root 2 | 1.4142 | 0.7071 |
| Resolution | Square-root 12 | 3.4641 | 0.2887 |

## Combine Uncertainty (RSS Method)

Individual uncertainty components may not be added. In order to combine the uncertainty components, the RSS method is used. This is assuming that the uncertainty components are random and independent.

Type A uncertainty components are added together using the RSS method.

$$u_{c_a} = \sqrt{u_{1_a}^2 + u_{2_a}^2 + u_{3_a}^2 + \cdots + u_{n_a}^2}$$

Type B uncertainty components are added together using the RSS method.

$$u_{c_b} = \sqrt{u_{1_b}^2 + u_{2_b}^2 + u_{3_b}^2 + \cdots + u_{n_b}^2}$$

The combined Type A and Type B components are then added to obtain the total combined uncertainty using the RSS method.

$$u_{c_{ab}} = \sqrt{u_{c_a}^2 + u_{c_b}^2}$$

An argument can be made that all Type A and Type B components can be combined at one time. If the components are combined separately in a methodical approach as shown, however, it is easier to troubleshoot calculation related errors later.

An example of using the RSS method for Type A uncertainty components follows.

| Parameter | Standard Uncertainty |
|---|---|
| Repeatability | 0.015 units |
| Reproducibility | 0.005 units |

So, the combined Type A uncertainty is:

$$u_{c_a} = \sqrt{0.015^2 + 0.005^2} = 0.015811$$

An example of using the RSS method for Type B uncertainty components follows.

| Parameter | Standard Uncertainty |
|---|---|
| Resolution | 0.001 units |
| Calibration | 0.0002 units |
| Temperature | 0.005 units |

So, the combined Type B uncertainty is:

$$u_{c_b} = \sqrt{0.001^2 + 0.0002^2 0.005^2} = 0.005103$$

The combined Type A and B uncertainty is:

$$u_{c_{ab}} = \sqrt{0.015811_{c_a}^2 + 0.005013_{c_b}^2} = 0.016614$$

## Expanded Uncertainty

Combined uncertainty is assigned a coverage factor k that denotes a degree of confidence interval associated with it. The GUM recommends k = 2 at 95% confidence interval. Combined uncertainty multiplied by the coverage factor: k is known as expanded uncertainty and denoted U.

$$U = k \cdot u_c$$

Various confidence interval values and their associated k values are shown on the following table.

| Coverage Factor (k) | Confidence Level |
|---|---|
| 1.000 | 68.27 |
| 1.645 | 90.00 |
| 1.960 | 95.00 |
| 2.000 | 95.45 |
| 2.576 | 99.00 |
| 3.000 | 99.73 |

Using the combined uncertainty sum from the previous example, the expanded uncertainty at k = 2, 95% confidence interval is:

$$U = 2(0.016614) = 0.033228$$

## Uncertainty Report

All the results of the measurement uncertainty determination are documented in a measurement uncertainty report. It is important to ensure that this document not only contains the calculations, but also the reasoning used to derive and justify the calculations. If software is used to determine measurement uncertainty, the software needs to be validated by an alternate calculation method. The uncertainty report needs to be well documented with appropriate comments where necessary so that someone other than the originator of the report can understand the reasoning behind the measurement uncertainty analysis.

The uncertainty report is a live document, such that, from time to time, it needs to be reevaluated. Examples prompting reevaluation are:

- Method changes.
- Operator changes.
- New equipment is used in the process.
- Equipment used is recalibrated.
- Different standards are used.

An example of an uncertainty report is shown in Figure 29.2.

| | | Uncertainty Report | | | | |
|---|---|---|---|---|---|---|
| | Company | A Good Calibration Laboratory | | | | |
| | Name | | | | | |
| | Parameter | Voltage | | | | |
| | Nominal or range | 10 Volts | | | | |
| | Primary equipment | 10 Volt Standard, Precision Voltmeter | | | | |
| | Personnel | I. M. A. Metrologist | | | | |
| | Date | 01/05/04 | | | | |
| | | | | | | |
| | Additional equipment | | | | | |
| | | Type A Uncertainty | | | | |
| | Uncertainty description | Uncertainty | Distribution | Divisor | Standard uncertainty | Variance |
| 1 | Repeatability | 6.46E-07 | normal | 1 | 6.46E-07 | 4.17E-13 |
| 2 | | | | | | |
| 3 | | | | | | |
| 4 | | | | | | |
| | | | Combined Type A Uncertainty | | 6.46E-07 | 4.17E-13 |
| | | | | | | |
| | | Type B Uncertainty | | | | |
| | Uncertainty description | Uncertainty | Distribution | Divisor | Standard uncertainty | Variance |
| 1 | Resolution (1E-7) | 1.00E-07 | Rectangular | 3.4641016 | 2.89E-08 | 8.33E-16 |
| 2 | 10 Volt standard (+/- 1E-10) | 1.00E-10 | Rectangular | 1.7320508 | 5.77E-11 | 3.33E-21 |
| 3 | Thermal stability (0.01 ppm/V/0C) | 2.00E-07 | U-shaped | 1.4142136 | 1.41E-07 | 2.00E-14 |
| 4 | Calibration (1.5E-8) k=2 | 1.50E-08 | 2U | 2 | 7.50E-09 | 5.63E-17 |
| | | | Combined Type B Uncertainty | | 1.4453E-07 | 2.09E-14 |

**Figure 29.2** A sample uncertainty report.

*continued*

*continued*

| Combined Uncertainty Results | | | | | |
|---|---|---|---|---|---|
| | TYPE A standard uncertainty | 6.46E-07 | Distribution | Divisor | 1/Divisor |
| | TYPE A variance | 4.17E-13 | Rectangular | 1.7321 | 0.5774 |
| | | | Triangular | 2.4495 | 0.4082 |
| | | | U-shaped | 1.4142 | 0.7071 |
| | TYPE B standard uncertainty | 1.45E-07 | Resolution | 3.4641 | 0.2887 |
| | TYPE B variance | 2.09E-14 | | | |
| | | | Coverage factor (k) | Confidence level | Coverage factor based on Eff. *df* |
| | | | 1.000 | 68.27 | |
| | TYPE AB (combined) standard uncertainty | 6.62E-07 | 1.645 | 90.00 | |
| | TYPE AB (combined) variance | 4.38E-13 | 1.960 | 95.00 | |
| | | | 2.000 | 95.45 | |
| | Effective degrees of freedom | | 2.576 | 99.00 | |
| | Coverage factor (k) | 2 | 3.000 | 99.73 | |
| | Expanded uncertainty | 1.32E-06 | | | |
| | *Comments* | | | | |
| | Repeatability | | | | |
| | | | 1 | | 10.0000007 |
| | | | 2 | | 10.0000000 |
| | | | 3 | | 10.0000006 |
| | | | 4 | | 10.0000020 |
| | | | 5 | | 10.0000018 |
| | | | 6 | | 10.0000006 |
| | | | 7 | | 10.0000003 |
| | | | 8 | | 10.0000016 |
| | | | 9 | | 10.0000005 |
| | | | 10 | | 10.0000003 |
| | | | 11 | | 10.0000010 |
| | | | 12 | | 10.0000001 |
| | | | 13 | | 10.0000002 |
| | | | Mean | | 10.0000007 |
| | | | Uncertainty (Std. Deviation) | | 6.4564E-07 |

**Figure 29.2** A sample uncertainty report.

Note that the comments are important part of the uncertainty report. In this example, the raw data for repeatability analysis are provided. Other information such as instruments used. The method of calculation or why a certain distribution was used over another should be included for future reference.

## Measurement Uncertainty Considerations

The following is a guideline of factors to consider when determining measurement uncertainty. It is by no means complete, but it should give the reader some thought when estimating measurement uncertainty of the parameters considered. Refer to other examples provided in the CD-ROM.

- **Dimensional (Caliper).** In a simple process of measuring a dimension with a caliper, the following factors may contribute to the measurement uncertainty.
  - Caliper resolution
  - Caliper calibration
  - Caliper accuracy (specification)
  - Operator
  - Method
  - Repeatability and reproducibility
  - Environment

- **Electrical (voltmeter).** In a simple process of measuring voltage with a DMM, the following factors may contribute to the measurement uncertainty.
  - DMM resolution
  - DMM calibration
  - DMM accuracy (specification)
  - Operator
  - Method
  - Repeatability and reproducibility
  - Environment

- **Pressure.** In a simple process of measuring pressure with a digital pressure indicator, the following factors may contribute to the measurement uncertainty.
  - Indicator resolution
  - Indicator calibration
  - Indicator accuracy (specification)
  - Sensor specifications
  - Operator
  - Method
  - Repeatability and reproducibility
  - Environment
  - Location

- **Temperature.**
  - Temperature readout resolution
  - Temperature readout calibration
  - Temperature readout accuracy (specification)
  - Operator
  - Method
  - Repeatability and reproducibility
  - Environment
  - Location
- **Mass.**
  - Scale resolution
  - Scale calibration
  - Scale accuracy (specification)
  - Mass weight specifications
  - Operator
  - Method
  - Repeatability and reproducibility
  - Environment
  - Location
- **Torque.**
  - Torque calibrator resolution
  - Torque calibrator calibration
  - Torque calibrator accuracy (specification)
  - Torques arm accuracy
  - Torque arm specification
  - Operator
  - Method
  - Repeatability and reproducibility
  - Environment
  - Location

## OTHER MEASUREMENT UNCERTAINTY CONSIDERATIONS

### Sensitivity Coefficients

*Sensitivity coefficients* are defined as "the differential change in the output estimate generated by a differential change in the input estimate divided by the change in that input estimate."

These derivatives describe how the output estimate (y) varies with changes in the values of the input estimates ($x_1, x_2,..., x_n$).

Sensitivity coefficients show how components are related to result. The sensitivity coefficient shows the relationship of the individual uncertainty component to the standard deviation of the reported value for a test item. The majority of sensitivity coefficients for Type B evaluations will be one with a few exceptions.

### Effective Degrees of Freedom

According to K.A. Brownlee, "Degrees of freedom for the standard uncertainty, u, which may be a combination of many standard deviations, is not generally known. This is particularly troublesome if there are large components of uncertainty with small degrees of freedom. In this case, the degrees of freedom is approximated by the Welch-Satterthwaite formula."

$$V_{eff} = \frac{u_c^4(y)}{\sum_{i=1}^{N} \frac{u_i^4(y_i)}{v_i}}$$

where: $v_i$ = degrees of freedom.
$u_c$ = combined uncertainty
$u_i$ = individual uncertainties
$V_{eff}$ = effective degrees of freedom

For further information, refer to ANSI/NCSL Z540-2-1997 Annex G.6.4.

### Correlation Coefficients

Some uncertainties may be correlated in their effect on the measurand. In that case, the correlation coefficients and covariance determination may be required. While it is not in the scope of this book to describe in detail how to determine correlation coefficients and covariances, it is important to point this out. Determination of correlated cefficients may be avoided by giving some thought to how the uncertainties operate.[5]

Here's an example. If a voltmeter is used to measure the voltage of a standard thermocouple and of a thermocouple being measured by the standard, the uncertainties contributed by the voltmeter are correlated if made on the same voltmeter range. The uncertainties contributed by the voltmeter for both the thermocouples (standard and measured) will be almost identical and cancel out.[5] Thus no determination of correlation coefficients is required.

If the measurements are made on a different range of the voltmeter, the uncertainties are partially correlated. The technician has three options:

1. If the uncertainties are relatively small, use the uncertainty for one of the measurements.

2. Calculate correlation coefficients.

3. Use the voltmeter on the same range for both the measurements, thereby canceling out the correlation effects.

For further information on correlated input quantities, refer to ANSI/NCSL Z540-2-1997, Chapter 5.

## MANAGING UNCERTAINTY

It is important to ensure that once the measurement uncertainty for a process (calibration or test) has been determined, that the data are not filed away. The work is not finished. Managing measurement uncertainty is a continuous process and not a one-time exercise.

When a manufacturer provides measurement uncertainty data for its product, it is specified under stated operating conditions. The manufacturer may specify 30-day, 90-day, one-year, or other periods for use under defined humidity and temperature conditions.

It is important to estimate measurement uncertainty under the laboratory's operating environment and not the manufacturer's stated conditions. From time to time, the uncertainty estimates need to be reevaluated to ensure that estimates have not changed significantly. As a minimum, whenever equipment is recalibrated or readjusted, an evaluation is necessary to ensure that its measurement uncertainty has not changed.

Various tools exist for managing measurement uncertainty data. Use of software minimizes calculation errors and helps in managing data from a computer workstation. Computer spreadsheet packages also help in easing data calculations.

When using automated tools and utilities for managing measurement uncertainty, it are important to note that the data are only as good as the user determined it to be. The user must make responsible decisions about the quality of the data and the uncertainty analysis, even when a computer is used to perform the calculations. Remember the old adage from the 1960s: garbage in, garbage out. While software tools help in automating calculations, they do not make a sound decision. That is the responsibility of the technician.

Follow good math practices. Ensure that consistent rounding number conventions are followed. Do not mix units in uncertainty budgets. State all parameters in one unit. If that is not possible, state the data in percentage or in parts per million (ppm).

## References

ANSI/NCSL. 1997. *ANSI/NCSL Z540-2-1997, U. S. Guide to the expression of uncertainty in measurement.* Boulder, CO: NCSL International. (Also called the GUM).

Bentley, Robert E. 2000. *Uncertainty in measurement: the ISO guide with examples.* Sydney: National Measurement Laboratory CSIRO.

Brownlee, K.A. 1960. *Statistical theory and methodology in science and engineering.* New York: John Wiley & Sons.

European Cooperation for Accreditation. 1999. EA-4/02, *Expression of the uncertainty of measurement in calibration*. Paris: European Cooperation for Accreditation.

NIST/SEMATECH. 1999. "2.5.7.1 Degrees of Freedom," e-Handbook of Statistical Methods. Gaithersburg, MD: NIST and Austin, TX: International SEMATECH. www.itl.nist.gov/div898/handbook/mpc/section5/mpc571.htm

# Part VI
# Measurement Parameters

**Chapter 30**  Introduction to Measurement Parameters
**Chapter 31**  DC and Low Frequency
**Chapter 32**  Radio Frequency and Microwave
**Chapter 33**  Mass and Weight
**Chapter 34**  Dimensional and Mechanical Parameters
**Chapter 35**  Other Parameters: Chemical, Analytical, Electro-Optical, and Radiation

# Chapter 30
# Introduction to Measurement Parameters

*When you can measure what you are speaking about, and express it in numbers, you know something about it; but when you cannot measure it, when you cannot express it in numbers, your knowledge is of a meager and unsatisfactory kind: it may be the beginning of knowledge, but you have scarcely, in your mind, advanced to the stage of science ...*

William Thomson, Baron Kelvin of Largs, 1883.

Part VI is about the process of obtaining the knowledge described by Lord Kelvin—the numbers. Measurements are what transform opinion about something into the facts of it.

This part is an overview of the major parameters measured by calibration laboratories, and some information on how the measurements of each are made. This part is not intended to be a detailed or complete review of each parameter. The measurement parameters are grouped into five major types, based on the divisions commonly found in the industry.

- DC and low frequency
- RF and microwave
- Physical measurements
- Dimensional and mechanical
- Chemical, electro-optical, analytical, and radiation

Each chapter discusses some important measurements that are made in each section. For each measurement, there is information about the relevant SI units, typical measurement and transfer standards, typical workload items, and some information about the parameter and how it is measured. Remember, though, the information here can only supplement—not replace—the information in other more detailed references that apply to specific areas. Table 30.1 lists the principal parameters that are measured in each measurement area, in no particular order.[1] Not all of these are discussed in this book.

**Table 30.1** The five major measurement parameters.

| DC and Low Frequency | RF and Microwave | Physical | Dimensional and Mechanical | Chemical, Analytical, Electro-optical, Radiation |
|---|---|---|---|---|
| Capacitance | Antenna gain | Fluid density | Acceleration | Acidity (pH) |
| Current | Attenuation | Mass | Acoustics | Biological properties |
| Electrical phase angle | Electromagnetic field strength | Pressure | Angle | Chemical properties |
| Energy | Frequency | Relative humidity | Flatness | Color |
| Frequency and time interval | Impedance | Temperature | Force | Conductivity |
| Impedance | Noise figure | Vacuum | Hardness | Ionizing radiation |
| Inductance | Phase | Viscosity | Length | Light intensity |
| Magnetics | Power | | Optical alignment | Light power |
| Power | Pulse rise time | | Roundness | Light spectral analysis |
| Resistance | Reflection coefficient (VSWR) | | Surface finish | Nuclear activity |
| Voltage | Voltage | | Volume | Optical density |
| | | | | Radiation dosimetry |

# Endnote

1. Based in part on NCSL RP-9, 7–8.

# References

NCSL. 1989. Recommended Practice RP-9: *Calibration laboratory capability documentation guideline.* Boulder, CO: National Conference of Standards Laboratories.

Today in Science. 2003. *Lord Kelvin, physicist.*
   www.todayinsci.com/K/Kelvin_Lord/Kelvin_Lord.htm

# Chapter 31
# DC and Low Frequency

It is probably safe to say that almost everyone has had some interaction with devices that use direct current or alternating current in the low frequency range. Batteries provide direct current to things such as automobiles, radios, and toys. Alternating current is provided by national power distribution systems to industrial plants, businesses, and individual homes. The spread of technology through the twentieth century carried these things to every place on Earth.

## DC AND AC

When a voltage or current is at a constant level over time (ignoring random noise) it is said to be direct. When measuring or referring to voltage, it is correct to use the term *direct voltage*, or DV. When measuring or referring to current, it is correct to use the term *direct current*, or DC. It is common, though, for people to use DC to refer to both current and voltage.

When a voltage or current changes magnitude in a periodic manner (ignoring random noise), it is said to be alternating. When measuring or referring to voltage, it is correct to use the term *alternating voltage*, or AV. When measuring or referring to current, it is correct to use the term *alternating current*, or AC. It is common, though, for people to use AC to refer to both current and voltage.

When using measuring instruments, it is often important to consider the input impedance of the measuring instrument. For voltage and current measurements, the input impedance appears as resistance in parallel with the measurement. It can become a significant factor when measuring low voltages or high resistance. For current measurements, the shunt impedance is in series with the measurement and can become a significant factor when measuring low currents.

## LOW FREQUENCY

Direct current is the absence of any periodic variation, or 0 Hz. For electronic metrology, low frequency AC is generally considered to be from a frequency greater than zero up to 100 kHz. There is some overlap with the RF area, because a number of AC meters can measure up to one MHz, and some up to 10 MHz or more. Traditionally this area also includes frequency standards; they commonly operate at frequencies of 1, 5, or 10 MHz (as well as others).

Frequency ranges of particular interest are:

- Power frequencies (mostly 45 to 65 Hz and 400 Hz) are used for distribution of energy. The most common are 50 Hz and 60 Hz. 400 Hz is common in aircraft, submarines, and some other applications because the higher frequency allows smaller—and therefore lighter—motors, transformers, and other components.

- Audio frequencies (20 Hz to 20 kHz) are defined by the average limits of human hearing. Everything below 20 Hz is called *infrasonic*, and everything above 20 kHz is called *ultrasonic*.

- Ultrasonics (above 20 kHz). Ultrasonic frequencies up to several megahertz are used in applications such as nondestructive testing, medical imagery, motion detection, and short-distance range finding.

It is useful to note two important distinctions here. *Electronics* is concerned with variations of voltage or current (the movement of electrons) in a conductor or semiconductor, and with propagation of electromagnetic waves through free space. *Acoustics* is specifically concerned with the propagation of pressure waves (sound) through a physical medium. The medium most commonly referred to is air, which gives rise to the common perception of acoustics as having to do only with hearing. Acoustic energy is also carried through solid materials and at much higher frequencies. Applications of ultrasonics were mentioned. There is also overlap from ultrasonics into electronics: surface acoustic wave (SAW) devices use mechanical vibration within a structure to control electronic properties. SAW devices are used as filters and for other applications. Without them, many modern electronic devices (such as mobile telephones) would be larger, heavier, and more complex than they are now. SAW devices are well above the frequency range of this section, though.

## MEASUREMENT PARAMETERS

Following is a discussion of the most important measurement parameters in the DC and low frequency area. This does not include all parameters. Also, it is not a complete discussion of them, especially considering that in many cases complete chapters of books have been written about each. The parameters are:

- Direct voltage
- Direct current
- Resistance
- Alternating voltage
- Alternating current
- Capacitance
- Inductance
- Time interval and frequency

- Phase angle
- Electrical power

## Direct Voltage

Voltage is the difference in electric potential between two points in a circuit. Direct voltage has been known since ancient times as static electricity. It is inherent in the structure of matter, as the basic unit appears to be the charge or potential of the electron. Now, the measurement of direct voltage is one of the most common tasks in metrology and can be done with very high accuracy. See Table 31.1 for details about direct voltage.

Measurements of direct voltage can be made to uncertainties of a few parts per million or better. Most measuring instruments measure voltage using the direct method, but differential and ratio methods are also used.

When using transfer standards such as electronic volt standards or saturated standard cells, the usual practice is to use them in groups and use the group average as the assigned value. All of these devices drift. Using a large number allows the drift rates and directions to be averaged, which results in a more stable and predictable value. Many laboratories use automated systems to run daily intercomparisons between the cells or electronic standards in the group. Specialized switch matrixes and software collect measurements and report on the state of individual units in the group, the group average and standard deviation, and the drift rate. When saturated standard cells are used, it is usual to have a group of up to 12 used as the working standard, a group of four as a transfer standard and another group of four as a check standard.[1] Electronic volt standards are commonly used in groups ranging from four to six. At least four are needed to maintain the laboratory's local value of the volt and provide redundancy.[2]

The volt is an SI derived unit, expressed in terms of the watt and the ampere. The primary standard for the representation of the volt is the Josephson junction array. The array is used while immersed in liquid helium. It is an intrinsic standard that uses a

Table 31.1 Direct voltage parameters.

| Parameter Name | Direct Voltage (DV or most commonly DC) |
| --- | --- |
| SI unit | Volt (watt / ampere) |
| Typical primary standards | Josephson junction array<br>Saturated standard cells |
| Typical transfer standards | Electronic volt standards<br>Saturated standard cells |
| Typical calibration standards | DV calibrator, multifunction calibrator |
| Typical IM&TE workload | Direct voltage function of multimeters, thermocouple meters, galvanometers, voltage dividers, multifunction calibrators |
| Typical measurement method | Direct, ratio, differential |
| Measurement considerations | Temperature (thermal emf)<br>Excessive source loading |

property of superconductivity (the Josephson effect) to represent the volt in terms of two constants of quantum physics. The defining equation of the Josephson constant is:

$$K_J = 2e/h$$

In this equation, $e$ is the elementary charge of the electron, and h is Planck's constant. Effective January 1, 1990, the CIPM defined the conventional value of $K_{J-90}$ to be 483,597.9 GHz/Volt. A plot of voltage versus current shows distinct steps; with an applied frequency of 10 GHz, each step is about 20 µV high.

An electronic volt standard uses a zener diode in a temperature-controlled oven. The important property of a zener diode is that when reverse-biased, it has a nearly constant voltage drop over a very wide current range. The voltage developed across the zener diode is filtered to remove noise and is usually amplified to the desired level. Most of these standards have a primary output of 10 V nominal, with the actual output known to a resolution of 0.1 mV. They often also have a secondary output that divides this down to 1 V or to 1.018 V for compatibility with older saturated standard cells. Zener-based direct voltage references are normally used in groups or at least four. The members of the group are intercompared regularly (weekly or daily) to provide a group average value. While these standards can usually source a current of several milliamps if necessary, they are not power supplies. Measurements are normally made using voltmeters with high input impedance or by using the differential technique.

Important performance characteristics of a zener-based electronic volt reference are stability, noise, predictability, and hysteresis. *Stability* and *noise* are short-term performance characteristics. *Predictability* is how well the future performance can be predicted from linear regression of historical data. *Hysteresis* is a measure of how closely the standard will return to its previous value after its power has been off.

Saturated standard cells were the first widely used stable and predictable direct voltage standards. Over the past 60 years, the most common type has been the Weston saturated mercury-cadmium standard cell that was widely used in all levels of metrology laboratories. (There also was an unsaturated standard cell that was used in portable IM&TE such as thermocouple potentiometers and differential voltmeters.) Because saturated standard cells have a temperature coefficient of about 50 ppm/°C, they are kept in controlled-temperature enclosures or oil baths to minimize this effect. The voltage available from a standard cell is approximately 1.018 V, and measurements are often made with a resolution of 0.1 µV. Because of the high internal resistance of a standard cell, the voltage cannot be measured by any method that draws significant current from it. Voltage measurements and comparisons made using standard cells are always performed using the differential technique with a null meter. At the null point, where opposing voltages are balanced, the current is virtually zero and therefore the load impedance is effectively infinite. Standard cells are now largely being replaced by zener-based direct voltage references in many general calibration laboratories.

Particular care must be taken to avoid the effects of thermoelectric potentials when measuring low voltage (10 V and less). A thermal emf will be produced at every connection and can affect the voltage measurement. In effect, each connection is a thermocouple. Table 31.2 lists some common connector material pairs and their thermoelectric potential.[3]

In Table 31.2, all of the metals are assumed to be clean, bright, and untarnished, except one. Copper oxide is the tarnish that forms on copper relatively quickly after its

Table 31.2 Thermoelectric effects from connector materials.

| Materials | Thermoelectric Potential |
|---|---|
| Copper – Copper | ≤ 0.2 µV/°C |
| Copper – Silver | 0.3 µV/°C |
| Copper – Gold | 0.3 µV/°C |
| Copper – Lead/Tin solder | 1 to 3 µV/°C |
| Copper – Nickel | 21 µV/°C |
| Copper – Copper oxide | 1000 µV/°C |

protective insulation is removed. From this, it should be obvious that it is very important for all copper wires and connection points to be regularly cleaned to keep them bright and shiny.

Each connection ideally should be at the same temperature, which means allowing time for them to stabilize after handling. Connections must be clean and tight. Connections should be made with low thermal emf alloys such as gold-plated tellurium copper or bright clean copper. Tarnished copper and nickel-plated connectors should be avoided because of the very high thermal emf potentials.[4] Residual thermal effects can be evaluated by making two sets of measurements. Measure the voltage, reverse the voltage sense leads at the meter and repeat the measurement, then average the values.[5]

## Direct Current

Electric current is the movement of electrons through a conductor. If the overall movement is in a single direction with no periodic reversal, it is direct current. Current is measured in amperes and is a SI base unit. See Table 31.3 for details about direct current.

For calibration purposes, direct current is measured using the indirect technique. The current is passed through a known resistance (called a *shunt*), and the voltage developed across the resistance is measured. The relationship is one of the arrangements of Ohm's law:

$$I = E / R$$

where I is the current in amperes, E is the voltage in volts, and R is the resistance in ohms.

Measurement problems arise when dealing with very small or very high currents. Very small currents, 100 pA or less, are difficult to measure because the input impedance of a typical laboratory digital multimeter is equal to or less than the resistance needed to generate a measurable voltage. Measurements in this area require a picoammeter. Very high currents are difficult to measure with a shunt and voltmeter because a very low resistance is necessary to avoid overheating, but the voltage developed may be too low to measure accurately with a normal digital multimeter. Measurements in this area require an electrometer or a nanovoltmeter.[6]

There are some meters that measure direct current by passing it through coils in the meter movement to deflect and indicating needle, but these are normally calibration workload items rather than measurement standards.

**Table 31.3** Direct current parameters.

| Parameter Name | Direct Current (DC) |
|---|---|
| SI unit | Ampere |
| Typical primary standards | Calculable capacitor, current balance |
| Typical transfer standards | Resistance and voltage standards, using Ohm's law |
| Typical calibration standards | DC calibrator, multifunction calibrator, transconductance amplifier |
| Typical IM&TE workload | Direct current function of multimeters, multifunction calibrators, current shunts, current sources |
| Typical measurement method | Indirect |
| Measurement considerations | Thermal effects, excessive source loading, very low or very high currents |

## Resistance

Resistance is the opposition to electric current. In most situations it is a property of all materials in varying degrees. Some materials exhibit no measurable resistance (superconductivity) at cryogenic temperatures. See Table 31.4 for detail about resistance.

The Ohm is a derived unit, equal to the voltage divided by current. This relationship, (R = E/I), is known as Ohm's law for direct current circuits.

The quantum Hall effect, discovered in 1980 by Klaus von Klitzing, is used to make a representation of the Ohm. The Hall effect, which was discovered in 1880, is exhibited when a semiconductor with a current passing through it and exposed to a magnetic field produces a voltage that is proportional to the strength and polarity of the field. (A Hall device is used in many current sensors and electronic compasses.) A quantum Hall effect (QHE) device is made so that electron flow is confined to an extremely thin layer, and it is operated at a temperature of less than 4 kelvin. Under these conditions, the voltage changes in steps instead of continuously as the magnetic field is varied. Von Klitzing discovered that the steps are multiples of a ratio of two fundamental values of physics, Plank's constant ($h$), and the elementary charge of the electron ($e$):

$$R_k = h/e^2$$

where $R_k$ is the von Klitzing constant. Effective January 1, 1990, the conventional value of $R_{k-90}$ is defined by the CIPM as 25,812.807 Ω. In use, the QHE device is connected in series with a resistor, usually 6400 Ω, and the ratio of the Hall voltage and the voltage across the resistor are determined. Ratio methods are then used to transfer the resistor value to the more common 1 Ω and 10 kΩ standard resistors.[7]

In most cases, the value of the Ohm is maintained using banks of 1 Ω and 10 kΩ standard resistors. There are several types: the Thomas One Ohm standard, the Reichsanstalt design for values below one Ohm, and the Rosa or NBS style for values above 10 Ω. These are made using special alloys, the oldest of which (Manganin) was developed in 1884.[8] Resistance values are transferred using these and other devices, such as the Hamon style transfer standard, in various values.

Table 31.4 Resistance parameters.

| Parameter Name | Resistance |
|---|---|
| SI unit | Ohm |
| Typical primary standards | Quantum Hall effect apparatus, calculable capacitor, standard resistors |
| Typical transfer standards | Standard resistors |
| Typical calibration standards | Digital multimeter, multifunction calibrator, standard resistors, resistance bridge |
| Typical IM&TE workload | Resistance function of multimeters, multifunction calibrators, resistors, current shunts |
| Typical measurement method | Ratio, transfer, indirect |
| Measurement considerations | Thermal effects, excessive current |

The value of standard resistors is normally transferred by ratio methods. One method is the use of any of several types of resistance bridge circuits. The other method compares the voltages across two resistors when they are connected in series and therefore have the same current flowing through them. In either case, one of the resistors is a known standard used as the reference for the ratio.

Multimeters typically use one of two different indirect measurement methods. One method passes a known current through the resistor and measures the voltage developed across it. This method is implemented in most digital multimeters. The other method places an ammeter and adjustable resistor in series with the unknown resistor and a fixed voltage across the combination. The ammeter is set to a reference with zero resistance and then used to measure the unknown. This method is implemented in most analog meters and some digital multimeters. With either method, the effect is using Ohm's law to determine resistance from values of voltage and current.

High-accuracy resistance measurement is performed using the four-wire measurement method. One pair of wires is used to connect the current that is passed through the resistor. The other pair of wires is used to measure the voltage across it. Advantages of the four-wire method are elimination of the test lead resistance, higher accuracy voltage measurements, and the possibility of measuring both voltage and current simultaneously.

When measuring low resistances (10 kΩ and less), particular care must be taken to avoid the effects of thermoelectric potentials. A thermal emf will be produced at every connection and can affect the voltage measurement. Each connection should ideally be at the same temperature, which means allowing time for them to stabilize after handling. Connections must be clean and tight. Connections should be made with low thermal emf alloys such as gold-plated tellurium copper or bright clean copper. Tarnished copper and nickel-plated connectors should be avoided because of the very high thermal emf potentials.[9] Residual thermal effects can be evaluated with the meter in DC volt mode. One method is to measure the voltage, reverse the voltage sense leads at the meter and repeat the measurement, then average the values.[10] Another method is to reverse the polarity of the current and repeat the measurement.[11] Some laboratory digital multimeters have measurement modes that use one of these methods automatically.

## Alternating Voltage

Alternating voltage changes value in a periodic manner. This makes it useful in many ways that direct voltage cannot be. In particular, alternating voltage can be stepped up or down using transformers, which makes it practical for long-distance energy transmission. See Table 31.5 for details about alternating voltage.

Alternating voltage metrology is based on the conversion of electrical power to heat. When a voltage is across a resistor, it converts the power to a proportional amount of heat. The relationship is a version of Ohm's law:

$$W = E^2/R$$

where $W$ is the power in watts, $E$ is the voltage, and $R$ is the resistance. A resistor produces a certain amount of heat when an unknown alternating voltage is applied. If a direct voltage is then applied and adjusted so that the same amount of heat is produced, then its value it must be equal to the alternating voltage. This is the basic principle of a thermal voltage converter.

The thermoelement in a thermal voltage converter is a fairly simple device. It consists of a resistive heater element and a thermocouple that is physically very close but electrically isolated. The assembly is enclosed in a glass envelope and a vacuum is created inside. There are different models to accommodate various voltage ranges. An ideal thermal voltage converter will have square-law response, so the direct voltage output represents the RMS value of the alternating voltage. Deviations from the ideal response are on the device's calibration report as an AC–DC difference. Another error, reversal error, is present due to different responses to +DC and −DC in the thermoelement. There some disadvantages to this type of thermal converter. There is some unit-to-unit variation, so a converter must be recalibrated if the thermoelement is replaced. They are delicate, and their output varies with increasing frequency. In addition, the thermocouple output is in millivolts, which can cause measurement difficulties. The devices also respond fairly slowly, which can be a disadvantage.[12]

A multijunction thermal converter of this type does exist, but is not widely used. This device uses multiple thermocouples in series to increase the DC output to around

**Table 31.5** Alternating voltage parameters.

| Parameter Name | Alternating Voltage (AV or most commonly AC) |
|---|---|
| SI unit | Volt (watt/ampere) |
| Typical primary standards | DC Volt, thermal voltage converter, micropotentiometer |
| Typical transfer standards | AC/DC thermal voltage converter |
| Typical calibration standards | AV calibrator, multifunction calibrator, DV reference, thermal converter |
| Typical IM&TE workload | Alternating voltage function of multimeters, multifunction calibrators, ratio transformers |
| Typical measurement method | Indirect, transfer, direct |
| Measurement considerations | Transfer error, frequency |

100 mV. It is limited in the amount of applied voltage and in frequency.[13] There is also some work on a thin-film multijunction thermal converter fabricated using microcircuit techniques, but this is not known to be in commercial use yet.

Another type of thermal AC/DC converter was developed by Fluke in the 1970s.[14] This sensor uses two resistor/transistor sets on a microcircuit. Each set is thermally isolated from the other. One resistor is the input; applied voltage heats the transistor next to it. That unbalances an amplifier that then produces a direct voltage to the other resistor. When the second transistor is at the same temperature as the first, the circuit is balanced and the amplifier is producing a direct voltage proportional to the input voltage. Major advantages of this sensor are a 2 V output, shorter response time, and lower DC reversal error.[15]

The thermal devices are used for the lowest uncertainty measurements in their range (50 ppm or less) and as transfer standards. Most meters do not use these methods. Analog and digital alternating voltage meters use one of the following several methods to make direct measurements.

- **High-speed sampling of the input and direct computation of the RMS value.** This method can be very accurate, but it requires an unvarying input at a frequency no higher than half the sampling rate. This method has minimum input frequency and amplitude limits.

- **Logarithmic amplifier.** This method uses a logarithmic amplifier and other circuitry to produce a converter output proportional to the RMS value of the input in real time. Many true RMS digital multimeters use this method. This method has minimum input frequency and amplitude limits.

- **Direct measurement of the average voltage and multiplying by a scale factor.** Unless the meter specification says it does true RMS measurement, this is probably what it does. The meter measures the average alternating voltage and multiplies it by 1.414. For a pure sine wave, that will make the meter read the same as an RMS responding meter. Any deviation from a sine wave will result in increased error. (The specifications for this type of meter will often say something like "average responding, RMS calibrated.")

At voltages over 1000 V, resistive voltage dividers or electrostatic voltmeters are commonly used.

## Alternating Current

Alternating current occurs when the direction of electron flow in a conductor changes direction in a periodic manner. Like alternating voltage, this makes it useful in many ways that direct current cannot be. In particular, alternating current can be stepped up or down using transformers, which makes it practical for long-distance energy transmission. When passing through a transformer, the resulting current ratio has an inverse relationship to the voltage ratio. If the voltage is doubled, the current is halved. The total energy remains the same, but transmission is more efficient because the power loss in a conductor with resistance is proportional to the square of the current. See Table 31.6 for details on alternating current.

**Table 31.6** Alternating current parameters.

| Parameter Name | Alternating Current (AC) |
|---|---|
| SI unit | Ampere |
| Typical primary standards | Direct current, thermal current converter |
| Typical transfer standards | AC/DC thermal current converter |
| Typical calibration standards | AC calibrator, multifunction calibrator, DC reference, thermal converter |
| Typical IM&TE workload | Alternating current function of multimeters, multifunction calibrators, current transformers, current and power meters |
| Typical measurement method | Indirect, transfer |
| Measurement considerations | Transfer error, frequency |

Alternating current metrology is based on the conversion of electrical power to heat. When a current is passed through a resistor, it converts the power to a proportional amount of heat. The relationship is a version of Ohm's law:

$$W = I^2/R$$

where W is the power in watts, I is the current, and R is the resistance. Except for being used to measure current, this is the same as a thermal voltage converter.

The thermoelement in a thermal current converter is identical to the one used in a thermal voltage converter. The difference is that a current shunt that has a very small AC/DC difference is connected in parallel with the thermoelement. This forms a current divider, limiting the current thought the thermoelement to its full-scale current when the shunt's rated current is applied. Instruments using Fluke's solid-state thermal converter are used with current shunts in the same manner. Note that for best results in either case, the shunts and AC/DC transfer standard should be calibrated together.[16]

The thermal devices are used for the lowest uncertainty measurements in their range (0.05% or less) and as transfer standards. Most meters do not use these methods. Analog and digital alternating current meters typically measure the voltage developed across an internal shunt, using one of the AV measurement methods described earlier.

To measure currents over 20 A, current transformer coils are generally used. The coil itself is the secondary coil of a transformer, sized to provide 5 amperes when the full rated current is passed through the primary. The primary is a cable loop connected to both sides of the current source and passing once through the center of the current transformer.[17]

## Capacitance

Capacitance is a property of a circuit or device that opposes a change in voltage. A capacitor can store a charge in the electric field of the dielectric (insulation) between its conductors. See Table 31.7 for details on capacitance.

The unit of capacitance, a farad, is equal to one coulomb of electric charge divided by one volt. A coulomb is the quantity of electricity moved in one second by a current

**Table 31.7** Capacitance parameters.

| Parameter Name | Capacitance |
| --- | --- |
| SI unit | Farad (derived from ampere and second) |
| Typical primary standards | Calculable capacitor, standard capacitors, current comparator |
| Typical transfer standards | Standard capacitors |
| Typical calibration standards | Standard capacitors, capacitance bridge, electronic impedance bridge |
| Typical IM&TE workload | Capacitance current function of multimeters, decade capacitors, tank level indicators or test sets |
| Typical measurement method | Ratio |
| Measurement considerations | Transfer error, temperature, frequency, shielding |

of one ampere. If a current of one ampere is applied to a one farad capacitor, the charge stored in it will change at the rate of one volt per second. For practical use, one farad is an extremely large value. Most capacitance measurements are in microfarads to picofarads.

A property of a capacitor in an AC circuit is its reactance $X_C$:

$$X_C = \frac{1}{2\pi f C}$$

where f is the frequency in hertz and C is the capacitance in farads.

The reactance of a capacitor can be measured in the same way as DC resistance—voltage across the capacitor divided by the current in the circuit, with the result expressed as ohms. If the AC frequency is known and the reactance is measured, the value of an unknown capacitor can be found by rearrangement of the defining equation for reactance.

At a standards laboratory, the farad may be realized by using a calculable capacitor. This device is made from a set of four long, parallel metal rods arranged so that when viewed from one end they are at the corners of a square. A short ground rod is fixed in the center at one end and a movable ground rod is inserted from the other end. The arrangement of the parallel rods provides a constant value of capacitance per meter of length. Changing the position of the movable ground changes the effective length and therefore changes the measured capacitance. The theoretical value can be calculated from the length and the speed of light in air.[18]

Capacitance measurements are generally made by the ratio method. A capacitance bridge is a frequently-used item in many calibration labs. The capacitance bridge includes high-stability reference capacitors in the ratio arm, and the range can be extended if necessary by substituting an external standard capacitor. Some types of ratio bridges can be used to compare a standard capacitor to a standard resistor, thereby comparing capacitive reactance to resistance.

Some electronic impedance bridges place the unknown capacitor in series with a known resistor and make a ratio measurement of the voltages across them. This ratio is also the ratio of the capacitive reactance to the known resistance, so the value of the capacitor can be both calculated and compared to a standard.

## Inductance

Inductance is a property of a circuit or device that opposes a change in current. An inductor can store energy in the magnetic field around its conductors. See table 31.8 for details about inductance.

The unit of inductance, a henry, is equal to one weber of magnetic flux divided by one volt. A weber is the amount of magnetic flux produced by a current that is changing amplitude at the rate of one volt per second.

A property of an inductor in an AC circuit is its reactance, $X_L$:

$$X_L = 2\pi f L$$

where f is the frequency in hertz and L is the inductance in henrys.

The reactance of an inductor can be measured in the same way as DC resistance—voltage developed across the inductor divided by the current through it, with the result expressed as ohms. If the AC frequency is known and the reactance is measured, the value of an unknown inductor can be found by rearrangement of the defining equation for reactance.

In principle, a calculable inductance standard can be made by constructing an extremely uniform solenoid. (A solenoid is a long wire coil, where the length of the coil is much greater than the radius of each turn.) The inductance is calculated from the length, radius, and number of turns. In practice, various problems make this not very practical, although a few have been made. Very precise standard capacitors are easier to make, so they are used with a ratio bridge to determine the values of standard inductors.[19] Most practical standard inductors are wound as toroids to minimize their physical size and their external magnetic field.

Inductance measurements are usually made by the ratio method. Inductors can be compared with each other, standard capacitors, or standard resistors.

**Table 31.8** Inductance parameters.

| Parameter Name | Inductance |
|---|---|
| SI unit | Henry (derived from ampere and second) |
| Typical primary standards | Standard inductor, inductance bridge, standard capacitor |
| Typical transfer standards | Standard inductors |
| Typical calibration standards | Standard inductors, inductance bridge, electronic impedance bridge |
| Typical IM&TE workload | Decade inductors, LCR meters, impedance bridges |
| Typical measurement method | Ratio |
| Measurement considerations | Transfer error, temperature, current limits, frequency, magnetic field shielding |

Some electronic impedance bridges place the unknown inductor in series with a known resistor and make a ratio measurement of the voltages across them. This ratio is also the ratio of the inductive reactance to the known resistance, so the value of the inductor can be both calculated and compared to a standard.

## Time Interval and Frequency

The second is the SI base unit of time interval. The hertz is the derived unit of frequency, with one hertz being one complete cycle per second. This makes the two unique in that by knowing one, the other is known automatically. Another unique attribute is the ability to measure these values with extremely high precision. In 1996 Michael Lombardi wrote, "Frequency and time interval can be measured with greater precision than all other physical quantities. In some fields of calibration, one part per million ($1 \times 10^{-6}$) is considered quite an accomplishment. In the world of frequency calibrations, measurements of one part per billion ($1 \times 10^{-9}$) are routine, and even one part per trillion ($1 \times 10^{-12}$) is commonplace.[20]" See Table 31.9 for details about time interval and frequency.

The SI definition of the second is the duration of 9,192,631,770 periods of radiation resulting from a pair of quantum transitions of the caesium 133 atom.[21] This is the basis of the international atomic time scale (TAI) maintained by the BIPM. Because it is based on quantum physics, this time scale is uniform—it does not vary. The time maintained by BIPM is based on the weighted average of over 200 atomic time standards operated by about 50 laboratories worldwide.[22] The definition of the second also means that a caesium beam frequency standard is an intrinsic standard and is assumed to be correct so long as it is operating correctly. (But they still need to be verified!)

**Table 31.9** Time interval and frequency parameters.

| Parameter Name | Time Interval / Frequency |
|---|---|
| SI unit | Second / hertz |
| Typical primary standards | Atomic frequency standards (caesium beam, caesium fountain, hydrogen MASER) |
| Typical transfer standards | Caesium beam frequency standards<br>Terrestrial or satellite RF signals. |
| Typical calibration standards | Caesium beam frequency standards<br>Quartz or rubidium oscillators disciplined to LORAN or GPS transmissions<br>Frequency and time interval counters, signal generators (low frequency, RF and microwave), function generators |
| Typical IM&TE workload | Frequency and time interval counters, signal generators (low frequency, RF and microwave), function generators, spectrum analyzers, navigation and communication equipment |
| Typical measurement method | Ratio (phase comparison)<br>Direct (time interval or frequency counter) |
| Measurement considerations | Phase noise |

Note: It is important not to confuse the SI unit of time interval, the second, with other common definitions of the second that are derived from astronomical observations. (See the section on time of day on page 340.)

It is probable that almost every electronic calibration laboratory has a frequency standard of some type. Frequency standards have a fixed output of at least one frequency, and units with several different outputs are common. Common frequencies are 1, 5 and 10 MHz, as well as 1 kHz and 1 Hz. A laboratory frequency standard can readily be compared to a national standard by any of several means. All of these use a radio signal as a transfer standard between the laboratory's frequency standard and the national metrology institute. Some examples follow:

- At least 17 countries provide HF radio broadcasts as time and frequency standards. The carrier frequencies of these broadcasts can be compared with a frequency standard. Although most of the transmitted frequencies have uncertainties of $10 \times 10^{-12}$ or better,[23] disturbances of the signal propagation through the atmosphere limit the available received accuracy to several parts in $10^7$.[24] Since these broadcasts are provided by the national agency responsible for time and frequency standards, they are traceable by definition.

- Several countries transmit standard signals in the LF or VLF radio bands. For example, in the United States, a 60 kHz signal is transmitted by NIST from WWVB in Fort Collins, Colorado. These signals can also be compared with a laboratory's frequency standard. Because of differences in propagation over the course of a day, best accuracy is obtained when the comparison is done over a full 24-hour period. Since these broadcasts are provided by the national agency responsible for time and frequency standards, they are traceable by definition. In a lot of cases these signals also carry time of day information that can be used by demodulating the signal. (As well as laboratory uses, this has spawned a wide variety of consumer timepieces equipped to receive these signals and commonly being advertised as atomic watches or clocks.)

- The U.S. Coast Guard operates the LORAN-C radio navigation network; the signal of that can be received throughout North America, most of the northern Pacific Ocean, and much of the North Atlantic Ocean. (Another LORAN-C network operates in northwestern Europe.) Timing signals from the LORAN system can be compared with a frequency standard with accuracy of several parts in $10^{11}$ or better under ideal conditions. The frequency of LORAN transmissions and the timing of the reference pulses are traceable to national standards. (Note that operation of the LORAN system was scheduled to be terminated at the end of 1999. It is still operating at the end of 2003, but its long-term future availability is undetermined.)

- Widespread use of navigation satellites—the NAVSTAR Global Positioning System (GPS) operated by the U.S. Department of Defense and the similar GLONASS system operated by Russia—have made an important improvement in comparison of frequency standards. The intended use of GPS is navigation. If at least four satellites are in view, a GPS receiver can determine three-dimensional position with respect to a model of the Earth's surface; some receivers can also display the velocity vector. If an antenna is in a permanent location and the receiver can be instructed to not update that position, then the same data can be used as a transfer standard for frequency, time interval, and time of day. If two laboratories have the same satellite in view, then the signals received from it can be compared and used as a transfer standard (the common-view technique).

- In the United States, a laboratory can subscribe to the Frequency Measurement and Analysis System (FMAS) provided by NIST. This is a leased GPS-based system installed at the laboratory by NIST and used to measure frequency standards and oscillators. This system is traceable to national standards both through the GPS system and because it is regularly monitored by NIST over a telephone link.[25] The common-view technique is used with this system. The frequency uncertainty can be about $1 \times 10^{-13}$ with a one day measurement time and possibly an order of magnitude better at one week.[26] The FMAS is able calibrate up to five oscillators simultaneously, although normally one of the five would be the laboratory reference standard.

The GPS system has a very high level of performance. (Sources of technical details are listed in the bibliography.) Each of the 24 satellites has at least four atomic frequency standards (two caesium and either two or three rubidium). GPS receivers intended for time and frequency measurements must be simultaneously tracking at least four satellites to obtain valid signals, and most are capable of tracking as many as 8 to 12. The receiver output is an average from all of the satellites in view, so in effect the signals of at least 16 atomic standards are being averaged. The receiver output is used to phase lock, or discipline, a frequency standard. There are a large number of companies that manufacture GPS-disciplined quartz or rubidium oscillators, although not all are suitable for frequency metrology. The main risk of using a satellite-based system is that position and timing accuracy for nonmilitary users can be intentionally degraded at any time by the Department of Defense. This degradation, called *Selective Availability* or *SA*, was turned off in May 2000, but can be turned on again any time it is needed for military requirements. When GPS is used as a transfer standard, SA appears as additional phase noise.

The frequency and time interval derived from the GPS system is traceable to the NIST master atomic clock array because the same signals are monitored by NIST and they provide correction data to the GPS control center.[27] (It is very important to note again that many GPS receivers are *not* suitable for this type of use.[28]) As well as using the signal directly, a laboratory must also obtain performance data from various sources and compare that with the historical performance of their system. These data are available from USNO, NIST, U.S. Coast Guard, BIPM, and others.[29] One of the corrections that the GPS control center applies to the system accounts for time errors due to general relativity—the first time this has been necessary outside the scope of a scientific study.[30] GPS also provides time of day information. What is actually transmitted is a value called GPS time, but the data also include the difference between that and universal coordinated time.

When calibrating a frequency standard, three methods are used. One method is to use a phase comparator to compare the unit under test to the standard. (Phase comparison is a ratio method). The frequency of the UUT is adjusted to minimize the phase shift over a given time period. The accuracy improves as the time period is increased. Another method is using a time interval counter. The counter is triggered by the start of a pulse from the frequency standard and stopped by the start of a pulse from the timebase under test. The time interval between them is measured, and the rate at which it changes is computed and used to adjust the UUT. A limitation is the time resolution of the time interval counter, but units with picosecond resolutions are available. The other method is direct measurement with a frequency counter. This is limited by the resolution and uncertainty of the counter, but counters with resolution of 0.001 Hz or better at 10 MHz are readily available.

Frequency calibrations can take considerable time, especially when calibrating a high quality oscillator and using LORAN or GPS as the transfer standard. The problem is noise. Atmospheric propagation effects add noise and phase uncertainty to both signals, with LORAN being affected more than GPS. The GPS signal will have significantly greater uncertainty if SA is turned on. Although the long-term stability of both systems is high, the noise creates a great deal of short-term instability. It is recommended that the measurement period for most oscillators should be at least 24 hours.[31] Caesium beam frequency standards and rubidium oscillators may require a measurement period of several days to a week to adequately verify their performance.

## Time of Day

It is important not to confuse the preceding information on time interval with the common use of time of day. The TAI time scale is based on the SI defined value of the second, which is a fixed time interval (or duration) and does not change. Time of day has many time scales, most of which are based on astronomical observations. The second is used as a unit in time of day, but different time scales use seconds of different duration (or size).

Terrestrial time, formerly called ephemeris time, is used mostly by astronomers. It uses a time scale based on the duration of the Earth's orbit around the sun. The ephemeris second is a fraction of the tropical year, using the interval between the spring equinoxes. The tropical year is defined as 31,556,925.9747 ephemeris seconds in duration.[32]

Many customary time scales are based on the mean solar day, which has 86,400 seconds. For example, this is the basis of legal time in the United States.[33] The duration of a day is determined by observations of the sun as it passes through the zenith. This means that the duration of the mean solar second is not a constant. It slowly increases because of short-term variations and gradual decrease in the Earth's rotational speed.

The seconds of atomic time, ephemeris time, and mean solar time are defined so that they were all equal on the starting instant of 1900.[34]

Universal time is used for navigation and for ordinary timekeeping. Universal time (UT) has three different components, of which only one is of practical importance to nonspecialists. UT1 is a time scale that is kept in step with the Earth's rotation as measured on the zero meridian at Greenwich, England. This means that the solar second is offset from the atomic time scale. Navigators use UT1 time to measure longitude.

Universal coordinated time (UTC) is a uniform atomic time scale, and the UTC second has the same duration as the TAI second. UTC is synchronized with UT1, but this means that UTC must be adjusted occasionally to keep the synchronization. Synchronizing the scales is accomplished by adding a leap second as needed to keep UTC within 0.9 s of UT1.[35] (Therefore, a given year may be one or two seconds longer than the year before or after.) The agencies that provide time of day standards around the world have agreed to keep their master clocks coordinated with UTC as determined by the BIPM. The UTC time scale was formerly called Greenwich Mean Time (GMT).

In the United States, UTC is maintained and provided by two sources. The U.S. Naval Observatory (USNO) provides UTC to the Department of Defense and the two major radio navigation systems—GPS and LORAN-C. If a GPS receiver is tracking at least four satellites, it is also receiving UTC time information that is normally within

350 nanoseconds (or better) of UTC as maintained by USNO.[36] This level of accuracy is necessary for the GPS system to realize its intended purpose as a navigation tool. For all other purposes, UTC is provided by NIST from its master clock. This may be important for a calibration laboratory that is performing calibration related to time of day in addition to frequency and time interval. If it is using the GPS system as a transfer standard, then frequency and time interval are traceable to the SI through NIST as discussed earlier. If it is using the time of day information, that is traceable to UTC and UT1 through USNO. As a practical matter, however, NIST and USNO keep their representations of UTC within a few nanoseconds of each other; and the differences between each and UTC are published monthly in BIPM Circular T.

## Phase Angle

Accurate measurement of electrical phase is important in areas as diverse as metering electrical energy, precise positioning of control devices, or safe navigation of aircraft. See table 31.10 for details on phase.

The SI unit that applies to phase measurements is the radian, but the commonly used unit of measure is the degree (°). The division of the circle into 360°, with 60 minutes per degree and 60 seconds to the minute originated in ancient Babylon. Now it is also common for subdivisions of the degree to be expressed in decimal form, 12.34°, for example. The relationship to radians is:

$$\text{Degree} = 1/360 \text{ circle} = \pi/180 \text{ rad}$$

$$\text{Minute} = 1/60 \text{ degree} = \pi/10\,800 \text{ rad}$$

$$\text{Second} = 1/60 \text{ minute} = \pi/648\,000 \text{ rad}$$

$$360° = 2\pi \text{ rad}.$$

A 1978 paper by R. S. Turgel and N. M. Oldham opens with the statement, "Phase measurements are essentially a determination of the ratio of two time intervals, and as

**Table 31.10** Phase angle parameters.

| Parameter Name | Phase |
|---|---|
| SI unit | Radian |
| Typical primary standards | High-precision phase calibration standard, resistor bridges, capacitive bridges |
| Typical transfer standards | Resistor bridges, capacitive bridges |
| Typical calibration standards | Phase calibration standard, phase angle voltmeter, synchro/servo simulator, ratio transformer |
| Typical IM&TE workload | Phase angle voltmeter, angle position indicator, phase meters, power analyzers, radio navigation equipment |
| Typical measurement method | Ratio |
| Measurement considerations | Phase noise, RMS voltmeter limits |

such are not dependent on any system of measurements."[37] There are two things that can be drawn from this.

- Phase measurements are made by the ratio method.
- There is no realizable standard radian, at least not in the same sense as we have realizable standards for units such as the meter or the volt.

Electrical phase angles can, of course, be generated and measured. There are also standards that generate phase relationships and methods for verifying those standards.

Signals used in phase standards are generated digitally and use digital timing to shift the phase. Display resolution is usually millidegrees (0.001°), but digital quantization may limit the resolution to something less than this. One commercial standard, for example, has an output that only increments in steps of $360/2^{18}$, or 0.001373°, due to binary counting in the digital circuits.[38] This must be accounted for in the uncertainty. IM&TE that is calibrated often has resolution of 1 or 10 millidegrees.

Phase standards are verified using two-arm resistive and capacitive bridges. One arm is connected to the reference output, the second arm is connected to the variable output, and the center of the bridge is connected to a voltmeter or oscilloscope. If the two arms are equal (a 1:1 ratio bridge) and the phase is set to 180.000°, the signal should cancel exactly. As the phase is varied in 1 m° steps above and below 180.000°, each of the output steps can be observed. Verification of phase performance is simply observing that all of the digital steps are present. When a 1:1 bridge is used and the signal amplitudes are equal, it is possible to separate the errors from the phase standard and those from the bridge. After the first set of readings are made, the inputs to the bridge are swapped and the measurement is repeated. The phase errors are calculated for each set. It has been shown that at any step, half the sum of the phase errors is the error of the phase standard and half the difference is the error in the bridge. This provides an absolute verification of the phase standard and the bridge at the same time.[39] Note that bridges made with other ratios (10:1 or 100:1, for example) are themselves verified by comparison to other well-characterized bridges.

A measurement problem may arise because the theoretical output of the bridge is zero or very close to it. This can be a problem for some RMS voltmeters. Some methods to work around this include using angle pairs further away from the null point, using unequal-signal amplitudes, or using a high-gain preamplifier.

## Electrical Power

Measurement of electric power (or energy, when integrated over time) is done so frequently that it is virtually invisible to most people—except when the electricity bill arrives at home or business for payment. See Table 31.11 for details on electrical power.

From Ohm's law we know that a DC power in watts is $P = E \times I$ where E is the voltage and I is the current. If an AC circuit had only pure resistance, the same relationship would be true, however, there is always at least a small amount of inductance or capacitance. This leads to a phase difference between voltage and current and therefore a difference between the apparent power (above) and the true power. True power is

$$P = EI \cos \theta$$

where θ is the phase angle between voltage and current.

Table 31.11 Electrical power parameters.

| Parameter Name | Power |
|---|---|
| SI unit | Watt |
| Typical primary standards | Voltage, current, resistance, time interval, phase sampling wattmeter, synthesized power source |
| Typical transfer standards | Voltage, current, resistance, time interval, phase, characterized watt-hour meter |
| Typical calibration standards | Voltage source, current source, AC/DC shunts, voltmeters, phase standard |
| Typical IM&TE workload | Watt-Hour meters, power meters, power analyzers, power supplies, load banks |
| Typical measurement method | Indirect |
| Measurement considerations | Safety, AC/DC transfer, measurement uncertainty—especially of small shunt voltages. |

One watt is one joule (J) of energy delivered in one second. If this is continued for an hour, 3600 J or one watt-hour of energy is delivered. In actual practice, the customary practical units are kilowatt-hour for selling energy and megawatt-hour for generating it. In any case, it can be seen that the accuracy of metering is important to an electric utility company.

There are several methods of measuring power or energy. All are indirect methods, except the first does make a direct measurement of voltage.

- Power alone (without considering time) can be found by placing an appropriate current shunt in the circuit and then measuring the voltages across the load and the shunt. The values are multiplied to find the power. If either the impedance or phase can be measured then true power can be found, otherwise only apparent power can be found.

- Electrodynamic and electrostatic methods use mechanical meters to multiply the voltage and current. An example of an electrodynamic meter is the familiar electric meter outside a home or business.

- Thermal methods, using either a thermoelement or a calorimeter.

- Electronic methods using a Hall effect sensor.

- Digital sampling of the current and voltage waveforms and integrating the measured values. This method is used by a wattmeter developed at NIST.[40] An advantage of this method is the capability to accurately measure distorted and noisy waveforms.

## Endnotes

1. Calhoun, 18.
2. Fluke, 7–9.
3. Adapted from Keithley, 3–4.
4. Fluke, 33-3.
5. Ibid, 8–8.
6. Keithley, 1-3 – 1-34.
7. See Fluke, 8-3 – 8-5; Calhoun, 33–34 and 50.
8. Fluke, 8–11.
9. Ibid, 33-3.
10. Ibid, 8-8.
11. Keithley, 3-19 – 3-20.
12. Fluke, 10-3 – 10-11; Calhoun, 159–160.
13. Fluke, 10-10.
14. Calhoun, 160, and Fluke, 10-6.
15. Fluke, 10-6 – 10-7.
16. Ibid, 10-9 – 10-10, and 11-11.
17. Calhoun, 174.
18. Fluke, 12-7 – 12-8; Calhoun, 99–100.
19. Ibid, 12-3 – 12-4); Calhoun, 115–116.
20. Lombardi, "Frequency calibration, part I," 2.
21. IUPAC 2003. *caesium* is the preferred spelling, but is commonly spelled *cesium* in the United States.
22. BIPM, 41–143 of the English-language edition.
23. NPL. www.npl.co.uk/time/time_trans.htm
24. Fluke, 14-6.
25. NIST. www.bldrdoc.gov/timefreq/service/fms.htm
26. Lombardi, Nelson et. al., 31.
27. Lombardi, 1999, 39.
28. A discussion of the possible problems is on the NIST internet page "Using a Global Positioning System (GPS) receiver as a NIST traceable frequency reference," www.boulder.nist.gov/timefreq/service/gpscal.htm
29. USNO GPS Timing Data and Information tycho.usno.navy.mil/gps_datafiles.html; NIST Global Positioning System (GPS) Data Archive; www.boulder.nist.gov/timefreq/service/gpstrace.htm; USCG Navigation Center GPS Information www.navcen.uscg.gov/gps/; BIPM Scientific Publications, Time Section www.bipm.org/en/publications/scientific/tai.html
30. Helfrick, 115.
31. Lombardi, "Frequency calibration, part I," 3.
32. Fluke, 14-3.
33. Levine, 55.
34. Fluke, 14-4.
35. Ibid, 14-4 – 14-5.
36. USNO.
37. Turgel and Oldham, 460.
38. Hess and Clarke, 39.
39. Hess and Clarke, 42; Clarke and Hess, 53.
40. Stenbakken, 2919.

# References

BIPM. 1998. *The International system of units (SI)*. 7th ed. Sevres, France: International Bureau of Weights and Measures. www1.bipm.org/en/publications/brochure

Calhoun, Richard. 1994. *Calibration & standards: DC to 40 GHz*. Louisville, KY: SS&S Inc.

Clarke, Kenneth K. and Donald T. Hess. 1990. "Phase measurement, traceability, and verification theory and practice." *IEEE Transactions on Instrumentation and Measurement*, 39 (February).

Fluke Corporation. 1994. *Calibration: Philosophy in practice*, 2nd ed. Everett, WA: Fluke Corporation.

Helfrick, Albert D. 2000. *Principles of avionics*. Leesburg, VA: Avionics Communications Inc.

Hess, Donald T., and Kenneth K. Clarke. 1995. "Phase standards: Design, construction, traceability and verification." *Cal Lab* (November–December).

IUPAC. 2003. "IUPAC Periodic Table of the Elements", version dated 7 November 2003. Research Triangle Park, NC: International Union of Pure and Applied Chemistry. www.iupac.org/reports/periodic_table

Keithley. 1998. *Low level measurement*. 5th ed. Cleveland, OH: Keithley Instruments.

Lombardi, Michael, Lisa Nelson, Andrew Novick, and Victor Zhang. 2001. "Time and frequency measurements using the Global Positioning system." *Cal Lab* (July-September): 26-33.

Lombardi, Michael. 1996. "An introduction to frequency calibration, part I." *Cal Lab* (January-February).

Lombardi, Michael. 1996. "An introduction to frequency calibration, part II." *Cal Lab* (March-April).

Lombardi, Michael. 1999. "Traceability in time and frequency metrology." *Cal Lab* (September-October): 33–40.

NIST. 2003. "NIST frequency measurement and analysis service." www.bldrdoc.gov/timefreq/service/fms.htm (viewed November 2003).

NPL. 2003. "Standard time and frequency transmissions." Teddington, England: National Physical Laboratory. www.npl.co.uk/time/time_trans.htm

Stenbakken, Gerard N. 1984. "A wideband sampling wattmeter." *IEEE Transactions on power apparatus and systems*, vol. PAS-103, (October): 2919–2926.

Turgel, Raymond S., and N. Michael Oldham. 1978. "High-Precision audio-frequency phase calibration standard." *IEEE Transactions on Instrumentation and Measurement*, vol. IM-27 (December): 460–464.

USNO. 2003. "GPS Time transfer performance." April 2003. Washington DC: U.S. Naval Observatory. tycho.usno.navy.mil/pub/gps/gpstt.txt

# Chapter 32
## Radio Frequency and Microwave

The radio frequency (RF) and microwave measurement area is concerned with the part of the electromagnetic spectrum where information can be transmitted through free space and received using the electrical or magnetic waves of the signal. Until the last quarter of the twentieth century, that would have been sufficient as a definition and would have covered the great majority of applications. Since then, the explosive growth of communications systems and digital information processing has greatly expanded the number and variety of products where RF and microwave engineering and measurement are important. Much of this is due to ever-increasing processing speed. This requires faster processors and logic and thus higher frequencies. This creates new challenges for circuit engineers because, for example, the printed circuit board traces and many electronic components are large compared to the signal wavelength and, therefore, have to be treated much differently than a wire with DC flowing in it.

Consider what has happened in the United States since 1975:

- In 1975 telephone service was from the phone company—there was only one company. The telephone was attached to the wall with a wire, and the concept of being able to buy (instead of lease) a telephone set was less than 10 years old. Now there are multiple phone companies, it is common to have local service from one and long-distance from another, and many users have at least one cordless telephone set that can be purchased just about anywhere.

- If you were out on the road and wanted to call someone, you had to find a coin telephone booth and a dime. Now you just pick up your personal cellular phone and place a call to anywhere in the world. In fact, in the United States the regional telecommunications companies are exiting the pay phone market.[1]

- Cable TV was something you had only if your community was out of reception range of a city with a TV station. Now cable TV is so widespread that the only places it is not available are so far out in the country it costs too much to run the optical fiber there. (But consumers can also have one or more direct satellite receivers.)

- Handheld calculators were still a new spin-off from the Apollo space program, and they were expensive. (A basic four-function unit with square root and memory cost about $80—about $280 in today's dollars after adjusting for inflation. In 2003 you can pick one up just about anywhere for less than $5.) As for business computers, most large companies and universities had one or two, and a few had several.

- In 1975 spread-spectrum communication technology and GPS were classified military technology, and frequencies in the gigahertz range were used almost exclusively by space systems, military, and law enforcement. Now applications such as home and office wireless telephones, wireless Internet ports, and wireless links between computers and their accessories use spread-spectrum technology at gigahertz frequencies; and you can buy a GPS receiver in your local Wal-Mart store.

- The January 1975 issue of *Popular Electronics* featured the first of a series of articles about the MITS Altair 8800 microcomputer. This kit is generally considered to be the first home computer and used the Intel 8080 microprocessor that had been invented the year before. (By late 1977, you could walk into an electronics store and buy a ready-to-use personal computer, such as a Commodore PET 2001, Apple II, or Radio Shack TRS-80. But the Internet would be still under the control of the Department of Defense and the National Science Foundation for another 10 years.) Now you can buy a handheld computer with as much power as room-filling business systems of 1975 and connect it wirelessly to the full range of offerings on the Internet.

- In 1975 there were virtually no consumer products (other than calculators and computers) that incorporated any kind of digital logic or processing. Now, only the very simplest appliances and toys do not have any microprocessors.

All of these new systems, and their continually increasing operating speeds, demand more of RF and microwave measurement systems and have required advances in related measurement sciences.

The International Telecommunication Union (ITU) Radiocommunication Sector manages the radio frequency part of the spectrum on a worldwide level. Within individual countries the radio spectrum is managed by the national government, usually in conformance to the ITU regulations and recommendations. In the United States, the responsible agencies are the National Telecommunications and Information Administration (NTIA), part of the U.S. Department of Commerce, and the Federal Communications Commission (FCC), which is an independent government agency.

## RF AND MICROWAVE FREQUENCIES

The general definitions of the terms used in this section are found in Federal Standard 1037C, *Telecommunications: Glossary of telecommunication terms*.[2]

- *Radio waves* are electromagnetic waves that are below an arbitrary limit of 3000 GHz (wavelength of 100 μm, in the infrared region).

- *Microwaves* are electromagnetic waves with frequencies from 1 GHz up to 300 GHz (wavelength from 30 cm to 1 mm).

- *Infrared* is the portion of the electromagnetic spectrum from 300 GHz up to the long-wavelength end of the visible light spectrum, approximately 0.7 μm.

These frequency boundaries are somewhat fuzzy, however. For example,

- Radio frequencies are not allocated below 9 kHz, but the U.S. Department of Defense has some transmitters that operate below 300 Hz.

- The NTIA frequency allocation chart shows the microwave region starting at 100 MHz.[3]

- There may be other names specific to an industry or company for various portions of the spectrum.

For calibration purposes, the bottom edge of the RF area is generally considered to be 100 kHz. The top edge is a moving target, regularly redefined as new or improved technology makes another advance. Optical fiber metrology (see Chapter 35) shares many of the same principles. An optical fiber is essentially a waveguide for photons and may be considered an analogue to a metallic waveguide for lower frequency electromagnetic waves.

For convenience, the RF spectrum is often divided into bands or groupings of frequencies. The divisions most commonly known to the public (in North America) are the AM and FM broadcast radio bands, 535 to 1605 kHz and 88 to 108 MHz, respectively. The most commonly used general divisions are shown in Table 32.1.[4]

The ITU defines 21 specific bands in the electromagnetic spectrum; the lowest starts at 3 Hz and the highest ends at 3000 EHz (Exahertz, 3000 x $10^{18}$ Hz, X-ray frequencies.) Other common methods of referring to groups of frequencies are the bands used by amateur radio operators and the bands used by shortwave radio listeners. An example is the amateur radio two-meter band, 144 to 148 MHz.

**Table 32.1** Common frequency bands and names.

| | | |
|---|---|---|
| 30 to 300 Hz | Extremely low frequency | ELF |
| 0.3 to 3 kHz | Super low frequency | SLF |
| 3 to 30 kHz | Very low frequency | VLF |
| 30 to 300 kHz | Low frequency | LF |
| 0.3 to 3 MHz | Medium frequency | MF |
| 3 to 30 MHz | High frequency | HF |
| 30 to 300 MHz | Very high frequency | VHF |
| 0.3 to 3 GHz | Ultra high frequency | UHF |
| 3 to 30 GHz | Super high frequency | SHF |
| 30 to 300 GHz | Extremely high frequency | EHF |

**Table 32.2** Common waveguide bands.

| Frequency (GHz) | Common Name |
|---|---|
| 8.2 to 12.4 | X |
| 12.4 to 18.0 | P, KU |
| 18.0 to 26.5 | K |
| 26.5 to 40.0 | R, KA |

## Waveguides

In the microwave area, frequencies are often grouped in bands that are associated with waveguide sizes. The most commonly used waveguide bands are shown in Table 32.2.[5]

There are many more waveguide sizes than this, of course, and many different names for each size. Waveguide systems are (or have been) used from 320 MHz to 325 GHz.[6] Waveguide band names are created by industry consensus or government action and have changed several times over the past 50 years. There are several overlapping systems for the bands, the waveguide pipe, and even the flanges at waveguide ends. Common designations in North America and the NATO area include consensus names such as in Table 32.2, the EIA WR numbers, the IEC R numbers, and British WG numbers. The U.S. military has both the JAN RG numbers and the MIL-W numbers, and, in many cases, each system has two different numbers for the same frequency range.[7] The actual interior dimensions of waveguides, however, do have a rational basis. The wide side of plain rectangular waveguide is approximately equal to the half-wavelength of the lowest frequency the waveguide is designed for; the narrow side is about half that dimension. This applies rectangular waveguide only; the physics are more complicated for ridged and circular waveguide. (See Chapter 2 of Adam's *Microwave Theory and Applications* for more details.)

Now that coaxial cable is being used to carry signals up to 65 GHz or more, the waveguide name system is becoming less well-known. Waveguide is still used, however, in radar and satellite transmitters—and other applications where there is a requirement to deliver high power to the antenna. Waveguide is also still used in metrology because some of the best measurement standards are waveguide types (rotary vane variable attenuators, for example.)

## MEASUREMENT PARAMETERS

Following is a discussion of the most important measurement parameters in the DC and low frequency area. This does not include all parameters. Also, it is not a complete discussion of them, especially considering that, in many cases, complete books or chapters of books have been written about each. The parameters are:

- RF power
- Attenuation or insertion loss
- Reflection coefficient or standing wave radio

- RF voltage
- Modulation
- Noise figure and excess noise ratio

## RF Power

Power is one of the fundamental measurements that are commonly made in any radio frequency system. It is usually one of the most important parameters specified by customers and designers. Power measurement is also fundamental to other measurements, such as attenuation and reflection coefficient. See Table 32.3 for more details on RF power.

The SI unit of RF and microwave power is the watt, a derived unit defined as joules/second. Because power measurements can cover several orders of magnitude, it is usual to express absolute power measurement as decibels relative to one milliwatt (dBm); the logarithmic scale makes the numbers easier to handle. Reference levels other than one milliwatt may also be used.

At DC and in low frequency AC measurements, power is determined by measuring some combination of voltage, current, and phase angle. For example, when determining the power available in a distribution system, an electrician will use instruments

**Table 32.3** RF power parameters.

| Parameter Name | RF Power |
|---|---|
| SI units | Watt (J/s). |
| Typical primary standards | Calorimeter; bolometer, direct voltage, AC/DC transfer, Dual six-port network analyzer. |
| Typical transfer standards | Thermistor mount (coaxial, waveguide). |
| Typical calibration standards | Thermistor mount (coaxial, waveguide). |
| Typical IM&TE workload | Thermistor mounts, thermocouple and diode-detector power sensors. |
| Typical measurement method | *Calorimeter.* Direct measurement of energy absorbed as heat. *Dual six-port.* S-parameter network analysis. *Bolometer.* Ratio (resistance and/or AC/DC power transfer). *Thermocouple.* Comparison to bolometer. *Diode-detector power sensors.* Comparison to bolometer. |
| Measurement considerations | Excessive power, electrostatic discharge, connector dimensions, connector torque. |
| Major uncertainty sources | *Coaxial connection repeatability.* Always use a torque wrench if the connector design allows it. *Reflection coefficient.* Affected by connector cleanliness, pin recession or protrusion dimensions, and overall connector wear. *Ambient temperature.* Standards and workload must be protected from drafts, and temperature changes during the calibration process. *Electromagnetic interference.* In some cases, the calibration may need to be done inside a screen room. |

that measure the voltage, current, and power factor (cosine of the phase) simultaneously. (See Chapter 31.) As the frequency increases to the RF region, these parameters become both more difficult to measure and less relevant to the average user. There are several considerations:

- The transmission line characteristics have an increasing effect on the signal and its measurement as frequency increases.

- Impedance becomes a complex vector value that cannot be easily determined and that varies with position on the transmission line.

- At RF and microwave frequencies the voltage and current parameters are meaningless to the practical user.

In an ideal transmission line with no loss, power is constant along the length because it is a product of voltage and current that is independent of position along the line.[8] Voltage measurements are still used (with a diode detector), but many RF power measurements use a thermal sensor, either a thermistor or a thermocouple. In either case, the measurement system usually displays the result as power or as a power-related computed value.

An RF power measurement system is composed of a sensor and a meter. Some low-level sensors include a 30 dB reference attenuator, and some high-power sensors include attenuators. All of these items must be calibrated before valid measurements can be made with them. Also, some types of power sensors must be used only with a specific power meter (by serial number)—watch out for those.

## Sensors and Their Meters

In most cases, what needs to be measured is the RMS or average power level. One way to quantify RMS power is to equate it to the heating effect produced by direct current into the same impedance. Calorimeters, thermistors, and thermocouples for RF and microwave power measurement use heat in some form and are inherently average-responding. Diode sensors, on the other hand, sense either average or peak voltage and use computation to derive the RMS equivalent.

### Thermistor Sensors

A thermistor sensor (or thermistor mount) is a type of bolometer. *Bolometer* is a term that is common in the (older) fundamental literature, but is seldom used now. It refers to two types of sensors—a barretter or a thermistor. In general, a bolometer is a type of sensor that absorbs the RF power and changes the resistance of the sensing element. The bolometer element is usually in one arm of a bridge circuit, which allows the resistance to be measured by conventional DC or low frequency AC methods. A bolometer is inherently average-responding. A *barretter* is a metallic resistive element, often platinum, that increases in resistance as temperature goes up. The temperature change is determined by measuring the resistance.[9] Barretters are no longer used so this term is rapidly becoming archaic.[10] A *thermistor* is a small bead of semiconductor material, with wires for connecting to a measuring circuit. The temperature change is determined by measuring the resistance. The resistance of a thermistor element decreases as the temperature goes up.

A thermistor is a temperature-sensitive resistor and, as such, its resistance can be measured using a DC bridge. (See resistance in Chapter 31.) Alternatively, a bias current in the bridge can warm the thermistor so some preselected resistance value and then the bridge can be balanced for a zero indication. If RF power is added to the thermistor through the sensor input, the bridge will become unbalanced. If the bias current is reduced to bring the bridge back into balance, the amount of power ($I^2R$) removed from the bridge by reducing the current is equal to the RF power absorbed by the thermistor. This is the DC substitution method; it is traceable to the SI base units of the ampere and second, and the derived unit of the watt. Power meters that use thermistor mounts use this measurement method. An example is the Agilent 432A power meter.

The principal advantage of a thermistor sensor is absolute RMS power measurement by the DC substitution technique. The main disadvantages (for some applications) are limited dynamic range and slow response time. In a calibration lab, principal uses of thermistor sensors are transfer standards from higher echelons, calibration of power sensors, measurement of absolute RF power, and calibration of the 50 MHz 1 mW reference output used on meters for thermocouple and diode sensors. This last application requires a thermistor sensor selected for low SWR and optimized for low frequency operation, such as the Agilent 478A Option H75.[11]

### Thermistor Sensor Meters

There are two general types (portable and laboratory) of RF power meters that use thermistor sensors. They both use versions of the self-balancing DC substitution bridge that was developed by the U.S. National Bureau of Standards (NBS) in 1957 and refined in the 1970s.[12] As discussed earlier, the amount of DC power in the bridge is adjusted to keep the resistance of the thermistor constant, and the amount of change is the measure of the absorbed RF power. This type of meter and sensor is the only one that can give an absolute power measurement.

### Thermocouple Sensors

A thermocouple sensor measures power by sensing the heating of a thermocouple junction. A thermocouple produces a voltage that has a known relationship to temperature difference between the junction and the measuring end. The meter measures the voltage and displays the result as power. Like thermistors, thermocouples are inherently average-responding.

Thermocouple sensors have several advantages over thermistor types. They have much lower sensitivity to ambient temperature changes. They are somewhat more sensitive, which gives more dynamic range. They also have lower input reflection coefficient, which reduces measurement uncertainty.

The principal disadvantage of thermocouple sensors is that they only make relative power measurements, not absolute. The measurements are all made relative to a reference. The meter for this type of sensor has a 50 MHz oscillator output on the front panel; the output power level is usually set to 1 mW ±0.4%. That output is used to set the sensor and meter to a reference level, and measurements are made relative to that setting. Naturally, this adds an uncertainty element to the measurements.

### Diode Sensors

A diode power sensor rectifies the RF and produces an output voltage. The meter measures the voltage, references it to current or impedance (depending on the design), and displays the result as power. A meter that uses a diode sensor may respond to the average voltage and be calibrated to display an equivalent RMS value, or it may respond to the peak waveform. A peak-responding meter may be calibrated to display an equivalent RMS value (the most common type) or it may display the actual peak power. The measurements are based on the AC volt and either the ampere or the ohm. Depending on the design, they use the relationship $P = E^2/R$ or $P = I^2/R$ to display the power up to −20 dBm, the square law region.[13] At higher levels, the performance transitions to a linear output region. The degree of linearity in this region is an important specification, particularly at higher power levels. Diodes used to be limited to a maximum input of about −20 dBm due to use in the square law region, but modern diode sensors are usable up to about +20 dBm.[14] The expanded range is possible because of digital storage of correction factors for linearity errors. Depending on the design, the corrections may be stored in the sensor (which makes them interchangeable between compatible meters) or in the meter. If linearity correction factors are stored only in the meter, then the sensor is not interchangeable, and the sensor and meter must be calibrated together.

Peak power measurements are common in pulse modulation applications such as radar, aircraft transponders, and digital communications. Diode sensors are commonly found in peak power meters and network analyzers. Limits are defined by low-level noise from the diode and meter, and damage due to excessive power input. Advantages of a diode sensor are high sensitivity, relatively simple circuit design, and (in newer models) up to 90 dB dynamic range.

The principal disadvantage of broadband diode sensors is that in most cases they only make relative power measurements, not absolute. The measurements are all made relative to a reference. The power meters for this type of sensor have a 50 MHz oscillator output on the front panel; the output power level is usually set to 1 mW ±0.4%. That output is used to set the sensor and meter to a reference level, and measurements are made relative to that setting. This type of system cannot be used for absolute power measurements. Accurate measurement of the power output (using a thermistor mount) is an important part of calibrating the meter. Naturally, this adds an uncertainty element to the measurements.

### Thermocouple and Diode Sensor Meters

Most common RF power meters use either thermocouple or diode sensors. All of them make relative power measurements. With the exception of one type (to be discussed), the reference is a reference oscillator in the power meter. The reference oscillator is usually designed to produce a 1.00 mW output at 50 MHz. The user of the meter connects the sensor and checks or sets this as the reference level before every use. Therefore, calibration of the reference output is a critical part of calibrating the power measurement system.

The other type of meter that uses a diode sensor does not require a reference power setting. This type of meter uses the diode as a voltage detector on the coupled output arm of a directional coupler. The diode, coupler arm, and reference termination are housed in a plug-in element. Each element is designed for a specific frequency and power range, and for either average or peak power. They are not broadband devices (other power meters are), but they are often more suitable for field service work. An advantage of this type of meter is that simply rotating the plug-in element 180° allows measurement of reverse power, which can be used to measure the SWR on the transmission line. This type of meter is lower accuracy overall and makes measurements relative to the calibration standards.

### Peak Sensor Meters

Older peak power meters use an average-responding sensor and divide the result by the pulse duty cycle to estimate the peak power. This is suitable only if the modulation waveform is a rectangular pulse with a known and constant duty cycle.[15] Recent meter designs have started to use advanced signal-processing to detect and analyze the actual modulation envelope of the signal, and to use time-gated systems that measure the power only during the on portion of a pulse.[16]

## Calibration of RF Power Sensors and Meters

### Power Sensors

The most important result of calibrating a power sensor is determination of the calibration factor. This value is essential for the sensor and meter to be properly used as a system. The calibration factor is the ratio of power absorbed by the sensing element to power available at the input plane of the sensor. It includes the two primary error sources—mismatch error and sensor efficiency. Mathematically,

$$K_b = \eta_e (1 - \rho_l^2)$$

where  $K_b$ =  calibration factor,

$\eta_e$ = effective efficiency, ratio of power absorbed by the sensing element to total incident power,

$\rho_l$ = reflection coefficient magnitude of the sensor.[17]

RF power sensors are calibrated by comparison to a transfer standard. There are two types of systems in common use. One is based on a DC or low-frequency AC bridge, and the other uses a well-matched RF power splitter. The purpose of calibration is to verify the RF performance by comparison to a standard with known characteristics and to determine a value for a calibration factor.

The bridge system is capable of the lowest uncertainty and direct traceability to national standards. A popular portable meter based on an automatic bridge design is the Agilent 432A, which has been in production for more than 30 years. As a standalone meter, its instrumentation uncertainty is ±1% of full scale in addition to the sensor uncertainty. But since the bridge voltages are available on the rear panel, higher-accuracy DC voltmeters can be used for the measurements to achieve instrumentation uncertainty closer to ±0.2% of reading.[18] A popular laboratory system is based on the NIST Type-IV bridge design, developed in the early 1970s. This system is more complex and expensive, but can be used (with purchase of appropriate transfer standards) to obtain direct traceability to national standards. In addition to this system, the laboratory needs a signal generator, power meters compatible with the workload, and a means of measuring reflection coefficient.

The power splitter system is based on a two-resistor power splitter (see Figure 32.1). It is simple, easy to use, broadband, and requires relatively little equipment. It is restricted to use with coaxial power sensors, but the majority of them are that type now. The basic components needed are a signal generator, power splitter, a power sensor and meter for measuring the applied power and possibly leveling the generator, a transfer

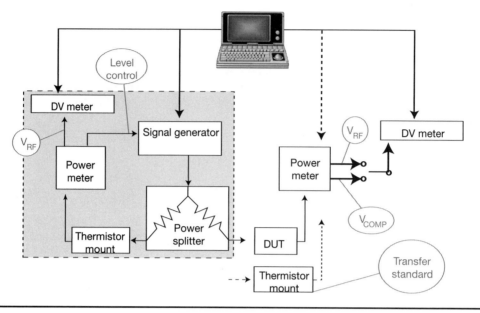

**Figure 32.1** Power sensor calibration system—block diagram.

standard sensor and meter, the sensor to be tested, and a means of measuring reflection coefficient. (The sensor to be tested can be used with its own meter provided the meter has been calibrated first.) A waveguide power sensor could be calibrated with the same type of system by replacing the two-resistor power splitter with a waveguide directional coupler and adjusting the readings to account for the coupling factor.

## Power Meters

The power meters have to be calibrated as well. Functionally, they are essentially DC or low-frequency AC voltmeters with a few additional special circuits. The most important parameters are range accuracy and linearity, and calibration factor accuracy. For meters that use thermocouple or diode sensors, the output level of the reference oscillator is a critical parameter.

## Attenuation, Insertion Loss Parameters

Attenuation is an important parameter of RF and microwave systems. It is double-edged: in some cases it is a desirable characteristic, and in others it is something to be minimized. Attenuation is normally measured using RF power and ratio methods. See Table 32.4 for details on attenuation or insertion loss.

Attenuation is a ratio. It is a decrease in power as a signal passes through a transmission line or other passive device. Insertion loss is an equivalent term. Attenuation is the opposite of the power gain provided by an amplifier. An attenuator is a device that is intended to reduce the input power by a predetermined ratio. Attenuation measurement values are typically reported in decibels (dB) due to the wide range of the numbers.[19]

**Table 32.4** Attenuation or insertion loss parameters.

| Parameter Name | Attenuation or Insertion Loss |
|---|---|
| SI units | (Ratio). |
| Typical primary standards | RF power standards, waveguide-beyond-cutoff (piston) attenuator, dual six-port network analyzer. |
| Typical transfer standards | Waveguide-beyond-cutoff (piston) attenuator, waveguide rotary vane attenuator. |
| Typical calibration standards | Waveguide-beyond-cutoff (piston) attenuator, waveguide rotary vane attenuator, precision coaxial attenuators, tuned RF attenuation measurement system, scalar network analyzer, vector network analyzer. |
| Typical IM&TE workload | Coaxial and waveguide attenuators (fixed or variable), tuned RF attenuation measurement system, signal generator output attenuators, receiver input attenuators. |
| Typical measurement method | Audio substitution attenuation measurement system, IF substitution attenuation measurement system, tuned RF attenuation measurement system, scalar network analyzer, vector network analyzer, RF power ratio. |
| Measurement considerations | Excessive power, electrostatic discharge, connector dimensions, connector torque. |
| Major uncertainty sources | *Coaxial connection repeatability.* Always use a torque wrench if the connector design allows it.<br>*Reflection coefficient.* Affected by connector cleanliness, pin recession or protrusion dimensions, and overall connector wear.<br>*Electromagnetic interference.* In some cases, the calibration may need to be done inside a screen room. |

Attenuation can be measured by either voltage or power measuring instruments. Attenuation is normally expressed in decibels, as the ratio of the output to the input voltage or power.

$$dB = 10 \log \frac{P_{out}}{P_{in}} \text{ or } dB = 20 \log \frac{V_{out}}{V_{in}}$$

If the output is less than the input, solving either of these relationships will result in a negative number, which is mathematically correct since attenuation is a negative gain. By convention, however, the negative sign is commonly dropped as long as the context makes it clear that attenuation is being discussed instead of gain.

Measurement of power is important to attenuation measurement, as the power ratio is often the most convenient to measure and use. Several different measurement methods are used, each with their own set of benefits. At low attenuation ratios (to about 40 dB) systems that use an audio substitution method are commonly used. A typical swept-frequency system that uses AF substitution is the Agilent 8757E scalar network analyzer. Intermediate Frequency (IF) substitution systems can measure attenuation ratios of 90 dB or more. A typical system that uses IF substitution is the TEGAM 8850-18, which is based on the model VM-7 30 MHz receiver.

## Reflection Coefficient, Standing Wave Ratio Parameters

The reflection coefficient or standing wave ratio (SWR) is another important parameter in RF systems. This needs to be minimized to improve power transfer and reduce the likelihood of equipment damage. SWR is a ratio measurement. See Table 32.5 for more details on reflection coeffiecient.

Reflection coefficient, SWR, and return loss are different ways of expressing the same parameter. Their relationships follow.[20]

Complex reflection coefficient = $\Gamma = \dfrac{V_{reflected}}{V_{incident}}$    (Has magnitude and phase)

where

Range 0 (no reflection) to 1 (total reflection) with phase 0 to ±180°

$V_{reflected}$ = voltage of the reflected signal

$V_{incident}$ = voltage of the incident signal

Scalar reflection coefficient = $\rho = |\Gamma|$    (Has magnitude only)

Range 0 (no reflection) to 1 (total reflection)

$\text{SWR} = \sigma = \dfrac{1+\rho}{1-\rho} = \dfrac{SWR-1}{SWR+1}$    (A unitless ratio)

Range 1 (no reflection) to ∞ (total reflection)

SWR (in dB) = 20 log σ

Return loss (dB) = −20 log ρ    Range ∞ (no reflection) to 0 (total reflection)

Mismatch loss (dB) = −10 log (1 − $\rho^2$)    Range 0 (no loss) to more than 50 dB loss

Impedance of a component being measured    Range 0 (short circuit) to ∞ (open circuit)

$Z_{DUT} = Z_0 \dfrac{1+\rho}{1-\rho}$

where $Z_0$ is the characteristic impedance of the system (normally 50Ω).

---

Note: The conversion from reflection coefficient to other values changes it from a vector with magnitude and phase to a scalar value with magnitude only. The loss of phase information increases measurement system uncertainty.

Table 32.5 Reflection coefficient, standing wave ratio parameters.

| Parameter Name | Reflection Coefficient |
|---|---|
| SI units | (Ratio). |
| Typical primary standards | Dual six-port network analyzer. |
| Typical transfer standards | Fixed reflection coefficient standards, sliding loads, sliding shorts, network analyzer calibration standards. |
| Typical calibration standards | Scalar and vector network analyzers, S-parameter test sets, SWR bridgesTypical IM&TE workload. |
| Typical IM&TE workload | All coaxial or waveguide RF and microwave devices. |
| Typical measurement method | Audio substitution attenuation measurement system, IF substitution attenuation measurement system, tuned RF attenuation measurement system, scalar network analyzer, vector network analyzer, RF power ratio. |
| Measurement considerations | |
| Major uncertainty sources | *Coaxial connection repeatability*. Always use a torque wrench if the connector design allows it. |

SWR is normally measured as a voltage ratio. SWR can also be measured as a power ratio, and it can be derived in terms of current or impedance ratios.[21]

Measurement of SWR is dependent on the ability to measure RF voltage or power, and attenuation. The ideal case is where the load absorbs all of the forward power and none is reflected. Imperfect impedance match, connector variations, transmission line discontinuities, or other problems cause reflected power, which is measured as reflection coefficient or SWR. Note: SWR may be measured in terms of either voltage (VSWR) or power (PSWR). The power ratio method is so rarely used now that it is increasingly common usage in the United States to drop the initial letter of the acronym. It is understood that SWR refers to the voltage ratio.[22]

SWR measurement uses attenuation or power measurement ratios and the theory of transmission lines. The measurement systems are initialized and verified with standard terminations: short-circuit, open-circuit, and standard mismatches with known reflection coefficients. (Open-circuit standards are generally not used in waveguide systems.) In a swept-frequency measurement system, attenuation measurements may be presented as a line graph of magnitude versus frequency or on a polar graph or a Smith chart.

Because the Smith chart is frequently used with measurements of reflection coefficient and other RF and microwave parameters, a few comments about it are appropriate. The Smith chart, invented in 1937 by Phillip H. Smith, is a useful tool for graphical analysis of and transformation between impedance and SWR. Full discussion of the Smith chart is beyond the scope of this book (it would take at least another chapter), but some useful references are listed in the references and bibliography. (See Stephen F. Adam, Agilent Technologies, and Philip H. Smith.) Figure 32.2 is an example of a Smith chart.[23] A printed chart can be plotted and evaluated manually, but more commonly now the chart is a display feature of measurement systems.

**360**  Part VI: Measurement Parameters

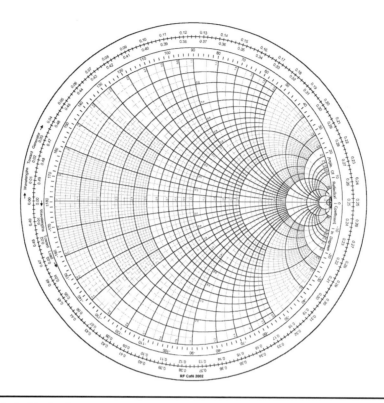

**Figure 32.2** Smith chart.

Some key features of the Smith chart are:

- All values on the Smith chart are normalized to the characteristic impedance of the system ($Z_0$) and the ½ wavelength of the signal frequency. This makes the graph independent of the actual impedance or frequency.
- The horizontal center line of the graph represents pure resistance.
- The circumference of the graph represents pure reactance. Positive values (above the resistance line) are inductive; negative values are capacitive.
- The intersection of the horizontal line (between 0° and ±180°) and the circle labeled 1.0 (the unit circle) represents the theoretically perfect complex impedance of the system—all resistance and no reactance.
- The circumference of the graph is ½ wavelength.
- When a point is plotted on the chart, the distance from the center represents the magnitude of the reflection coefficient. The center is minimum reflection since it represents $Z_0$, and the circumference is complete reflection, as from a perfect open or short.
- When a point is plotted on the chart, a line drawn from the center through that point and continuing to the edge will show the phase angle of the reflection coefficient.

- There are three scales around the circumference of the graph. The innermost one is the phase angle of the reflection coefficient. The other two are graduated in fractions of a wavelength towards the load (inner) and the generator (outer).

- Some printed versions of the Smith chart include additional horizontal scales above or below the graph. Most commonly, these are used to quickly plot or determine values such as standing wave ratio, or return loss in decibels, as well as other values. Automated systems can make these transformations internally.

In common metrology applications, it does not make much difference whether voltage or power is measured to determine SWR. That is because we have the capability to make the measurements at the reference planes of the devices being measured. In other applications, though, it does make a difference. If the measurement is made at the source end of a transmission line, the measurement will be correct only if a power ratio measurement is used. A voltage or current based measurement is only correct at the load or at exact half-wavelength intervals from it along the transmission line. At any other point, it is useful only as a relative indicator of load match—to adjust the load for a minimum—because the actual displayed value has very little meaning.[24] A common situation where this type of measurement may be made is making measurements of a transmitter antenna.

## RF Voltage Parameters

RF voltage measurement is used in two major ways. Many instruments use voltage measurements internally, even if they display a measurement result in other terms. If an instrument uses a diode type sensor, it is probably making voltage-based measurements. The other principal method is the use of a thermal voltage converter for high-accuracy AC/DC transfer measurement at radio frequencies. See Table 32.6 for more details on RF voltage.

**Table 32.6** RF voltage parameters.

| Parameter Name | RF Voltage |
| --- | --- |
| SI units | Volt (W/A). |
| Typical primary standards | RF voltage comparator. |
| Typical transfer standards | Thermal voltage converter. |
| Typical calibration standards | Thermal voltage converter. |
| Typical IM&TE workload | Signal generators, RF voltmeters. |
| Typical measurement method | AC/DC substitution at radio frequencies. |
| Measurement considerations | Excessive power, frequency limits, AC-DC difference. |
| Major uncertainty sources | *Coaxial connection repeatability.* Always use a torque wrench if the connector design allows it.<br>*Reflection coefficient.* Affected by connector cleanliness, pin recession or protrusion dimensions, and overall connector wear.<br>*Ambient temperature.* Standards and workload must be protected from drafts and temperature changes during the calibration process. |

As discussed in the section on RF power, many voltage measurements are commonly made using diode detectors. The detectors are commonly part of measurement systems for other parameters such as power, attenuation, or reflection coefficient. An important manual use was measuring SWR by means of voltage measurements along a slotted line, but now this is rarely done in the majority of calibration labs.

The most important manual measurement method of measuring RF voltage uses thermal voltage converters (TVC). Thermal voltage converters are used to transfer direct current or voltage values to alternating current or voltage. There are two types in wide commercial use. The most common is a TVC that uses a vacuum thermoelement and is usable up to 100 MHz. The other type, called a micropotentiometer, uses a ceramic disk resistor and is usable up to 1 GHz.[25] These devices have measurement uncertainties of about 1%. The operating principle is that an AC RMS level is equal to a DC level if it produces the same heating effect. A newer device, with uncertainties in the ppm range, is the multi-junction thin film thermal voltage converter system developed by Sandia National Laboratory and NIST.

## Modulation Parameters

Modulation is an essential property for communication because it is the modulation of a carrier signal that carries the information. The simplest form of modulation is turning the carrier signal on and off, but modern communication systems require much more complex methods. See Table 32.7 for more details on modulation.

Modulation is the area where low frequency AC and RF-microwave come together. A modulated signal is a single-frequency signal, usually in the RF-microwave region, that is being altered in a known manner in order to convey information. The modulation source can be audio, video, digital data, or the simple on-off action of a Morse code key. The modulation always adds energy to the carrier and increases either its

**Table 32.7** Modulation parameters.

| Parameter Name | Modulation |
|---|---|
| SI units | Ratio. |
| Typical primary standards | Derived from mathematical description of the modulation type and applied to RF and microwave primary standards. |
| Typical transfer standards | Voltage, frequency, power, mathematics. |
| Typical calibration standards | RF, microwave and audio signal generators, signal analyzers, spectrum analyzers, modulation theory. |
| Typical IM&TE workload | Test sets for RF and microwave communication systems, radar systems, radio navigation systems. |
| Typical measurement method | Comparison. |
| Measurement considerations | |
| Major uncertainty sources | *Coaxial connection repeatability.* Always use a torque wrench if the connector design allows it.<br>*Reflection coefficient.* Affected by connector cleanliness, pin recession or protrusion dimensions, and overall connector wear. Linearity of audio detection system. |

power or its bandwidth. The modulation can alter the carrier by changing amplitude (AM), frequency (FM), phase (PM), or any combination. For example, one of the signals transmitted by an aviation very high frequency omnidirectional range (VOR) navigation base station is an amplitude-modulated signal at the assigned frequency. The modulating signal is an audio frequency that is itself frequency-modulated with a second audio frequency.

Radar systems and microwave communications systems use pulse modulation, which is a form of amplitude modulation. Pulse modulation can be as simple as on–off, but most systems now use methods to vary the pulse position, width, amplitude, or frequency as part of the method of carrying information.

Modulation is a ratio: 0% is the absence of a modulating signal, and 100% is the maximum amount that can be applied without causing distortion of the signal. It is sometimes called the modulation index, which is a number between 0 and 1. Modulation, along with frequency and power, is one of the regulated parameters of a radio, television, or other transmitter.

Modulation is frequently measured by analyzers designed for that purpose, such as AM/FM modulation meters or video modulation analyzers. Amplitude modulation can be measured using an oscilloscope. Frequency modulation can be measured with a spectrum analyzer, using the Bessel null technique. Some modulation sources, such as the VOR signal mentioned earlier, require specialized test equipment.

## Noise Figure, Excess Noise Ratio Parameters

Noise is an important parameter to measure in communication systems. All electronic devices generate some amount of noise from the random motion of electrons, which is related to heat. The total amount of self-generated noise in a system sets a limit to the weakest signal that can be detected. See Table 32.8 for details on noise.

**Table 32.8** Noise figure, excess noise ratio parameters.

| Parameter Name | Noise Temperature |
|---|---|
| SI units | Kelvin. |
| Typical primary standards | Primary thermal-noise standards, total-power radiometer. |
| Typical transfer standards | Thermal noise standard. |
| Typical calibration standards | Thermal noise standard. |
| Typical IM&TE workload | Thermal noise standards, amplifiers, noise figure meters. |
| Typical measurement method | Comparison. |
| Measurement considerations | Temperature, bandwidth, connector quality. |
| Major uncertainty sources | *Reflection coefficient.* Affected by connector quality, condition and cleanliness, pin recession or protrusion dimensions, and overall connector wear.<br>*Coaxial connection repeatability.* Always use a torque wrench if the connector design allows it.<br>*Source stability.* |

The thermal noise (Johnson noise) in a resistor is

$$N = kTB$$

where N is the noise power in watts, $k$ is Boltzmann's constant, T is the absolute temperature in kelvin, and B is the bandwidth of the measurement system.

The thermal noise source calibration system used at standards laboratories has two resistive noise sources. One is immersed in liquid nitrogen (77.5 K or –195.7 °C) and the other is used at laboratory ambient temperature. (Some systems may operate the second source at 373 K or 100 °C).[26] For convenience in calculations, the laboratory ambient is usually taken to be a conventional value of 290 K (16.8 °C) and labeled $T_0$.

The value reported by NIST is the available noise temperature, defined as the available noise power spectral density at the measurement plane, divided by Boltzmann's constant. The temperature is reported in kelvins. For noise temperatures over $T_0$, the excess noise ratio (ENR) is also reported.

## MEASUREMENT METHODS

There are three important methods used for measuring RF and microwave parameters: spectrum analysis, scalar network analysis, and vector network analysis. While they are not measurement parameters, the understanding of these methods is important because of their common use and wide application.

### Spectrum Analysis

Like an oscilloscope, a spectrum analyzer measures signal amplitude on the vertical scale. The horizontal scale, however, is calibrated in terms of frequency instead of time. A spectrum analyzer displays the individual frequency components of a signal source.

The essential parts of a spectrum analyzer are a swept-frequency local oscillator, a superheterodyne receiver, and a display unit.[27] The local oscillator sweep is coupled to the horizontal axis of the display, and the receiver output is coupled to the vertical axis.

Spectrum analysis is not exclusive to RF—there are models that start in the subaudio range as well as ones going up to 110 GHz or more. The measurement from a spectrum analyzer is voltage-based, but the display is normally referenced to a power level and scaled in decibels.

### Scalar Network Analysis

A scalar network analyzer, like a spectrum analyzer, has a display that shows amplitude versus frequency. There are more inputs, though, and the principal use is to make ratio measurements. The essential parts of a scalar network analyzer are a swept-frequency signal source, a multi-input broadband receiver, a signal splitter or directional bridge, diode detectors, and a display unit.

The sweep ramp of the signal source also controls the horizontal display of the network analyzer. The RF output of the source is connected in different ways, depending on the application.

- To measure a two-port device such as an attenuator, the swept RF output goes through a power splitter. One side of the splitter is connected to the reference input of the analyzer. The other side of the splitter is connected to the measurement input of the analyzer through the device to be tested. Before use, the system is set up by measuring with a direct connection to set the zero level and possibly with a check standard to verify performance.

- To measure a one-port device such as a termination, the swept RF output goes through a bridge or directional coupler. The system is set up by using short and open standards to set the zero return loss level and possibly verified by using mismatch standards. The device under test is then attached to the measurement port and measured.

Because of the diode detectors, the scalar network analyzer is a broadband device. This is either an advantage or not a problem for many measurements, but does have some limitations. The sensitivity is generally limited to about −60 dB and cannot be improved by averaging.[28]

A scalar network analyzer requires a normalization procedure whenever it is started and whenever the test setup or parameters are changed. The normalization (often called *self-calibration*) is done using high-quality transmission and reflection standards. The scalar network analyzer computer makes and stores a set of response measurements, which are used as a reference for measurements of the devices being tested. This minimizes the effect of any power versus frequency variations. Because the measurements are scalar quantities, the error model cannot account for any other source of systematic error.[29]

---

### Scalar and Vector: What's the Difference?

A scalar is a real number. In measurement, a *scalar* is a quantity that only has a magnitude. The majority of the numbers encountered are scalars. For example, if an object is moved, the measurement of the distance is a scalar.

A *vector* is a quantity that has both magnitude and a direction. If plotted on a graph, a vector starts at a defined point (its origin), has a length equal to the magnitude, and is at a specific angle to the reference line. Continuing the example above, a vector would describe not only how far the object was moved but also in what direction.

A scalar network analyzer measures magnitude only. The reflection coefficient (or return loss or VSWR) adds uncertainty. Its phase with respect to the incident signal changes with frequency and is unknown. So, at a given frequency, the actual magnitude is uncertain but somewhere in the range

(measured value ± reflection coefficient)

This uncertainty is the reason why the reflection coefficient of RF and microwave components, connections, transmission lines, and systems should be as low as possible.

A vector network analyzer measures magnitude and phase at each frequency point. The measured value has greatly reduced uncertainty because both the magnitude and phase of the reflection coefficient is known.

*continued*

*continued*

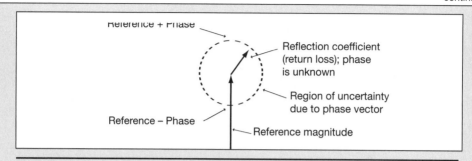

**Figure 32.3** A scalar network analyzer measures magnitude only. The reflection coefficient phase varies with frequency and causes an area of uncertainty.

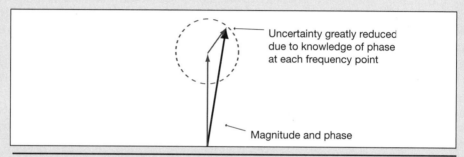

**Figure 32.4** A vector network analyzer measures magnitude and phase at each frequency, so magnitude uncertainty is reduced.

To help understand the difference, look at measurements of attenuation made by scalar- and vector-swept frequency measurement systems. See Figures 32.3 and 32.4.

As shown in the figures, the scalar measurement has an uncertainty circle (the dashed line) around it. This defines the uncertainty introduced because the reflection coefficient of the system is unknown. At any given frequency, the reflection coefficient could be anywhere from completely added to the magnitude to completely subtracted from it, but the exact value is unknown. A vector measurement system includes phase in the measurement and therefore produces a more exact result. There is still some residual uncertainty, but it is one or two orders of magnitude less than a scalar measurement.

## Vector Network Analysis

A vector network analyzer (VNA) makes the same kind of measurements as a scalar analyzer, but in a manner that preserves the phase relationships of the signals. The essential parts of a vector network analyzer are similar to a scalar analyzer, but there are important differences. The major differences are use of narrow-band tuned receivers and the addition of an S-parameter test set or a transmission-reflection test set.

Conceptually, the S-parameter test set contains two dual directional couplers, four diode sensors, and some high-speed switching devices. (Actual implementation is more complex, of course!) The signal from the source is sent in one direction, and measurements are made of the four coupled outputs. The signal direction is reversed and the measurements are repeated. The values are then combined mathematically to determine the four S-parameters of the device under test.[30] Some units use a transmission-reflection (T/R) test set. It is very similar to a S-parameter test set, except that the signal only goes one direction. This makes it simpler because there is only one dual directional coupler, half as many detectors, and no need for the switching arrangement. However, the T/R test set is limited to one-port measurements.[31]

A vector network analyzer requires a normalization procedure whenever it is started and whenever the test setup or parameters are changed. The normalization (often called *self-calibration*) is done using very high-quality transmission and reflection standards, along with a mathematical model of each standard. The VNA computer makes a set of measurements, compares them to the known model of the standard, and stores the difference as a correction. This minimizes the effect of any systematic error sources within the analyzer. Because the measurements are vector quantities, the error model accounts for all of the major sources of systematic error.[32]

Some important advantages of vector network analysis are:[33]

- Using a tuned-receiver design provides better sensitivity and more dynamic range than simple diode detection.

- Phase information is preserved during the detection process. This results in lower measurement uncertainty. It also allows more complete error correction to compensate for the test setup.

- Dynamic range can be improved by averaging, since the averaging is done with the vector data. Another advantage is the ability to display the data in forms that may be more useful than an amplitude-versus-frequency display. For example, data can also be displayed on a polar graph or a Smith chart.

## S-parameters

S-parameter measurements are developed from the fundamental theorems of electronic networks. Recall that if a component or system is modeled as a single block with no knowledge of what is inside it, its behavior can be characterized by measuring the response of its outputs when various inputs are applied. In a simple case such as a coaxial attenuator, the device has two ports, which are called the input and output, or 1 and 2 for convenience. Each port has two nodes, usually labeled a and b. A node is a point for measuring voltage or current passing into or out of the network. Figure 32.5 is a diagram of a basic two-port network.

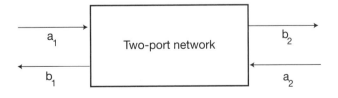

**Figure 32.5** A general two-port network. This can be used to represent a device such as a coaxial attenuator.

The operation of a vector network analyzer is based on measurements of a two-port network and application of the mathematics of vector measurements and scattering or s-parameters. A basic understanding of those subjects is essential to understanding the VNA and using it effectively. The references by Adam and Agilent have more detailed information.[34]

A vector quantity is one that has both magnitude and phase information. (If the phase is not measured, then the quantity is a scalar.) By definition, the VNA makes vector measurements. Also, all of the raw measurements are voltages. S-parameters are a set of measurements of energy flow into and out of a device when the device is terminated in its characteristic impedance ($Z_0$). S-parameters are relatively easy to measure at RF and microwave frequencies; directly represent values that are used by metrologists, technicians, and engineers; and are used to model components in design software.

S-parameters are usually illustrated using a network flow graph diagram, a visual method of analyzing voltage traveling waves. A flow graph separates each port (input or output) into two *nodes*, labeled $a_n$ and $b_n$. Node $a$ always represents waves entering the network and node $b$ always represents waves leaving the network. The number n is the port number that the node represents. All nodes are connected with directional lines that illustrate the possible signal flow paths. A network with N ports has 2N nodes and $N^2$ s-parameters. For example, a three-port network (such as a directional coupler) has three ports (input, output, coupled output) and nine s-parameters.

In the Figure 32.6, $a_1$, $b_1$, $a_2$ and $b_2$ are the nodes. Node $a_1$ represents the incident wave entering port 1, and node $b_1$ represents the reflected wave leaving port 1. Nodes $a_2$ and $b_2$ represent the same things for port 2. The four s-parameters are $s_{21}$, $s_{11}$, $s_{12}$, and $s_{22}$. The convention for the subscript numbers is that the first digit is the port number of the output node, and the second digit is the port number of the input node.

An advantage of s-parameters is that they directly represent the transmission and reflection values that are used in metrology. Any one s-parameter can be measured from the values of one $a$ node and one $b$ node, provided the other $a$ node value is zero. If the device is terminated in its $Z_0$ impedance, this is assumed to be true. The relationships are shown below.

Input reflection coefficient $\qquad s_{11} = \dfrac{b_1}{a_1}$ when $a_2 = 0$

Forward transmission ratio $\qquad s_{21} = \dfrac{b_2}{a_1}$ when $a_2 = 0$

Reverse transmission ratio $\qquad s_{12} = \dfrac{b_1}{a_2}$ when $a_1 = 0$

Output reflection coefficient $\qquad s_{22} = \dfrac{b_2}{a_2}$ when $a_1 = 0$

All of the s-parameters are voltage ratios. The input and output reflection coefficients are the same as G, described earlier. The forward and reverse transmission ratios can be converted to attenuation (or gain) as described earlier:

$$dB = 20 \log \frac{V_{out}}{V_{in}}$$

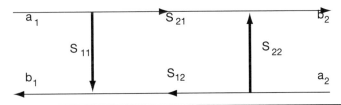

**Figure 32.6** S-parameter flowgraph of a two-port network.

A new type of VNA has recently been introduced—a four-port VNA. Design and test engineers use this in developing equipment that uses differential inputs and outputs. A differential input is one that has both sides of the input or output floating—not connected to circuit common, or ground. This has a number of technical advantages that are not relevant here. High-speed data communication systems are working at RF frequencies, where RF behavior is important, so there is increasing use of RF and microwave test tools. An example of where the four-port VNA might be used is in design and testing of a high-speed serial data bus to connect a peripheral device such as a video camera to a computer. The four port VNA can operate as a two-port described above, using only four s-parameters. In the four-port mode, it requires 16 s-parameters (eight each for differential mode and common mode) to fully describe the device under test.[35]

## Endnotes

1. Hochberg, radio broadcast; Luke, C1.
2. FS 1037C, www.its.bldrdoc.gov/fs-1037
3. NTIA, www.ntia.doc.gov/osmhome/allochrt.pdf
4. Adapted from Shrader.
5. Adapted from Adam, 51 – 74; Saad, 21 – 92; and Shrader, 627 – 631.
6. Adam, 58–59 reprints a table of waveguide sizes and frequency ranges. The original table was from Microwave Development Laboratories (MDL) (Needham Heights, Massachusetts, USA; www.mdllab.com). This table also indicates some of the band names that were originally assigned by MDL and which are now so embedded in the industry that few are aware of their origins.
7. EIA is the Electronics Industries Alliance. IEC is the International Electrotechnical Commission. JAN refers to the US military Joint Army-Navy specifications system; A useful reference for other current waveguide designations can be found at Agilent Technologies. In their web site (www.agilent.com) search for "Agilent Waveguide Overview."
8. Adam 1969, 211–212.
9. Laverghetta, 34.
10. Agilent 2003b, 5.
11. Agilent 2003a, 26.
12. NBS is now NIST; Larsen 1976. 343–347.
13. Carr 1996, chapter 24.
14. Agilent 2003b, 19–50.
15. Agilent 2002, 4-23.
16. Agilent 2002, 4-24 – 4-30.

17. Agilent 2003a, 18-19; and Agilent 2003c, 5-19.
18. Agilent 2003b, 8.
19. ANSI/IEEE 474-1973.
20. Adam, 21-25; and Agilent 2000a, 10.
21. Carr, 493.
22. Agilent 2003a, 31.
23. The Smith chart is copyright by Analog Instruments Company, New Providence NJ. This image is courtesy of RF Café (www.rfcafe.com).
24. Carr, 495.
25. DeWitt 2003.
26. Laverghetta, 142.
27. Carr, 535.
28. Agilent 2000b, 6. (AN-1287-2).
29. Agilent 2002, 1-54.
30. For a comprehensive review of S-parameters and their measurement, see chapter 6.1 of Adam (1969). Other useful references are Agilent AN 1287-1/2/3 and AN 154.
31. Agilent, AN 1287-2b.
32. Agilent 2002, 1-48 – 1-77.
33. Agilent, AN 1287-2b.
34. Adam, 86–106 and 351–457; Agilent 2000a, 1–14; and Agilent 2000c, 1–44.
35. Agilent Technologies. 2002b, 1–16.

# References

Adam, Stephen F. 1969. *Microwave theory and applications*. Englewood Cliffs, NJ: Prentice-Hall.

Agilent Technologies. 2000a. *Application Note 1287-1: Understanding the fundamental principles of vector network analysis*. Publication number 5965-7707E. Palo Alto, CA: Agilent Technologies.

Agilent Technologies. 2000b. Application Note 1287-2: *Exploring the architectures of network analyzers*. Publication number 5965-7708E. Palo Alto, CA: Agilent Technologies. (Available

Agilent Technologies. 2000c. *Application Note AN 154: S-Parameter design*. Publication number 5952-1087. Palo Alto, CA: Agilent Technologies.

Agilent Technologies. 2002a. "Power Measurement Basics." *Agilent 2002 Back to Basics Seminar*. Publication number 5988-6641EN. Palo Alto, CA: Agilent Technologies.

Agilent Technologies. 2002b. *Application Note 1382-7: VNA-based system tests the physical layer*. Publication number 5988-5075EN. Palo Alto, CA: Agilent Technologies.

Agilent Technologies. 2003a. *Application Note 1449-1: Fundamentals of RF and Microwave Power Measurements (Part 1): Introduction to power, history, definitions, international standards and traceability*. Publication number 5988-9213EN. Palo Alto, CA: Agilent Technologies.

Agilent Technologies. 2003b. *Application Note 1449-2: Fundamentals of RF and Microwave Power Measurements (Part 2)—Power sensors and instrumentation*. Publication number 5988-9214EN. Palo Alto, CA: Agilent Technologies.

Agilent Technologies. 2003c. *Application Note 1449-3: Fundamentals of RF and Microwave Power Measurements (Part 3): Power measurement uncertainty per international guides*. Publication number 5988-9215EN. Palo Alto, CA: Agilent Technologies.

ANSI/IEEE 474-1973. *Specifications and test methods for fixed and variable attenuators, DC to 40 GHz*. New York: Institute of Electrical and Electronic Engineers. ts.nist.gov/ts/htdocs/230/233/calibrations/Electromagnetic/pubs/IEEE-474.pdf

Carr, Joseph J. 1996. *Elements of electronic instrumentation and measurement*. 3rd ed. Upper Saddle River, NJ: Prentice-Hall.

DeWitt, Peter. 2003. Telephone conversation with Graeme C. Payne, 12 September.
FS 1037C. 1996. Federal Standard 1037C, *Telecommunications: Glossary of telecommunication terms*. Washington: General Services Administration. www.its.bldrdoc.gov/fs-1037
Hochberg, Adam. 2001. "Analysis: How Bell South's plan to get out of the pay phone industry will affect the lower-income and those in need of emergency assistance." National Public Radio, *All Things Considered*. Broadcast date 14 February 2001.
Larsen, Neil T. 1976. "A New Self-Balancing DC-Substitution RF Power Meter." *IEEE Transactions on Instrumentation and Measurement*. vol. IM-25 (December):343-347.
Laverghetta, Thomas S. 1981. *Handbook of microwave testing*. Dedham, MA: Artech House.
Luke, Robert. 2003. "Era ends for pay phones." *Atlanta Journal-Constitution*. (27 December).
NTIA. 1996. *United States frequency allocations chart*. National Telecommunications and Information Administration. Internet, www.ntia.doc.gov/osmhome/allochrt.pdf (July 2003)
Saad, Theodore S., Robert C. Hansen, and Gershon J. Wheeler. 1971. *Microwave engineer's handbook*. vol. 1. Dedham, MA: Artech House.
Shrader, Robert L. 1980. *Electronic communication*. 4th ed. New York: McGraw-Hill.

# Chapter 33

## Mass and Weight

Although measurements have been performed in one form or another for as long as human beings have been on Earth, metrology, as a branch of physical science, is relatively young. The development of metrology was made possible in part by its universal applicability in all fields of science, acting as a cohesive element, and in the same time by its specificity as an independent and indispensable science.

Operating with standards, measuring units, and their dissemination from national level down to the consumer level, metrology also includes the design and manufacturing of measuring devices as well as the accuracy of performed measurements. As much as it is regarded as a physical science, metrology differs from other sciences by the implications concerning rules and regulations to be followed. From that results the dual function of Metrology: scientific and legal.

Due to the developments of both scientific and legal metrology, two organizations were created: International Bureau of Weights and Measures, BIPM (from the French Bureau International des Poids et Mesures), which, according with 1875 Treaty of Meter, covers the scientific activity, and International Organization of Legal Metrology, OIML (from the French *Organisation International de Metrologie Legale*), which coordinates, since its creation in 1956, the legal activity.

One of the main objectives of metrology is to eliminate the trade barriers generated by nonuniform units of measures, terminology, standards, and so on, and to create a favorable climate for a harmonious and rational worldwide market.

Mass is one of the base quantities in classical mechanics. The concept of mass constitutes a universal characteristic of bodies. As a physical quantity, the mass could be determined in relationship with other physical quantities. Such relationships are Newton's second law of motion and the law of universal gravitation. The concept of mass was introduced by Newton in his second law of motion, which can be stated: The rate of change of the velocity of a particle, or its acceleration, is directly proportional with the resultant of all external forces exerted on the particle and is in the same direction as the resultant force; their ratio is constant.

Experimentally, it was found that if the same force is applied to different bodies, they show different accelerations. Therefore, it was concluded that the accelerations of the bodies are generated not only by the external forces, but also by a physical property, characteristic for each body, the mass.

The mathematical expression of Newton's second law of motion is:

$$\Sigma F = m \times a$$

where:     $\Sigma F$ = summation of all the external forces applied to the particle,

$a$ = the acceleration of the particle,

$m$ = the mass of the particle.

From this equation we can write:

$$m = \frac{F}{a}$$

The mass is numerically determined by the ratio between force and acceleration.

Newton's law of gravitation, published by him in 1686, may be stated: Every particle of matter in the universe attracts every other particle with a force, which is directly proportional to the product of the masses of the particles and inversely proportional to the square of the distance between them. Or in a mathematical form:

$$F = k \frac{m_1 m_2}{r^2}$$

where:     $F$ = the attraction force between the two bodies $m_1$ and $m_2$

$m_1, m_2$ = the masses of the two bodies

$r$ = the distance between the centers of the two bodies

$k$ = gravitational constant ($k = 6.670 \times 10''n$ N m2 kg'2 )

An immediate application of the universal attraction to be observed on Earth is the weight of the bodies. The force with which an object is attracted to the Earth is called its *weight*. Weight is different from mass, which is a measure of the inertia an object displays. Although they are different physical quantities, mass and weight are closely related.

The weight of an object is the force that causes it to be accelerated when is dropped. All objects in free fall near the Earth surface have a downward acceleration of approximately $g = 9.8$ m/s$^2$, the acceleration of gravity. If the object's mass is $m$, then the downward force on it, which is the weight $w$, can be found from the second law of motion (equation 1) where $F=W$ and $a = g$.

Evidently we have:

$$W = mg$$

or Weight = mass times acceleration of gravity.

The weight of any object is equal to its mass multiplied by the acceleration of gravity. Since g is a constant at any specific location near the Earth's surface, the weight w of an object is always directly proportional to its mass $m$: a large mass is heavier than a small one. The gravitational acceleration varies with latitude (because the Earth is not a perfect sphere), elevation, and local variations in subsurface density. So, the weight, when measured with a sufficiently sensitive scale, will vary from one location to another; but the mass of an item remains constant.

The mass of an object is a more fundamental property than its weight, because its mass, when at rest, is constant and is the same everywhere in the universe, whereas the gravitational force on it depends upon its position relative to the Earth or to some other astronomical body (at the Earth's poles $g$ = 9.832 17 m/s², which is the maximum value, and at the equator $g$ = 9.780 39 m/s², which is the minimum value). Mass remains a constant quantity only in classical mechanics. Modern physics, particularly the theory of relativity, shows that the mass of a body increases with the increase of its velocity. The variation of mass with velocity follows the formula:

$$m = \frac{m_0}{\sqrt{1-\frac{v^2}{c^2}}}$$

The symbols have the following meanings:

$m_0$ = mass measured when object is at rest (rest mass),

$m$ = mass measured when the object is in relative motion,

$v$ = velocity of relative motion,

$c$ = velocity of light.

Relativistic mass increases only at velocities approaching that of light. For velocities smaller than $10^6$ m/s, the difference between $m$ and $m_0$ is insignificant.

## UNITS OF MEASURES FOR MASS

In the metric system we normally indicate mass in kilograms whereas the unit of measure for force is *newton*. In the customary (British) system, the unit of mass is the *slug* and the unit force is the *pound* (lb). A body with a mass of 1 slug experiences an acceleration of 1 ft/s² when a force of 1 lb works on it. Accordingly:

$$1 \text{ lb} = 1 \text{ slugft/s}^2$$

The newton and slug are not familiar units because in every day life weights rather than masses are specified in both the metric and the customary system. For example, we ask for 1 kg of apples, not 9.8 newtons of apples, and similarly, we ask for 8 lb of oranges and not ¼ slug of oranges. To avoid general misconception, the following stipulations should be taken into consideration:

- In the metric system, the unit for mass is called *kilogram* and the unit for force is called *newton*.

- In the British system, the unit for mass is called *slug*, and the unit for force is called *pound*.

Customarily and erroneously, when the British system is employed, people use the unit pound to designate the mass of an object. Customarily and erroneously, the term *weight* is used to designate the mass of an object. We say, "the weight of the watermelon was . . .". or "the person's weight is . . . ". In the technical sense, the term *weight* of a body means the force, which applied to that body, would give it an acceleration equal to the local force of gravity.

Some of the most-used conversion factors between the metric system and the customary system, in mass field, are shown in Table 33.1.

**Table 33.1** Most-used conversion factors.

| | | |
|---|---|---|
| 1 pound avoirdupois (lb avdp) | = | 0.453 592 37 kilograms (kg) |
| | = | 7 000 grains |
| | = | 1.215 troy or apothecaries pounds |
| 1 ounce avoirdupois (oz avdp) | = | 28.350 grams (g) |
| | = | 437.5 grains |
| | = | 0.911 troy or apothecaries ounces |
| 1 pound troy or apothecaries | = | 373.242 grams (g) |
| (lb t or lb ap) | = | 5 760 grains |
| | = | 0.823 avoirdupois pound |
| 1 ounce troy or apothecaries | = | 31.103 grams |
| (oz t or oz ap) | = | 480 grains |
| | = | 1.097 avoirdupois ounces |
| 1 kilogram (kg) | = | 2.205 pounds |
| 1 carat (c) | = | 200 milligrams (mg) |
| | = | 3.086 grains |
| 1 gram (g) | = | 0.035 avoirdupois ounce |
| | = | 15.432 grains |
| 1 grain | = | 64.798 91 milligrams (mg) |
| 1 metric ton (t) | = | 2 204.623 pounds |
| | = | 1.102 net tons |
| 1 ton (net or short) | = | 2 000 pounds |
| | = | 0.907 metric ton |
| 1 slug | = | 14.6 kilograms (kg) |

## Official Definition of the Kilogram

The first CGPM in 1889 sanctioned the international prototype of the kilogram and declared, "This prototype shall henceforth be considered to be the unit of mass." The third CGPM in 1901, in a declaration intended to end the ambiguity that existed as to the meaning of the word *weight* in popular usage, confirmed that the kilogram is the unit of mass; it is equal to the mass of the international prototype of the kilogram.

The United States is in possession of two national prototype kilograms identified as K20 and K4, made of the same alloy of 10% platinum and 90% iridium as the international prototype. Also the international prototype, as well as the national prototypes, is right circular cylinders having the height equal with the diameter and equal with 39 mm (1.535″). The material density is 21.5 g/cm$^3$. The international prototype is kept at the BIPM at the Pavillon de Breteuil, Sevres, near Paris. The national prototypes of meter and kilograms are kept at NIST, the highest national entity in metrology, located in Gaithersburg, Maryland.

## Definitions and Conventions

The following definitions might better help the reader in his or her use of mass standards and terminology.

**Apparent mass.** See conventional mass.

**Conventional mass.** Sometimes called mass in air, or apparent mass; this is the value of the mass required to balance the object in conventionally chosen conditions of air density of 1.2 kg/m$^3$, temperature of 20 °C, atmospheric pressure of 760 mm Hg. If the mass required to balance the object is made out of brass with a conventional density of 8390.94 kg/m$^3$ at 20 °C, we obtain the conventional mass (or apparent mass) versus brass of that object.

If the mass required to balance the object is made out of stainless steel with a conventional density of 8000 kg/m$^3$ at 20 °C, we obtain the conventional mass (or apparent mass) versus stainless steel of that object.

Note: The word *mass* comes from the Latin word *massa* meaning bulge of shapeless material, which, in turn, comes from the Greek *massein* or *maza*. Other older and different meanings may be assumed, including the religious meanings of the word; also new meanings are used today such as mass media or mass production.

**Mass comparator.** A weighing device used only as an intermediary between weights with a known value (standards) and weights with unknown value with the purpose of determining the unknown value. Note: Any scale or balance may be used as a mass comparator.

**Scale.** A measuring device designed to evaluate the mass of the objects; in every day vocabulary, a scale is used to determine the weight of an object. A scale is usually of industrial precision.

**True mass.** The mass of an object determined in vacuum. The true mass of an object can be calculated from the conventional mass by applying the buoyancy correction. True mass value is usually determined for special applications such as in nuclear field, biotechnology, and so on; implies weights of ASTM class 1.

**Conventional reference conditions.** Through international consensus, for the purpose of uniformity and reproducibility of the measurements, the conventional reference conditions are:

- Reference temperature          20 °C
- Reference atmospheric pressure  760 mm Hg
- Reference altitude             Sea level
- Reference latitude             45° northern hemisphere

These are also known as normal reference conditions or standard reference conditions.

**Mass.** The state scalar physical quantity expressing both the inertial property of the matter and the property of the matter to generate a field of attraction forces (gravitational field). This is also called *true mass* or *mass in vacuum*. Numerically the mass is expressed as the ratio between a force a material object and its resulting acceleration. Here are some examples using the definitions just discused.

The difference between the buoyant effect acting on a Pt_Ir standard and that acting on a stainless steel weight is approximately 94 mg. (in other words 150 mg – 55.814 mg = 94.186 mg). In a primary standards laboratory, differences such as these are significant and must be accounted and corrected for on the balances used for calibration. This requires knowledge of the density of the unknown mass $\Delta_a$ and the density of air at the time of the measurement.

$$\Delta_a \text{ or } \Delta_n = \text{air density or normal air density}$$

Or, the mass of the stainless steel object is about 1 kg + 94 mg. The true mass of the stainless steel object is about 1 kg + 94 mg (true mass = mass). The vacuum mass of the stainless steel object is about 1 kg + 94 mg (vacuum mass = mass). The apparent mass of the stainless steel kilogram is about 1 kg + 0.0 mg when measured against Pt_Ir standards in air of density 1.2 mg/cm$^3$ at a temperature of 20 °C. The conventional mass of the stainless steel kilogram is about 1 kg + 0.0 mg when measured againstPt_Ir standards in air of density 1.2 mg/cm$^3$ at a temperature of 20 °C (conventional mass = apparent mass).

## CALIBRATION STANDARDS

Calibration standards in mass measurement are the mass measures called (improperly) weights, represent in the SI as the materialization of the mass unit, the kilogram, its decimal multiples and submultiples, and also their half and double values. The pound unit, in the pound-inch system, derives its value from the kilogram, since there is no international pound mass standard. In the United States, the avoirdupois, troy or apothecary pound are used. The avoirdupois pound or ounces set of weights is found both in decimal and fractional multiples and submultiples.

Both the SI and pound-inch systems are using the following denominations for weights: 5321, 5221, and 52111. Denominations of weights are selected in order to cover the most possible nominal values with a minimum number of weights. This is particularly important when someone needs to calibrate customary weights using metric standards; in this case because of the conversion factors, the equivalent number of weights composing the standard, is larger than initially designed, therefore the uncertainty of the calibration will be larger versus a one-to-one calibration.

There are three major classifications of weights in use in the United States concentrated in documents generated by three organizations.

1. NIST with the following documents:

    - Circular 547, section 1, 1954

        "Precision Laboratory Standards of Mass and Laboratory Weights"

        Note: This classification was withdrawn.

    - Circular 3, 1918

        "Design and Test of Standard of Mass"

        Note: This classification was withdrawn.

- NIST Handbook 105-1, revised 1990

  "Specifications and Tolerances for Reference Standards and Field Standards Weights and Measures"

  1. Specifications and Tolerances for Field Standards Weights (NIST Class F),

  Note: NIST, Weights and Measures Division currently uses this classification for the states laboratory program, in legal metrology.

2. American Society for Testing Materials, with the following document:

  - ASTM E617/97

  "Standard Specification for Laboratory Weights and Precision Mass Standards"

3. International Organization for Legal Metrology

  - OIML R111/1994

  "Weights of classes E1, E2, F1, F2, M1, M2, M3"

## Material

The following materials are most used in the manufacturing of weights.

- Platinum-Iridium (90 percent Platinum, 10 percent Iridium) used for manufacturing the international prototype and national standards. Density: 21.5 g/cm$^3$
- Stainless steel, density 8.0 g/cm$^3$
- Tantalum, density 16.0 g/cm$^3$
- Aluminum, density 2.7 g/cm$^3$
- Brass, density 8.4 g/cm$^3$
- Cast iron, density 7.8 g/cm$^3$

## Design

Weights design varies with the nominal value and application. Small weights, from 1 mg to 500 mg, are leaf shaped with a vertical lip to accommodate handling using fine tip nonmetallic tweezers. Fractional pounds (for example, 0.005 lb) may be shaped as a dish. Larger weights, up to 50 kg and 100 lb, are usually of cylindrical or rectangular shape with or without an adjusting cavity. The adjusting cavity is sealed either with a threaded screw and aluminum seal or a threaded screw with a knob.

## Maximum Permissible Error

Each weight within a class is assigned a maximum permissible error (MPE) (or tolerance). Unless it is specified otherwise, the maximum permissible error is assigned both plus and minus sign.

For example: ASTM Class 6, 1 kg weight has ± 100 mg MPE, meaning the weight value may be anywhere between 1kg − 100 mg and 1 kg + 100 mg values.

## Handling, Storage, Packing, Shipping

- Weights of ASTM classes 5 or smaller should not be handled with bare hands, due to possible contamination that will change the mass. There are special nonmetallic, nonmagnetic tweezers to handle small and big weights up to 100g. Special lifters are used for bigger weights with cushioning material at the point of contact. Chains with soft material sleeves or high-density nylon belts are used for lifting large weights.

- Weights are usually offered as a set. Precision weights are provided in cases made of wood or plastic material with seats for each weight. All weights need to be stored in places free of dust or any other contamination with temperature and humidity not to exceed prescribed limits. In general a temperature of 20 °C ± 4 °C (68 °F ± 7 °F) is considered acceptable for storage.

- Weights of small denomination are packed in their cases so they are not loose or lost in transportation. Precision weights are wrapped individually with dry, wax-free paper and placed in their case seats. Cases are wrapped individually with one-inch thick foam or other similar wrapping material and placed in cardboard boxes. Weights up to 10 kg or 25 lb are wrapped individually and should not touch each other. Large weights are packed in specially made wooden boxes with one-inch thick cushion material inside.

- It is preferable that precision weights be hand carried. In general, boxes containing weights should be treated as fragile materials and sensitive equipment for transportation.

## CLASSIFICATION OF WEIGHING DEVICES (SCALES, BALANCES, MASS COMPARATORS)

Weighing devices are the measuring devices exclusively designed to determine the mass of an object. The classification of weighing devices could be accomplished according to the following criteria:

- The mass measuring technique
- The extent of operator participation in the weighing process
- Installation
- The method of obtaining the equilibrium position
- The type of the load-receiving element: pan, platform, bucket, rail, hook, conveyer belt, and so on

## The Mass Measuring Technique

According to this criterion we have direct weighing devices and indirect weighing devices. Weighing devices that directly determine the mass of an object are those devices employing levers in their construction. The mass is determined by the equality between the moments created by the measurand and the standard weight. This type of weighing devices is called *balances* if they have a single lever, with equal or unequal arms. If they have multiple lever(s), they are called *scales* and are used for weighing voluminous and heavy objects (for example, 50,000 lb).

Weighing devices that indirectly determine the mass of an object are based on one the following principles:

- Elastic element (flexure, torsion, compression, traction, and so on)
- Hydraulic element (variation of a liquid pressure)
- Electronic (capacitance, resistance, strain gauge, and so on)

## Operator Participation in the Weighing Process

According to this criterion we have manual, semiautomated, and automated weighing devices. A manual-weighing device is that type where the operator does all three main operations: loading, weighing, and unloading. Examples include equal arm balances, platform scales, and truck scales. A semiautomated weighing device is loaded and unloaded manually, but the weighing operation is automated. The most common are those scales with a dial or digital indicator. An automated weighing device has all three operations: loading, weighing, and unloading completed without the operator intervention. Examples include the belt conveyer scale and the new generation of electronic balances.

## Installation

According to this criterion weighing devices are classified as portable and stationary (stable).

## Method of Obtaining the Equilibrium Position

Based on this criterion we have two types of weighing devices:

- Those with constant equilibrium position. In other words, two indicators—one fixed and one mobile or two mobile indicators.
- Those with variable equilibrium position. In other words, the indicator moves in front of a graduated scale; *semiautomated* weighing devices.

### Scales

Weighing devices employing multiple levers in their design are called *scales*. There are a large variety of scales, from which we choose to mention the following categories: platform scales, either portable of built in floor, vehicle scales, and railroad scales.

The design of scales has changed very much from levers and knife-edge joints to levers with flexure joints and, lately, to levers with load cells or just load cells (see

Figure 33.1). Nowadays a scale has three general components: the load-receiving element, the load cell, and an indicating device. Large capacity scales with load cell elements are achieving higher accuracy than traditional scales employing beams.

## Balances and Mass Comparators

For over three thousand years the design of balances did not change significantly, mainly because the requirements were unchanged. The last half of twentieth century changed the situation, and new balances and mass comparators based on electromagnetic force compensation emerged to satisfy new technologies, such as computers, biotechnology and drug manufacturing, sophisticated and more accurate weaponry, and so on.

## Electromagnetic Force Compensation Mass Comparators

These weighing devices are called mass comparators as they perform a comparison between a known weight piece (standard weight) and an unknown weight piece (unknown) in order to determine the mass of the later. Such devices are designated for high accuracy mass determinations in the national and primary calibration laboratories.

These weighing devices employ flexure joints in their design, replacing the knife-edge joints. There are no moving parts, making these devices practically maintenance free. Although the electromagnetic force compensation principle was known well before, it is the last generation of mass comparators, created in the last decade, that achieved the accuracy of former equal arm balances combined with electro-optical systems. Another significant advantage of the new mass comparators is the possibility of computer interfacing. Using the computer to record data eliminates one source of errors (writing the wrong value) and gives the possibility of instantaneous statistical analysis of real time data, so critical for making the right decisions.

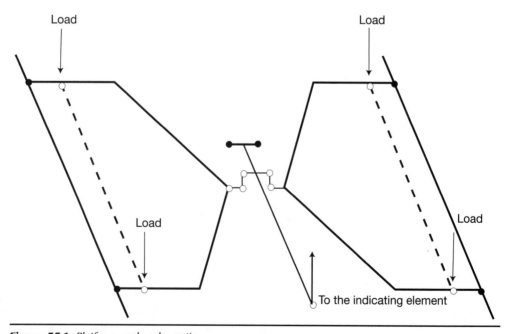

**Figure 33.1** Platform scale schematic.

## Weighing Methods

Selecting the weighing method for your application depends on a series of factors such as:

- Required level of accuracy and precision
- Available equipment
- Available trained operators
- Cost effectiveness

Sometimes a lower level procedure may be sufficient for a specific application, rather than a laborious state-of-the-art intercomparison. Having the entire palette of weighing methods available will offer a better opportunity for selecting the most appropriate one.

## Direct Reading on an Equal Arm Balance

Using an equal arm balance for simple weighing could be accomplished by loading one pan with the measurand (unit to be measured or unknown weight), and the other pan with the necessary weights, until the balance is brought in a normal equilibrium position. This is a lower level weighing method for applications that do not require special care with regard of accuracy. In this case the mass of the measurand is equal with the total mass of the added weights.

## Direct Reading on Single Pan Optical-Mechanical and Electronic Balances

The value of the measurand, in this case the unknown mass, is determined by placing it on the pan of the balance, releasing the pan, and reading the value of the mass on the balance display. Certainly, we have to verify if the display shows zero when the balance is unloaded.

Single pan optical-mechanical scales provide apparent mass value, depending on balance adjustment. Both single pan optical-mechanical and electronic scales rely on internal weights of the balance. For that reason the internal weights should be tested periodically. Also the optical scale should be tested for the same reason.

## Substitution Method

The substitution method, known also as the Borda method, eliminates the systematic error induced in the weighing process by the unequal arms of the balance beam. The substitution method consists of the following: the measurand is placed on the left pan of the balance while on the right pan we place tare weights until the balance remains in the normal equilibrium position or close to this position. We then replace the measurand with known weights, adding them until obtaining the same equilibrium position as previously. The mass $m_X$ of the measurand is equal with $m_P$, the mass of the known weights. The substitution method is largely employed in most of the laboratories performing mass measurements.

In the single substitution method the sequence is as follows:

- Standard, S, observation O1
- Unknown, X, observation O2
- Unknown plus sensitivity weight, X + sw, observation O3

The measurement sequence in double substitution is as follows:

- Standard, S, observation O1
- Unknown, X, observation O2
- Unknown plus sensitivity weight, X + sw, observation O3
- Standard plus sensitivity weight, S + sw, observation O4

**Transposition Method**

The transposition method, or Gauss method, after the German mathematician and astronomer Carl Friedrich Gauss, consists of the following:

The measurand is placed on the left pan of the balance, while on the right pan we place known weights (standard) until the balance remains in the normal equilibrium position or close to this position; observation O1. We permute the measurand and the standard, adding known weights (standards) to the left pan, while the measurand is placed on the right pan; observation O2. A sensitivity weight is added to the left pan; observation O3.

## Alternative Solutions

There are many options available for an alternative method of defining a kilogram, but three are at the forefront.

1. **A new physical mass.** The current standard mass is 90% platinum and 10% iridium. This mixture is especially stable and dense. A dense mass is preferred because more porous metals would absorb gases and be harder to clean, both affecting mass. Because of these properties, a new physical standard mass would most likely be made of the same materials, just with better manufacturing techniques and closer tolerances.

2. **The Avogadro project.** The Avogadro project can be summed up by the following. Avogadro's number is a natural physical constant. It is to molecular science what pi is to trigonometry. As stated in *Comprehensive Mass Metrology*, "By definition, an Avogadro number of Carbon-12 atoms weigh exactly 12 grams. As such, the kilogram could be redefined as the mass of 1000/12 x (Avogadro's number)." To use this method, though, Avogadro's number would need to be found more precisely. In determining the constant more precisely, the preferred method has been to use a highly polished silicon crystal sphere of 93.6 mm diameter, with a roundness in the range of 60 nanometers. Roundness refers to the amount of change in radius from point to point on the sphere. The less change, the rounder the sphere is.

    A crystal structure is used because crystals are the most stable structure. For example, the crystal structure of carbon is diamonds, while its other forms of shale and graphite are much weaker. Silicon is used because it has a well-known crystal structure and is stable.

    The idea is that by knowing exactly what atoms are in the crystal, how far apart they are and the size of the ball, the number of atoms in the ball can be calculated. That number then becomes the definition of a kilogram.

3. **The Watt Balance approach.** The Watt Balance approach is described in VIM as the comparison of the weight caused by a test mass to "a vertical force generated by induction, with a current injected in a moving coil surrounding a fixed permanent magnet." The system works on the same principle as an electromagnet. It uses electrical force to pull on an object and checks how much electrical force, or current, it takes to counteract the force gravity has on the object. Current is relatively easy to measure, and the amount of current it takes can be set to the new standard.

# References

Harris, Georgia L. 1995, *Western Regional Assurance Program workbook*. Gaithersburg, MD: NIST, Office of Weights and Measures.

Harris, Georgia L. (editor and lead). 2003. NIST Special Publication 1001, *Basic mass metrology*, CD-ROM interactive. Gaithersburg, MD: NIST.

Harris, Georgia L., and Jose A. Torres. 2003. NIST IR 6969, *Selected laboratory and measurement practices, and procedures, to support basic mass calibration*. Gaithersburg, MD: NIST.

Hazarian, Emil. 2003. "Elements of Measurement Techniques," MSQA Program, QAS 516 Measurement and Testing Techniques, Supplement. Dominguez Hills: California State University.

Hazarian, Emil. 2002. Mass Measurement Techniques IV. "Scales Uncertainty." Tutorial. Anaheim: Measurement Science Conference.

Hazarian, Emil. 1994. *Techniques of mass measurement workbook*. Los Angeles: Technology Training, Inc.

Hazarian, Emil. 1992. "Some Characteristics of Weighing Devices" Technical paper at Measurement Science Conference. Anaheim.

ISO. 1993. *International vocabulary of basic and general terms in metrology (VIM)*. Geneva. ISO.

Jones, Frank E., and Randall M. Schoonover. 2002. *Handbook of mass measurement*. Boca Raton, FL: CRC Press.

Kochsiek, M., and M. Gläser, (editors). 2000. *Comprehensive mass metrology*. Berlin. Wiley-VCH.

Mettler-Toledo. 2000. *Glossary of weighing terms: A practical guide to the terminology of weighing*. With the assistance of Prof. Dr. M. Kochsiek, Physikalisch-Technische Bundesanstalt, PTB. English Translation by J. Penton. Berlin: Mettler-Toledo.

NBS Metric Information. 1976. Letter Circular LC 1071, *Factors for high-precision conversion*. Gaithersburg, MD: National Bureau of Standards.

Prowse, David D. 1995. *The calibration of balances*. Australia: Commonwealth Scientific and Industrial Research Organization.

# Chapter 34
# Dimensional and Mechanical Parameters

Between 1889 and 1960, the definition of the meter has been based on the international prototype of platinum-iridium, the so-called archive meter. In 1960, the 11th CGPM replaced the old definition with a new one based on the wavelength of the krypton-86 atomic radiation. In 1983, the 17th CGPM, in order to satisfy the industry's growing demand, an increase in the precision of materialization of the meter was considered a necessity, therefore a new definition of the meter was adopted. It is, "the meter is the length of the path traveled by light in vacuum during a time interval of 1/299 792 458 of a second." (17th CGPM, 1983, Resolution 1)

Length, surface, or angle do not differ only from the quality point of view, they differ quantitatively as well. They could be smaller or larger and we can ascertain how small or how large they are. In other words we can measure them. Therefore, we can say that these geometrical quantities are part of the greater category of physical quantities.

From the metrological point of view, geometrical quantities could be classified, with respect of their dimension, as follows:

- Length (dimension L)
- Plane surface (dimension $L^2$)
- Plane angle (dimension $L^0$)
- Geometric dimensioning and tolerances (dimension L or $L^0$)
- Convergence of optical systems (dimension $L^{-1}$)

This classification does not include the volume (dimension $L^3$), which constitutes a separate branch in metrology.

## Conversion Factors

1 inch = 0.0254 meters

1 foot = 0.3048 meters

1 yard = 0.9144 meters

## LENGTH MEASURES

The length measures category may be subdivided into categories.

Length measures such as:

- Graduated measures (graduated lines, tapes, and so on.)
- End measures (slide calipers, gage blocks)
- Combined measures (rules, tapes)
- Radiation measures (Kr 86 radiation)

Angular measures:

- Graduated angular measures (protractor)
- End-angular measures (angular gage blocks, calipers, polygons)

Plane measures:

- Plan meters
- Area measuring machines

Geometric dimensioning and tolerances measures:

- Straightness measures (knife edge rules)
- Flatness measures (surface plates, autocollimators)
- Roughness measures (specimen blocks, profilometers)

Length measuring instruments of general designation for measuring:

- Length
- Angle
- Surface
- Geometric dimensioning
- Convergence of optical systems

Length measuring instruments of special designation for measuring:

- Gears
- Threads
- Conicity
- Coatings' thickness
- Wires
- Railroads dislevement

- Trees (dendrometry or hypsometry)
- Human body (anthropometry)

Choosing the design principle as criteria, the length measuring instruments may be classified as:

- Mechanical
- Optical
- Combined optical and mechanical
- Pneumatic
- Hydraulic
- Interferential
- Electronic
- Magnetic
- Photoelectric
- Ultrasound
- Laser
- X-rays
- Nuclear radiation

## ANGLE MEASUREMENT—DEFINITIONS

**Angle (geometric).** The figure formed by two straight lines (sides) emanating from one point (vertex), or by two or more planes.

**Slope (m).** The inclination of a straight line versus the horizontal plane. The slope could be evaluated through trigonometric tangent of the angle. Note: The slope is usually expressed in mm/m (inch/foot) for machinist levels, or in percentage for road construction. For example 15% slope corresponds to a level difference of 15 feet for 100 feet road horizontal projection.

**Conicity.** The angular size expressed by the ratio between the difference of two diameters, reduced to unit, and the axial distance between those diameters.

### Angle Units of Measure

**Radian (rad).** The ratio between the length of the arc subtended by the central angle of a circle and the radius of that circle. The radian is the supplementary unit of measure for a plane angle in SI. Considering that the plane angle is expressed as a ratio between two lengths, the supplementary unit, radian, is a dimensionless derived unit. The right angle has $\pi/2$ radians.

**Sexagesimal degree.** The sexagesimal system divides the circle into 360 degrees (°), each degree is divided into 60 minutes of arc ('), and each minute is divided into 60 seconds of arc ("). The right angle has 90°.

**Centesimal grade.** The centesimal system divides the circle into 400 grades ($^g$), each grade is divided into 100 minutes ($^c$), and each minute is divided into 100 seconds ($^{cc}$). The right angle has 100$^g$.

**Millieme or mil system.** Using as an angular unit a thousandth of a radian. The length of a circle is 2B radians, or 6.283185. radians, or 6,283 milliradians.

**The practical mil system.** Using as an angular unit the radian divided, for practical purposes, into 6,400 equal portions.

**Rimailho millieme.** Using as an angular unit the radian divided exactly into 6,000 portions. The conversion factors between the above angular units are given in Table 34.1.

## Angle Measurement—Instrumentation

**Line graduated measures.** Graduated discs of circular shape, semicircular, or a quarter of a circle with divisions of 1° or 1$^g$. A vernier scale is often used for finer divisions. This type of measures are used for educational purpose, in drafting, cartography, or inside optical instruments such as rotary tables, dividing heads, theodolites, goniometers, and so on.

**Table 34.1** Conversion factors for plane angle units.

| Units | (rad) | (L) | (°) | (') | (") | ($^g$) | (m) | ($m_R$) |
|---|---|---|---|---|---|---|---|---|
| (rad) | 1 | 2/π | 180/π | 10800/π | 648000/π | 200/π | 3200/π | 3000/π |
| (L) | π/2 | 1 | 90 | 5400 | 324000 | 100 | 1600 | 1500 |
| (°) | π/180 | 1/90 | 1 | 60 | 3600 | 10/9 | 160/9 | 50/3 |
| (') | π/10800 | 1/5400 | 1/60 | 1 | 60 | 1/54 | 8/27 | 5/18 |
| (") | π/648000 | 1/324000 | 1/3600 | 1/60 | 1 | 1/3240 | 2/405 | 1/216 |
| ($^g$) | π/200 | 1/100 | 9/10 | 54 | 3240 | 1 | 16 | 15 |
| (m) | π/3200 | 1/1600 | 9/160 | 27/8 | 405/2 | 1/16 | 1 | 15/16 |
| ($m_R$) | π/3000 | 1/1500 | 3/50 | 18/5 | 216 | 1/15 | 16/15 | 1 |

Where: rad = radian

L = right angle (quadrant)

(°) = sexagesimal degree

(') = sexagesimal minute

(") = sexagesimal second

($^g$) = centesimal grade

(m) = the practical mil unit

($m_R$) = the Rimailho millieme

**End measures.** Angle gages usually employ integer values of angles. They are used in mechanical workshops mostly for the verification of cutting tools angles. The angle gages are made of tool or hardened steel, or have the active surfaces plated with metallic carbide. The angle gage accuracy is verified either with fixed taper gages or other angle measurement instruments.

**Angle squares.** Angle squares are materializing unique plane angle values, usually 90°. Although there are angle squares of 30°, 45°, 60°, or 120°. Angle squares are used along with machine tools for verification of external or internal angles, or for angular tracing (marking). They are made of tool steel entirely or of regular steel having the active measuring surfaces hardened. The measuring surfaces flatness error is less than 0.01mm (0.0004inch).

Angle squares are classified as a function of the longer side length, in four classes of accuracy, in respect to the perpendicularity error. The permissible errors are given by the following formulas:

$$E_1 = \pm\left(0.002 + \frac{h}{100}\right)$$

$$E_2 = \pm\left(0.02 + \frac{h}{100}\right)$$

where $h$ (expressed in meters), is the length of the longer side of the square.

The angle accuracy, for a right angle square is verified by the two square method and the three-square method. Both methods are comparing the squares to be verified between them versus a standard right angle square. Also an angle tester instrument may be employed for the same purpose.

**Taper gages.** These are fixed angular measures usually fabricated for standard dimensions. The most utilized systems in taper gage are the metric (1:20 taper) and the Morse (0.625 = 5/8 inch per foot). Taper gages are used either for the verification of internal conicity, called taper plugs, or for the verification of external conicity, in which case they are called sleeve gages. In both cases, taper gages are made for the verification of a specific reference diameter.

**Angle gage blocks.** These blocks are the most common angular measuring devices. In fact, angle gage blocks are end measures for a plane angle, having the shape of a rectangular prism. They are usually made of tempered steel.

The main components of an angle gage block are as follows:

- Measuring surfaces, producing the active angle
- Nonmeasuring surfaces
- The angle gage block may employ one active angle or more. The size of the active angle of an angle gage block may vary between 10" (seconds of arc) and 100° (degrees of arc) combined in sets having different configuration depending upon manufacturer and destination. For angles bigger than 100°, the angle gage blocks are combined suitably using special accessories.

Angle gage blocks are classified by one of the principal U.S. manufacturer in three classes of accuracy:

- Laboratory master, with an accuracy grade of ± ¼ second of arc
- Inspection master, with an accuracy grade of ± ½ second of arc
- Tool room, with an accuracy grdea of ± 1 second of arc

In the metric system there are also three classes of accuracy:

- Class 0, with an admissible angular deviation of ± 3 sec. of arc
- Class 1, with an admissible angular deviation of ± 10 sec. of arc
- Class 2, with an admissible angular deviation of ± 30 sec. of arc

The number of angle gage blocks in a set varies from six to 16 pieces.

The calibration of angle gage blocks is to accomplish directly by using standard goniometers or by comparison with other standard angle measures with the assistance of an autocollimator.

The angle gage blocks have multiple applications such as:

- Direct measuring of angles or angular variation
- Calibration of angular instrumentation (mechanical and optical protractor, and so on)
- Inspecting rotary tables and dividing heads

**Polygons.** Polygons are angle end measuring devices made of metal or optical glass (quartz) having three to 72 active surfaces, therefore angles magnitude between 120° and 5°. The polygons are standard measures of angles used for verification and calibration of angle measurement instrumentation. They are manufactured in two classes: reference class, with an accuracy grade of ± 0.25 sec. of arc, and calibration class, with an accuracy grade of ± 0.5 sec. of arc. The maximum allowable error of calibration varies between 0.05 second and 3 seconds of arc.

## Complex Angle Measurement Instrumentation

**Mechanical protractor.** This is one of the most common angle measuring devices in a workshop. The main components of a mechanical protractor are a semicircular sector, divided in 180° with a resolution of 1°, and a mobile ruler provided with an indicator.

**Mechanical protractor with vernier.** Has a similar design with the mechanical protractor although the resolution is improved by the vernier to 5 minutes of arc. The vernier is shaped as a circular sector divided in 12 even intervals on either side of "0" division. The value of division of the vernier is given by:

$$\frac{1°}{12'} = \frac{60'}{12'} = 5'$$

**Optical protractor.** This is very similar to the mechanical protractor with vernier using an optical system for reading. The smallest value of division is 5 minutes of arc.

**Sine bars and plates.** Both sine bars and sine plates are based on trigonometrical function of sine:

$$\sin\alpha = \frac{h}{L}$$

where $h$ is the gage block height and $L$ is the fixed length between the two axes of the rolls.

The sine plates are similar with sine bars having the reference surface bigger. The angle generated by the sine bars and plates is from 0 to 60°. The distance between the two rolls varies from 100mm to 200mm (5 inches to 10 inches). The reference surface is up to 25 x 220mm (1 x 10.75 inches) for sine bars and up to 500 x 1200mm (12 x 24 inches) for sine plates.

The angle accuracy depends on the accuracy of the distance between the axes of the two rolls, the diameter of the rolls and the combined height of the gage blocks. The achieved accuracy is within seconds of arc.

**Tangent bar.** The tangent bar has also two rolls, but of different diameters. Between the two rolls the distance $L$ is determined by gage blocks.

The desired angle is given by formula:

$$tg\frac{\alpha}{2} = \frac{D-d}{D+d+2L}$$

The obtained angle accuracy is within seconds of arc and depends on the accuracy of the two rolls, gage blocks, and the parallelism between top and bottom surfaces of the bar.

# LEVELS

**Levels with vials and bubble.** This type of level, invented in 1666 by Thévenot, is mostly used for determining small angles of deviation from horizontal and vertical direction. This type of level has a vial with a curvature given by the radius $R$. The curvature is obtained by lapping, for precision levels, or by bending the glass tubing, for lower precision levels. The inside liquid utilized is ethanol (ethyl alcohol) with a bubble of the respective liquid. The vial is graduated outside within bubble motion range. The graduations are equidistant. If we consider the level

$$\sin\alpha = \frac{d}{radian}$$

inclination $\alpha$, the bubble moves with interval $d$. The corresponding inclination angle will be:

Customarily, the angle is expressed by mm/m or inch/foot. Therefore:

$$\sin\alpha = \frac{1}{1000}\frac{d}{radian} \quad mm/m\,(inch/foot)$$

The value of division could vary from 0.01mm/m to 20mm/m or higher, in metric system, or from 0.0005 inch/foot to 1 inch/foot or higher, in English system.

There are few distinct designs of vial levels differentiated by shape and scope:

- **Cylindrical level.** The vial is mounted inside a cylindrical tube. It is used for verifying the horizontality of V-shaped guides (grooves).

- **Precision block level.** The bottom is the active surface with a good flatness, also V-shaped for checking cylindrical surfaces.

- **Block level with micrometer.** Inside the block, the vial has an articulation on one end, the other end being adjusted with a micrometer.

- **Frame level.** Frame levels are equipped with cross-vials. The V-shape accommodates cylindrical surfaces.

- **Level with microscope.** In actuality this is a combination between a level and an optical protractor.

- **Level with coincidence.** This type of level covers a range between $\pm 10$mm/m, in metric system, or $\pm 1/8$ inch/foot in English system. The value of division achieved is 0.01mm/m (2 seconds of arc).

- **Levels with flexible tubing.** This type of levels employs the principle of communicating vessels. The main components are two graduated vessels with a bottom flange as a reference surface. Flexible tubing connects the two vessels. These levels are mainly used in construction or big workshops for mounting components at the same level.

**Electronic levels.** In an electronic level, an inductive transducer determines the angular position of a pendulum versus the case level. The value of division varies between 0.05mm/m and 0.01mm/m, with a corresponding range of $\pm 0.75$mm/m and $\pm 0.15$mm/m. The electronic circuits are battery powered. In the English system the value of division varies from $5 \times 10^{-6}$ inch/inch to $1 \times 10^{-4}$ inch/inch with a corresponding range of $1 \times 10^{-4}$ inch/inch to $2 \times 10^{-3}$ inch/inch. The high precision makes the electronic levels an indispensable tool in the aircraft industry, thermal, hydro or nuclear power plants, etc.

**Autocollimators.** Autocollimators are optical instruments for measuring small angles, such as the deviation from rectilinear, flatness, for alignment of different components, for calibrating angular polygons, and angle gage blocks. The value of division for an autocollimator varies between 0.1 second of arc and 10 seconds of arc for a corresponding range of 5 minutes of arc to 30 minutes of arc. The autocollimators are calibrated with standard plane angle generator, using the tangent bar principle.

**Optical dividing heads.** These are manufactured for a range of 360° and a value of division varying from 1 second of arc to 60 seconds of arc. The calibration of optical dividing heads is performed with standard polygons and autocollimators.

**Optical rotary tables.** Similar with optical dividing heads, although these have T-grooves for easier mounting of parts and may have a value of division of 0.5 second of arc.

**Electronic encoders.** Electronic encoders are not measuring devices, although they are used in conjunction with appropriate devices, and can be used for measuring and recording circular divisions. Electronic encoders are using circular capacity transducer and may have an accuracy of 1 second of arc.

**Theodolites.** These are instruments capable of measuring angles in a horizontal or vertical plane, within a range of 360° or 400$^g$. The value of division varies from 1 second of arc to 1 minute of arc.

**Goniometers.** Goniometers are instruments capable of measuring angles in a horizontal plane, also used in topography along with a leveling staff.

Both theodolites and goniometers are operated mounted on a strong tripod.

# Chapter 35

# Other Parameters: Chemical, Analytical, Electro-optical, and Radiation

As explained in Chapter 30, this part can only supplement the information found in greater detail in other documents, books, reference manuals, and so on. With a focus on being general in nature, the hope is to explain some of the areas that most are unfamiliar with, have limited usage in the metrology community, or give knowledge to help the reader understand areas that are very focused in their nature and sometimes not even considered when referring to calibration or IM&TE. They should be considered, though. For example, chemistry has been around as long as measurement. Alchemists and wizards had a need to measure their potions and concoctions to ensure the desired results. Too much eye of newt and there might be more than enough toil and trouble to go around. The mere mention of optics brings a myriad of ideas to mind; including deep space exploration, the Hubble telescope, and, possibly, the local surveying crew with their pole and transit.

## OPTICS

The first subject discussed is the history behind optics, and how their importance continues into our future. Let's begin with some basics. Figures 35.1, 35.2, 35.3, 35.4, and 35.5. show optical principles.

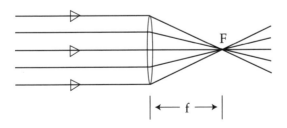

**Figure 35.1** Converging (positive) lens. Convex lenses are those that are wider in the center than they are on the top and bottom. They are used to converge light.

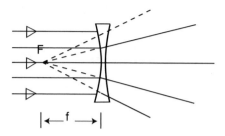

**Figure 35.2** Diverging (negative) lens. Concave lenses are those that are wider at the top and bottom and narrower in the center. They are used to diverge light.

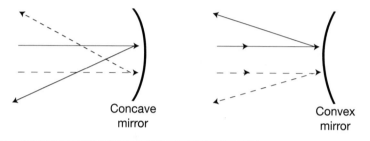

**Figure 35.3** Concave mirrors reflect light inward, and convex mirrors reflect light outward.

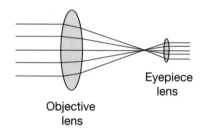

**Figure 35.4** Refracting telescopes use lenses to focus light.

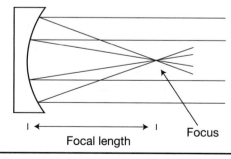

**Figure 35.5** Reflecting telescopes use mirrors to focus light.

## Optics

A grunt. A thump on the chest of his eldest son. The sweep of an arm from a rock by a river to a distant tree. Some prehistoric father had just described a property line. In doing so, he tapped a wellspring of dispute, lawsuits and war that still flows. He also invented the title business and the surveying profession.

Some 200,000 years later, the earliest grants of land in America still involved the broadest sort of property description. But, as the populations grew and land was subdivided, the necessity arose for more precise descriptions of property. At first, the old world practice of "metes and bounds" sufficed. Property was described in distances from landmark to landmark. To conform those descriptions, men were employed to survey, define and mark boundaries. George Washington, this land's first county surveyor, was among them.

**A grand scale.** Metes and bounds were adequate in the original 13 colonies, Kentucky, Tennessee and parts of other eastern states. But the vast western territories received a simple scheme of grand scale. So the Congress of a young United States approved the subdivision of public lands into a grid of 36-square-mile townships, and square-mile sections. The corners of each were to be measured from the intersection of north/south meridian and east/west baselines.

To that end, teams of surveyors were dispatched by the U. S. Coast and Geodetic Survey. For the better part of 100 years, they took the measure of the land, using instruments that had been invented in the 16th and 18th centuries and techniques that evolved as they went along. The results were less than perfect. Today's land surveyors spend much of their time finding and correcting the errors of their forerunners in the interest of clearing title. Of course, today's surveyor has the advantage of soaring technology. Surveying instruments have changed more in the past 10 years than in the previous 200. And those changes will ultimately impact the way that property is measured, described and held.

**Little change.** The instruments used to survey America changed little from the early 1800s well into the twentieth century. An 1813 surveying text notes that, in New England, most work was done with a magnetic compass and a surveyor's chain.

The compass, invented in 1511, was in wide use until 1894. The chain was invented in 1620 by Edmund Gunter, an Englishman. It was made of 100 iron or steel links and was 66 feet long. Eighty chains made up one mile. Ten square chains made one acre. Gunter's chain was in universal use until the steel tape measure replaced it in the last decades of the 18th century.

The transit was first made in 1831 by Philadelphian William J. Young. It was an adaptation of the theodolite invented in 1720 by John Sisson of England. Sisson had combined a telescope (invented circa 1608), a vernier—a device for subdividing measurements by 10ths (1631)—and a spirit level (1704) into a single instrument. Young's improvement was to permit the telescope to revolve, or transit, upon its axis—a useful feature when prolonging straight lines or taking repeated readings to confirm accuracy. Improved versions of Young's transit were still in use for land surveying in the 1950s and are still broadly used in the construction trade.

*continued*

*continued*

**Inaccuracies.** Early surveys were often grossly inaccurate. The iron chains stretched with use. An error of one link (about 8 inches) in 3 to 5 chains was considered normal. The magnetic compass was a major source of error. It is subject to daily, annual and lunar variations in the earth's magnetic field, solar magnetic storms, local attractions and static electricity in the compass glass. Optical glass varied in quality. There were no standards for equipment and many manufacturers. And, of course, no way to re-calibrate equipment damaged 100 miles from nowhere.

Survey procedures were often less than precise. If a tree blocked a line of sight, a surveyor might sight to the trunk, walk around it and approximate the continuing line. One modern text referring to 19th century surveys cautions that "No line more than one-half mile can be regarded as straight." Of course, precision seemed unimportant when the land seemed endless and, at $1.25 an acre, cheap. Especially to a surveyor who was paid by the mile. Other factors contributing to inaccuracy included a lack of supervision, a shortage of trained surveyors, an abundance of hostile Indians, bears, wolves, wind, rain, snow, burning sun and rugged terrain.

**Technology soars.** The technological boom of the past 15 years has greatly increased both the accuracy and precision of land surveys. In the 1860s an error of only 1 in 1,500 feet was considered highly accurate, (*even though it worse than the dimensional accuracy of the Pyramids in Egypt* – editor). By today's standards, an error of 1 in 10,000 feet is reasonably accurate and measurements accurate to 1 in 1,000,000 feet are possible. A 19th century compass measurement that came within 60 seconds was acceptably accurate. Modern electronic instruments are accurate to within one second (a second is 1/3,600 of a degree).

Until recently, modern surveying was accomplished using manual devices known as Theodolites along with plumb bobs and both measuring tapes and rods. The Theodolite is an optical instrument that works similar to a set of binoculars in that focusing the lenses will provide an approximate verification of distance. The instrument also typically has an angle function to allow a fairly good reading of the angle, or elevation. More exact distance measurements are obtained by use of the calibrated measuring tape and the rod. The tape can be very long, even into the hundreds of feet or meters. The Theodolite is mounted on a tripod and the plumb bob is used for leveling of the instrument.

The most modern versions of the Theodolites can include lasers and/or GPS – allowing much more precise measurements to be made.

The most stunning breakthrough of modern technology has been in the measurement of distance. Electronic Distance Meters (EDMs) have replaced the steel tapes. EDMs operate on the basis of the time it takes a signal to travel from an emitter to a receiver, or to reflect back to the emitter. Short range EDMs use infrared signals. EDMs designed for distances from 2 to 20 miles use microwaves. "They are accurate to within 3 millimeters on a clear day and adjust for atmospheric haze distortion and curvature of the earth," says Duke Dutch of Hadco Instruments, a major southern California distributor of survey equipment. Surveyor's transits now incorporate digital electronics that read down to one second.

*continued*

Instead of jotting his calculations and notes in a field book, today's surveyor plugs an electronic field book into his electronic transit. The electronic field book is a magnetic tape recorder with a digital display and keyboard. It automatically records each observation made by the transit. Before the electronics boom, surveyors used a bulky plane table to manually plot maps in the field. Today's surveyor can hook his electronic field book to a computer. Special coordinate geometry software speeds the process of checking observations and, through interface with a plotter, making maps. During the past several years, the EDM, electronic transit and electronic field book have been combined into a single unit called a "total station." "Surveyors who don't have the latest technology simply won't survive," observes Dutch. "They won't be competitive. These instruments cut the time and cost of a survey by as much as 40 percent."

**A Coming Change?** The almost microscopic accuracy of electronic surveying equipment, when combined with advances in astronomy and satellite technology, may soon change the way property is described. It may even stimulate basic changes in title law, according to Paul Cuomo, section chief with the Orange County, California, Surveyor's Office and treasurer of the California Land Surveyors' Association. "The basic problem," Cuomo says, "is the perpetuation of monuments." Under current laws, surveyors' monuments denoting corners have priority in the determination of boundaries. In most cases, the boundaries defined by a monument will stand even if the original survey was in error and a new survey demonstrates that fact. If old corners are obliterated, they may be reconstructed from physical evidence or testimony. If they are lost entirely, as is often the case, a new survey must retrace them from some known monument.

Monuments are plowed under, washed away, rotted away or simply moved as often as not. And sometimes they cannot be found because they were misplaced in the first place. (An error of one degree by a 19th century surveyor translates into 90 feet at one mile.) Cuomo foresees a day when boundaries will be precise and permanent—with or without monuments. "All it would take is to tie all future surveys to the State Plane Coordinates System," he says. That system is a nationwide grid of survey stations established over the past 40 years. Each station's location was precisely plotted astronomically. Each is in sight of others, so triangulations are convenient.

"Any survey of record tied to this system could be recreated on the ground, exactly, forever. All that would be required would be a record of the northing and easting," says Cuomo.

In surveyor's parlance, that means a notation of the exact longitude and latitude of at least one corner of the property. That notation would be include in the property description in the deed. Formerly, such a notation required time-consuming observations and calculations that were prohibitively expensive. The advent of rapid operating, optically superior and highly precise electronic survey devices changes the picture. "It will take major changes in the law," says Cuomo. "And we'll have to get everyone to agree: the courts, the lawyers, the title companies. But it will happen someday." And, when it does, the line that someone's ancestors drew from the rock by the river to the distant tree will hold. Forever.[1]

Since the previous material was written (1986) an even greater change has come to surveying. Although optical transits are still used, many survey systems now use GPS receivers, technology that was in its infancy in the mid-1980s. The GPS system allows determination of three-dimensional position (latitude, longitude and elevation) completely independently of other references. The normal accuracy is to within 10 meters, but that can be improved by several methods. Combined with optical triangulation and electronic distance measuring, this discipline has been taken to a new level.

## COLORIMETRY

### New NIST Color Reference More Than a Shade Improved

A new reference instrument for measuring the surface color of materials with high accuracy has been developed by the National Institute of Standards and Technology (NIST) Optical Technology Division, which plans to offer a calibration service for 0 degrees/45 degrees industrial color standards starting in January 2003. Because color often plays a major role in the acceptability of a product, this service is designed to meet a demand for improved measurements and standards to enhance the color matching of products.

The new reference colorimeter measures with the best possible accuracy a non-fluorescent sample's spectral reflectance properties, from which color quantities are calculated. The instrument design can perform measurements at all possible combinations of illumination and viewing angles, which is important for accurate image rendering. In addition, the standard 0 degrees/45 degrees geometry (illumination at 0 degrees and viewing at 45 degrees) is highly automated through the use of a sample wheel with a capacity of 20 samples.

The new calibration service will be NIST's first for color measurement in many years, a response to needs articulated in recent reports of the Council for Optical Radiation Measurements. This new service complements ongoing services in reflectance, transmittance, and specular gloss. Industrial customers are expected to send samples (typically colored tiles) to NIST for measurement, and then use these samples as standards to calibrate their own instruments. Users then typically convert a spectral reflectance measurement into the color coordinate system used by that particular industry.[2]

### Ionizing Radiation

Ionizing radiation sources can be found in a wide range of occupational settings, including health care facilities, research institutions, nuclear reactors and their support facilities, nuclear weapon production facilities, and other various manufacturing settings, just to name a few. These radiation sources can pose a considerable health risk to affected workers if not properly controlled. This page provides a starting point for technical and regulatory information regarding the recognition, evaluation, and control of occupational health hazards associated with ionizing radiation.

Ionizing radiation is radiation that has sufficient energy to remove electrons from atoms. In this document, it will be referred to simply as radiation. One source of

*continued*

radiation is the nuclei of unstable atoms. For these radioactive atoms (also referred to as radionuclides or radioisotopes) to become more stable, the nuclei eject or emit subatomic particles and high-energy photons (gamma rays). This process is called radioactive decay. Unstable isotopes of radium, radon, uranium, and thorium, for example, exist naturally. Others are continually being made naturally or by human activities such as the splitting of atoms in a nuclear reactor. Either way, they release ionizing radiation. The major types of radiation emitted as a result of spontaneous decay are alpha and beta particles, and gamma rays. X rays, another major type of radiation, arise from processes outside of the nucleus.

### Alpha Particles

Alpha particles are energetic, positively charged particles (helium nuclei) that rapidly lose energy when passing through matter. They are commonly emitted in the radioactive decay of the heaviest radioactive elements such as uranium and radium as well as by some manmade elements. Alpha particles lose energy rapidly in matter and do not penetrate very far; however, they can cause damage over their short path through tissue. These particles are usually completely absorbed by the outer dead layer of the human skin and, so, alpha emitting radioisotopes are not a hazard outside the body. However, they can be very harmful if they are ingested or inhaled. Alpha particles can be stopped completely by a sheet of paper.

### Beta Particles

Beta particles are fast moving, positively or negatively charged electrons emitted from the nucleus during radioactive decay. Humans are exposed to beta particles from manmade and natural sources such as tritium, carbon-14, and strontium-90. Beta particles are more penetrating than alpha particles, but are less damaging over equally traveled distances. Some beta particles are capable of penetrating the skin and causing radiation damage; however, as with alpha emitters, beta emitters are generally more hazardous when they are inhaled or ingested. Beta particles travel appreciable distances in air, but can be reduced or stopped by a layer of clothing or by a few millimeters of a substance such as aluminum.

### Gamma Rays

Like visible light and X rays, gamma rays are weightless packets of energy called photons. Gamma rays often accompany the emission of alpha or beta particles from a nucleus. They have neither a charge nor a mass and are very penetrating. One source of gamma rays in the environment is naturally occurring potassium-40. Manmade sources include plutonium-239 and cesium-137. Gamma rays can easily pass completely through the human body or be absorbed by tissue, thus constituting a radiation hazard for the entire body. Several feet of concrete or a few inches of lead may be required to stop the more energetic gamma rays.

*continued*

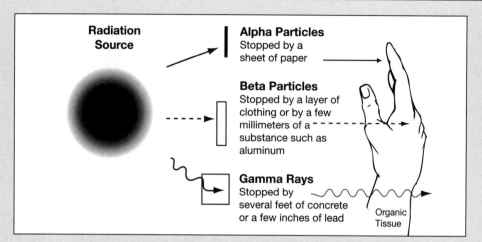

Penetrating powers of Alpha and Beta Particles, and Gamma Rays.

### X Rays

X rays are high-energy photons produced by the interaction of charged particles with matter. X rays and gamma rays have essentially the same properties, but differ in origin; i.e., X rays are emitted from processes outside the nucleus, while gamma rays originate inside the nucleus. They are generally lower in energy and therefore less penetrating than gamma rays. Literally thousands of X ray machines are used daily in medicine and industry for examinations, inspections, and process controls. X rays are also used for cancer therapy to destroy malignant cells. Because of their many uses, x rays are the single largest source of manmade radiation exposure. A few millimeters of lead can stop medical X rays.

### Sources of Radiation

**Natural Radiation.** Humans are primarily exposed to natural radiation from the sun, cosmic rays, and naturally occurring radioactive elements found in the earth's crust. Granite, for example, usually contains small grains of minerals that include radioactive elements. Radon, which emanates from the ground, is another important source of natural radiation. Cosmic rays from space include energetic protons, electrons, gamma rays, and X rays. The primary radioactive elements found in the earth's crust are uranium, thorium, and potassium, and their radioactive derivatives. These elements emit alpha and beta particles, or gamma rays.

**Manmade Radiation.** Radiation is used on an ever-increasing scale in medicine, dentistry, and industry. Main users of manmade radiation include: medical facilities such as hospitals and pharmaceutical facilities; research and teaching institutions; nuclear reactors and their supporting facilities such as uranium mills and fuel preparation plants; and Federal facilities involved in nuclear weapons production as part of their normal operation.

*continued*

*continued*

> Many of these facilities generate some radioactive waste; and some release a controlled amount of radiation into the environment. Radioactive materials are also used in common consumer products such as digital and luminous-dial wristwatches, ceramic glazes, artificial teeth, and smoke detectors.
>
> **Health Effects of Radiation Exposure**
>
> Depending on the level of exposure, radiation can pose a health risk. It can adversely affect individuals directly exposed as well as their descendants. Radiation can affect cells of the body, increasing the risk of cancer or harmful genetic mutations that can be passed on to future generations; or, if the dosage is large enough to cause massive tissue damage, it may lead to death within a few weeks of exposure."

## OPTICAL RADIATION

Optical radiation is all around us every day. Generally we are not harmed by it. Optical radiation is typically measured by wavelength, rather than by frequency. For example, the red laser pointer that is often used in meetings has a wavelength of approximately 632.8nm (Nanometers).

Optical radiation is really light. Visible light occupies a very small portion of the light spectrum, yet we also are often aware of light effects outside the visible spectrum. For example, we apply sunscreen lotion to cut down on the harmful UV (ultra violet) rays from the sun. The UV rays are a lower wavelength than visible, typically in the 200nm–400nm range.

Light measurements are performed using devices known as optical detectors and power meters. The detectors have specified wavelength ranges and power ranges. Measuring the output of a home lightbulb would only require a low power detector coupled to a power meter, operating in the visible range. Measuring the output of a high-powered laser would require a high power detector and power meter.

The power meters typically use electronic calibration methods, as they are electronic. The detectors require calibration using a certified light source and system known as a monochromator. The light source can be one of two types, either the pulse type, which is a specific wavelength or very small band, or a continuous wave that has a very wide wavelength range. The continuous wave light sources for the monochromator system then require selectable mirrors inside the monochromator for wavelength selection. The wavelength must be selected, in order to plot the calibration response of the detector. Detector response is typically plotted as power at given wavelengths. There are several detector types, based on the material used for what is known as the active surface. That is the portion that is actually excited by the light. Each material has a different response range.

A device that is often coupled with a detector for actual use is the integrating sphere. The name is descriptive, although some other shapes are also used. The integrating sphere is used to ensure all the electrons from the light source are applied to the active surface.

Generally, there are three different negative effects that can occur with optical radiation, either in use or during calibration. These effects are diffusion, reflection, and refraction. *Diffusion* is simply the effect of electrons escaping from the focused beam that is being sent to the receiving device. The easy way to demonstrate the effect is to shine a flashlight on an object, and along the beam path have a dark background. If you can see the actual beam of light, then the beam is experiencing diffusion. If the beam cannot be seen, and only the impact spot of the light can be seen, then there is no diffusion. *Reflection* is exactly what the term suggests. The beam or some portion is reflected back toward the source. In the microwave realm this is known as VSWR. Refraction is the dissemination of some of the electrons at an angle that is not desired. The easy way to demonstrate refraction is to get a half-full glass of water, put a pencil in the glass, and note that part of the pencil is exactly as expected and part appears to be bent in a different direction.

Sometimes the directed and amplified light, often meaning lasers, have specific safety precautions. This description will address lasers, but any directed and amplified light should be treated with the utmost regard for safety. There are five classes of lasers, 1 to 5. Class one is very low power, such as the supermarket scanner or the laser pointer. An accidental glance into one, such as at the supermarket checkout, will not typically damage the eyes, but is never recommended, as prolonged exposure to even the low-power laser can damage eyes. Class 3 and above require special safety equipment, beginning with goggles. The two most important safety precautions for Class 3 and above are *always* wear your goggles, and *never* get in front of the beam. Class 3 can cause serious damage to a body, and Class 4 can sever body parts. Class 5 is the level that was developed for the "Star Wars" defense research project and is unlikely to be seen outside that environment in the foreseeable future.

---

### Percent of Hydrogen (pH)

**The water molecule.** All substances are made up of millions of tiny atoms. These atoms form small groups called molecules. In water, for example, each molecule is made up of two hydrogen atoms and one oxygen atom. The formula for a molecule of water is $H_2O$. "H" means hydrogen, "2" means 2 hydrogen atoms, and the "O" means oxygen.

**Acids and bases in water.** When an acid is poured into water, it gives up H (hydrogen) to the water. When a base is poured into water, it gives up OH (hydroxide) to the water.[3]

### A Bit of History

It all began with food. Food is truly of the essence, and good food is highly valued but too rarely found. In earlier years people used to taste the food in order to establish the quality of a product. Working for the food industry could be a hard job. Little did they know at the time that the pH of food / drinks can often yield information about its state such as whether fruit is fresh, or not or whether wine will taste sweet or bitter. Some of them were lucky to work with pleasant products like wine or juices and they were happy. However some of them were a bit less lucky (vinegar?) and they must have been sad. Not only to make the sad ones happy but to make the happy ones even happier, science had to think up something called potentiometry, which then enabled people to

*continued*

directly measure the pH instead of tasting. Not only was this the birth of proper quality control in the food industry but in conjunction with further developments allowed many other industries to grow.

The magic term "pH"...Is a liquid (or solid for that matter) acidic like lemon juice or is it basic (sometimes also termed alkaline, although this actually refers to the presence of alkali ions) like bleach? How do we know? How do we measure? Does anyone care? Whether a substance is acidic or basic all depends on a single ion: $H_3O+$, the hydronium ion. If the hydronium ion is present at a concentration higher than 0.0000001 mol/l ($10^{-7}$ mol/l) we are talking about an acidic solution. The higher the concentration of the hydronium ion the more acidic the solution is. On the other hand concentrations below $10^{-7}$ mol/l $H_3O+$ lead to a basic solution. Well, these numbers are terribly long and difficult to handle, which is probably what Mr. Sørensen thought when he devised the pH term and pH scale in 1909. pH simply stands for the negative logarithm of the hydronium ion concentration. So a concentration of $10^{-7}$ mol/l $H_3O+$ means a pH value of 7. Likewise, a $H_3O+$ concentration of 0.1 mol/l would give a pH of 1. A difference of 1 in pH therefore means that the hydronium ion concentration has changed by a factor of ten! A solution at pH 6 has ten times more hydronium ions than one at pH 7. Don't forget: a solution is acidic if the pH value is below 7; basic if the pH value is above 7; neutral if the pH value is exactly 7.

### Modern pH measurement

**The method: potentiometry.** Did that word scare you? Don't let it. All potentiometry does is to measure (meter) the voltage (potential) caused by our friend, the hydronium ion: $H_3O+$. However this new method gave accuracy, reliability and faster results than the taste of any human being. It also saved some people from early death and gave the rest of the chemical industry a chance to prosper. Finally, scientists all over the world could measure things which were previously unmeasurable.

**Tools for measuring pH.** As mentioned, potentiometry is a measurement of voltage. The tools used for this are: a pH meter (to accurately measure and transform the voltage caused by our hydronium ion into a pH value); a pH electrode (to sense all the hydronium ions and to produce a potential); a reference electrode (to give a constant potential no matter what the concentration of our hydronium ion is).

*The pH meter*
Basically, a pH meter measures the potential between our pH electrode (which is sensitive to the hydronium ions) and the reference electrode (which doesn't care what's in the solution).

*The pH electrode*
The pH electrode's potential changes with the $H_3O+$ ion concentration in the solution. A pH electrode is built as follows: The clever bit is that the pH electrode only senses the hydronium ions. This means that any voltage produced

*continued*

*continued*

is from hydronium ions only. This way we can relate the potential directly to the hydronium concentration. Pretty neat.

*The reference electrode*
The reference electrode supplies a "constant" value against which we measure the potential of the pH electrode. That's the funny thing about potentials, they have to be in pairs to produce a voltage. A reference electrode is built as follows:

*The classical set-up*
The classical set-up for measuring pH consisted of a pH meter, a pH electrode and a reference electrode:

*The modern set-up*
Although you can perfectly measure pH using the "classical" set-up, it was soon realized that the two electrodes could be built into the same probe (although there still are two totally separate electrodes). This is nowadays called the combined pH electrode, which is, of course, much more practical."

## Calibration of pH Instruments

Calibration of pH meters depends on their construction. In many cases, the meter and probe are separate items and the meter can use interchangeable probes. The meter can be calibrated electrically, since it is basically a millivolt meter. The probe itself—and the entire instrument if the probe and meter are a single unit—must be standardized using wet chemistry, immediately before each use.

A pH measuring system is a classic example of a measuring system that must be standardized before every use, even if it has a separate calibrated meter. To calibrate a pH measuring system, the operator needs at least two buffer solutions and some distilled water. The buffer solutions are liquids that are known to have a specific pH because of their chemical composition. They may be purchased commercially in sealed single-use packages or prepared by a chemistry lab immediately before use. (Standard chemistry handbooks have recipes for buffer solutions.) One of the buffers will have a pH of 7.0, and the other has a pH near the expected measured value. The process is: rinse the probe in distilled water; put it in the 7.0 buffer and verify the reading; rinse the probe in clean distilled water; put it in the other buffer and verify the reading; and finally rinse the probe again. Buffer solutions should be discarded after use, as carbon dioxide from the air dissolves into the solution and changes the pH over time.

Two notes about pH probes. First, they have a limited lifetime and should be considered a consumable item. Second, the design of some probes requires that they be stored wet, usually in distilled water. Other designs may be stored dry—read the instructions to be sure.

## Endnotes

1. *Backsights Magazine*, www.surveyhistory.org/the_changing_chains.htm
2. NIST, ts.nist.gov/ts/htdocs/230/233/calibrations/optical-rad/cal.op.htm
3. www.miamisci.org/ph/hoh.html

## References

Nadal, Maria. 2003. Colorimetry. Gaithersburg, MD: NIST
www.nist.gov/public_affairs/update/upd20021216.htm#Colorimetry
NIST, 2003. Calibration Services. Gaithersburg, MD: NIST.
www.ts.nist.gov/ts/htdocs/230/233/calibrations/optical-rad/cal.op.htm
Surveyors Historical Society. 1986. "The Changing Chains" *Backsights Magazine*. Reprinted from *Reflections* (Autumn), a publication of First American Title Insurance Company.
www.surveyhistory.org/the_changing_chains.htm
U.S. Department of Labor, Occupational Safety & Health Administration,
www.osha.gov/SLTC/radiationionizing

# Chapter 36
# Getting Started

Part VII is an attempt to show how calibration and metrology functions are managed. There are no hard and fast rules for conducting a successful calibration laboratory, metrology department, or combination of the two. Nor is there a silver bullet for eliminating the alligators and dragons that too frequently populate our work environments. There are, however, tried-and-true policies and procedures that have been successfully used to help managers and supervisors get the most out of their resources.

Different strategies and ideas that have been incorporated into various work environments, from large calibration facilities to one-man calibration companies, are shown. They have been used in calibration labs, metrology sections, groups and departments, machine shops, production lines, and the vast array of calibration functions located around the world. How one incorporates them into their work environment is determined by the requirements of the metrology organization.

Within the upcoming chapters, you will find ideas on improving customer service, using metrics, preventive maintenance programs, and how to use surveys to improve the organization's policies and procedures. We also discuss different ideas on how to manage the workflow, starting from when IM&TE comes in the front door, all the way to going back to the customer, budgeting and resource management, vendors and suppliers, housekeeping and safety, and professional associations.

One topic that should be discussed—small enough in content but broad enough that it must be brought to light—is ethics. One of the best sets of guidelines can be found in every ASQ certification brochure. It is reprinted here.

> ASQ Code of Ethics: "To uphold and advance the honor and dignity of the profession, and in keeping with high standards of ethical conduct, I acknowledge that I:
>
> **Fundamental Principles**
>
> I. Will be honest and impartial; will serve with devotion my employer, my clients, and the public.
>
> II. Will strive to increase the competence and prestige of the profession.

*continued*

*continued*

> III. Will use my knowledge and skill for the advancement of human welfare and in promoting the safety and reliability of products for public use.
>
> IV. Will earnestly endeavor to aid the work of the Society.
>
> **Relations With the Public**
>
> 1.1 Will do whatever I can to promote the reliability and safety of all products that come within my jurisdiction.
>
> 1.2 Will endeavor to extend public knowledge of the work of the Society and its members that relates to the public welfare.
>
> 1.3 Will be dignified and modest in explaining my work and merit.
>
> 1.4 Will preface any public statements that I may issue by clearly indicating on whose behalf they are made.
>
> **Relations With Employers and Clients**
>
> 2.1 Will act in professional matters as a faithful agent or trustee for each employer or client.
>
> 2.2 Will inform each client or employer of any business connections, interests, or affiliations that might influence my judgment or impair the equitable character of my services.
>
> 2.3 Will indicate to my employer or client the adverse consequences to be expected if my professional judgment is overruled.
>
> 2.4 Will not disclose information concerning the business affairs or technical processes of any present or former employer or client without his or her consent.
>
> 2.5 Will not accept compensation from more than one party for the same service without the consent of all parties. If employed, I will engage in supplementary employment of consulting practice only with the consent of my employer.
>
> **Relations With Peers**
>
> 3.1 Will take care that credit for the work of others is given to those to whom it is due.
>
> 3.2 Will endeavor to aid the professional development and advancement of those in my employ or under my supervision.
>
> 3.3 Will not compete unfairly with others; will extend my friendship and confidence to all associates and those with whom I have business relations.

The various standards discuss the requirements for maintaining client confidentiality, and good business practices dictate this as an unspoken rule. A company, department, or laboratory must keep the secrets that are entrusted to it and conduct its business in a professional manner, which assumes that its conduct and ethics are above reproach.

Not only is this good business, but also should be a carryover from the way calibrations are performed, data collected, and traceability ensured. Professional conduct and ethics form the foundation for any business that maintains a reputation for honesty, integrity, and truth in their dealings with their customers.

Naturally, one would not find all the answers to their management questions in these chapters. The ideas and suggestions given here are to provide direction. Good managers are not born. They are molded, taught, and shaped by their experiences and training. An old axiom goes: "It's better (and faster) to learn from other's mistakes, than to make them yourself." These chapters are in the same vein.

# Chapter 37
# Best Practices

*Originality is nothing but judicious imitation. The most original writers borrowed one from another. The instruction we find in books is like fire. We fetch it from our neighbor's, kindle it at home, communicate it to others, and it becomes the property of all.*

Voltaire

Such is the case with best practices. These are a compilation of success stories, policies, procedures, work instructions, military axioms, learning experiences, and, in some cases, just good old common sense. If they can help anyone improve their process, production, profits, or performance . . . then the goals of this chapter have been met.

It is doubtful that any of the suggestions, policies, or practices that will be related here are original to the author who suggested them. The originality came in how they were applied to the particular situation that created their need. The phrase, "Improvise, adapt, and overcome" has been attributed to the United States Marine Corps. It could also apply to best practices. Use them when needed or mix and match them to an organization's situation.

Paul Arden wrote in his book *It's Not How Good You Are, It's How Good You Want To Be*:

> Do not covet your ideas. Give away everything you know, and more will come back to you . . . remember from school other students preventing you from seeing their answers by placing their arm around their exercise book or exam paper? It is the same at work, people are secretive with ideas. "Don't tell them that, they'll take the credit for it." The problem with hoarding is you end up living off your reserves. Eventually you'll become stale. If you give away everything you have, you are left with nothing. This forces you to look, to be aware, to replenish. Somehow the more you give away the more comes back to you. Ideas are open knowledge. Don't claim ownership. They're not your ideas anyway, they're someone else's. They are out there floating by on the ether. You just have to put yourself in a frame of mind to pick them up.

Arden's bit of knowledge could be used either in the preface or conclusion to this handbook. In writing this handbook, the authors are sharing their ideas, programs, procedures, and solutions in the hope that the metrology and calibration community can continue to grow and improve in the years to come. By sharing knowledge, experiences, failures and successes, we hope to reduce the need to continually recreate the wheel when good wheels are already out there being used by fellow practitioners.

We hope as future revisions to this handbook occur, more best practices are added and the sharing of ideas and programs continue. That is not to say you should give away the farm by telling trade secrets or patented ideas, but it is common knowledge that anyone can calibrate more items in a given period of time by doing like items, instead of having to change set-ups and standards several times a day. Is this a trade secret, someone's common sense, a written policy or procedure, or just a good business practice? One guess is as good as any other. But we practice doing like items on a daily basis and reap the benefits. So, it is hoped that the reader gains something from the first installment of best practices.

## CUSTOMER SERVICE (LAB LIAISONS)

Who are your customers? What is the difference between good customer service and bad customer service? What are they paying you to do for them? What should they expect in return for their money? How do you know if you're providing the level of customer service they expect? A *customer* is a person who buys, especially regularly at the same place. A *service* is an act of assistance or benefit to another or others. By definition, one could say that *customer service* is an act of assistance to a person who buys regularly at the same place. Nothing in this definition mentions quality or timeliness. There is also no mention of error free, satisfaction, most bang for the buck, or best price. But if the customer never returns, something was obviously missing.

An awful lot of questions hang on two simple words—customer service. If the answers were simple, then there would be no need for this section. But the answers are not simple. It's complicated, sometimes overanalyzed, and can make the difference between a company staying in business and closing its doors forever.

Most of us have been on both sides of this issue: we've been the customer and dealt with customers. We all know how good it feels to walk away from a satisfactory encounter at a store, shop, or establishment after we've made a purchase. We also know how upset we get when the experience wasn't up to par. We may not always remember the good exchanges, but we have a hard time forgetting the bad ones. Let's explore ways to take care of our customers and keep them coming back for more quality service.

Customer service is not only a good idea from a buyer's point of view; it's also a requirement in different standards. ANSI/ISO 17025-1999 states in paragraph 4.7, Service to the client: "The laboratory shall afford clients or their representatives cooperation to clarify the client's request and to monitor the laboratory's performance in relation to the work performed, provided that the laboratory ensures confidentiality to other clients."

Such cooperation may include: (a) providing the client or the client's representative reasonable access to relevant areas of the laboratory for the witnessing of tests and/or calibrations performed for the client; and (b) preparation, packaging, and dispatch of test and/or calibration items needed by the client for verification purposes.

Clients value the maintenance of good communication, advice, and guidance in technical matters, and opinions and interpretations based on results. Communication with the client, especially in large assignments, should be maintained throughout the work. The laboratory should inform the client of any delays or major deviations in the performance of the tests and/or calibrations.

Laboratories are encouraged to obtain other feedback, both positive and negative, from their clients (for example, client surveys). The feedback should be used to improve the quality system, testing and calibration activities, and client service.

BSR/ISO/ASQ Q10012:2003 states in paragraph 5.2, Customer focus: "The management of the metrological function shall ensure that:

- Customer measurement requirements are determined and converted into metrological requirements,

- The measurement management system meets the customers' metrological requirements, and

- Compliance to customer-specified requirements can be demonstrated."

To paraphrase both of these standards:

- Cooperate with the client and monitor the labs performance for the required work.

- Allow clients access to see how their items are calibrated, while maintaining any confidentiality issues.

- Keep the client informed of deviations or delays, while asking for positive and negative feedback.

- The client's needs are determined and converted into the requirements of the IM&TE calibration.

- Ensure the quality system meets clients' requirements while providing data in the form of records, certificates, and data.

If all of these requirements are fulfilled, are you guaranteed a happy customer? Is that all you have to do to keep them satisfied? During a presentation at a NIST seminar in November of 2002, the author made the following statement: "The customer is always right. I don't believe that to be true. The customer is not always right—but . . . they are always the customer!" This is not to say that calibration facilities have to cater to clients that do not know what they are talking about or do not understand uncertainty or the time it takes to complete a complex calibration, but the customer still is the source of income for many companies, and their idiosyncrasies, lack of knowledge, and/or limited understanding of metrology must be factored into any equation. Honest, intelligent communication with calibration customers about their capabilities, scope, and products can only enhance the calibration facilities ability to provide satisfactory service. The calibration facility is no better or worse than anybody else in the business community. It must find a way to deal with its customers in a professional manner, providing the type of service for which the customers are willing to pay.

It's a good idea to maintain an up-to-date customer database with contact points, telephone numbers, and email addresses. If lab technicians have to contact their customer prior to them picking up the equipment, can they do so in an expeditious manner?

Do they have the correct contact point? Do they need to know the owner or user of the IM&TE? Depending on the function, calibration lab, or metrology department, the user could be either or both the user and owner.

For example, the USAF PMEL had Owning Work Center (OWC) monitors assigned as the liaisons between the different squadrons (or work centers) and their supporting PMELs. If the OWC monitor was not the user of the IM&TE, he or she knew whom the user's supervisor was in order to allow notification of IM&TE being out of tolerance or when approval was required for limited calibrations. The relationship between the PMEL scheduler and the OWC monitor was crucial from both perspectives. The OWC monitors received monthly updates to their master IM&TE listings; they were the ones who delivered and picked-up the IM&TE on a regular basis and usually knew who to contact for limited calibrations or who to notify when their test equipment was out of tolerance. The PMEL scheduler had a contact point for doing business and was the interface between the OWC and PMEL's management and calibration technicians.

In most cases, PMEL schedulers represented the calibration function to those they supported. The attitude and customer service displayed by the scheduler represented all of the individuals working behind the scenes. Such is the case with most laboratories and calibration functions. The person who talks to the customer represents the organization's supervision, technicians, and, possibly, its quality assurance personnel. When a less than positive attitude is presented to the customer, it represents the entire company, group, department, or laboratory to the customer. This is why the hiring of competent and professional staff that will be dealing with clients, customers, or the public in general is so important.

To maintain a good working relationship with customers in a company setting, lab, division, or group, liaisons have filled the gap in some organizations. The liaison is the contact person for questions about the IM&TE, and, also, in some situations, delivers the test equipment to the calibration function when applicable. Good liaisons need to receive training in what they are responsible for, how to properly maintain their IM&TE, and what the various listings and terminology mean in the metrology world. One cannot expect to effectively communicate with the customer if one perceives them to be speaking in a different language. Traceability, uncertainty, 4:1 ratios, NIST, IM&TE, calibration, and reproducibility terms may seem like Greek to the untrained person. An orientation session can help improve vocabulary, while giving the technician the opportunity to answer the customers' questions and get a better feel for their needs and expectations.

Here are some suggestions on how to keep customers coming back for more:

- Send them a list of their items supported, with the next due date. (They were asked what kind of calibration interval they wanted to have, right?)

- Maintain a website with a frequently asked questions section.

- Keep this area up-to-date with the customer's actual questions. Sometimes it's easy to forget that new customers (hopefully there are some) are always coming through the door, and communicating at the lowest level can be a wise decision.

- Have orientation sessions or tours of your facilities. Open houses can be a way to draw in new customers who are curious about what you do or how you do it.

- Participate in local events to get help make your name or business more available to those who would not normally associate with calibration or metrology. Is this really a part of customer service? If the customer doesn't come to you, you have to go to them. Once you have someone as your customer, then you need to shift gears to keep them as your customer.

- Benchmark to identify and adapt best industry practices.

- Survey customers for valuable feedback information.

- Make available audits accomplished by accreditation bodies.

Timely information, quality service, accurate data, and a friendly smile (or voice over the phone) can go a long ways in keeping customers happy, satisfied, and coming back for more.

Let's spend some time looking at the other side of the customer service coin. Does the organization have a program for addressing customer complaints? Is it written out in the quality system or just a form located at the front desk that is never used? Paragraph 4.8 of ANSI/ISO 17025-1999 states, "The laboratory shall have a policy and procedure for the resolution of complaints received from clients or other parties. Records shall be maintained of all complaints and of the investigations and corrective actions taken by the laboratory (see also 4.10)."

An old adage goes, "the squeaky wheel gets the grease." Is this true in an organization? Does a person have to continually complain to be heard? Or does the organization take both suggestions and complaints seriously? An in-house lab may need to pay special attention to this if it has a captive audience who has no other recourse than to use its services. The internal customers are the unheard masses that have nowhere to turn if their IM&TE is not up to speed, reliable, or producing the results they need. They do not have the choice of going to another vendor. They are stuck with one organization for their calibration requirements! Don't they deserve the same quality service as the commercial customer? Absolutely!

## USING METRICS FOR DEPARTMENT/LABORATORY MANAGEMENT

Generally speaking, humans are visual beings. We see most things as a picture, in color, with shape and form. Place a bunch of numbers in front of us, and we have to start thinking. When a supervisor asks for data, which is easier to analyze a column of 17 numbers or a graph with 17 bars of different heights, sorted by whatever the common factor used to collect the data? For most of us, the graph gives us immediate recognition of what is important, what is not, and where to focus our attention.

When we present data or give a presentation, we are either trying to sell it, persuade our audience, or entertain them. Metrics are a form of communication. They are visual, immediately tell a story, and help get your point across in a medium that is easy to understand and comprehend.

Metrics can help an organization forecast its workload, show production trends, and repair problems. How often IM&TE is out of tolerance can help determine future calibration intervals, and production totals can graphically show who should be promoted or given additional responsibilities.

Here are some examples of metrics. Keep in mind that they have been applied to a specific work environment, and when used outside of that situation, may not give the same results. When forecasting for future workload, make a list of all the items that are due for calibration over the next 12 months. Sort the list by the next calibration due dates. Once that is accomplished, total the number of calibrations due for each of the next 12 months. Then place those numbers into a graph. The graph might look something like Figure 37.1.

Figure 37.1 shows a projected forecast for the next year. The straight line is the average of all the months added together and divided by twelve. As seen in the graph, October, January, May, and June could be smoothed out with increases in the other months to balance the workload. Generally speaking, each item in the inventory that shows up in this graph could have its calibration date moved back to accommodate the smoothing of the forecast. None of the due dates could be moved forward since their items would then be overdue for calibration. Naturally, there would be a mix of calibration intervals within this inventory and, because different groups of IM&TE would be coming due at different rates in the coming years, this one-time smoothing of the peaks and valleys would need to be repeated on a regular basis.

To help in getting an accurate forecast of the coming workload, experience shows that a precalculated increase in the numbers would more accurately reflect the actual workload coming in the door. This number is found by comparing actual workload against forecasted workload over the previous year's collected data. With new items continuously being added to the inventory, items requiring repair also need calibration, and other cases where calibrations were performed out of cycle, the actual number of IM&TE requiring calibration each month was significantly higher than the forecast for the next 12 months. By analyzing the historical data and projecting the difference into the yearly forecast, it is easier to forecast the future calibration workload.

Another metric that has plays a significant impact on the workload is the compilation of out-of-tolerance items versus calibrations performed. As in most cases where ugly things stay in your memory longer than beautiful ones, it is found that actual numbers versus perceived data provide a far more accurate idea of which IM&TE was reliable compared to others that needed their calibration interval reduced. Usually

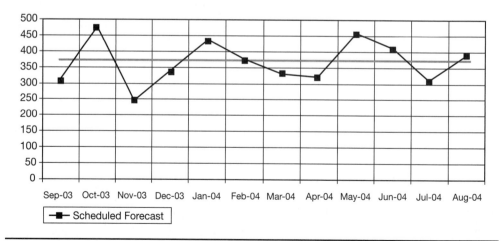

**Figure 37.1** Scheduled forecast of calibrations due.

managers and supervisors are focused on the one or two bad players that required longer time to repair, adjust, or calibrate after being down for maintenance. When, in fact, many more items passed calibration without any adjustment, but were easily forgotten in the daily grind to produce a quality product. By tallying the yearly pass rate for each type of equipment and setting limits as to when it would increase, decrease, or allow the calibration interval to remain the same, an organization can continue riding the cutting edge instead of slipping on the ragged edge with IM&TE requiring adjustment and repair before they come due for calibration. The pass rate can be figured for each item by dividing the number of items found out of tolerance during their scheduled calibration by the total number of like items receiving calibration and then multiplying by 100. If the pass rate exceeds 95 percent, then that type of equipment is eligible for an increase in its calibration interval (95 percent and 98 percent receive different increase rates). If the type falls below 95 percent, then there is discussion concerning if the calibration interval would remain the same or if the time between calibrations should be decreased. Special circumstances should be considered. Many types of equipment may have their calibration intervals remain the same even though they had less than a 95 percent pass rate.

Most numbers, of course, can work for or against an organization. As an example, let's look at production numbers. Dick and Jane both work in the same area, calibrating the same type of equipment. Dick produces 45 units a month, while Jane turns out 25 units. At first glance, one might think either that Dick is an outstanding producer or that Jane needs additional training or assistance. By analyzing the production numbers, it is found that Dick is putting in 10 additional hours of work a week (two hours a day) because he was single and wanted higher production totals. Jane is actually calibrating more of the items that have longer calibration procedures and is waiting for Dick to let her use the standards they have to share. It is possible to equalize these factors by giving them a weight or factor in determining the value a technician is worth. As an example, one might give a weight factor of one to a multimeter calibration while a spectrum analyzer might receive a weight factor of seven, based on both the time required for calibration and/or adjustment and the sophistication of technology and training to use the specific IM&TE.

Different types of calibration functions will post different types of metrics. The question might be what to post and what to keep secret. One may or may not wish to embarrass liaisons that do not get their IM&TE to them in a timely manner. Labs cannot post or publish confidential information on their customers, but they can post generic information about the number of items they support or calibrate without breaking confidentiality. Production totals, pass rates, overdue rates, trends in growth, productivity, or forecasts can all help an organization see where they are or where they wish to go.

Another metric that may be overlooked for different reasons is turnaround time. The customer wants its IM&TE back as quickly as possible. Calibration technicians (or supervisor or manager) may be graded on their ability to satisfy customers, and this is a valid indicator as to how responsive they are to customer needs. Like any metric, though, this one also has its good and bad points. One must consider the time IM&TE wait for parts, the adjustment to be made, and re-calibration if required. This adds to the overall time in the hands of the technician, when in fact there was nothing anyone could do while the item sat waiting for something or someone else to do their job. These factors should be added to the equation for an accurate, competitive look at a laboratory's actual turnaround time.

## PREVENTIVE MAINTENANCE PROGRAMS

"You can pay me now or pay me later." Have you heard those words before? Was it from a TV commercial or did your father utter those words while showing you how to fix your car in the family garage? Or did you hear it from your supervisor the first time you started working with IM&TE? The fact is, it's true.

The optimum word here is *preventive*. The person accomplishing work is trying to prevent something negative from happening by doing something positive first. Invest some resources, some time and money to preclude spending large amounts of time and money in the future. Preventive maintenance (PM) can really make a difference.

Just because the equipment is a solid-state device doesn't mean it will last forever. The same is true for mechanical, dimensional, or any other category of IM&TE. Would a technician put away gage blocks without cleaning them and coating them with a protective film? At the very least that's being preventive. We change the oil and oil filter in our cars on a regular basis. Why? Because it will save time and money on the investment in transportation.

Most of us do not work in a clean environment and neither does our equipment. Over time, air filters are contaminated with dust and dirt, then heat starts to build up inside the equipment, and it's not long before there is an equipment breakdown. Organizations should perform preventive maintenance on high-use items at least once a year, and, in circumstances called for by the manufacturer, even more often. Returning to the automobile example, in many cases the owner's manual has different maintenance schedules for different types of operation. In some work environments, PMs are scheduled during routine calibration, or if the item does not require calibration, scheduled on a regular basis. If one is only cleaning filters, vacuuming the inside of an item, or lubricating bearings, it is another opportunity to check the unit for smooth operation and proper function.

Here are some areas that might require a technician's attention. Any IM&TE that has a fan, usually has an air filter that needs to be checked. If the environment where the unit is used contains above average contamination, it would be appropriate to check it more often than one used in a cleaner area. If an item has ball bearings, an armature, or any type of moving parts that might require lubrication or replacement of parts, a regular check on their condition could save a breakdown when the item is most needed. Murphy's Law, "If anything can go wrong it will," is alive and well in the metrology world. Not only will it go wrong, but it will go wrong at the most inopportune time. Usually an item is not used until it is most needed, which means it has not been used for an extended period of time. When taken off the shelf and fired up, it has a better chance of breaking than if it has been used on a regular basis. The common factors causing this are dried-up grease or lubricants in the unit, a clogged fan filter, or mysterious dust bunnies have built a nest inside and the first time it's turned on sparks fly and electronic components short out. There are many reasons why IM&TE break down, but a reliable PM program can help prevent this from happening if performed on a regular basis.

Another benefit of having a PM program is the availability of common parts being on hand. If filters are changed, bearings lubricated, and special parts cleaned on a regularly scheduled basis, technicians are more likely to have the required parts, components, or supplies in stock. It is one thing to have something break down when most needed and quite another to have to add days to the down time because the required parts or supplies have to be researched and then ordered.

An important part of any PM program is having the correct service manuals on hand. They often give specific instructions on how to perform the PMs, along with a recommended list of required parts. The manufacturer usually knows more about what can go wrong, what areas need special attention, and which parts or supplies should be on hand, and this eliminates having to reinvent the wheel for every preventive maintenance situation.

Writing the instructions for performing PM inspections into procedures if service manuals are not readily available is advisable. This precludes having to memorize them, makes them available for training new personnel, and allows for updates and improvements as equipment and procedures are changed.

If performing a preventive maintenance inspection on a calibrated item, you should perform an 'As Found' calibration prior to the PM. Cleaning parts, replacing components, or changing the settings on internal adjustments could lead to the wrong conclusions when calibrating the unit without prior clarification. (Some light cleaning may be needed first to protect your own measurement standards; for example, cleaning grease and dirt from the anvil and spindle faces of a micrometer before letting it near your gage blocks.) Generally, most quality systems require documentation when repairs, adjustments, or calibration are performed on IM&TE. This is also true for preventive maintenance inspections. Even if nothing is replaced or adjusted, a record that the PM was performed as scheduled should be documented. When it was accomplished, who did it, and the final results should also be in the record.

Some IM&TE require that periodic checks be performed with check standards or self-tests be run on a regular basis. These should also be documented for easy retrieval to see what has been done on a particular piece of IM&TE. Schedules for accomplishing use of check standards would normally require updating of the computer system that generates the schedule. This would automatically update the database for recordkeeping purposes, but if a customer is performing the self-tests, or auto-calibrations, one needs to have a system in place for identifying who is performing the task, when it was accomplished, and the final results. As with any PM program, self-diagnostics and the use of check standards are activities that can catch a small problem before it becomes a big problem and save both time and money in the long run. Documenting when they are done, who did them, and the final results can also save big bucks during the life of any company's IM&TE.

## SURVEYS AND CUSTOMER SATISFACTION

How does an organization know if its customers are satisfied? If they only use the organization's service once and never return, that could be one way of knowing they are not satisfied. There could be other reasons too. There has to be a less ambiguous way to know. There is . . . ask them!

Everyone has seen the customer surveys next to the cash register or by the door exits at most stores. Has anyone ever filled one out, sent it in, or even thought of letting the store know you were not satisfied? Or better yet, let the store know that it received exceptional service from the summer helpers who only work a 20-hour week?

Unless we have our own business and understand the importance of knowing how our customers feel, we usually do not care. We receive surveys in the mail, as pop-ups on the internet, and during dinner in the evenings when the phone rings. Why are we being asked for our opinions? Does someone really care? Yes, they do. Surveys give important information about our needs, how they can be satisfied, and what we most want out of our relationship with the seller. We are willing to spend our money. The seller is willing to take it. The seller needs to know what we are willing to buy and for how much; while we want to know what we can get for the least amount of money. The quickest and easiest way of exchanging that information is by using a survey.

When most of us hear the word *survey*, we think of someone asking us questions about things we are not interested in or it concerns something we have already purchased (a car, appliance, or furniture). If we're satisfied, we may just throw the survey in the trash. If not, there is more of a chance that we complete and return it. What if we received the survey when we received the product? Can we realistically complete it at that time without knowing the quality of the service? How do we know if the item will even work properly till we return to work or home? How about getting the survey one week after picking up the product? Each situation has different circumstances and should probably be analyzed by different types of survey.

If an organization wants to hear from the disgruntled customer who visits its establishment, it might wish to have customer satisfaction cards at the front door or reception desk. If it wants to know the quality of the IM&TE it is sending back to the customer, maybe it would be better to send a survey back with the item after performing calibrations and/or repairs. Maybe a customer satisfaction form available online would provide an easier, more reliable form of passing on expectations to the customer. Whichever way they are used, surveys can be an invaluable way of communicating with those who like the service given and those who don't. The first will be returning on a regular basis, the latter won't. An organization needs to focus on the latter to ensure they become the first. One unhappy customer can cost a company many future customers simply through word of mouth. And it could be something as simple as the receptionist having a bad day or the telephone operator accidentally hanging up on the customer.

Once you have a completed survey in hand, what can you do with it? A couple of ideas come to mind. One could perform root cause analysis (big words for simply asking why five times) on the problems identified. But what do you do if there are multiple problems? Possibly sort them by importance (to the customer, not the company; by the number of clients concerned about a particular topic; or possibly by how often they reoccur within the survey. Some quality systems require preventive and/or corrective action plans to be in place. It would be appropriate to identify problem(s) in your corrective action plan, give a timeline for solving the problem, and follow up to see if the solution(s) implemented succeeded in satisfying the client. Documenting the problems, solution implementation, and final results can only make process improvements easier, while showing your customers that the investment in completing the surveys was well worth the time and effort.

Sometimes a survey will show that some customers have unrealistic expectations. Handling those expectations may require a follow-up conversation to determine if they really want that or if they simply do not know what is realistically possible. If the laboratory has a web page, it can be an opportunity to educate customers about which expectations are realistic and which ones are not.

Organizations with a captive customer base still need to check the pulse of their customers on a regular basis. Are they providing customers the quality service they require? How do they know? Are customer satisfaction forms available? Is there a formal complaint system in place for their use? One might think that with a captive audience there is little or no need for surveys or complaint forms. Nothing could be further from the truth. Everyone has a boss, and negative feedback from customers won't help an individual's career or an organization's bottom line. But more importantly, the quality of the product an organization provides its customers should be foremost on its mind. Their ability to do their job could be determined in a big part by the quality of the IM&TE they use. The safety of flight, purity of drug manufacturing, or traceability of product could hinge on their willingness to produce a quality product, whether it has a captive audience or not.

The adage "The squeaky wheel gets the grease" has never been more applicable than when it comes to customer complaints. Being proactive in asking for their comments, suggestions, or ideas, can only increase customers' opinions of an organization's department, group, or calibration lab. Everyone likes to feel wanted, and when sincerely asked for opinions or comments, it is much easier to relate small or irrelevant problems instead of waiting for them to grow into large problems that take time away when you can least afford it. By encouraging customer feedback when they are satisfied, instead of when they are not, you increase the customers' willingness to look at your service with an open mind and possibly receive information that has been overlooked previously. In retail sales, it has been known for decades that while a happy customer may only tell two or three other people, an unhappy customer will tell an average of 10 other people. It would not be surprising if this is found to be true of metrology customers as well, so it is definitely beneficial to make all of your customers happy ones.

## References

ANSI/ISO 17025-1999, *American National Standard—General requirements for the competence of testing and calibration laboratories.* Milwaukee: ASQ Quality Press.

ANSI/ISO/ASQ Q9001-2000, *American National Standard—Quality management systems—Requirements.* Milwaukee: ASQ Quality Press.

ANSI/ASQ M1-1996, *American National Standard for calibration systems.* Milwaukee: ASQC Quality Press.

ANSI/NCSL Z540-1-1994, *American National Standard for calibration—Calibration laboratories and measuring and test equipment—general requirements.* Boulder, CO: National Conference of Standards Laboratories.

Arden, Paul. 2003. *It's not how good you are, its how good you want to be: The world's best selling book.* New York: Phaidon Press.

Bertermann, Ralph E. 2002. *Understanding current regulations and international standards; Calibration compliance in FDA regulated companies.* Mt. Prospect, IL: Lighthouse Training Group.

BSR/ISO/ASQ Q10012:2003(E), *Measurement management systems—Requirements for measurement processes and measuring equipment.* Milwaukee. ASQ Quality Press.

Bucher, Jay L. 2000. *When your company needs a metrology program, but can't afford to build a calibration laboratory . . . What can you do?* Boulder, CO: National Conference of Standards Laboratories.

www.dartmouth.edu/~ogehome/CQI/PDCA.html
deming.eng.clemson.edu/pub/tutorials/qctools/qct.htm
FDA Backgrounder, 1999, Updated August 5, 2002.
   www.fda.gov/opacom/backgrounders/miles.html
FDA. U.S. Government. 2001. Code of Federal Regulations, Title 21, Volume 8, Part 820—
   *Quality system regulation, Sec. 820.72 Inspection, measuring, and test equipment.* Revised April 1, 2001. Rockville, MD: U.S. Government Printing Office.
Hirano, Hiroyuki. 1995. *5 Pillars of the visual workplace: The sourcebook for 5S implementation.* (Bruce Talbot, translator.) Shelton, CT: Productivity Press.
Kimothi, S. K., 2002, *The uncertainty of measurements, physical and chemical metrology impact and analysis.* Milwaukee: ASQ Quality Press.
National Conference of Standards Laboratories. 1999. Recommended Practice RP-6: *Calibration control systems for the biomedical and pharmaceutical industry.* Boulder, CO: NCSL International.
NIST Special Publication 811–1995 Edition, *Guide for the use of the international system of units (SI)*; Barry N. Taylor, Physics Laboratory, National Institute of Standards and Technology, Gaithersburg, MD 20899-0001, (Supersedes NIST Special Publication 811, September 1991).
Pinchard, Corinne. 2001. *Training a calibration technician . . . In a metrology department?* Boulder, CO: National Conference of Standards Laboratories.
Praxiom Research Group Limited, 9619 - 100A Street, Edmonton, Alberta, T5K 0V7, Canada
projects.edtech.sandi.net/staffdev/tpss99/processguides/brainstorming.html
U.S. Food and Drug Administration—Center for Devices and Radiological. 1999. *Medical device quality systems manual: A small entity compliance guide,* 1st ed. (Supersedes the Medical Device Good Manufacturing Practices [GMP] Manual). www.fda.gov/cdrh/ dsma/gmp_man.html

# Chapter 38

## Personnel Organizational Responsibilities

Fulfillment of organizational objectives requires personnel be assigned specific duties / functions within an organization. Assignment of responsibilities is necessary to ensure that personnel are held accountable for completion of tasks within the allotted time using a finite set of resources. It is not only necessary that individuals be cognizant of their job duties, but that other personnel, internal as well as external to the organization, are clear as to whom they should consult for specific matters. Without a clear understanding of the duties and responsibility individuals play in an organization, chaos often results.

Many quality standards have requirements for defining the responsibility for personnel within the metrology calibration testing organization. BSR/ISO/ASQ Q10012: 2003, *Measurement Management System—Requirements for Measurement Processes and Measuring Equipment*, section 6.1.1. requires, "the management of the metrological function shall define and document the responsibilities of all personnel assigned to the management system." ANSI/ISO 17025-1999, *General Requirements for the Competence of Testing and Calibration Laboratories*, section 4.1.4 states that a calibration laboratory shall,

- "Have managerial and technical personnel with the authority and resources needed to carry out their duties . . .

- Specify the responsibility, authority and interrelationships of all personnel who manage, perform or verify work affecting the quality of the tests and/or calibrations

- Provide adequate supervision of testing and calibration staff, including trainees, by persons familiar with the test and/or calibration methods and procedures . . .

- Have technical management which has overall responsibility for the technical operations . . .

- Appoint a member of staff as quality manager (however named) who, irrespective of other duties and responsibilities, shall have defined responsibility and authority for ensuring that the quality system is implemented and followed at all times . . ."

These and other quality standards are absolute regarding the necessity that personnel within the organization have their responsibilities clearly defined and documented. The following are common positions within the metrology/calibration/testing organization.

## LABORATORY MANAGER

The Laboratory Manager, often referred to as the department manager, has overall responsibility for all operations of a calibration/testing laboratory. JobGenie at www.stepfour.com/jobs gives the following job description for a department manager;

> Directs and coordinates, through subordinate supervisors, department activities in commercial, industrial, or service establishment: Reviews and analyzes reports, records, and directives, and confers with supervisors to obtain data required for planning department activities, such as new commitments, status of work in progress, and problems encountered. Assigns, or delegates responsibility for, specified work or functional activities and disseminates policy to supervisors. Gives work directions, resolves problems, prepares schedules, and sets deadlines to ensure timely completion of work. Coordinates activities of department with related activities of other departments to ensure efficiency and economy. Monitors and analyzes costs and prepares budget, using computer. Prepares reports and records on department activities for management, using computer. Evaluates current procedures and practices for accomplishing department objectives to develop and implement improved procedures and practices. May initiate or authorize employee hire, promotion, discharge, or transfer. Workers are designated according to functions, activities, or type of department managed.

## TECHNICAL MANAGER

From ISO IEC 17025 we can conclude that the technical manager has overall responsibility for the technical operations of the calibration/testing laboratory. Technical operations typically include the following:

- Validity of calibration/testing methodologies per application requirements
- Development and validation of mathematical algorithms/computational calculations
- IM&TE selection per application requirements
- Interpretation of measurement data
- Root cause analysis of technical discrepancies/abnormalities

## QUALITY MANAGER

From ANSI/ISO 17025-1999 we can conclude that the quality manager has overall responsibility for assuring the calibration/testing laboratory's quality system is being implemented and is being monitored for compliance. The quality manager is responsible for making sure quality-related discrepancies are identified and documented and that appropriate corrective actions plans are formulated and implemented.

## CALIBRATION TECHNICIAN

The ASQ Certified Calibration Technician (CCT) brochure defines a certified calibration technician as one who, "... tests, calibrates, maintains, and repairs electrical, mechanical, electromechanical, analytical, and electronic measuring, recording, and indicating instruments and equipment for conformance to established standards."

JobGenie at www.stepfour.com/jobs gives the following job description for a calibration laboratory technician:

> Tests, calibrates, and repairs electrical, mechanical, electromechanical, and electronic measuring, recording, and indicating instruments and equipment for conformance to established standards, and assists in formulating calibration standards: Plans sequence of testing and calibration procedures for instruments and equipment, according to blueprints, schematics, technical manuals, and other specifications. Sets up standard and special purpose laboratory equipment to test, evaluate, and calibrate other instruments and test equipment. Disassembles instruments and equipment, using handtools, and inspects components for defects. Measures parts for conformity with specifications, using micrometers, calipers, and other precision instruments. Aligns, repairs, replaces, and balances component parts and circuitry. Reassembles and calibrates instruments and equipment. Devises formulas to solve problems in measurements and calibrations. Assists engineers in formulating test, calibration, repair, and evaluation plans and procedures to maintain precision accuracy of measuring, recording, and indicating instruments and equipment.

## CALIBRATION ENGINEER

A calibration engineer is quite simply an engineer whose main duties are in support of calibration activities. Normally the use of an engineering title implies that an individual has successfully completed a bachelor of science degree in an engineering discipline such as electronic/electrical, mechanical, systems, or other technical field such as chemistry, physics, or mathematics. The following is a work elements job description excerpt from an aeronautical company for a calibration engineer:

- Analyze inspection, measuring, and test equipment (IM&TE) to determine the calibration requirements. Determine the functions to be verified and their specifications, the methods to be used, and the measurement standards required.
- Prepare and test new calibration procedures.
- Analyze requirements for measurement standards and make recommendations.
- Perform engineering and statistical analyses of the equipment and historical data to determine appropriate calibration intervals to meet reliability goals.

- Assist in maintaining a laboratory quality management system that is registered to ISO 9001:2000. [Or] Assist in implementing laboratory accreditation to ANSI/ISO 17025-1999 and conformance to other standards.

- Communicate with other departments in the company, with equipment manufacturers and vendors, and with representatives of regulatory agencies regarding test, measurement and calibration.

- Analyze, evaluate, and document measurement uncertainty and laboratory calibration capability.

- Prepare laboratory documentation (calibration procedures, quality procedures, analysis spreadsheets, reference documents, and others) in a variety of formats including but not limited to word processor documents, HTML files, portable document format files, databases, or visual presentations.

- Document laboratory best practices and assist in training calibration technicians to implement them. Document and implement quality management system procedures and methods, and assist in training staff in them.

- Provide engineering analysis and guidance for metrology issues.

- Apply knowledge of national and international standards documents to the calibration business. Relevant standards include, but are not limited, to ISO 9000 series, ISO 17025, ANSI/NCSL Z540-1 and -2, NCSL Recommended Practices, and BSR/ISO/ASQ Q10012:2003.

- Develop and implement measurement process improvements based on appropriate statistical methods and proficiency studies.

- Define, develop, and implement data and record requirements for calibration laboratory information database systems and automated calibration systems.

- Develop and implement web-based systems for accessing laboratory documents and resources on the laboratory intranet.

## METROLOGIST

According to the United States Department of Labor's Office of Administrative Law Judges Law Library, a metrologist:

> Develops and evaluates calibration systems that measure characteristics of objects, substances, or phenomena, such as length, mass, time, temperature, electric current, luminous intensity, and derived units of physical or chemical measure: Identifies magnitude of error sources contributing to uncertainty of results to determine reliability of measurement process in quantitative terms. Redesigns or adjusts measurement capability to minimize errors. Develops calibration methods and techniques based on principles of measurement science, technical analysis of measurement problems, and accuracy and precision requirements. Directs engineering, quality, and laboratory personnel in design, manufacture, evaluation, and calibration of measurement standards, instruments, and test systems to ensure selection of approved instrumentation. Advises others on methods of resolving measurement problems and exchanges

information with other metrology personnel through participation in government and industrial standardization committees and professional societies.

## LOGISTICAL SUPPORT PERSONNEL

Logistical personnel provide the resources necessary to perform tasks in nontechnical support of calibration/testing activities such as administration, pickup and delivery, and so on. The following are common positions supporting calibration/testing activities:

- **Scheduler.** Identifies workflow requirements and assigns jobs according to worker availability, expertise, seniority, job classification, and preferences. Often required to adjust schedules to meet urgent customer demands and/or emergency situations. Some organizations have found that it is beneficial for the person in this role to have at least a basic knowledge of the types of calibration or testing work performed in the organization. That knowledge improves the scheduler's understanding of what is and is not possible and facilitates his or her acting as an interface between the technical staff and the customers.

- **Material handler.** Receives, processes, and delivers incoming IM&TE and/or packs and ships outgoing IM&TE. May be required to relocate IM&TE within a corporation. Makes IM&TE pickups and deliveries, as required.

- **Document control administrator.** Duties often include: creating, importing, editing, reviewing, approving, issuing, and processing requests for changes or to withdraw or purge documents associated with calibration/testing laboratory operations.

- **General administration.** Duties often include administrative tasks supporting the financial and logistical activities of a calibration/testing laboratory such as parts ordering, updating vendor information, accounts payable, accounts receivable, completing shipping documentation, and so on.

Personnel titles often vary considerably between different organizations, as does the scope of personnel duties and assigned responsibilities. Sometimes one person may fill two or more of these roles. Regardless of these differences, quality standards throughout the world, as well as good business practices, dictate that personnel positions should be defined in terms of their duties, responsibilities, interactions, and authorities. Within the metrology/calibration/testing organization, personnel duties and responsibilities are required not only to be clearly defined but also documented as a proviso for accreditation as well as ISO certification.

## References

ANSI/ISO 17025-1999, *American National Standard—General requirements for the competence of testing and calibration laboratories.* Milwaukee: ASQ Quality Press.

BSR/ISO/ASQ Q10012:2003(E), *Measurement management systems—Requirements for measurement processes and measuring equipment*. Milwaukee: ASQ Quality Press.

Fluke Corporation. 1994. *Calibration: philosophy in practice*. 2nd ed. Everett, WA: Fluke Corporation.

ISO. 1993. *International vocabulary of basic and general terms of metrology (VIM)*. Geneva: ISO.

Kimothi, S. K. 2002. *The uncertainty of measurements, physical and chemical metrology impact and analysis*. Milwaukee: ASQ Quality Press.

# Chapter 39
## Process Workflow

One size does not fit all when it comes to how IM&TE should flow through a facility for calibration. Depending on how the operation is set-up, staffed, managed, or controlled, there are many ways to get the IM&TE in and back out with efficient, economical processes. By giving the reader a few examples of how this process workflow operates in different environments and organizations, it is hoped improvements can be made with minimum impact on their current operations or cost in terms of money and/or resources.

One of the authors (Bucher) manages a metrology department for a biotechnology company. Each of the calibration technicians has responsibility for specific facilities, areas, and departments. They are taught the most efficient way to schedule their workload, and over a period of time, make improvements to the system that allows them flexibility, innovation, and efficiency in the scheduling and management of their workload. The following is a more detailed example of how they schedule their work and accomplish their responsibilities.

After producing a list of calibrations that will be due in the next 30 days, they sort the list by type of equipment and the location of that equipment. By calibrating 'like items' as much as possible, they reduce the time it normally takes to change calibration procedures, standards, and the calibration set-up. Except for a couple of types of items that are calibrated in the technician's work area, everything else is calibrated 'on-site', meaning, in the actual environment where it is used. This translates to taking everything needed to calibrate a particular piece of IM&TE to the location where the item is used. With limited resources to maintain duplicate sets of standards, the efficient use of available standards is critical to the success of the department. By coordinating with the other calibration technicians, time and money are saved in the sharing and use of department standards.

One might believe that calibrating 'like items' is common sense or taught throughout the metrology community as a standard practice. Many years of experience in the field of metrology have shown that nothing can, or should be taken for granted. The calibration of 'like items' refers to setting up your work to produce (calibrate) the maximum number of items with the minimum amount of time and effort. By calibrating all the pH meters due calibration during a specific period of time, as an example, in the next two weeks, in a specific facility or area; you might greatly reduce the time it takes to set-up for the calibrations (standards, procedures, forms, calibration labels, cables, buffers, etc.). By reducing the time for set-up, teardown, changing of standards, accessories, forms, and procedures, your production increases while your customer receives a faster turn around time.

By allowing the calibration technicians to set his or her own schedule, the opportunity for boredom is reduced exponentially. They decide how to best fit the most calibrations into their workday, while keeping repetitive work to a minimum (doing like items is encouraged, but dozens of the same calibration can get anybody down!) The needs of the company and their customers come first; but the flexibility of doing their own scheduling have paid dividends in higher productivity, outstanding customer service, and most importantly, a calibration technician that has learned to think-on-their-feet, become more self-sufficient, and requires minimal supervision. In reality, a manager/supervisor's 'dream come true'.

Once the scheduling of work is accomplished, the technician prepares for the calibration of the IM&TE. This includes getting the appropriate calibration procedure; reading it to ensure the proper standards, accessories, forms, and labels are available; and collecting everything and setting up for the actual calibration. Since they use a paperless system, a laptop computer is utilized for the collection of the calibration data. They must ensure the laptop is charged up, the correct electronic form is available, and then they proceed to where the IM&TE is located. If they are doing 'like items', they might collect all of the day's workload into one area, or move the cart on which all of their required items are transported, from room to room, calibrating each item where it is actually used by the customer.

Once the calibration is complete, to include recording the last calibration date, the old calibration sticker is removed and a new one placed on the unit. One of the policies that are in place includes leaving the IM&TE in better condition than it was found. This has helped increase the usable life span of the IM&TE, raised the status of the metrology department in the eyes of the customer, and helped them become proactive in identifying problems before they impact the use of the test equipment. By cleaning filters, vacuuming the internal areas of some items, and inspecting for leaks, fluids, and missing parts or components, they have saved valuable time in either eliminating or reducing the number of future repairs. This could also include replacing common user replaceable parts such as batteries, front replaceable bulbs, etc. This also includes inspecting the equipment from a safety aspect before leaving the lab, i.e. checking for frayed power cords, inoperative safety locks, proper functioning of all lamps or indicating devices, proper wiring of the power cord and fuse system to include being able to remove the fuse using the fuse cap when an inherently safe fuse holder is not in use, and so on.

The technician completes the calibration record and stores it for review by the supervisor, and then the unit is returned to the customer. Depending on the type of system that is in place, a second set of eyes may be needed on documentation before it can be signed off. The technician must also update the software system to show the unit has been calibrated, with comments, standards used, and any repair costs that might have been incurred. Remember: The job isn't finished till the paperwork is done is as applicable in metrology and calibration as any other industry. Without the system being updated for the next calibration due date, nobody would know that the calibration ever occurred. This is also the perfect time to record problems, discrepancies, or adjustments that were made, and track any trends, both good and bad, on that particular type of equipment.

If a quality assurance program of some type is used in an organization, it would probably be incorporated after the technician completes the calibration and before the IM&TE is returned to the user or customer. Some organizations perform a quality inspection of IM&TE as the items are received. This could have a twofold benefit. It might identify problems with the equipment that would preclude it receiving a timely calibration (missing parts, cables, manuals, and so on) and the customer could be contacted in a timely manner to resolve the problem. Also, if the customer actually delivers the items to the organization and problems are found, the technician could coordinate resolutions or take the unit back without having to make additional trips. Incoming inspections are also a good time for capturing warranty dates, lease return dates, and any type of information that could accurately update your database and help keep track of information that could be used in the future. In cases where multiple accessories are included with the incoming IM&TE, it is often advantageous to take a digital picture of the items and archive with the unit's incoming inspection history, if possible. This practice is visual insurance to protect both the customer and the calibration practitioner.

Within any organization that performs calibrations, there will be certain accessories needed for completing calibrations, including adapters, cables, buffers, standards, work benches, and so on. Two schools of thought come to mind for the process of distributing these assets. One idea is to have a central location for these items. As the items are needed, the technician retrieves and uses them, and returns them at the end of the calibration or the end of the workday. Another idea is to give each technician the minimum number of accessories needed to do the job and have the more expensive standards or accessories available at a central location. Both have their benefits and drawbacks. It is more expensive to purchase duplicate cables, loads, standards, and so on, but it can save valuable time in getting and returning them. Also, if an accessory is used only on rare occasions, it is a waste of resources to purchase one for each technician to have available when needed. If a special setup is in place for ease in calibrating a large variety of equipment (for example, an oscilloscope package or microwave system), it would be counterproductive to remove a cable or load to complete another calibration while that system sat idle waiting for the return of the removed accessory. This has been observed in more than one organization. Time is money for most calibration functions, and the availability of resources is a double edged sword. It costs money to purchase standards and accessories. It also costs money (in wasted time) for technicians to be idle waiting to perform calibrations. Efficient scheduling of standards and the calibration of like items can help reduce both wasted time and duplication of standards.

A commercial calibration function might use the following processes in its workflow. The lab receives a customer inquiry:

- Determine if it is a qualifying job.
  - Typical job—qualifies.
  - Nontypical job. Obtain verbal and written (e-mail, fax, mail) details, loading/unloading, discuss additional cost, provide an estimated uncertainty to see if will satisfy the potential client, obtain management approval for overtime, extra labor force, and so on.

- Schedule the job tentatively.

- Requestor needs to provide a purchase order in a time limit (for example: two days) to firm up the schedule. Purchase order will also contain all discussed details the requestor desires.

- Schedule the job.

- Job arrives at the calibration laboratory.

- Items with paperwork passport are unloaded and given a unique nonrepetitive number in the logbook. This number will follow the job at every step and will be the main identifier.

- Items are unpacked.

- Items are prepared for calibration: paperwork, cleaning, temperature balance, and so on.

- Items are calibrated using: method, standards, environment, operator.

- Necessary calculations are performed.

- Calibration certificate containing found measured error and uncertainty and all other ANSI/ISO 17025-1999 requirements is written.

- Customer called with cost, pick-up instructions.

- Send necessary billing information to the business office.

- Business office confirms job was paid.

Measurement or calibration activities are characterized as being routine or nonroutine. Routine activities generally:

- Have established precedents.

- Do not require deviating from normal workflow patterns or established laboratory policies.

- Have available all necessary support equipment, documentation, and technical know-how to successfully perform them.

- Produce data consistent with the activity, and results in an equipment status, which is acceptable for a given application.

An activity that deviates from this list is considered nonroutine. Furthermore, any activity can be treated as nonroutine until it is determined to be otherwise. Nonroutine activities involving workflow, equipment status, and laboratory policy or practice issues should be addressed by the calibration laboratory manager. Nonroutine activities requiring internal or external technical support should be addressed by the technical manager. This includes the use of alternative test methodologies, equipment, and specifications. Nonroutine activities may result in a corrective action form (CAF) being generated.

Figure 39.1 is an example of a business process interaction diagram similar to that used by one calibration laboratory that is registered to ISO 9001:2000. It shows the core

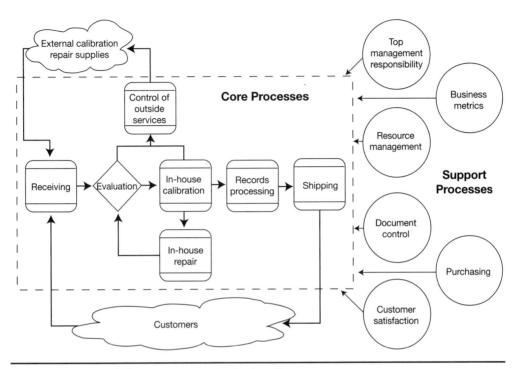

**Figure 39.1** A sample business process interaction diagram.

business processes and their main interactions, and other processes needed to support them and the quality management system.

The core processes for this laboratory are:

- **Receiving.** Unpacking shipped items, entering receipt data for them and customer-delivered items into the database.

- **Evaluation.** Determine the type of service required. If there is nothing that says otherwise, the assumption is that routine calibration is required and the item is staged on the ready-to-work shelves. Some items may be sent directly to external suppliers.

- **Calibration.** Calibrate the equipment and place it on the completed-work shelf. If more than minor repair is needed, transfer it to the repair process.

- **Records processing.** Verify that all database entries have been made, any other records are updated as appropriate, and then apply the appropriate calibration label.

- **Shipping.** Notify local customers that equipment is ready for pickup. Package, prepare shipping documents, and dispatch to other customers.

- **Control of outside services.** As needed, manage interactions with other calibration and repair providers.

- **In-house repair.** As needed, repair equipment and then return it to the beginning of the calibration process. (All repaired items are calibrated before they are released.)

This is a top-level view. This level of detail does not show exceptions to the normal flows, such as out-of-tolerance conditions or in-place calibrations. Those details are shown in quality procedures for each of the core processes. A laboratory information database system is used to manage and record the workflow including: special instructions for certain equipment, ensuring that the current calibration procedure is used and that all measurement standards are within their due dates, and automatically printing calibration certificates and labels. The database automatically records data such as calibration time per unit, days for outside supplier service, and even the temperature and humidity at the start of a calibration.

In this laboratory, if a calibrated IM&TE item is repaired in-house, it is always calibrated before return to the customer. This is a very important action for two reasons. First, passing the calibration procedure validates the repair action and serves as the quality assurance for repair. Second, and as important, it allows the laboratory to justify excluding manufacturer operating and service manuals from the document control system. This can be done because the performance of the equipment is verified using the calibration procedure, which is a controlled document.

# Chapter 40
# Budgeting and Resource Management

From one author's point of view, the greatest angst among managers comes from either training, or budgeting and resource management. Training was discussed in Chapter 15. Like training, maintaining an accurate, up-to-date budget is not only critical to staying out of the red on your balance sheet, but by being proactive in meeting your needs, one can anticipate extra expenses before they actually occur.

Resource management is not only a part of any calibration function, but also a requirement listed in BSR/ISO/ASQ Q10012:2003, *Measurement Management Systems – Requirements for Measurement Processes and Measuring Equipment*. Section 6, Resource management, covers both human resources (responsibilities of personnel) and competence and training. More on this topic will be discussed later.

**Budget.** What comes to mind when this word is spoken, written, or mentioned in any way, shape, or form? Money comes to mind. How to spend it also appears. By definition, *budget* is an itemized summary of probable expenditures and income for a given period. Generally speaking, budgets are a way to show where the organization expects to spend its resources for a given period of time. Most calibration functions have to pay their staff, purchase parts, maintain a facility, and keep their standards in a calibrated state. All of these items require money. Whether you receive the funds from outside sources or internally from a company, it is expected that a valid estimate be provided on a regular basis of the expenses expected to occur. Companies typically make budget forecasts within the realm of various assumptions and management directives. Assumptions may presume some inflationary figure or expected growth rate, possible personnel workforce reductions, increase efficiencies, and so on. Directives from senior management are often in the guise of mandates for keeping expenditures flat, in other words previous expenditures = projected forecast, or reducing them by some apportionment.

There are many ways to accomplish budget forecasting that is both accurate and reliable. Historical data are usually a good starting point. By maintaining records of past expenses, it is possible to accurately forecast future needs. Knowing when standards will come due for calibration and what it cost previously to have the calibrations accomplished will usually allow a close estimate. Some organizations automatically include small inflation increases, while others do not.

If an organization includes the cost of replacing parts in the IM&TE it supports as part of its budget, computer software can provide this information if it is available. Compiling the cost of spare parts over a specific period of time can provide accurate data from a historical perspective. This can be done on a quarterly, semi-annual, or yearly basis, depending on how the budget is set up.

Depending on the situation, the following areas might have to be covered in a budget: salaries and wages, bonuses, general calibration supplies (cables, loads, filters, lubricants, fuses, and so on), dues and subscriptions (for professional organizations), telephone calls, hotel and lodging (for conferences, over night on-site calibrations, and so on), airfare, transportation costs (taxis or rental cars), business meals, training (seminars, conferences, schooling), repair expenses on IM&TE and/or standards, educational reimbursement (for those lucky enough to have company support in this area), having standards calibrated off-site or at a higher echelon laboratory, or the cost of maintaining a company van or truck. The list could be a lot longer depending on responsibilities, size, and location.

As previously mentioned, resource management can be an integral part of one's responsibilities. This could include being responsible for all personnel assigned to the calibration system in the organization. These responsibilities could be defined in an organization chart, individual job descriptions, quality management system manual, or work instruction and/or procedures.

A current job description for each person assigned to the calibration function should be in the individual's training or personnel folder. This provides an accurate gage for what each is responsible for and can be used to guide supervisors in providing the proper training when needed. During some types of audits in a calibration facility, an auditor will use the individual's training record or job description to ascertain if he or she is qualified to perform a specific task. Some regulated agencies require documented training, education, or qualification for an individual to perform repair and/or calibration on IM&TE. It can also be used to determine if personnel are qualified for promotions, bonuses, additional training, seminars, or conferences.

Maintaining an up-to-date job description and training record should be the responsibility of the individual more so than the supervisor. After all, it's their career that is dependent on those records, and they should play a major role in keeping them accurate and current. In some large organizations, each department (such as the calibration lab) may have a person specifically assigned to maintain training records as an additional duty. This can work well if a person's training record is not part of the main personnel file, otherwise privacy issues can be an issue. The U.S. Air Force is an example of one organization that does this.

# References

BSR/ISO/ASQ Q10012:2003(E), *Measurement management systems—Requirements for measurement processes and measuring equipment.* Milwaukee: ASQ Quality Press.

# Chapter 41
# Vendors and Suppliers

Does your organization have procedures in place for selecting a qualified supplier? Or do you ask your second cousin on your mother's side if Uncle Harry can get what you need? Most standards and requirements recognize this problem and address it in their manuals.

ANSI/ISO 17025 has two specific paragraphs covering this topic. Paragraph 4.5 Subcontracting of tests and calibrations states:

- "When a laboratory subcontracts work . . . this work shall be placed with a competent subcontractor. A competent subcontractor is one that, for example, complies with this International Standard for the work in question.

- The laboratory shall advise the client of the arrangement in writing and, when appropriate, gain the approval of the client, preferably in writing.

- The laboratory is responsible to the client for the subcontractor's work, except in the case where the client or a regulatory authority specifies which subcontractor is to be used.

- The laboratory shall maintain a register of all subcontractors that it uses for tests and/or calibrations and a record of the evidence of compliance with this International Standard for the work in question.

Note: This register is frequently referred to as an approved vendor list." The AVL, besides documenting logistical information (address, contacts, and so on) and listing compliance/accreditation status, also lists any qualifiers such as 'Authorized for Repair Only,' 'Authorized for Dimensional Calibrations Only,' and so on. Laboratory personnel use the AVL to select vendors for external work. Vendors not appearing on the AVL would require the quality manager to authorize them before work could be done.

Paragraph 4.6, Purchasing services and supplies, states:

- "The laboratory shall have a policy and procedure(s) for the selection and purchasing of services and supplies it uses that affect the quality of the tests and/or calibrations. Procedures shall exist for the purchase, reception and storage of reagents and laboratory consumable materials relevant for the tests and calibrations.

- The laboratory shall ensure that purchased supplies and reagents and consumable materials that affect the quality of tests and/or calibrations are not used until they have been inspected or otherwise verified as complying with standard specifications or requirements defined in the methods for the tests and/or calibrations concerned. These services and supplies used shall comply with specified requirements. Records of actions taken to check compliance shall be maintained.

- Purchasing documents for items affecting the quality of laboratory output shall contain data describing the services and supplies ordered. These purchasing documents shall be reviewed and approved for technical content prior to release. Note: The description may include type, class, grade, precise identification, specifications, drawings, inspection instructions, other technical data including approval of test results, the quality required and the quality system standard under which they were made.

- The laboratory shall evaluate suppliers of critical consumables, supplies and services which affect the quality of testing and calibration, and shall maintain records of these evaluations and list those approved."

BSR/ISO/ASQ Q10012:2003 states, in paragraph 6.4, Outside suppliers, "The management of the metrological function shall define and document the requirements for products and services to be provided by outside suppliers for the measurement management system. Outside suppliers shall be evaluated and selected based on their ability to meet the documented requirements. Criteria for selection, monitoring and evaluation shall be defined and documented, and the results of the evaluation shall be recorded. Records shall be maintained of the products or services provided by outside suppliers."

NCSL RP-6 lists requirements in several places. In paragraph 5.5, Supplier control, it states: "Biomedical or pharmaceutical organizations, using outside supplier maintenance and calibration services, must ensure that the supplier's calibration control system complies with all the organization's requirements. To this end, they are advised to establish an agreement with outside supplier service organizations to:

- Utilize approved procedures outlining methodology used in maintaining and calibrating the client's measurement and test equipment;

- Provide a certified report of calibration, complete with all necessary supporting documentation;

- Perform the maintenance and calibration activities meeting the organization's specific requirements;

- Provide a copy of a quality manual or alternate documentation of their quality system;

- Provide objective evidence of quality records, environmental control, equipment history files, and any other relevant quality materials."

NCSL RP-6, Paragraph 5.5.1, Supplier specifications, states: "The organization should provide the supplier with a specification document that can be used to ensure that calibration activities and desired results match the organization's requirements. This document may include but is not limited to:

- Type of calibration service;
- Applicable documents, technical reference, or standard;
- Equipment specification;
- Service performance requirements;
- Physical, environmental, and data-format requirements;
- Safety and biohazard requirements;
- Confidentiality and security requirements;
- Certificate of calibration requirements."

Finally, RP-6 paragraph 5.5.2, Supplier audits, states: "Each organization is responsible for conducting a documented audit of all suppliers on a periodic basis that attests the capabilities of the supplier and the competencies of all supplier personnel. Accredited suppliers may not require periodic audits when the contracting organization's regulatory requirements are met by the accreditation process."

Some organizations or regulatory agencies require parts only be replaced by the OEM, while others only stipulate that the part is a direct and equal replacement of the part being replaced. Depending on the quality system an organization conforms to, the cost could be a factor in what to procure or not affect the choice at all. If you must use the OEM, consider comparing prices at different locations. Another consideration might be in purchasing a service contract with an equipment vendor. For a fixed amount of money, all repairs, including parts and labor, for a predetermined amount of time, usually a year, can be arranged. If the IM&TE is getting old but is still serviceable, a service contract can save money in the long run. Money is saved by not having to purchase new equipment, a set amount can be budgeted, and all costs are borne by the vendor. Plus, the supplier usually provides some type of warranty on the replaced parts and labor. Also, some vendors provide discounts if the total amount for all the service contracts exceeds a predetermined limit. This is something to consider when looking at the option of using service contracts.

Market consolidation and contraction over the past decade has resulted in some manufacturers becoming the sole source for the type of products they make and the associated parts and service. In many cases the manufacturers in these situations continue to provide fast, efficient, quality, and reasonably-priced service. In some cases, however, the sole-source equipment manufacturer may be a handicap to providing quality service to your customer. There are cases where any rational supplier qualification scheme would exclude the manufacturer as a supplier for one or more reasons, but you are forced to deal with them anyway because there is no alternative. The laboratory must have a process for dealing with cases like this. A primary recommendation is to fully document every interaction with the supplier. The documentation can be useful in dealing with their complaint process, although a realistic view is that the typical calibration laboratory has little or no leverage in dealing with a huge sole-source corporation. If necessary, you can also use the documentation as reference material in dealing with your customers and/or your auditors and assessors.

## References

ANSI/ISO 17025-1999, *American National Standard—General requirements for the competence of testing and calibration laboratories.* Milwaukee: ASQ Quality Press.

BSR/ISO/ASQ Q10012:2003(E), *Measurement management systems—Requirements for measurement processes and measuring equipment.* Milwaukee: ASQ Quality Press.

National Conference of Standards Laboratories. 1999. Recommended Practice RP-6: *Calibration control systems for the biomedical and pharmaceutical industry.* Boulder, CO: NCSL International.

# Chapter 42
# Housekeeping and Safety

On a list of most important items, where is safety found? At the top, part way down, or close to the bottom? Does money make a difference in an organization's philosophy about safety? Should it?

It's easy to see where this is going, but do you really care? You should! No matter what type of calibration performed or in what environment it is accomplished, safety should be the driving force behind every task. Calibrations can be reaccomplished, and money replaced. A person's eye, finger, limb, or life cannot.

Good housekeeping and safety should go together. It is hard to have one without the other. Different industries have their special hazards, and most are explained during orientation or safety briefings. If you work around high voltage or current, you should not wear wire rim glasses, rings, watches, jewelry, or anything that could conduct electricity. Some organizations use the rule "no metal above the waist." When working with machinery, loose clothing or long hair can get a person in trouble very quickly, as can inattentive work habits. The authors have tried to put together a list of items that specific industries should be cautious of when working with or around IM&TE.

## All Industries

- Everything should have a place to be when it is not in use. There are many ways to accomplish this, some organizations have had good results from adopting the 5S methodology.[1]

- Know what personal protective equipment (PPE) is required and when to use it. Use your PPE when required.

- Provide safe and appropriate storage for chemicals. Do not store oxidizers with combustibles or flammables.

- Know where the exit, fire alarm, and fire extinguisher are. Know how to use the fire extinguisher.

## Biotechnology and Pharmaceutical Industry

- Always wear safety glasses around chemicals and reagents.

- Wear gloves at all times, not knowing what chemicals or reagents have come in contact with the IM&TE being calibrated and/or repaired.

- Most companies require a smock or lab coat to be worn when in a lab or working with IM&TE.

- Special considerations when working with radioactive material or isotopes, including gloves, eye protection, lab coats as a minimum.

- Know where the nearest emergency eyewash and emergency shower are and how to use them. Ensure that they are always in proper working order.

## Electronics or High Voltage/Current Industries

- Remove rings, watches, jewelry, chains, metal rim eyewear—anything that could conduct current or voltage. Do not wear metal above the waist.

- Do not work inside energized equipment unless there is no alternative. If you are working with high voltage or high current, have a qualified safety observer nearby.

- Everyone in the laboratory should be qualified in emergency first aid and cardio-pulmonary resuscitation (CPR).

- Calibration procedures should be designed that performance verification of the unit under test can be performed with all covers installed, using only the standard external inputs and outputs.

- The high voltage calibration area should have limited access, so people not working on that equipment can't accidentally wander in. (This will help reduce accidental shocks.) A good practice is to have a visible indicator, such as a flashing light, so other people know when the high voltage is energized.

- If the laboratory has a repair area, it should have limited access, so people not working on that equipment can't accidentally wander in. (This may help reduce accidental burns.) The soldering station needs to have an air filtering or exhaust system to eliminate the fumes.

## Airline Industry

- Be sure the IM&TE meets all requirements before it is released. It will be used to perform maintenance on an airframe, engine, or aircraft component, and the performance of those items can directly affect flight safety.

- If the workload includes aircraft system test sets, the laboratory will require 400 Hz single-phase and/or three-phase electrical power. The receptacles for this power should be clearly marked to indicate that they are not the normal utility power frequency.

## Physical-Dimensional Calibration

- Treat torque testers with respect. Although slight, there is a possibility that a torque wrench or a part of the tester might break and fly off in a random direction.

- When performing pressure calibrations, first be sure all of the hydraulic piping, hoses and couplings are in excellent condition and properly fastened.

- A lot of things are heavy. Lift them properly to protect your back. Handle them with care to protect the equipment. Be especially careful placing things on or removing them from a granite or steel surface plate so you don't damage the top of the surface plate.

- Don't stare into a laser beam. It may be the last thing you see.

- Anything used for calibration of oxygen or breathing air equipment should be in a separate room dedicated for that purpose. That area, and the equipment in it, must be totally free of contamination from oil and other petroleum products.

- If you have a mercury manometer (a primary standard used for absolute pressure calibration), it must be in a room separate from everything else. That room must have a separate ventilation system, a mercury vapor detector and alarm, and a mercury spill cleanup kit.

- Treat thermal standards with care. They can cause burn or freeze injuries, depending on the temperature.

## Computer Industry

- Follow the One Hand Rule. If you work with high voltage circuits, you should keep one hand in your pocket or behind your back, to prevent yourself from bridging the circuit with both hands.

- Use the buddy system. Never engage in work with hazardous material, hazardous electrical potentials, or work having a less than remote possibility of injury due to falls, burns, contact with machinery, and so on, without somebody close at hand that can render assistance in case of an accident.

- Never cheat safety interlocks unless specifically authorized by the OEM as required to service its equipment.

- Any hazardous work should require engineering safeguards to prevent unauthorized contact or interference. Some examples of engineering safeguards include posting signs, roping off areas, and posting personnel to prevent access.

- Laboratory personnel should be aware of emergency numbers, location of emergency exits and emergency equipment, evacuation procedures, and so on.

- Monthly self-assessments are encourage in order to identify possible hazardous conditions as frayed power cords, emergency exits/equipment access blocked, hazardous chemical not stored correctly, and so on. A checklist is recommended to document compliance and to note discrepancies. All safety discrepancies are escalated to highest priority level for corrective action.

## Endnote

1. Hirano. Based on Japanese words that begin with S, the 5 S philosophy focuses on effective work place organization and standard work procedures. 5S simplifies your work environment, reduces waste and nonvalue activity while improving quality efficiency and safety (sort–Seiri, set in order—*Seiton*, shine—*Seiso*, standardized—*Seiketsu*, and sustain—*Shitsuke*).

## Reference

Hirano, Hiroyuki. 1995. *5 Pillars of the visual workplace: The sourcebook for 5S implementation.* (Bruce Talbot, translator.) Shelton, CT: Productivity Press.

# Appendix A
## Professional Associations

If this handbook does not have the answer to your calibration or metrology question, where does one turn for more information? There are numerous organizations available to help answer questions or point in the right direction. A compilation with contact numbers and addresses is furnished for the convenience of the reader, with the understanding that this is not an endorsement of any specific organization or company. The sharing of nonproprietary information throughout the metrology community is one of the hallmarks that help in keep industry, both public and private, on the cutting edge with the latest technology.

### NATIONAL CONFERENCE OF STANDARDS LABORATORIES INTERNATIONAL

2995 Wilderness Place, Suite 107
Boulder, CO 80301-5404
Tel: 303-440-3339
Fax: 303-440-3384

Established in 1961, NCSL International is a professional association for individuals engaged in all spheres of international measurement science. In addition to providing valuable real-time professional career support and advancement opportunities, NCSL International sponsors an annual technical Annual Workshop & Symposium with panels, exhibits, and individual presentations to provide a forum for attendees to exchange information on a wide variety of measurement topics, including implementing national and international standards, achieving laboratory accreditation, new measurement technology, advances in measurement disciplines, laboratory management procedures and skills, equipment management, workforce training, and new instrumentation.

NCSL International was formed in 1961 to promote cooperative efforts for solving the common problems faced by measurement laboratories. Today, NCSL International has over 1500 member organizations from academic, scientific, industrial, commercial, and government facilities around the world. This wide representation of experience provides members a rich opportunity to exchange ideas, techniques, and innovations with others engaged in measurement science.

NCSL International is a nonprofit organization, whose membership is open to any organization with an interest in the science of measurement and its application in research, development, education, or commerce. Its vision is to promote competitiveness and success of NCSL International members by improving the quality of products and services through excellence in calibration, testing, and metrology education and training. The mission of NCSL International is to advance technical and managerial excellence in the field of metrology, measurement standards, conformity assessment, instrument calibration, as well as test and measurement, through voluntary activities aimed at improving product and service quality, productivity, and the competitiveness of member organizations in the international marketplace.

NCSL International accomplishes its mission through activities whose purposes are to:

- Promote voluntary and cooperative efforts to solve common problems faced by its member organizations.

- Promote and disseminate relevant information that is important to its member organizations.

- Formulate consensus positions of the membership when requested by outside organizations and government bodies that will serve all or segments of the member organizations.

- Advance the state-of-the-art in metrology and related activities in both the technical and the management area.

- Provide liaison with technical societies, trade associations, educational institutions, and other organizations or activities that have common interests.

- Assess metrology requirements and develop uniform, recommended practices related to the activities of the membership.

- Provide a forum to accomplish the objectives of NCSL through conferences, regional and sectional meetings, committee activities, and publications.

- Serve as an effective channel to assist various national laboratories disseminate information to metrological communities, and to collect and present information to strengthen and improve national measurement systems and the horizontal linkages between these systems.

## AMERICAN SOCIETY FOR QUALITY

600 North Plankinton Avenue
Milwaukee, WI 53203
North America: 800-248-1946, Fax: 414-272-1734
International: 414-272-8575

The American Society for Quality (ASQ) is the world's leading authority on quality since 1946. The 104,000-member professional association creates better workplaces and communities worldwide by advancing learning, quality improvement, and knowledge exchange to improve business results. By making quality a global priority, an organizational imperative, and a personal ethic, ASQ is the community for everyone who seeks technology, concepts, or tools to improve themselves and their world.

## Changing the World

A world of improvement is available through ASQ, providing information, contacts and opportunities to make things better in the workplace, in communities, and in people's lives.

## An Impartial Resource

ASQ makes its officers and member experts available to inform and advise the U.S. Congress, government agencies, state legislatures, and other groups and individuals on quality-related topics. ASQ representatives have provided testimony on issues such as training, healthcare quality, education, transportation safety, quality management in the federal government, licensing for quality professionals, and more.

Send e-mail to customer service. A Customer Care representative will respond as soon as possible, usually within one business day. Send your message to help@asq.org.

## The Measurement Quality Division of ASQ

Members include quality and instrument specialists who develop, apply, and maintain the calibration of measuring equipment and systems, and quality engineers and educators concerned with measurement process capability.

The Measurement Quality Division (MQD) supports, assists, and guides ASQ members and others in the measurement field in the application of both established and innovative tools of measurement and quality. The goal is to improve measurement-based decisions in laboratory, calibration, manufacturing, and management processes at all levels of accuracy.

The MQD supports standards development, disseminates measurement and quality-related information, offers technical support, provides education, sponsors research, fosters professional interaction in the measurement of quality and the quality of measurement, and emphasizes the importance of measurements in the quality process.

# THE INSTRUMENTATION, SYSTEMS, AND AUTOMATION SOCIETY (ISA)

67 Alexander Drive
Research Triangle Park, NC 27709 USA
Phone: 919-549-8411, FAX: 919-549-8288
E-Mail: info@ISA.org, www.isa.org

ISA was founded in 1945 as the Instrument Society of America, with a focus on industrial instrumentation. It now has over 39,000 members in 110 countries. ISA is a nonprofit professional society for people in automation and control systems. It promotes innovation and education in application and use of automation and control systems, and in their theory, design and manufacture. ISA has annual conferences and technical training, and is a publisher of books, magazines, technical standards, and recommended practices.

## INSTITUTE OF ELECTRICAL AND ELECTRONIC ENGINEERS (IEEE)

www.ieee.org

IEEE Instrumentation and Measurement Society (ewh.ieee.org/soc/im)

## THE AMERICAN SOCIETY FOR NONDESTRUCTIVE TESTING

www.asnt.org/home.htm

The American Society for Nondestructive Testing, Inc., (ASNT) is the world's largest technical society for nondestructive testing (NDT) professionals. Through its organization and membership, it provides a forum for exchange of NDT technical information; NDT educational materials and programs; and standards and services for the qualification and certification of NDT personnel. ASNT promotes the discipline of NDT as a profession and facilitates NDT research and technology applications.

ASNT was founded in 1941 (under the name of The American Industrial Radium and X-Ray Society) and currently boasts an individual membership of nearly 10,000 and a corporate membership of about 400 companies. The society is structured into local sections (or chapters) throughout the world. There are over 75 local sections in the United States and 12 internationally.

## THE AMERICAN SOCIETY OF TEST ENGINEERS

www.astetest.org

The ASTE is a all volunteer, nonprofit corporation with members in 22 states and Canada, including several active chapters. The ASTE is dedicated to the quality, integrity, and advancement of the test engineering profession.

## THE NATIONAL SOCIETY OF PROFESSIONAL ENGINEERS (NSPE)

www.nspe.org

The National Society of Professional Engineers (NSPE) is the only engineering society that represents individual engineering professionals and licensed engineers across all disciplines. Founded in 1934, NSPE strengthens the engineering profession by promoting engineering licensure and ethics, enhancing the engineer image, advocating and protecting PEs' legal rights at the national and state levels, publishing news of the profession, providing continuing education opportunities, and much more.

# Appendix B
## ASQ and Certification

### CERTIFIED CALIBRATION TECHNICIAN INFORMATION

The Certified Calibration Technician . . .
. . . tests, calibrates, maintains, and repairs electrical, mechanical, electromechanical, analytical, and electronic measuring, recording, and indicating instruments and equipment for conformance to established standards.

### Education and/or Experience

You must have five years of on-the-job experience in one or more of the areas of the Certified Calibration Technician body of knowledge. If you are now or were previously certified by ASQ as a Quality Engineer, Quality Auditor, Reliability Engineer, Software Quality Engineer, or Quality Manager, experience used to qualify for certification in these fields applies to certification as a calibration technician. If you have completed a degree* from a college, university, or technical school with accreditation accepted by ASQ, part of the five-year experience requirement will be waived, as follows (only one of these waivers may be claimed):

Diploma from a technical, trade, or military school—two years waived

Associate degree—two years waived

Bachelor's degree—two years waived

Master's or doctorate degree—two years waived

* Degrees/diplomas from educational institutions outside the United States must be equivalent to degrees from U.S. educational institutions.

### Proof of Professionalism

Proof of professionalism may be demonstrated in one of three ways:

1. Membership of ASQ, an international affiliate society of ASQ, or another society that is a member of the American Association of Engineering Societies or the Accreditation Board for Engineering and Technology

2. Registration as a Professional Engineer

3. The signatures of two persons—ASQ members, members of an international affiliate society, or members of another recognized professional society—verifying that you are a qualified practitioner of the quality sciences

## Examination

Each certification candidate is required to pass a written examination that consists of multiple-choice questions that measure comprehension of the body of knowledge. The Calibration Technician examination is a one-part, 125-question, four-hour exam, and is offered in the English language only.

# BODY OF KNOWLEDGE

The topics in this body of knowledge include additional detail in the form of subtext explanations and the cognitive level at which the questions will be written. This information will provide useful guidance for both the Examination Development Committee and the candidates preparing to take the exam. The subtext is not intended to limit the subject matter or be all-inclusive of what might be covered in an exam. It is intended to clarify the type of content to be included in the exam. The descriptor in parentheses at the end of each entry refers to the highest cognitive level at which the topic will be tested. A complete description of cognitive levels is provided at the end of this document.

Note: Regarding IM&TE, the Test Specification Committee that created this body of knowledge recognizes that different industries and branches of the military use various descriptors and abbreviations to refer to the units being calibrated. To avoid confusion, the committee decided to use the term *IM&TE* as the most globally descriptive term. This term will be used in both the BOK and the examination itself.

## I. General Metrology (30 Questions)

A. **Base SI units.** Describe and define the seven base units: meter, kilogram, second, ampere, kelvin, candela, mole. (Comprehension) Note: The application of these units is covered in I.B., I.C., and I.E.

B. **Derived SI units.** Define and calculate various derived units, including degree, ohm, pascal, newton, joule, coulomb, hertz, and so on. (Analysis)

C. **SI multipliers and conversions.** Define and apply various multipliers (for example, zeta, kilo, deci, centi, milli) and convert between them (for example, mega to kilo, micro to milli). (Application)

D. **Fundamental constants.** Recognize various fundamental constants and identify their standard symbols and common applications, such as $c$ (velocity or speed of light in a vacuum), $g$ (gravitational constant), $R$ (universal gas constant), and so on. (Knowledge) *Note:* The values or formulas for calculating these constants will not be tested.

E. **Common measurements.** Describe and apply IM&TE in measuring the following: temperature, humidity, pressure, torque, force, mass, voltage/current/resistance, time/frequency, linear displacement, and so on. (Evaluation)

F. **Principles and practices of traceability.** Identify various aspects of traceability, including traceability through commercial and national laboratories and international metrology organizations. (Comprehension)

G. **Types of measurement standards.** Recognize and distinguish between various types of standards, including primary, reference, working, intrinsic, derived, consensus, transfer, and so on. (Application)

H. **Substitution of calibration standards.** Determine when and how calibration standards are substituted based on measurement requirements, equipment availability, equipment specifications, and so on. (Application)

## II. Measurement Systems (25 Questions)

A. **Measurement methods.** Describe and use various measurement methods, including direct, indirect, ratio, transfer, differential, and substitution, in other words, replacing a reference standard device with a unit under test (UUT). (Evaluation)

B. **Measurement data.** Identify and respond to various measurement data considerations, including readability, integrity, confidentiality, resolution, format, suitability for use, and so on. (Analysis)

C. **Characteristics of measurements.** Define and distinguish between various measurement characteristics, including variability, sensitivity, repeatability, bias, linearity, stability, reproducibility, and so on. (Comprehension) Note: The application of these characteristics is covered in VI.A.3. and VI.B.

D. **IM&TE specifications.** Describe and use IM&TE specifications in terms of common descriptors, for example, percent of full scale (FS), percent of range, parts per million (ppm) of reading, and number of counts. (Application)

E. **Primary error sources.** Identify and correct for various types of error sources that can affect measurement uncertainty, including drift, bias, operator error, environment, and so on. (Evaluation)

F. **Measurement systems and capabilities.** Describe and distinguish between measurement systems and measurement capabilities. (Comprehension)

G. **Measurement assurance programs (MAPs).** Identify and describe basic concepts of MAPs, including interlaboratory comparisons, proficiency tests, gage R&R studies, and so on. (Comprehension)

## III. Calibration Systems (25 Questions)

A. **Calibration procedures.** Identify and define common components of calibration procedures, such as required equipment, ambient conditions, revisions, equipment listing, environmental restraints, and so on. (Comprehension)

B. **Calibration methods.** Define and use common calibration methods, including spanning, nulling, zeroing, linearization, and so on. (Application)

C. **Industry practices and regulations.**

   1. *Industry-accepted practices.* Recognize various sources of industry-accepted metrology and calibration practices (for example, published, manufacturer, ANSI). (Comprehension)

   2. *Directives and mandates.* Define and describe different types of calibration directives such as state and federal regulations, traceability, and other requirements mandated by legal metrology and guidance from national or international standards, and identify which rules or conventions take precedence in various situations. (Application)

D. **Control of the calibration environment.** Define and describe various environmental parameters for humidity, dust levels, electrostatic discharge (ESD), temperature, vibration, and so on, and their influence on the calibration function. (Application)

E. **Calibration processes for IM&TE.**

   1. *Process flow.* Identify and describe the basic flow of IM&TE throughout the calibration process. (Comprehension)

   2. *Logistical information.* Identify various aspects of IM&TE logistical information, such as equipment identification, ownership, service history, process tracking, and so on. (Comprehension)

   3. *Roles and responsibilities.* Identify various roles and responsibilities of staff such as technical manager, scheduler, quality manager, technician, and so on. (Comprehension)

   4. *Scheduling.* Describe various IM&TE scheduling considerations, including calibration intervals, recalls, how overdue schedules are determined, steps in the notification process, and so on. (Knowledge)

F. **Manual and automated calibration.** Recognize various issues related to developing, validating, and using both manual and automated calibration processes, including software-driven processes. (Comprehension)

G. **Systems records and records management.** Identify the importance of maintaining document control, confidentiality, and integrity in relation to various records (for example, training records, audit results, uncertainty budgets, customer data) in both electronic and hard-copy formats. (Comprehension)

H. **Reporting results.** Identify and distinguish between various types of calibration results reports, including certificates, test reports, labels, reports of nonconforming calibration, and so on. (Application)

## IV. Applied Math and Statistics (20 Questions)

A. **Technical and applied mathematics (Application).**

1. *Scientific and engineering notation.* Express a floating point number in scientific and engineering notation.

2. *English/metric conversions.* Convert various units of measurement between English and metric units, including length, area, volume, capacity, and weight.

3. *Ratios.* Express ratios in terms of parts per million (ppm), percentage, decibels (dB), and so on.

4. *Linear interpolation and extrapolation.* Interpret tables and graphs to determine intermediate and extrapolated values.

5. *Rounding, truncation, and significant figures.* Round and truncate a given number to a specified number of digits.

6. *Number bases.* Convert numbers between various number bases (for example, decimal, binary, octal, hexadecimal).

7. *Volume and area.* Calculate volume and area of various geometric shapes (for example, cube, sphere, pyramid, cylinder).

8. *Angle conversions.* Convert between various angular units (for example, degrees, radians).

9. *Graphs and plots.* Determine the slope, intercept, and linearity of data sets.

B. **Applied statistics.**

1. *Basic statistical tools.* Define and use basic statistics such as measures of central tendency (mean, standard deviation, and so on), sample vs. population, degrees of freedom, and so on. (Application)

2. *Common distributions.* Classify data distributions as being normal, rectangular, triangular, or U-shaped. (Application)

3. *Descriptive statistics.* Calculate the variance, root mean square (rms), root sum square (rss), and standard error of the mean (SEM) for a data set. (Application)

4. *Sampling issues.* Recognize various terms, including acceptance sampling, sample size, sufficient number of points, and so on. (Knowledge)

## V. Quality Systems and Standards (15 Questions)

**A. Quality management systems.**

1. *System components.* Define and distinguish between various components of a quality system, including organizational leadership, market and customer focus, organizational performance measures and analysis, employee training and development, continuous improvement models, and so on. (Application)

2. *Procedures.* Identify various methods and tools used in the development, validation, improvement, and review of a quality system, including mission and goals, strategic planning, cross-functional teams, and so on. (Comprehension)

**B. The seven quality control tools.** Select and apply the basic quality tools: cause and effect diagrams, flowcharts/process maps, check sheets, Pareto diagrams, scatter diagrams, control/run charts, and histograms. (Analysis)

**C. Quality audits.** Define basic audit types (for example, internal, external, product, process) and roles (for example, auditor, auditee, client), and identify basic components of an audit (for example, audit plan, audit purpose, audit standard) and describe various auditing tools (for example, checklist, final report). (Comprehension)

**D. Preventive and corrective action.**

1. *Process improvement techniques.* Determine and select areas for improvement using various quality tools (for example, PDCA, confidence checks, brainstorming, mistake-proofing, fishbone diagram). (Application)

2. *Nonconforming material identification.* Determine conformance status and apply various methods of identifying and segregating nonconforming IM&TE materials. (Evaluation)

3. *Impact assessment of nonconformances.* Define and use various tools (for example, reverse traceability, customer notification, product recall, calibration standard evaluation, root-cause analysis) in response to out-of-tolerance conditions for IM&TE. (Application)

**E. Supplier qualification and monitoring.** Identify various activities used to qualify, monitor, and sustain approved suppliers. (Knowledge)

**F. Professional conduct and ethics.** Identify appropriate behaviors, such as those listed in the ASQ Code of Ethics, for various situations requiring ethical decisions. (Application)

**G. Occupational safety requirements.**

1. *Hazards and safety equipment.* Identify potential hazards within the working environment (for example, ventilation, mercury, lighting, soldering) and describe the proper use of personal protective equipment (PPE). (Knowledge)

2. *Hazardous communications (HAZ-COM).* Identify and interpret various HAZ-COM directives (for example, right-to-know (RTK), material safety data sheet (MSDS), material labeling). (Comprehension)

3. *Housekeeping.* Describe the importance of good housekeeping tools and methods (for example, maintenance, cleaning). (Knowledge)

H. **Quality standards and guides.** Explain the benefits and importance of the following in relation to calibration:

1. Quality standards such as ISO/IEC 17025, ANSI/NCSL Z540-1-1994, ISO/IC 10012, ISO 9000- 2000, and so on.;

2. Quality guides such as GUM, ANSI/NCSL Z540-2-1997, VIM, and so on.;

3. Accreditation and registration boards such as NVLAP, A2LA, IAS, LAB, RAB, IRCA, and so on. (Comprehension)

## VI. Uncertainty (10 Questions)

A. **Uncertainty budget components.** Identify various Type A and Type B uncertainty components, including environment, human factors, methods and equipment, item under test, reference standards, materials, and so on, and identify the key elements of developing an uncertainty budget. (Application)

B. **Uncertainty management.** Define basic terms, such as guardbanding, test uncertainty ratio (TUR), test accuracy ratio (TAR), bias, error, percent of tolerance, and so on. (Knowledge)

C. **Uncertainty determination and reporting.** Identify and use various methods to determine and report measurement uncertainty, including combined and expanded uncertainty, weighted factors, explanatory graphics, coverage factors, confidence levels, effective degrees of freedom, and so on. (Application)

# SIX LEVELS OF COGNITION BASED ON BLOOM'S TAXONOMY(1956)

In addition to content specifics, the subtext detail also indicates the intended complexity level of the test questions for that topic. These levels are based on Levels of Cognition (from Bloom's Taxonomy, 1956) and are presented in rank order, from least complex to most complex.

**Knowledge Level.** (Also commonly referred to as recognition, recall, or rote knowledge.) Be able to remember or recognize terminology, definitions, facts, ideas, materials, patterns, sequences, methodologies, principles, and so on.

**Comprehension Level.** Be able to read and understand descriptions, communications, reports, tables, diagrams, directions, regulations, and so on.

**Application Level.** Be able to apply ideas, procedures, methods, formulas, principles, theories, and so on, in job-related situations.

**Analysis.** Be able to break down information into its constituent parts and recognize the parts' relationship to one another and how they are organized; identify sublevel factors or salient data from a complex scenario.

**Synthesis.** Be able to put parts or elements together in such a way as to show a pattern or structure not clearly there before; identify which data or information from a complex set are appropriate to examine further or from which supported conclusions can be drawn.

**Evaluation.** Be able to make judgments regarding the value of proposed ideas, solutions, methodologies, and so on, by using appropriate criteria or standards to estimate accuracy, effectiveness, economic benefits.

# Appendix C
# Acronyms and Abbreviations

| Acronym | Meaning | More Information |
|---|---|---|
| A2LA | The American Association for Laboratory Accreditation | www.a2la.org |
| A. M. | *ante meridian* (Latin) —between midnight and midday | |
| AC | Alternating current | |
| AF | Audio frequency | |
| AFC | Automatic frequency control | |
| AGC | Automatic gain control | |
| AIAG | Automotive Industry Action Group | www.aiag.org |
| ALC | Automatic level control | |
| AM | Amplitude modulation | |
| ANOVA | Analysis of variance | |
| ANSI | American National Standards Institute, Inc. | www.ansi.org |
| API | Application programming interface (software) | |
| APLAC | Asia-Pacific Laboratory Accreditation Cooperation | www.ianz.govt.nz/aplac |
| ARFTG | IEEE Automatic RF Techniques Group | www.arftg.org |
| ASCII | American Standard Code for Information Interchange (data processing) | |
| ASIC | Application-specific integrated circuit | |
| ASQ | American Society for Quality | www.asq.org |
| ASQ CCT | ASQ Certified Calibration Technician | www.asq.org |
| ASQ MQD | ASQ Measurement Quality Division | www.asq.org |
| ATE | Automated (or Automatic) test equipment | |
| ATM | Asynchronous transfer mode (data communications) | |
| ATM | Automatic teller machine (banking) | |
| AV | Alternating voltage | |

| Acronym | Meaning | More Information |
|---|---|---|
| AVC | Automatic volume control | |
| AWG | American wire gage | |
| BASIC | Beginner's All-purpose Symbolic Instruction Code (software) | |
| BCD | Binary-coded decimal | |
| BIOS | Basic input/output system (computers) | |
| BIPM | International Bureau of Weights and Measures | www.bipm.fr |
| BS | British standard | |
| BSI | British Standards Institution | www.bsi.org.uk |
| CAD | Computer-aided design | |
| CAM | Computer-aided manufacturing | |
| CASE | Coordinating Agency for Supplier Evaluation | www.caseinc.org |
| CCT | Certified Calibration Technician | www.asq.org/cert/types/cct |
| CD | Compact disk | |
| CDMA | Code division multiple access (cellular telephones) | |
| CENAM | National Center for Metrology (Mexico) | www.cenam.mx |
| cGLP | current Good Laboratory Practices | |
| cGMP | current Good Manufacturing Practices | |
| CGPM | General Conference on Weights and Measures | www.bipm.fr |
| CIPM | International Committee of Weights and Measures | www.bipm.fr |
| CMM | Component Maintenance Manual (aviation) | |
| CMM | Coordinate measuring machine | |
| CODATA | Committee on Data for Science and Technology | www.codata.org |
| CPR | Calibration Problem report | |
| CPR | Cardio-pulmonary resuscitation | |
| CRT | Cathode-ray tube | |
| DAC | Data acquisition and control | |
| DAC | Digital to analog converter | |
| DARPA | Defense Advanced Research Projects Agency (USA) | |
| dB | Decibel | |
| dBm | Decibels relative to 1 milliwatt in a specified impedance | |
| DC | Direct current | |
| DEW | Directed-energy weapon | |
| DEW | Distant early warning | |
| DIN | German Institute for Standardization | www.din.de |
| DKD | German Calibration Service (accreditation body) | www.dkd.info |
| DMAIC | Define, measure, analyze, improve, control | |

| Acronym | Meaning | More Information |
|---|---|---|
| DMM | Digital multimeter | |
| DOD | Department of Defense (USA) | www.dod.gov |
| DTMF | Dual tone, multiple frequency | |
| DUT | Device under test | |
| DV | Direct voltage | |
| DVD | Digital video disk or digital versatile disk | |
| DVM | Differential voltmeter | |
| DVM | Digital voltmeter | |
| EA | European Cooperation for Accreditation | www.european-accreditation.org |
| EDI | Electronic Data Interchange | |
| EIA | Electronic Industries Alliance | www.eia.org |
| EMF | Electromotive force, voltage | |
| EMI | Electromagnetic interference | |
| EMU | Electromagnetic units (obsolete) | |
| EN | European standard | |
| ESD | Electrostatic discharge | |
| ESU | Electrostatic units (obsolete) | |
| EUROMET | European Collaboration in Measurement Standards | www.euromet.org |
| EUT | Equipment under test | |
| FAA | Federal Aviation Administration (USA) | www.faa.gov |
| FAQ | Frequently-asked questions | |
| FAX | Facsimile (transmission of images by telephone) | |
| FDA | Food and Drug Administration (USA) | www.fda.gov |
| FFT | Fast fourier transform (mathematics) | |
| FM | Frequency modulation | |
| FO | Fiber optics | |
| FS | Federal standard (USA) | |
| FTP | File transfer protocol (Internet) | |
| GIDEP | Government-Industry Data Exchange Program (USA) | http://www.gidep.org |
| GLONASS | Global navigation satellite system (Russia) | |
| GMT | Greenwich mean time—obsolete, see UTC | |
| GPETE | General purpose electronic test equipment | |
| GPIB | General purpose interface bus (Tektronix term for IEEE-488) | |
| GPS | Global positioning system (satellites) (short form of NAVSTAR GPS) | gps.losangeles.af.mil |
| GSM | Global system for mobile (telecommunications) | |
| GUM | *Guide to the Expression of Uncertainty in Measurement* | |

# 466  *Appendix C*

| Acronym | Meaning | More Information |
|---|---|---|
| HIPOT | High potential (a type of electrical test) | |
| HPGL | Hewlett-Packard graphics language (plotters, printers) | |
| HPIB | Hewlett-Packard interface bus (Hewlett-Packard term for IEEE-488) | |
| HPML | Hewlett-Packard multimeter language | |
| HTML | Hypertext markup language | www.w3.org |
| HTTP | Hypertext transfer protocol (what makes the www work!) | www.w3.org |
| HV | High voltage | |
| HVAC | Heating, ventilation, and air conditioning | |
| IAF | International Accreditation Forum Inc. | www.accreditationforum.com |
| IAQG | International Aerospace Quality Group | www.iaqg.sae.org |
| IAS | International Accreditation Service, Inc. | www.iasonline.org |
| IATF | International Automotive Task Force | |
| IC | Integrated circuit | |
| ICC | International Code Council | www.iccsafe.org |
| IEC | International Electrotechnical Commission | www.iec.ch |
| IEEE | Institute of Electrical and Electronics Engineers | www.ieee.org |
| IFCC | International Federation of Clinical Chemistry and Laboratory Medicine | www.ifcc.org |
| ILAC | International Laboratory Accreditation Cooperation | www.ilac.org |
| ILC | Inter-laboratory comparison | |
| IM&TE | Inspection, measuring, and test equipment | |
| IMEKO | International Measurement Confederation | |
| IR | Infrared | |
| ISA | Instrumentation, Systems & Automation Society | www.isa.org |
| ISO | Insurance Services Office (insurance risk ratings) | www.iso.com |
| ISO | Is *not* an acronym for International Organization for Standardization | www.iso.org |
| IST | International Steam Table | |
| IT | Information technology | |
| ITU | International Telecommunications Union | www.itu.int |
| IUPAC | International Union of Pure and Applied Chemistry | www.iupac.org |
| IUPAP | International Union of Pure and Applied Physics | www.iupap.org |
| JAN | Joint Army-Navy (USA) | |
| JIT | Just-in-time | |
| JUSE | Union of Japanese Scientists and Engineers | |
| KCDB | Key comparison database | kcdb.bipm.org |
| LAN | Local area network | |
| LASER | Light amplification by stimulated emission of radiation | |

| Acronym | Meaning | More Information |
|---|---|---|
| LCR | Inductance-capacitance-resistance (meter) from the circuit symbols | |
| LED | Light-emitting diode | |
| LORAN | Long-range aid to navigation | |
| LSD | Least significant digit | |
| MAP | Measurement Assurance Program | |
| MASER | Microwave amplification by stimulated emission of radiation | |
| MIL-HDBK | Military handbook (USA) | |
| MIL-PRF | Military performance-based standard (USA) | |
| MIL-STD | Military standard (USA) | |
| MOD | Ministry of Defence (UK) | |
| MRA | Mutual recognition arrangement | www.bipm.org/en/convention/mra |
| MSA | Measurement system analysis | |
| NACC | North American Calibration Committee | |
| NACLA | National Cooperation for Laboratory Accreditation | www.nacla.net |
| NAFTA | North American Free Trade Agreement | |
| NAPT | National Association for Proficiency Testing | www.proficiency.org |
| NASA | National Aeronautics and Space Administration (USA) | www.nasa.gov |
| NATO | North Atlantic Treaty Organization | www.nato.int |
| NAVSTAR GPS | Navigation system with time and ranging global positioning system | gps.losangeles.af.mil |
| NBS | National Bureau of Standards - see NIST | www.nist.gov |
| NCO | Non-Commissioned Officer (military rank) | |
| NCSL | National Conference of Standards Laboratories (see NCSLI) | www.ncsli.org |
| NCSLI | NCSL International | www.ncsli.org |
| NDE | Nondestructive evaluation | www.ndt-ed.org |
| NDT | Nondestructive testing | www.ndt-ed.org |
| NIST | National Institute of Standards and Technology (USA) | www.nist.gov |
| NMI | National Metrology Institute | |
| NORAMET | North American Cooperation in Metrology | |
| NPL | National Physical Laboratory (UK) | www.npl.co.uk |
| NRC | Nuclear Regulatory Commission (USA) | www.nrc.gov |
| NTSC | National Television System Committee (TV format in USA, Canada, Japan) | |
| NVLAP | National Voluntary Laboratory Accreditation Program | www.ts.nist.gov/ts/htdocs 210/214/214.htm |
| OEM | Original equipment manufacturer | |

| Acronym | Meaning | More Information |
|---|---|---|
| OIML | International Organization of Legal Metrology | www.oiml.org |
| OOT | Out of tolerance | |
| P. M. | *post meridian* (Latin)—between midday and midnight | |
| PAL | Phase alternation by line (TV format in Europe, China) | |
| PARD | Periodic and random deviation | |
| PAVM | Phase angle voltmeter | |
| PC | Personal computer | |
| PC | Printed circuit (wiring board) | |
| PCS | Personal communications services (telecommunications, USA) | |
| PDCA | Plan, do, check, act (the Deming or Shewhart cycle) | |
| PDF | Portable document format (documents; trademark of Adobe Corp.) | |
| PDSA | Plan, do, study, act (the Deming or Shewhart cycle, another version) | |
| PM | Phase modulation | |
| PM | Preventive maintenance | |
| PME | Precision measuring equipment | |
| PME | Professional military education (U.S. Armed Forces) | |
| PMEL | Precision measurement equipment laboratory | |
| PMET | Precision measuring equipment and tooling | |
| PO | Petty Officer (Naval rank) | |
| PO | Purchase order | |
| POTS | Plain old telephone service (a basic subscriber line with no extra features) | |
| PPE | Personal protective equipment | |
| PPM | Parts per million (preferred usage is "parts in $10^6$") | |
| PT | Proficiency test | |
| PTB | Physikalisch-Technische Bundesanstalt (German NMI) | www.ptb.de |
| PXI | Compact PCI extensions for Instruments | www.pxisa.org |
| QA | Quality assurance | |
| QC | Quality control | |
| QHE | Quantum Hall effect | |
| QMS | Quality management system | |
| QS | Quality system (as in QS-9000) | |
| R&R | Repeatability and reproducibility | |
| RADAR | Radio detection and ranging | |

| Acronym | Meaning | More Information |
|---|---|---|
| RAM | Random-access memory | |
| RF | Radio frequency | |
| RFID | Radio frequency identification (tags) | |
| RMS | Root-mean-square | |
| ROM | Read-only memory (computers) | |
| RP | Recommended practice (as in NCSL RP-1) | |
| RS | Recommended standard (as in EIA RS-232) | |
| RTD | Resistor temperature device | |
| RTF | Rich text format (documents) | |
| SAE | Society of Automotive Engineers | www.sae.org |
| SAW | Surface acoustic wave | |
| SCPI | Standard commands for programmable instrumentation | scpiconsortium.org |
| SI | International system of weights and measures | www.bipm.fr |
| SONAR | Sound navigation and ranging | |
| SOP | Standard operating procedure | |
| SPC | Statistical process control | |
| SPETE | Special purpose electronic test equipment | |
| SQC | Statistical quality control | |
| SQL | Structured query language | |
| SQUID | Superconducting quantum interference device | |
| SRM | Standard reference material | ts.nist.gov/ts/htdocs/230/232/232.htm |
| Sr. NCO | Senior NCO (military rank) | |
| SWR | Standing wave ratio | |
| TAG | Technical Advisory Group (ISO) | |
| TAG | Truck Advisory Group (AIAG) | |
| TAI | International atomic time | www.bipm.org/en/committees/cc/cctf |
| TAR | Test accuracy ratio | |
| TCP/IP | Transfer control protocol / Internet protocol | |
| TDMA | Time division multiple access (cellular telephones) | |
| TDR | Time-domain reflectometer | |
| TELCO | Telephone company | |
| THD | Total harmonic distortion | |
| TI | Test item | |
| TMDE | Test, measurement, and diagnostic equipment | |
| TQL | Total quality leadership | |
| TQLS | Total quality lip service (humor) | |
| TQM | Total quality management | |
| TRMS | True RMS | |

| Acronym | Meaning | More Information |
|---|---|---|
| TUR | Test uncertainty ratio | |
| TVC | Thermal voltage converter | |
| UMTS | Universal mobile telecommunications system | |
| USB | Universal serial bus | www.usb.org |
| USNO | U.S. Naval Observatory | www.usno.navy.mil |
| UT1 | Universal time scale 1 | www.bipm.org/en/committees/cc/cctf |
| UTC | Universal coordinated time (formerly GMT): time of day at 0 deg. Longitude | www.bipm.org/en/committees/cc/cctf |
| UUC | Unit under calibration | |
| UUT | Unit under test | |
| VIM | ISO International Vocabulary of Basic and General Terms in Metrology | |
| VME | (No particular meaning) | www.ee.ualberta.ca/archive/vmefaq.html |
| VNA | Vector network analyzer | |
| VOM | Volt-Ohm meter | |
| VTVM | Vacuum tube voltmeter | |
| VXI | VME extensions for instrumentation | www.vxibus.org |
| WAN | Wide area network | |
| W-CDMA | Wideband code division multiple access (cellular telephones) | |
| WPAN | Wireless personal area network (range approx. 10 m) | |
| WWV | NIST time and frequency radio transmitter in Ft. Collins, Colorado | |
| WWVB | NIST 60 kHz digital time code radio transmitter in Ft. Collins, Colorado | |
| WWVH | NIST time and frequency radio transmitter in Kauai, Hawaii | |
| WWW | World-wide web (Internet) | www.w3.org |
| Z | Symbol for impedance | |
| ZULU | (U.S. military) indicates reference to UTC time | |
| ZULU | Phonetic for the last letter of the alphabet | |

# Appendix D
## Glossary of Terms

### INTRODUCTION

This glossary is a quick reference to the meaning of common terms. It is a supplement to the VIM, GUM, NCSL Glossary, and the information in the other references listed at the end.

In technical, scientific and engineering work (such as metrology) it is important to correctly use words that have a technical meaning. Definitions of these words are in relevant national, international and industry standards; journals; and other publications, as well as publications of relevant technical and professional organizations. Those documents give the intended meaning of the word, so everyone in the business knows what it is. *In technical work, only the technical definitions should be used.*

Many of these definitions are adapted from the references. In some cases several may be merged to better clarify the meaning or adapt the wording to common metrology usage. The technical definitions may be different from the definitions published in common grammar dictionaries. However, the purpose of common dictionaries is to record the ways that people actually use words, not to standardize the way the words should be used. *If a word is defined in a technical standard, its definition from a common grammar dictionary should never be used in work where the technical standard can apply.*

Terms that are not in this glossary may be found in one of these primary references:

1. ISO. 1993. *International vocabulary of basic and general terms in metrology* (called the VIM); BIPM, IEC, IFCC, ISO, IUPAC, IUPAP, and OIML. Geneva: ISO.

2. ANSI/NCSL. 1997. ANSI/NCSL Z540-2-1997, *U. S. Guide to the expression of uncertainty in measurement* (called the GUM). Boulder, CO: NCSL International.

3. NCSL. 1999. *NCSL Glossary of metrology-related terms.* 2nd ed. Boulder, CO: NCSL International.

Some terms may be listed in this glossary in order to expand on the definition, but should be considered an *addition to* the references listed above, *not a replacement* of them. (It is assumed that a calibration or metrology activity owns copies of these as part of its basic reference material.) Most acronyms and abbreviations are listed in Appendix C.

# GLOSSARY

**Accreditation (of a laboratory)**—Formal recognition by an accreditation body that a calibration or testing laboratory is able to competently perform the calibrations or tests listed in the accreditation scope document. Accreditation includes evaluation of both the quality management system and the competence to perform the measurements listed in the scope.

**Accreditation body**—An organization that conducts laboratory accreditation evaluations in conformance to ISO Guide 58.

**Accreditation certificate**—Document issued by an accreditation body to a laboratory that has met the conditions and criteria for accreditation. The certificate, with the documented measurement parameters and their best uncertainties, serves as proof of accredited status for the time period listed. An accreditation certificate without the documented parameters is incomplete.

**Accreditation criteria**—Set of requirements used by an accrediting body that a laboratory must meet in order to be accredited.

**Accuracy (of a measurement)**—Accuracy is a qualitative indication of how closely the result of a measurement agrees with the true value of the parameter being measured. (VIM, 3.5) Because the true value is always unknown, accuracy of a measurement is always an estimate. An accuracy statement by itself has no meaning other than as an indicator of quality. It has quantitative value only when accompanied by information about the uncertainty of the measuring system.
*Contrast with*: accuracy (of a measuring instrument)

**Accuracy (of a measuring instrument)**—Accuracy is a qualitative indication of the ability of a measuring instrument to give responses close to the true value of the parameter being measured. (VIM, 5.18) Accuracy is a design specification and may be verified during calibration.
*Contrast with:* accuracy (of a measurement)

**Assessment**—Examination typically performed on-site of a testing or calibration laboratory to evaluate its conformance to conditions and criteria for accreditation.

**Best measurement capability**—For an accredited laboratory, the best measurement capability for a particular quantity is "the smallest uncertainty of measurement a laboratory can achieve within its scope of accreditation when performing more-or-less routine calibrations of nearly ideal measurement standards intended to define, realize, conserve, or reproduce a unit of that quantity or one or more of its values; or when performing more-or-less routine calibrations of nearly ideal measuring instruments designed for the measurement of that quantity." (EA-4/02) The best measurement capability is based on evaluations of actual measurements using generally accepted methods of evaluating measurement uncertainty.

**Bias**—Bias is the known systematic error of a measuring instrument. (VIM, 5.25) The value and direction of the bias is determined by calibration and/or gage R&R studies. Adding a correction, which is always the negative of the bias, compensates for the bias.
*See also:* correction, systematic error

**Calibration**—(1). (See VIM 6.11 and NCSL pages 4–5 for primary and secondary definitions.) *Calibration* is a term that has many different—but similar—definitions. It is the process of verifying the capability and performance of an item of measuring and test equipment by comparison to traceable measurement standards. Calibration is performed with the item being calibrated in its normal operating configuration—as the normal operator would use it. The calibration process uses traceable external stimuli, measurement standards, or artifacts as needed to verify the performance. Calibration provides assurance that the instrument is capable of making measurements to its performance specification when it is correctly used.

The result of a calibration is a determination of the performance quality of the instrument with respect to the desired specifications. This may be in the form of a pass/fail decision, determining or assigning one or more values, or the determination of one or more corrections.

The calibration process consists of comparing an IM&TE unit with specified tolerances, but of unverified accuracy, to a measurement system or device of specified capability and known uncertainty in order to detect, report, or minimize by adjustment any deviations from the tolerance limits or any other variation in the accuracy of the instrument being compared.

Calibration is performed according to a specified documented calibration procedure, under a set of specified and controlled measurement conditions, and with a specified and controlled measurement system.

Notes:

- A requirement for calibration does *not* imply that the item being calibrated can or should be adjusted.

- The calibration process *may* include, if necessary, calculation of correction factors or adjustment of the instrument being compared to reduce the magnitude of the inaccuracy.

- In some cases, minor repair such as replacement of batteries, fuses, or lamps, or minor adjustment such as zero and span, may be included as part of the calibration.

- Calibration does *not* include any maintenance or repair actions except as just noted.
  *See also:* performance test, calibration procedure
  *Contrast with:* calibration (2) and repair

**Calibration**—(2 A) Many manufacturers *incorrectly* use the term calibration to name the process of alignment or adjustment of an item that is either newly manufactured or is known to be out of tolerance, or is otherwise in an indeterminate state. Many calibration procedures in manufacturers' manuals are actually factory alignment procedures that only need to be performed if a UUC is in an indeterminate state because it is being manufactured, is known to be out of tolerance, or after it is repaired. When used this way, *calibration* means the same as alignment or adjustment, which are repair activities and excluded from the metrological definition of calibration.

(B) In many cases, IM&TE instruction manuals may use calibration to describe tasks normally performed by the operator of a measurement system. Examples include performing a self-test as part of normal operation or performing a self-calibration (normalizing) a measurement system before use. When calibration is used to refer to tasks like this, the intent is that they are part of the normal work

done by a trained user of the system. These and similar tasks are excluded from the metrological definition of calibration.
*Contrast with:* calibration (1)
*See also:* normalization, self-calibration, standardization

**Calibration activity or provider**—A laboratory or facility—including personnel—that perform calibrations in an established location or at customer location(s). It may be external or internal, including subsidiary operations of a larger entity. It may be called a calibration laboratory, shop, or department; a metrology laboratory or department; or an industry-specific name; or any combination or variation of these.

**Calibration certificate**—(1) A calibration certificate is generally a document that states that a specific item was calibrated by an organization. The certificate identifies the item calibrated, the organization presenting the certificate, and the effective date. A calibration certificate should provide other information to allow the user to judge the adequacy and quality of the calibration. (2) In a laboratory database program, a certificate often refers to the permanent record of the final result of a calibration. A laboratory database certificate is a record that cannot be changed; if it is amended later a new certificate is created.
*See also:* calibration report

**Calibration procedure**—A calibration procedure is a controlled document that provides a validated method for evaluating and verifying the essential performance characteristics, specifications, or tolerances for a model of measuring or testing equipment. A calibration procedure documents one method of verifying the actual performance of the item being calibrated against its performance specifications. It provides a list of recommended calibration standards to use for the calibration; a means to record quantitative performance data both before and after adjustments; and information sufficient to determine if the unit being calibrated is operating within the necessary performance specifications. A calibration procedure always starts with the assumption that the unit under test is in good working order and only needs to have its performance verified.
Note: A calibration procedure does not include any maintenance or repair actions.

**Calibration program**—A calibration program is a process of the quality management system that includes management of the use and control of calibrated inspection, and test and measuring equipment (IM&TE), and the process of calibrating IM&TE used to determine conformance to requirements or used in supporting activities.
   A calibration program may also be called a *measurement management system* (ISO 10012:2003).

**Calibration report**—A calibration report is a document that provides details of the calibration of an item. In addition to the basic items of a calibration certificate, a calibration report includes details of the methods and standards used, the parameters checked, and the actual measurement results and uncertainty.
*See also:* calibration certificate

**Calibration seal**—A calibration seal is a device, placard, or label that, when removed or tampered with, and by virtue of its design and material, clearly indicates tampering. The purpose of a calibration seal is to ensure the integrity of the calibration. A calibration seal is usually imprinted with a legend similar to "Calibration Void if Broken or Removed" or "Calibration Seal—Do Not Break or Remove." A calibration seal provides a means of deterring the user from tampering

with any adjustment point that can affect the calibration of an instrument and detecting an attempt to access controls that can affect the calibration of an instrument.
Note: A calibration seal may also be referred to as a tamper seal.

**Calibration standard**—(See VIM, 6.1 through 6.9, and 6.13, 6.14; and NCSL pages 36–38.) A calibration standard is an IM&TE item, artifact, standard reference material, or measurement transfer standard that is designated as being used only to perform calibrations of other IM&TE items. As calibration standards are used to calibrate other IM&TE items, they are more closely controlled and characterized than the workload items they are used for. Calibration standards generally have lower uncertainty and better resolution than general-purpose items. Designation as a calibration standard is based on the use of the specific instrument, however, not on any other consideration. For example, in a group of identical instruments, one might be designated as a calibration standard while the others are all general purpose IM&TE items. Calibration standards are often called measurement standards.
See also: standard (measurement)

**Combined standard uncertainty**—The standard uncertainty of the result of a measurement, when that result is obtained from the values of a number of other quantities. It is equal to the positive square root of a sum of terms. The terms are the variances or covariances of these other quantities, weighted according to how the measurement result varies with changes in those quantities. (GUM, 2.3.4)
See also: expanded uncertainty

**Competence**—For a laboratory, the demonstrated ability to perform the tests or calibrations within the accreditation scope and to meet other criteria established by the accreditation body. For a person, the demonstrated ability to apply knowledge and skills.
Note: The word *qualification* is sometimes used in the personal sense, since it is a synonym and has more accepted usage in the United States.

**Confidence interval**—A range of values that is expected to contain the true value of the parameter being evaluated with a specified level of confidence. The confidence interval is calculated from sample statistics. Confidence intervals can be calculated for points, lines, slopes, standard deviations, and so on. For an infinite (or very large compared to the sample) population, the confidence interval is:

$$CI = \bar{x} \pm t\, \frac{s}{\sqrt{n}} \quad \text{or} \quad CI = p \pm \sqrt{\frac{p(1-p)}{n}}$$

where  $CI$ is the confidence interval,

$n$ is the number of items in the sample,

$p$ is the proportion of items of a given type in the population,

$s$ is the sample standard deviation,

$\bar{x}$ is the sample mean, and

$t$ is the Student's T value for $\alpha/2$ and $(n-1)$ ($\alpha$ is the level of significance).

**Correction (of error)**—A correction is the value that is added to the raw result of a measurement to compensate for known or estimated systematic error or bias. (VIM, 3.15) Any residual amount is treated as random error. The correction value is equal to the negative of the bias. An example is the value calculated to compensate for the calibration difference of a reference thermometer or for the calibrated offset voltage of a thermocouple reference junction.
*See also:* bias, error, random error, systematic error

**Corrective action**—Corrective action is something done to correct a nonconformance when it arises, including actions taken to prevent reoccurrence of the nonconformance.
*Compare with:* preventive action

**Coverage factor**—A numerical factor used as a multiplier of the combined standard uncertainty in order to obtain an expanded uncertainty. (GUM, 2.3.6) The coverage factor is identified by the symbol $k$. It is usually given the value 2, which approximately corresponds to a probability of 95 percent for degrees of freedom > 10'.

**Deficiency**—Nonfulfillment of conditions and/or criteria for accreditation, sometimes referred to as a nonconformance.

**Departure value**—A term used by a few calibration laboratories to refer to *bias, error or systematic error*. The exact meaning can usually be determined from examination of the calibration certificate.

**Equivalence**—(A) Acceptance of the competence of other national metrology institutes (NMI), accreditation bodies, and/or accredited organizations in other countries as being essentially equal to the NMI, accreditation body, and/or accredited organizations within the host country. (B) A formal, documented determination that a specific instrument or type of instrument is suitable for use in place of the one originally listed, for a particular application.

**Error (of measurement)**—(See VIM, 3.10, 3.12–3.14; and NCSL pages 11–13.) In metrology, error (or measurement error) is an estimate of the difference between the measured value and the probable true value of the object of the measurement. The error can never be known exactly; it is always an estimate. Error may be systematic and/or random. Systematic error (also known as bias) may be corrected.
*See also:* bias, correction (of error), random error, systematic error

**Gage R&R**—Gage repeatability and reproducibility study, which (typically) employs numerous instruments, personnel, and measurements over a period of time to capture quantitative observations. The data captured are analyzed statistically to obtain best measurement capability, which is expressed as an uncertainty with a coverage factor of $k = 2$ to approximate 95 percent. The number of instruments, personnel, measurements, and length of time are established to be statistically valid consistent with the size and level of activity of the organization.

**GUM**—An acronym commonly used to identify the ISO *Guide to the Expression of Uncertainty in Measurement*. In the United States, the equivalent document is ANSI/NCSL Z540-2-1997, *U. S. Guide to the Expression of Uncertainty in Measurement*.

**Hipot (test)**—Hipot is an acronym for high potential (voltage). A Hipot test is a deliberate application of extreme high voltage, direct or alternating, to test the insulation system of an electrical product well beyond its normal limits. An accepted guideline for the applied value, is double the highest operating voltage plus one kilovolt. Current through the insulation is measured while the voltage is applied. If the current exceeds a specified value, a failure is indicated. Hipot testing is normally done during research and development, factory production and inspection, and sometimes after repair. A synonym is dielectric withstand testing.

A high potential tester normally has meters to display the applied voltage and the leakage current at the same time. *Caution!* Hipot testing involves lethal voltages. *Caution!* Hipot testing is a potentially destructive test. If the insulation system being tested fails, the leakage creates a path of permanently lowered resistance. This may damage the equipment and may make it unsafe to use. Routine use of Hipot testing must be carefully evaluated.

Note: Hypot® is a registered trademark of Associated Research Corp. and should not be used as a generic term.

**IM&TE**—The acronym IM&TE refers to inspection, measuring, and test equipment. This term includes all items that fall under a calibration or measurement management program. IM&TE items are typically used in applications where the measurement results are used to determine conformance to technical or quality requirements before, during, or after a process. Some organizations do not include instruments used solely to check for the presence or absence of a condition (such as voltage, pressure, and so on) where a tolerance is not specified and the indication is not critical to safety.

Note: Organizations may refer to IM&TE items as MTE (measuring and testing equipment), TMDE (test, measuring, and diagnostic equipment), GPETE (general-purpose electronic test equipment), PME (precision measuring equipment), PMET (precision measuring equipment and tooling), or SPETE (special purpose electronic test equipment).

**Insulation resistance (test)**—An insulation resistance test provides a qualitative measure of the performance of an insulation system. Resistance is measured in megohms. The applied voltage can be as low as 10 volts DC, but 500 or 1000 volts are more common. Insulation resistance can be a predictor of potential failure, especially when measured regularly and plotted over time on a trend chart. The instrument used for this test may be called an insulation resistance tester or a megohmmeter.

An insulation tester displays the insulation resistance in Megohms and may display the applied voltage.

Note: Megger® is a registered trademark of AVO International and should not be used as a generic term.

**Interlaboratory comparison**—Organization, performance, and evaluation of tests or calibrations on the same or similar items or materials by two or more laboratories in accordance with predetermined conditions.

**Internal audit**—A systematic and documented process for obtaining audit evidence and evaluating it objectively to verify that a laboratory's operations comply with the requirements of its quality system. An internal audit is done by or on behalf of the laboratory itself, so it is a first-party audit.

**International Organization for Standardization (ISO)**—An international nongovernmental organization chartered by the United Nations in 1947, with headquarters in Geneva, Switzerland. The mission of ISO is "to promote the development of standardization and related activities in the world with a view to facilitating the international exchange of goods and services, and to developing cooperation in the spheres of intellectual, scientific, technological and economic activity." The scope of ISO's work covers all fields of business, industry and commerce except electrical and electronic engineering. The members of ISO are the designated national standards bodies of each country. (The United States is represented by ANSI.)
*See also:* ISO

**International System of Units (SI)**—A defined and coherent system of units adopted and used by international treaties. (The acronym SI is from the French *Système International*.) SI is international system of measurement for all physical quantities. (Mass, length, amount of substance, time, electric current, thermodynamic temperature, and luminous intensity.) SI units are defined and maintained by the International Bureau of Weights and Measures (BIPM) in Paris, France. The SI system is popularly known as the metric system.

**ISO**—*Iso* is a Greek word root meaning equal. The International Organization for Standardization chose the word as the short form of the name, so it will be a constant in all languages. In this context, ISO is not an acronym. (If the acronym was based on the full name were used, it would be different in each language.) The name also symbolizes the mission of the organization—to equalize standards worldwide.

**Level of confidence**—Defines an interval about the measurement result that encompasses a large fraction $p$ of the probability distribution characterized by that result and its combined standard uncertainty, and $p$ is the coverage probability or level of confidence of the interval. Effectively, the coverage level expressed as a percent.

**Management review**—The planned, formal, periodic, and scheduled examination of the status and adequacy of the quality management system in relation to its quality policy and objectives by the organization's top management.

**Measurement**—A set of operations performed for the purpose of determining the value of a quantity. (VIM, 2.1)

**Measurement system**—A measurement system is the set of equipment, conditions, people, methods, and other quantifiable factors that combine to determine the success of a measurement process. The measurement system includes at least the test and measuring instruments and devices, associated materials and accessories, the personnel, the procedures used, and the physical environment.

**Metrology**—Metrology is the science and practice of measurement (VIM, 2.2).

**Mobile operations**—Operations that are independent of an established calibration laboratory facility. Mobile operations may include work from an office space, home, vehicle, or the use of a virtual office.

**Natural (physical) constant**—A natural constant is a fundamental value that is accepted by the scientific community as valid. Natural constants are used in the basic theoretical descriptions of the universe. Examples of natural physical constants important in metrology are the speed of light in a vacuum (c), the triple point of water (273.16 K), the quantum charge ratio (*h/e*), the gravitational constant (G), the ratio of a circle's circumference to its diameter (π), and the base of natural logarithms (*e*).

**NCSL international**—Formerly known as the National Conference of Standards Laboratories (NCSL). NCSL was formed in 1961 to "promote cooperative efforts for solving the common problems faced by measurement laboratories. NCSL has member organizations from academic, scientific, industrial, commercial and government facilities around the world. NCSL is a nonprofit organization, whose membership is open to any organization with an interest in the science of measurement and its application in research, development, education, or commerce. NCSL promotes technical and managerial excellence in the field of metrology, measurement standards, instrument calibration, and test and measurement."

**Nondestructive Testing (NDT)**—Nondestructive testing is the field of science and technology dealing with the testing of materials without damaging the material or impairing its future usefulness. The purposes of NDT include discovering hidden defects, quantifying quality attributes, or characterizing the properties of the material, part, structure or system. NDT uses methods such as X-ray and radioisotopes, dye penetrant, magnetic particles, eddy current, ultrasound, and more. NDT specifically applies to physical materials, not biological specimens.

**Normalization, Normalize—**
*See:* self-calibration

**Offset**—Offset is the difference between a nominal value (for an artifact) or a target value (for a process) and the actual measured value. For example, if the thermocouple alloy leads of a reference junction probe are formed into a measurement junction and placed in an ice point cell, and the reference junction itself is also in the ice point, then the theoretical thermoelectric emf measured at the copper wires should be zero. Any value other than zero is an offset created by inhomogeneity of the thermocouple wires combined with other uncertainties.
*Compare with:* bias, error

**On-site operations**—Operations that are based in or directly supported by an established calibration laboratory facility, but actually perform the calibration actions at customer locations. This includes climate-controlled mobile laboratories.

**Performance Test**—A performance test (or performance verification) is the activity of verifying the performance of an item of measuring and test equipment to provide assurance that the instrument is capable of making correct measurements when it is properly used. A performance test is done with the item in its normal operating configuration. A performance test is the same as a calibration (1).
*See also:* calibration (1)

**Policy**—A policy defines and sets out the basic objectives, goals, vision, or general management position on a specific topic. A policy describes what management intends to have done regarding a given portion of business activity. Policy statements relevant to the quality management system are generally stated in the quality manual. Policies can also be in the organization's policy/procedure manual.
*See also:* procedure

**Precision**—Precision is a property of a measuring system or instrument. Precision is a measure of the repeatability of a measuring system—how much agreement there is within a group of repeated measurements of the same quantity under the same conditions. (NCSL, page 26) Precision is not the same as accuracy. (VIM, 3.5)

**Preventive action**—Preventive action is something done to prevent the possible future occurrence of a nonconformance, even though such an event has not yet happened. Preventive action helps improve the system.
*Contrast with*: corrective action

**Procedure**—A procedure describes a specific process for implementing all or a portion of a policy. There may be more than one procedure for a given policy. A procedure has more detail than a policy but less detail than a work instruction. The level of detail needed should correlate with the level of education and training of the people with the usual qualifications to do the work and the amount of judgment normally allowed to them by management. Some policies may be implemented by fairly detailed procedures, while others may only have a few general guidelines.
*Calibration*: see calibration procedure.
*See also*: policy

**Proficiency testing**—Determination of laboratory testing performance by means of interlaboratory comparisons.

**Quality manual**—The quality manual is the document that describes the quality management policy of an organization with respect to a specified conformance standard. The quality manual briefly defines the general policies as they apply to the specified conformance standard and affirms the commitment of the organization's top management to the policy. In addition to its regular use by the organization, auditors use the quality manual when they audit the quality management system. The quality manual is generally provided to customers on request. Therefore, it does not usually contain any detailed policies and never contains any procedures, work instructions, or proprietary information.

**Random error**—Random error is the result of a single measurement of a value, minus the mean of a large number of measurements of the same value. (VIM, 3.13) Random error causes scatter in the results of a sequence of readings and, therefore, is a measure of dispersion. Random error is usually evaluated by Type A methods, but Type B methods are also used in some situations.
*Note*: Contrary to popular belief, the GUM specifically does not replace random error with either Type A or Type B methods of evaluation. (3.2.2, note 2)
*See also*: error
*Compare with*: systematic error

**Repair**—Repair is the process of returning an unserviceable or nonconforming item to serviceable condition. The instrument is opened, or has covers removed, or is removed from its case and may be disassembled to some degree. Repair includes adjustment or alignment of the item as well as component-level repair. (Some minor adjustment such as zero and span may be included as part of the calibration.) The need for repair may be indicated by the results of a calibration. For calibratable items, repair is always followed by calibration of the item. Passing the calibration test indicates success of the repair.
*Contrast with*: calibration (1), repair (minor)

**Repair (minor)**—Minor repair is the process of quickly and economically returning an unserviceable item to serviceable condition by doing simple work using parts that are in stock in the calibration lab. Examples include replacement of batteries, fuses, or lamps; or minor cleaning of switch contacts; or repairing a broken wire; or replacing one or two in-stock components. The need for repair may be indicated by the results of a calibration. For calibratable items, minor repair is always followed by calibration of the item. Passing the calibration test indicates success of the repair.

*Minor repairs are defined as* repairs that take no longer than a short time as defined by laboratory management, and where no parts have to be ordered from external suppliers, and where substantial disassembly of the instrument is not required.

*Contrast with:* calibration (1), repair

**Reported value**—One or more numerical results of a calibration process, with the associated measurement uncertainty, as recorded on a calibration report or certificate. The specific type and format vary according to the type of measurement being made. In general, most reported values will be in one of these formats:

- *Measurement result and uncertainty.* The reported value is usually the mean of a number of repeat measurements. The uncertainty is usually expanded uncertainty as defined in the GUM.

- *Deviation from the nominal (or reference) value and uncertainty.* The reported value is the difference between the nominal value and the mean of a number of repeat measurements. The uncertainty of the deviation is usually expanded uncertainty as defined in the GUM.

- *Estimated systematic error and uncertainty.* The value may be reported this way when it is known that the instrument is part of a measuring system and the systematic error will be used to calculate a correction that will apply to the measurement system results.

**Round robin**—*See:* Interlaboratory Comparison

**Scope of accreditation**—For an accredited calibration or testing laboratory, the scope is a documented list of calibration or testing fields, parameters, specific measurements, or calibrations and their best measurement, uncertainty. The scope document is an attachment to the certificate of accreditation and the certificate is incomplete without it. Only the calibration or testing areas that the laboratory is accredited for are listed in the scope document, and only the listed areas may be offered as accredited calibrations or tests. The accreditation body usually defines the format and other details.

**Self-calibration**—Self-calibration is a process performed by a user for the purpose of making an IM&TE instrument or system ready for use. The process may be required at intervals such as every power-on sequence; or once per shift, day, or week of continuous operation; or if the ambient temperature changes by a specified amount. Once initiated, the process may be performed totally by the instrument or may require user intervention and/or use of external calibrated artifacts. The usual purpose is accuracy enhancement by characterization of errors inherent in the measurement system before the item to be measured is connected.

Self-calibration is *not* equivalent to periodic calibration (performance verification) because it is not performed using a calibration procedure and does not meet the metrological requirements for calibration. Also, if an instrument requires self-calibration before use, then that will also be accomplished at the start of a calibration procedure. Self-calibration may also be called normalization or standardization.
*Compare with:* calibration (2.B)
*Contrast with:* calibration (1)

**Specification**—In metrology, a specification is a documented statement of the expected performance capabilities of a large group of substantially identical measuring instruments, given in terms of the relevant parameters and including the accuracy or uncertainty. Customers use specifications to determine the suitability of a product for their own applications. A product that performs outside the specification limits when tested (calibrated) is rejected for later adjustment, repair, or scrapping.

**Standard (document)**—A standard (industry, national, government, or international standard; a *norme*) is a document that describes the processes and methods that must be performed in order to achieve a specific technical or management objective, or the methods for evaluation of any of these. An example is ANSI/NCSL Z540-1-1994, a national standard that describes the requirements for the quality management system of a calibration organization and the requirements for calibration and management of the measurement standards used by the organization.

**Standard (measurement)**—A standard (measurement standard, laboratory standard, calibration standard, reference standard; an *étalon*) is a system, instrument, artifact, device, or material that is used as a defined basis for making quantitative measurements. The value and uncertainty of the standard define a limit to the measurements that can be made: a laboratory can never have better precision or accuracy than its standards. Measurement standards are generally used in calibration laboratories. Items with similar uses in a production shop are generally regarded as working-level instruments by the calibration program.

*Primary standard.* Accepted as having the highest metrological qualities and whose value is accepted without reference to other standards of the same quantity. Examples: triple point of water cell and caesium beam frequency standard.

*Transfer standard.* A device used to transfer the value of a measurement quantity (including the associated uncertainty) from a higher level to a lower level standard.

*Secondary standard.* The highest accuracy level standards in a particular laboratory, generally used only to calibrate working standards. Also called a *reference standard*.

*Working standard.* A standard that is used for routine calibration of IM&TE.

The highest level standards, found in national and international metrology laboratories, are the realizations or representations of SI units.
*See also:* calibration standard

**Standard operating procedure (SOP)**—A term used by some organizations to identify policies, procedures, or work instructions.

**Standard reference material**—A standard reference material (SRM) as defined by NIST "is a material or artifact that has had one or more of its property values certified by a technically valid procedure, and is accompanied by, or traceable to, a certificate or other documentation which is issued by NIST. . . . Standard reference materials are . . . manufactured according to strict specifications and certified by NIST for one or more quantities of interest. SRMs represent one of the primary vehicles for disseminating measurement technology to industry."

**Standard uncertainty**—The uncertainty of the result of a measurement, expressed as a standard deviation. (GUM, 2.3.1)

**Standardization**—
See: self-calibration.

**Systematic error**—A systematic error is the mean of a large number of measurements of the same value minus the (probable) true value of the measured parameter. (VIM, 3.14) Systematic error causes the average of the readings to be offset from the true value. Systematic error is a measure of magnitude and may be corrected. Systematic error is also called *bias* when it applies to a measuring instrument. Systematic error may be evaluated by Type A or Type B methods, according to the type of data available.
Note: Contrary to popular belief, the GUM specifically does not replace systematic error with either Type A or Type B methods of evaluation. (3.2.3, note)
See also: bias, error, correction (of error)
Compare with: random error

**Test accuracy ratio**—(1) In a calibration procedure, the test accuracy ratio (TAR) is the ratio of the accuracy tolerance of the unit under calibration to the accuracy tolerance of the calibration standard used. (NCSL, page 2)

$$TAR = \frac{UUT\_tolerance}{STD\_tolerance}$$

The TAR must be calculated using identical parameters and units for the UUC and the calibration standard. If the accuracy tolerances are expressed as decibels, percentage, or another ratio, they must be converted to absolute values of the basic measurement units. (2) In the normal use of IM&TE items, the TAR is the ratio of the tolerance of the parameter being measured to the accuracy tolerance of the IM&TE.
Note: TAR may also be referred to as the accuracy ratio or (incorrectly) the uncertainty ratio.

**Test uncertainty ratio**—In a calibration procedure, the test uncertainty ratio (TUR) is the ratio of the accuracy tolerance of the unit under calibration to the uncertainty of the calibration standard used. (NCSL, page 2)

$$TUR = \frac{UUT\_tolerance}{STD\_uncert}$$

The TUR must be calculated using identical parameters and units for the UUC and the calibration standard. If the accuracy tolerances are expressed as decibels, percentage, or another ratio, they must be converted to absolute values of the basic measurement units.

Note: The uncertainty of a measurement standard is not necessarily the same as its accuracy specification.

**Tolerance**—A tolerance is a design feature that defines limits within which a quality characteristic is supposed to be on individual parts; it represents the maximum allowable deviation from a specified value. Tolerances are applied during design and manufacturing. A tolerance is a property of the item being measured.
*Compare with:* specification, uncertainty

**Traceable, traceability**—Traceability is a property of the result of a measurement, providing the ability to relate the measurement result to stated references, through an unbroken chain of comparisons each having stated uncertainties. (VIM, 6.10) Traceability is a demonstrated or implied property of the result of a measurement to be consistent with an accepted standard within specified limits of uncertainty. (NCSL, pages 42–43)

The stated references are normally the base or supplemental SI units as maintained by a national metrology institute; fundamental or physical natural constants that are reproducible and have defined values; ratio type comparisons; certified standard reference materials; or industry or other accepted consensus reference standards.

Traceability provides the ability to demonstrate the accuracy of a measurement result in terms of the stated reference.

Measurement assurance methods applied to a calibration system include demonstration of traceability. A calibration system operating under a program controls system only implies traceability.

Evidence of traceability includes the calibration report (with values and uncertainty) of calibration standards, but the report alone is not sufficient. The laboratory must also apply and use the data.

A calibration laboratory, a measurement system, a calibrated IM&TE, a calibration report, or any other thing is not and cannot be traceable to a national standard. Only the result of a specific measurement can be said to be traceable, provided all of the conditions just listed are met.

Reference to a NIST test number is specifically *not* evidence of traceability. That number is merely a catalog number of the specific service provided by NIST to a customer so it can be identified on a purchase order.

**Transfer measurement**—A transfer measurement is a type of method that enables making a measurement to a higher level of resolution than normally possible with the available equipment. Common transfer methods are differential measurements and ratio measurements.

**Transfer standard**—A transfer standard is a measurement standard used as an intermediate device when comparing two other standards. (VIM, 6.8) Typical applications of transfer standards are to transfer a measurement parameter from one organization to another, from a primary standard to a secondary standard, or from a secondary standard to a working standard in order to create or maintain measurement traceability. Examples of typical transfer standards are DC volt sources (standard cells or zener sources), and single-value standard resistors, capacitors, or inductors.

**Type A evaluation (of uncertainty)**—Type A evaluation of measurement uncertainty is the statistical analysis of actual measurement results to produce uncertainty values. Both random and systematic error may be evaluated by Type A methods. (GUM, 3.3.3 through 3.3.5) Uncertainty can only be evaluated by Type A methods if the laboratory actually collects the data.

**Type B evaluation (of uncertainty)**—Type B evaluation of measurement uncertainty includes any method except statistical analysis of actual measurement results. Both random and systematic error may be evaluated by Type B methods. (GUM, 3.3.3 through 3.3.5) Data for evaluation by Type B methods may come from any source believed to be valid.

**Uncertainty**—Uncertainty is a property of a measurement result that defines the range of probable values of the measurand. Total uncertainty may consist of components that are evaluated by the statistical probability distribution of experimental data or from assumed probability distributions based on other data. Uncertainty is an estimate of dispersion; effects that contribute to the dispersion may be random or systematic. (GUM, 2.2.3) Uncertainty is an estimate of the range of values that the true value of the measurement is within, with a specified level of confidence. After an item that has a specified tolerance has been calibrated using an instrument with a known accuracy, the result is a value with a calculated uncertainty.
*See also:* Type A evaluation, Type B evaluation

**Uncertainty budget**—The systematic description of known uncertainties relevant to specific measurements or types of measurements, categorized by type of measurement, range of measurement, and/or other applicable measurement criteria.

**UUC, UUT**—The unit under calibration or the unit under test—the instrument being calibrated. These are standard generic labels for the IM&TE item that is being calibrated, which are used in the text of the calibration procedure for convenience. Also may be called device under test (DUT) or equipment under test (EUT).

**Validation**—Substantiation by examination and provision of objective evidence that verified processes, methods, and/or procedures are fit for their intended use.

**Verification**—Confirmation by examination and provision of objective evidence that specified requirements have been fulfilled.

**VIM**—An acronym commonly used to identify the ISO *International Vocabulary of Basic and General Terms in Metrology*. (The acronym comes from the French title.)

**Work Instruction**—In a quality management system, a work instruction defines the detailed steps necessary to carry out a procedure. Work instructions are used only where they are needed to ensure the quality of the product or service. The level of education and training of the people with the usual qualifications to do the work must be considered when writing a work instruction. In a metrology laboratory, a calibration procedure is a type of work instruction.

# References

**National or international standards and related documents**
ANSI/ASQC M1-1996, *American National Standard for calibration systems*. Milwaukee: ASQC Quality Press.

ANSI/ISO/ASQ Q9000-2000, *Quality management systems—Requirements*. Milwaukee: ASQ Quality Press.

ANSI/NCSL Z540-2-1997, *U.S. Guide to the expression of uncertainty in measurement* (called the GUM). Boulder, CO: NCSL International. (Also called the GUM).

ANSI/NCSL Z540-1-1994, *American National Standard for Calibration: Calibration laboratories and measuring and test equipment—general requirements*. Boulder, CO: NCSL International.

BSR/ISO/ASQ Q10012:2003(E), *Measurement management systems—Requirements for measurement processes and measuring equipment*. Milwaukee: ASQ Quality Press.

European Cooperation for Accreditation. 1999. EA-4/02, *Expression of the uncertainty of measurement in calibration*. Paris: European Cooperation for Accreditation.

ISO. 1993. *International vocabulary of basic and general terms of metrology (VIM)*. Geneva: ISO.

NCSL. 1999. *NCSL glossary of metrology related terms*, 2nd ed. Boulder, CO: NCSL.

NCSL. 1995. *Handbook for the application of ANSI/NCSL Z540-1-1994*. Boulder, CO: NCSL.

NCSL. 1994. *NCSL glossary of metrology related terms*. Boulder, CO: NCSL.

NCSL. 1990. Recommended Practice RP-3, *Calibration procedures*. Boulder, CO: NCSL.

NIST. *NIST Policy on traceability*, www.nist.gov/traceability

U. S. Department of Defense. 2000. MIL-HDBK-1839A, *Calibration and measurement requirements*. Washington, DC: Government Printing Office.

U. S. Department of Defense. 2000. MIL-STD-1839C *Standard Practice for Calibration and Measurement Requirements*. Washington, DC: Government Printing Office.

U. S. Department of Defense. 1997. MIL-PRF-38793B *Performance Specification—Technical Manuals: Calibration procedures—Preparation*. Washington, DC: Government Printing Office.

U. S. Department of Defense. 1992. MIL-STD-1309D *Definitions of terms for testing, measurement and diagnostics*. Washington, DC: Government Printing Office.

**Other publications:**

AIAG (Automotive Industry Action Group). 2002. MSA-3: *Measurement systems analysis (MSA)*, 3rd ed. Southfield MI: AIAG.

American Association for Laboratory Accreditation (A2LA). 2003. *Mandatory guidance on editorial principles for calibration scopes of accreditation*. Frederick, MD: A2LA.

Delta Air Lines. 2003. *Electronic calibration laboratory glossary*. Atlanta: Delta Air Lines.

Fluke Corporation. 1994. *Calibration: philosophy in practice*. 2nd ed. Everett, WA: Fluke Corporation.

IAS (International Accreditation Service). 2002. IAS Calibration Laboratory, *Accreditation program: definitions*. Whittier, CA: IAS.

Kimothi, S.K. 2002. *The uncertainty of measurements, physical and chemical metrology impact analysis*. Milwaukee: ASQ Quality Press.

Kurtz, Max. 1991. *Handbook of applied mathematics for engineers and scientists*. New York: McGraw-Hill.

Lamprecht, James L. 1996. *ISO 9000 implementation for small business*. Milwaukee: ASQC Quality Press.

Page, Stephen B. 1998. *Establishing a system of policies and procedures*. Mansfield, OH: BookMasters Inc.

Pennella, C. Robert. 2004. *Managing the metrology system*. 3rd ed. Milwaukee: ASQ Quality Press.

Wadsworth Jr., Harrison M. 1990. *Handbook of statistical methods for engineers and scientists*. New York: McGraw-Hill.

# Appendix E
## Common Conversions

Note: Table is alphabetical within 'FROM/Name' column.

| TO CONVERT | FROM | | TO | | MULTIPLY BY |
|---|---|---|---|---|---|
| | SYMBOL | NAME | SYMBOL | NAME | |
| electric current | abA | abampere (EMU current, obsolete) | A | ampere | 10 |
| charge, electric, electrostatic, quantity of electricity | abC | abcoulomb (EMU charge, obsolete) | C | coulomb | 10 |
| electric capacitance | abF | abfarad (EMU capacitance, obsolete) | F | farad | 1000000000 |
| electric inductance | abH | abhenry (EMU inductance, obsolete) | H | henry | 1000000000 |
| electrical conductance | ℧ | abmho (EMU conductance, obsolete) | S | siemens | 1000000000 |
| electric resistance | abΩ | abohm (EMU resistance, obsolete) | Ω | ohm | 1000000000 |
| electric potential difference, electromotive force | abV | abvolt (EMU voltage, obsolete) | V | volt | 100000000 |
| area, plane | acre | acre (U.S. survey) | ha | hectare | 0.404687261 |
| area, plane | acre | acre (U.S. survey) | m² | square meter | 4046.87261 |
| volume, capacity | acre - ft | acre-foot (U.S. survey) | m³ | cubic meter | 1233.489238 |
| charge, electric, electrostatic, quantity of electricity | A - h | ampere hour | C | coulomb | 3600 |
| magnetomotive force | | ampere turn | A | ampere | 1 |
| magnetic field strength | | ampere-turn per inch | A / m | ampere per meter | 39.37007874 |
| magnetic field strength | | ampere-turn per meter | A / m | ampere per meter | 1 |
| length | A° | angstrom | m | meter | 1E-10 |

| TO CONVERT | FROM | | TO | | MULTIPLY BY |
|---|---|---|---|---|---|
| | SYMBOL | NAME | SYMBOL | NAME | |
| length | A° | angstrom | nm | nanometer | 0.1 |
| area, plane | a | are | $m^2$ | square meter | 100 |
| length | ua | astronomical unit | m | meter | 1.49598E+11 |
| pressure, stress | atm | atmosphere (std) | bar | bar | 1.01325 |
| pressure, stress | atm | atmosphere (std) | kPa | kilopascal | 101.325 |
| pressure, stress | atm | atmosphere (std) | Pa | pascal | 101325 |
| pressure, stress | at | atmosphere (technical) | kPa | kilopascal | 98.0665 |
| pressure, stress | at | atmosphere (technical) | Pa | pascal | 98066.5 |
| pressure, stress | bar | bar | kPa | kilopascal | 100 |
| pressure, stress | bar | bar | Pa | pascal | 100000 |
| area, plane | b | barn | $m^2$ | square meter | 1E-28 |
| volume, capacity | bbl | barrel (oil, 42 U.S. gallons) | $m^3$ | cubic meter | 0.158987295 |
| volume, capacity | bbl | barrel (oil, 42 U.S. gallons) | L | liter | 158.9872949 |
| activity | Bq | becquerel | $s^{-1}$ | per second (disintegration) | 1 |
| electric current | Bi | biot (see also abampere) | A | ampere | 10 |
| energy | $Btu_{IT}$ | Btu (International Table) | J | joule | 1055.055853 |
| energy | $Btu_{th}$ | Btu (thermochemical) | J | joule | 1054.35 |
| energy | Btu | Btu (mean) | J | joule | 1055.87 |
| energy | Btu | Btu (39 degree Fahrenheit) | J | joule | 1059.67 |
| energy | Btu | Btu (59 degree Fahrenheit) | J | joule | 1054.8 |
| energy | Btu | Btu (60 degree Fahrenheit) | J | joule | 1054.68 |
| conductivity, thermal | $Btu_{IT}$ - ft / (h - $ft^2$ - °F) | Btu (International Table) foot per hour square foot degree Fahrenheit | W / (m-K) | watt per meter kelvin | 1.73073466637139 |

## Common Conversions

| TO CONVERT | FROM | | TO | | MULTIPLY BY |
|---|---|---|---|---|---|
| | SYMBOL | NAME | SYMBOL | NAME | |
| conductivity, thermal | $Btu_{th}$ - ft / (h - ft$^2$ - °F) | Btu (thermochemical) foot per hour square foot degree Fahrenheit | W / (m-K) | watt per meter kelvin | 1.72957677165354 |
| conductivity, thermal | $Btu_{IT}$ - in / (h - ft$^2$ - °F) | Btu (International Table) inch per hour square foot degree Fahrenheit | W / (m-K) | watt per meter kelvin | 0.144227889 |
| conductivity, thermal | $Btu_{th}$ - in / (h - ft$^2$ - °F) | Btu (thermochemical) inch per hour square foot degree Fahrenheit | W / (m-K) | watt per meter kelvin | 0.144131398 |
| conductivity, thermal | $Btu_{IT}$ - in / (s - ft$^2$ - °F) | Btu (International Table) inch per second square foot degree Fahrenheit | W / (m-K) | watt per meter kelvin | 519.2203999 |
| conductivity, thermal | $Btu_{th}$ - in / (s - ft$^2$ - °F) | Btu (thermochemical) inch per second square foot degree Fahrenheit | W / (m-K) | watt per meter kelvin | 518.8730315 |
| conductivity, thermal | $Btu_{IT}$ / ft$^3$ | Btu (International Table) per cubic foot | J / m$^3$ | joule per cubic meter | 37258.94581 |
| conductivity, thermal | $Btu_{th}$ / ft$^3$ | Btu (thermochemical) per cubic foot | J / m$^3$ | joule per cubic meter | 37258.94581 |
| heat capacity, entropy | $Btu_{IT}$ / °F | Btu (International Table) per degree Fahrenheit | J / K | joule per kelvin | 1899.100535 |
| heat capacity, entropy | $Btu_{th}$ / °F | Btu (thermochemical) per degree Fahrenheit | J / K | joule per kelvin | 1897.83 |
| heat capacity, entropy | $Btu_{IT}$ / °R | Btu (International Table) per degree Rankine | J / K | joule per kelvin | 1899.100535 |
| heat capacity, entropy | $Btu_{th}$ / °R | Btu (thermochemical) per degree Rankine | J / K | joule per kelvin | 1897.83 |

| TO CONVERT | FROM | | TO | | MULTIPLY BY |
|---|---|---|---|---|---|
| | SYMBOL | NAME | SYMBOL | NAME | |
| heat flow rate | $Btu_{IT}/h$ | Btu (International Table) per hour | W | watt | 0.29307107 |
| heat flow rate | $Btu_{th}/h$ | Btu (thermochemical) per hour | W | watt | 0.292875 |
| irradiance, heat flux density, heat flow rate / area | $Btu/(h\cdot ft^2)$ | Btu per hour square foot | $W/m^2$ | watt per square meter | 3.154590745 |
| coefficient of heat transfer | $Btu_{IT}/(h\cdot ft^2\cdot {}^\circ F)$ | Btu (International Table) per hour square foot degree Fahrenheit | $W/(m^2\,K)$ | watt per square meter kelvin | 5.678263341 |
| coefficient of heat transfer | $Btu_{th}/(h\cdot ft^2\cdot {}^\circ F)$ | Btu (thermochemical) per hour square foot degree Fahrenheit | $W/(m^2\,K)$ | watt per square meter kelvin | 5.674464474 |
| coefficient of heat transfer | $Btu_{IT}/(s\cdot ft^2\cdot {}^\circ F)$ | Btu (International Table) per second square foot degree Fahrenheit | $W/(m^2\,K)$ | watt per square meter kelvin | 20441.74803 |
| coefficient of heat transfer | $Btu_{th}/(s\cdot ft^2\cdot {}^\circ F)$ | Btu (thermochemical) per second square foot degree Fahrenheit | $W/(m^2\,K)$ | watt per square meter kelvin | 20428.07211 |
| heat flow rate | $Btu_{th}/min$ | Btu per minute (thermochemical) | W | watt | 17.5725 |
| specific heat capacity, specific entropy | $Btu_{IT}/(lbm\cdot {}^\circ F)$ | Btu (International Table) per pound mass degree Fahrenheit | $J/(kg\text{-}K)$ | joule per kilogram kelvin | 4186.8 |
| specific heat capacity, specific entropy | $Btu_{th}/(lbm\cdot {}^\circ F)$ | Btu (thermochemical) per pound mass degree Fahrenheit | $J/(kg\text{-}K)$ | joule per kilogram kelvin | 4183.99895 |

| TO CONVERT | FROM | | TO | | MULTIPLY BY |
|---|---|---|---|---|---|
| | SYMBOL | NAME | SYMBOL | NAME | |
| specific heat capacity, specific entropy | $Btu_{IT}$ / (lbm - °R) | Btu (International Table) per pound mass degree Rankine | J / (kg-K) | joule per kilogram kelvin | 4186.8 |
| specific heat capacity, specific entropy | $Btu_{th}$ / (lbm - °R) | Btu (thermochemical) per pound mass degree Rankine | J / (kg-K) | joule per kilogram kelvin | 4183.99895 |
| energy per mass, specific energy | $Btu_{IT}$ / lb | Btu (International Table) per pound-mass | J / kg | joule per kilogram | 4183.99895 |
| energy per mass, specific energy | $Btu_{th}$ / lb | Btu (thermochemical) per pound-mass | J / kg | joule per kilogram | 2326 |
| energy, molar | Btu / lb - mol | Btu per pound-mole | J / kmol | joule per kilomole | 2326 |
| molar entropy, molar heat capacity | Btu / (lb - mol - °F) | Btu per pound-mole degree Fahrenheit | J / (kmol-K) | joule per kilomole kelvin | 4186.8 |
| heat flow rate | $Btu_{IT}$ / s | Btu (International Table) per second | W | watt | 1055.055853 |
| heat flow rate | $Btu_{th}$ / s | Btu (thermochemical) per second | W | watt | 1054.35 |
| energy per area | $Btu_{IT}$ / ft² | Btu (International Table) per square foot | J / m² | joule per square meter | 11356.52668 |
| energy per area | $Btu_{th}$ / ft² | Btu (thermochemical) per square foot | J / m² | joule per square meter | 11348.92895 |
| heat flow per area | $Btu_{IT}$ / (ft² - h) | Btu (International Table) per square foot hour | W / (m² K) | watt per square meter kelvin | 3.154590745 |
| heat flow per area | $Btu_{th}$ / (ft² - h) | Btu (thermochemical) per square foot hour | W / (m² K) | watt per square meter kelvin | 3.152480263 |

| TO CONVERT | FROM | | TO | | MULTIPLY BY |
|---|---|---|---|---|---|
| | SYMBOL | NAME | SYMBOL | NAME | |
| heat flow per area | $Btu_{th} / (ft^2 \cdot min)$ | Btu (thermochemical) per square foot minute | $W/(m^2 K)$ | watt per square meter kelvin | 189.1488158 |
| heat flow per area | $Btu_{IT} / (ft^2 \cdot s)$ | Btu (International Table) per square foot second | $W/(m^2 K)$ | watt per square meter kelvin | 11356.52668 |
| heat flow per area | $Btu_{th} / (ft^2 \cdot s)$ | Btu (thermochemical) per square foot second | $W/(m^2 K)$ | watt per square meter kelvin | 11348.92895 |
| heat flow per area | $Btu_{th} / (in^2 \cdot s)$ | Btu (thermochemical) per square inch second | $W/(m^2 K)$ | watt per square meter kelvin | 1634245.768 |
| mass | bu | bushel (barley) | kg | kilogram | 21.8 |
| mass | bu | bushel (corn, shelled) | kg | kilogram | 25.4 |
| mass | bu | bushel (oats) | kg | kilogram | 14.5 |
| mass | bu | bushel (potatoes) | kg | kilogram | 27.2 |
| mass | bu | bushel (soybeans) | kg | kilogram | 27.2 |
| volume, capacity | bu | bushel (U.S.) | $m^3$ | cubic meter | 0.03523907 |
| volume, capacity | bu | bushel (U.S.) | L | liter | 35.23907017 |
| mass | bu | bushel (wheat) | kg | kilogram | 27.2 |
| energy | $Cal_{IT}$, kcal | Calorie (nutrition, International Table) (kilocalorie) | J | joule | 4186.8 |
| energy | $Cal_{th}$, kcal | Calorie (nutrition, thermochemical) (kilocalorie) | J | joule | 4184 |
| energy | $cal_{mean}$ | calorie (mean) | J | joule | 4.19002 |
| energy | $cal_{15C}$ | calorie (15 °C) | J | joule | 4.1858 |
| energy | $cal_{20C}$ | calorie (20 °C) | J | joule | 4.1819 |
| energy | $cal_{IT}$ | calorie (International Table) | J | joule | 4.1868 |
| energy | $cal_{th}$ | calorie (thermochemical) | J | joule | 4.184 |

| TO CONVERT | FROM | | TO | | MULTIPLY BY |
|---|---|---|---|---|---|
| | SYMBOL | NAME | SYMBOL | NAME | |
| conductivity, thermal | $cal_{th}$ / (cm - s - °C) | calorie (thermochemical) per centimeter second degree Celsius | W / (m-K) | watt per meter kelvin | 418.4 |
| energy per mass | $cal_{IT}$ / g | calorie (International Table) per gram | J / kg | joule per kilogram | 4186.8 |
| energy per mass | $cal_{th}$ / g | calorie (thermochemical) per gram | J / kg | joule per kilogram | 4184 |
| specific heat capacity, specific entropy | $cal_{IT}$ / (g - °C) | calorie (International Table) per gram degree Celsius | J / (kg-K) | joule per kilogram kelvin | 4186.8 |
| specific heat capacity, specific entropy | $cal_{th}$ / (g - °C) | calorie (thermochemical) per gram degree Celsius | J / (kg-K) | joule per kilogram kelvin | 4184 |
| specific heat capacity, specific entropy | $cal_{IT}$ / (g - K) | calorie (International Table) per gram degree kelvin | J / (kg-K) | joule per kilogram kelvin | 4186.8 |
| specific heat capacity, specific entropy | $cal_{th}$ / (g - K) | calorie (thermochemical) per gram degree kelvin | J / (kg-K) | joule per kilogram kelvin | 4184 |
| energy, molar | $cal_{th}$ / mol | calorie (thermochemical) per mole | J / mol | joule per mole | 4.184 |
| molar entropy, molar heat capacity | $cal_{th}$ / (mol - °C) | calorie (thermochemical) per mole degree Celsius | J / (mol-K) | joule per mole kelvin | 4.184 |
| heat flow rate | $cal_{th}$ / min | calorie (thermochemical) per minute | W | watt | 0.069733333 |
| heat flow rate | $cal_{th}$ / s | calorie (thermochemical) per second | W | watt | 4.184 |
| heat energy per area | $cal_{th}$ / $cm^2$ | calorie (thermochemical) per square centimeter | J / $m^2$ | joule per square meter | 41840 |
| heat flow rate per area | $cal_{th}$ / $cm^2$ - min | calorie (thermochemical) per square centimeter minute | W / $m^2$ | watt per square meter | 697.3333333 |

| TO CONVERT | FROM | | TO | | MULTIPLY BY |
|---|---|---|---|---|---|
| | SYMBOL | NAME | SYMBOL | NAME | |
| heat flow rate per area | $cal_{th}$ / $cm^2$ - s | calorie (thermochemical) per square centimeter second | W / $m^2$ | watt per square meter | 41840 |
| luminance | cd / $in^2$ | candela per square inch | cd / $m^2$ | candela per square meter | 1550.0031 |
| luminous intensity | | candle | cd | candela | 1 |
| luminous Intensity | cp | candle power | cd | candela | 1 |
| mass | | carat (metric) | g | gram | 0.2 |
| mass | | carat (metric) | kg | kilogram | 0.0002 |
| mass | | carat (metric) | mg | milligram | 200 |
| temperature | °C | Celsius (interval) degree | K | kelvin | 1 |
| temperature | °C | Celsius (temperature) degree | K | kelvin | formula: $t_K$ = $t_{°C}$ + 273.15 |
| temperature | °C | centigrade (interval) degree | °C | Celsius (interval) degree | 1 |
| temperature | °C | centigrade (temperature) degree | °C | Celsius (temperature) degree | 1 |
| pressure, stress | cm Hg (°C) | centimeter mercury (0 °C) | Pa | pascal | 1333.221913 |
| pressure, stress | cm Hg (°C) | centimeter mercury (0 °C) | kPa | kilopascal | 1.333221913 |
| pressure, stress | cm Hg | centimeter mercury (conventional) | Pa | pascal | 1333.224 |
| pressure, stress | cm Hg | centimeter mercury (conventional) | kPa | kilopascal | 1.333224 |
| pressure, stress | cm $H_2O$ (4 °C) | centimeter water (4 °C) | Pa | pascal | 98.06375414 |
| pressure, stress | cm $H_2O$ | centimeter water (conventional) | Pa | pascal | 98.0665 |
| viscosity, dynamic | cP | centipoise | Pa-s | pascal second | 0.001 |
| viscosity, kinematic | cSt | centistokes | $m^2$ / s | per second square meter | 0.000001 |
| length | ch | chain (U.S. survey) | m | meter | 20.11684023 |
| area, plane | cmil | circular mil | $m^2$ | square meter | 5.06707E-10 |
| area, plane | cmil | circular mil | $mm^2$ | square millimeter | 0.000506707 |

| TO CONVERT | FROM | | TO | | MULTIPLY BY |
|---|---|---|---|---|---|
| | SYMBOL | NAME | SYMBOL | NAME | |
| resistance, thermal | clo | clo | K-m$^2$ / W | kelvin square meter per watt | 0.155 |
| volume, capacity | | cord | m$^3$ | cubic meter | 3.624556364 |
| volume, capacity | ft$^3$ | cubic foot | m$^3$ | cubic meter | 0.028316847 |
| volume / time, (flowrate) | cfm | cubic foot per minute | m$^3$ / s | cubic meter per second | 0.000471947 |
| volume / time, (flowrate) | cfm | cubic foot per minute | L / s | liter per second | 0.471947443 |
| volume / time, (flowrate) | ft$^3$ / s | cubic foot per second | m$^3$ / s | cubic meter per second | 0.028316847 |
| volume, capacity | in$^3$ | cubic inch | m$^3$ | cubic meter | 1.63871E-05 |
| volume / time, (flowrate) | in$^3$ / min | cubic inch per minute | m$^3$ / s | cubic meter per second | 2.73118E-07 |
| volume, capacity | mi$^3$ | cubic mile (international) | km$^3$ | cubic kilometer | 4.168181825 |
| volume, capacity | mi$^3$ | cubic mile (international) | m$^3$ | cubic meter | 4168181825 |
| volume, capacity | yd$^3$ | cubic yard | m$^3$ | cubic meter | 0.764554858 |
| volume / time, (flowrate) | yd$^3$ / min | cubic yard per minute | m$^3$ / s | cubic meter per second | 0.012742581 |
| volume, capacity | cup | cup (U.S. liquid) | m$^3$ | cubic meter | 0.000236588 |
| volume, capacity | cup | cup (U.S. liquid) | L | liter | 0.236588237 |
| volume, capacity | cup | cup (U.S. liquid) | mL | milliliter | 236.5882365 |
| activity | Ci | curie | Bq | becquerel | 37000000000 |
| mass | | dalton | kg | kilogram | 1.66054E-27 |
| permeability | darcy | darcy | m$^2$ | square meter | 9.86923E-13 |
| time | d | day (24h) | h | hour | 24 |
| time | d | day (24h) | s | second | 86400 |
| time | d | day (ISO) | s | second | 86400 |
| time | d | day (sidereal) | s | second | 86164.09057 |
| electric dipole moment | D | debye | C - m | coulomb meter | 3.33564E-30 |
| resistance, thermal | °F-h / Btu $_{IT}$ | degree Fahrenheit hour per Btu (International Table) | K / W | kelvin per watt | 1.895634241 |
| resistance, thermal | °F-h / Btu $_{th}$ | degree Fahrenheit hour per Btu (thermochemical) | K / W | kelvin per watt | 1.896903305 |
| resistance, thermal | °F-s / Btu $_{IT}$ | degree Fahrenheit second per Btu (International Table) | K / W | kelvin per watt | 0.000526565 |

| TO CONVERT | FROM | | TO | | MULTIPLY BY |
|---|---|---|---|---|---|
| | SYMBOL | NAME | SYMBOL | NAME | |
| resistance, thermal | °F-s / Btu $_{th}$ | degree Fahrenheit second per Btu (thermochemical) | K / W | kelvin per watt | 0.000526918 |
| insulance, thermal | °F - ft$^2$ - h / Btu $_{IT}$ | degree Fahrenheit square foot hour per Btu (International Table) | K m$^2$ / W | kelvin square meter per watt | 0.176110184 |
| insulance, thermal | °F - ft$^2$ - h / Btu $_{th}$ | degree Fahrenheit square foot hour per Btu (thermochemical) | K m$^2$ / W | kelvin square meter per watt | 0.176228084 |
| resistivity, thermal | °F - ft$^2$ - h / (Btu $_{IT}$ - in) | degree Fahrenheit square foot hour per Btu (International Table) inch | K-m / W | kelvin meter per watt | 6.933471799 |
| resistivity, thermal | °F - ft$^2$ - h / (Btu $_{th}$ - in) | degree Fahrenheit square foot hour per Btu (thermochemical) inch | K-m / W | kelvin meter per watt | 6.933471799 |
| angle, plane | ° | degree of angle | rad | radian | 0.017453293 |
| velocity, angular | deg / s | degree per second | rad / s | radian per second | 0.017453293 |
| acceleration, angular | deg / s$^2$ | degree per second squared | rad / s$^2$ | radian per second squared | 0.017453293 |
| mass / length | | denier | kg / m | kilogram per meter | 1.11111E-07 |
| mass / length | | denier | g / m | gram per meter | 0.000111111 |
| mass / length | | denier | mg / m | milligram per meter | 0.111111111 |
| force | dyn | dyne | N | newton | 0.00001 |
| moment of force, torque, bending moment | dyn-cm | dyne centimeter | N-m | newton meter | 0.0000001 |
| pressure, stress | dyn / cm$^2$ | dyne per square centimeter | Pa | pascal | 0.1 |
| electric capacitance | abF | EMU capacitance (abfarad) | F | farad | 1000000000 |
| electric current | abA | EMU current (abampere) | A | ampere | 10 |
| electric potential difference, electromotive force | abV | EMU electric potential (abvolt) | V | volt | 100000000 |
| electric inductance | abH | EMU inductance (abhenry) | H | henry | 1000000000 |
| electric resistance | abΩ | EMU resistance (abohm) | Ω | ohm | 1000000000 |

# Common Conversions

| TO CONVERT | FROM | | TO | | MULTIPLY BY |
|---|---|---|---|---|---|
| | SYMBOL | NAME | SYMBOL | NAME | |
| energy | erg | erg | J | joule | 0.0000001 |
| power | erg / s | erg per second | W | watt | 0.0000001 |
| power density, power / area | erg / cm$^2$ | erg per square centimeter | W / m$^2$ | watt per square meter | 0.001 |
| electric current | statA | ESU current (statampere) | A | ampere | 3.33564E-10 |
| electric capacitance | statF | ESU capacitance (statfarad) | F | farad | 1.11265E-12 |
| inductance, electrical | statH | ESU inductance (stathenry) | H | henry | 8.98755E+11 |
| electric resistance | stat$\Omega$ | ESU resistance (statohm) | $\Omega$ | ohm | 8.98755E+11 |
| electric potential difference, electromotive force | statV | ESU electric potential (statvolt) | V | volt | 299.792458 |
| temperature | °F | Fahrenheit (interval) degree | °C | Celsius (interval) degree | 0.555555556 |
| temperature | °F | Fahrenheit (interval) degree | K | kelvin | 0.555555556 |
| temperature | °F | Fahrenheit (temperature) degree | °C | Celsius (temperature) degree | formula: $t_{°C} = (t_{°F} - 32) / 1.8$ |
| temperature | °F | Fahrenheit (temperature) degree | K | kelvin | formula: $t_K = (t_{°F} + 459.67) / 1.8$ |
| charge, electric, electrostatic, quantity of electricity | faraday | faraday (based on ) carbon 12 | C | coulomb | 96485.3415 |
| length | fathom | fathom | m | meter | 1.8288 |
| length | fermi | fermi | fm | femtometer | 1 |
| length | fermi | fermi | m | meter | 1E-15 |
| length | ft | foot | ft | foot, U.S. survey | 1.000002 |
| length | ft | foot | m | meter | 0.3048 |
| velocity / speed | ft / h | foot per hour | m / s | meter per second | 8.46667E-05 |
| velocity / speed | ft / min | foot per minute | m / s | meter per second | 0.00508 |
| velocity / speed | ft / s | foot per second | km / h | kilometer per hour | 1.09728 |
| velocity / speed | ft / s | foot per second | m / s | meter per second | 0.3048 |
| acceleration, linear | ft / s$^2$ | foot per second squared | m / s$^2$ | meter per second squared | 0.3048 |
| energy | ft - pdl | foot poundal | J | joule | 0.04214011 |
| energy | ft - lbf | foot pound-force | J | joule | 1.355817948 |

| TO CONVERT | FROM | | TO | | MULTIPLY BY |
|---|---|---|---|---|---|
| | SYMBOL | NAME | SYMBOL | NAME | |
| moment of force, torque, bending moment | ft - lbf | foot pound-force (torque) | N-m | newton meter | 1.355817948 |
| energy density | ft - lbf / ft$^3$ | foot pound-force per cubic foot | J / m$^3$ | joule per cubic meter | 47.88025898 |
| power | ft - lbf / h | foot pound-force per hour | W | watt | 0.000376616 |
| power | ft - lbf / min | foot pound-force per minute | W | watt | 0.022596966 |
| power | ft - lbf / s | foot pound-force per second | W | watt | 1.355817948 |
| energy per area | ft - lbf / ft$^2$ | foot pound-force per square foot | J / m$^2$ | joule per square meter | 14.59390294 |
| power density, power / area | ft - lbf / ft$^2$ s | foot pound-force per square foot second | W / m$^2$ | watt per square meter | 14.59390294 |
| pressure, stress | ft H$_g$ | foot mercury (conventional) | kPa | kilopascal | 40636.66752 |
| pressure, stress | ft H$_g$ | foot mercury (conventional) | Pa | pascal | 40.63666752 |
| pressure, stress | ft H$_2$O | foot water (conventional) | kPa | kilopascal | 2.98906692 |
| pressure, stress | ft H$_2$O | foot water (conventional) | Pa | pascal | 2989.06692 |
| pressure, stress | ft H$_2$O (39.2 °F) | foot water (39.2 °F) | kPa | kilopascal | 2.988983226 |
| pressure, stress | ft H$_2$O (39.2 °F) | foot water (39.2 °F) | Pa | pascal | 2988.983226 |
| length | ft | foot, U.S. survey | m | meter | 0.30480061 |
| illuminance | fc | footcandle | lx | lux | 10.76391042 |
| luminance | fL | footlambert | cd / m$^2$ | candela per square meter | 3.4262591 |
| acceleration, linear | Gal | gal (galileo) | cm / s$^2$ | centimeter per second squared | 1 |
| acceleration, linear | Gal | gal (galileo) | m / s$^2$ | meter per second squared | 0.01 |
| volume, capacity | gal | gallon (Imperial) | m$^3$ | cubic meter | 0.00454609 |
| volume, capacity | gal | gallon (Imperial) | L | liter | 4.54609 |
| volume, capacity | gal | gallon (U.S. liquid) | m$^3$ | cubic meter | 0.003785412 |
| volume, capacity | gal | gallon (U.S. liquid) | L | liter | 3.785411784 |

## Common Conversions

| TO CONVERT | FROM | | TO | | MULTIPLY BY |
|---|---|---|---|---|---|
| | SYMBOL | NAME | SYMBOL | NAME | |
| volume / time, (flowrate) | gal / d | gallon (U.S. liquid) per day | $m^3/s$ | cubic meter per second | 4.38126E-08 |
| volume / time, (flowrate) | gal / d | gallon (U.S. liquid) per day | L / s | liter per second | 4.38126E-05 |
| volume / time, (flowrate) | gpm | gallon (U.S. liquid) per minute | $m^3/s$ | cubic meter per second | 6.30902E-05 |
| volume / time, (flowrate) | gpm | gallon (U.S. liquid) per minute | L / s | liter per second | 0.063090196 |
| volume / energy | gal / (hp-h) | gallon per horsepower hour | $m^3/J$ | cubic meter per joule | 1.41009E-09 |
| volume / energy | gal / (hp-h) | gallon per horsepower hour | L / J | liter per joule | 1.41009E-06 |
| magnetic flux density, induction | γ | gamma | nT | nanotesla | 1 |
| magnetic flux density, induction | g | gamma | T | tesla | 0.000000001 |
| magnetic flux density, induction | G | gauss | T | tesla | 0.0001 |
| magnetomotive force | Gi | gilbert | A | ampere | 0.795774715 |
| volume, capacity | gi | gill (Imperial) | $m^3$ | cubic meter | 0.000142065 |
| volume, capacity | gi | gill (Imperial) | L | liter | 0.142065313 |
| volume, capacity | gi | gill (U.S.) | $m^3$ | cubic meter | 0.000118294 |
| volume, capacity | gi | gill (U.S.) | L | liter | 0.118294118 |
| angle, plane | gon | gon | ° | degree of angle | 0.9 |
| angle, plane | gon | gon | rad | radian | 0.015707963 |
| angle, plane | grad | grad | ° | degree of angle | 0.9 |
| angle, plane | grad | grad | rad | radian | 0.015707963 |
| angle, plane | grade | grade | ° | degree of angle | 0.9 |
| angle, plane | grade | grade | rad | radian | 0.015707963 |
| mass | gr | grain | kg | kilogram | 6.47989E-05 |
| mass | gr | grain | mg | milligram | 64.79891 |
| density, mass / volume | gr / gal | grain per gallon (U.S. liquid) | $kg/m^3$ | kilogram per cubic meter | 0.017118061 |
| density, mass / volume | gr / gal | grain per gallon (U.S. liquid) | mg / L | milligram per liter | 17.11806105 |
| pressure, stress | $g_f/cm^2$ | gram force per square centimeter | Pa | pascal | 98.0665 |
| density, mass / volume | $g/cm^3$ | gram per cubic centimeter | $kg/m^3$ | kilogram per cubic meter | 1000 |
| acceleration, linear | $g_n$ | gravity, standard acceleration due to | $m/s^2$ | meter per second squared | 9.80665 |

| TO CONVERT | FROM | | TO | | MULTIPLY BY |
|---|---|---|---|---|---|
| | SYMBOL | NAME | SYMBOL | NAME | |
| area, plane | ha | hectare | $hm^2$ | square hectometer | 1 |
| area, plane | ha | hectare | $m^2$ | square meter | 10000 |
| power | hp | horsepower (550 foot pound-force per second) | W | watt | 745.6998716 |
| power | hp | horsepower (boiler, approx 33470 Btu per hour) | W | watt | 9809.5 |
| power | hp | horsepower (electric) | W | watt | 746 |
| power | hp | horsepower (metric) | W | watt | 735.4988 |
| power | hp | horsepower (U.K.) | W | watt | 745.7 |
| power | hp | horsepower (water) | W | watt | 746.043 |
| time | h | hour | s | second | 3600 |
| time | h | hour (sidereal) | s | second | 3590.17044 |
| mass | | hundredweight, long | kg | kilogram | 50.80234544 |
| mass | cwt | hundredweight, short (100 lb) | kg | kilogram | 45.359237 |
| mass | cwt | hundredweight, short (100 lb) | lb | pound-force (avoirdupois) | 100 |
| length | in | inch | cm | centimeter | 2.54 |
| length | in | inch | m | meter | 0.0254 |
| length | in | inch | mm | millimeter | 25.4 |
| pressure, stress | in Hg (0C) | inch mercury, 0 °C | Pa | pascal | 3386.383659 |
| pressure, stress | in Hg (0C) | inch mercury, 0 °C | kPa | kilopascal | 3.386383659 |
| pressure, stress | in Hg (60F) | inch mercury, 60 °F | Pa | pascal | 3376.846044 |
| pressure, stress | in Hg (60F) | inch mercury, 60 °F | kPa | kilopascal | 3.376846044 |
| pressure, stress | in Hg | inch mercury (conventional) | Pa | pascal | 3386.38896 |
| pressure, stress | in Hg | inch mercury (conventional) | kPa | kilopascal | 3.38638896 |
| speed, velocity | in / s | inch per second | m / s | meter per second | 0.0254 |
| acceleration, linear | in / $s^2$ | inch per second squared | m / $s^2$ | meter per second squared | 0.0254 |
| moment of section | $in^4$ | inch to the fourth power | $m^4$ | meter to the fourth power | 4.16231E-07 |
| moment of force, torque, bending moment | in - oz | inch ounce-force (torque) | N-m | newton meter | 0.007061552 |

## Common Conversions

| TO CONVERT | FROM SYMBOL | FROM NAME | TO SYMBOL | TO NAME | MULTIPLY BY |
|---|---|---|---|---|---|
| moment of force, torque, bending moment | in - lbf | inch pound -force (torque) | N-m | newton meter | 0.112984829 |
| pressure, stress | in $H_2O$ (39.2F) | inch water (39.2 °F) | Pa | pascal | 249.0819355 |
| pressure, stress | in $H_2O$ (60F) | inch water (60 °F) | Pa | pascal | 248.8432087 |
| pressure, stress | in $H_2O$ | inch water (conventional) | Pa | pascal | 249.08891 |
| length | in | inch, U.S. survey | m | meter | 0.025400051 |
| density, luminous flux | Jy | jansky | $W/(m^2 Hz)$ | watt per square meter hertz | 1E-26 |
| surface tension | $J/m^2$ | joule per square meter | $N/m$ | newton per meter | 1 |
| per length | K | kayser | $m^{-1}$ | per meter | 100 |
| temperature | K | kelvin | °C | Celcius (temperature) degree | formula: $t/°C = T/K - 273.15$ |
| energy | kcal $_{IT}$ | kilocalorie (International Table) | J | joule | 4186.8 |
| energy | kcal $_{th}$ | kilocalorie (thermochemical) | J | joule | 4184 |
| energy | kcal $_{mean}$ | kilocalorie (mean) | J | joule | 4190.02 |
| heat flow rate | kcal $_{th}$ / min | kilocalorie (thermochemical) per minute | W | watt | 69.73333333 |
| heat flow rate | kcal $_{th}$ / s | kilocalorie (thermochemical) per second | W | watt | 4184 |
| force | kgf | kilogram force | N | newton | 9.80665 |
| moment of force, torque, bending moment | kgf - m | kilogram-force meter | N-m | newton meter | 9.80665 |
| pressure, stress | kgf / $cm^2$ | kilogram-force per square centimeter | kPa | kilopascal | 98.0665 |
| pressure, stress | kgf / $cm^2$ | kilogram-force per square centimeter | Pa | pascal | 98066.5 |
| pressure, stress | kgf / $m^2$ | kilogram-force per square meter | Pa | pascal | 9.80665 |
| pressure, stress | kgf / $mm^2$ | kilogram-force per square millimeter | Pa | pascal | 9806650 |
| pressure, stress | kgf / $mm^2$ | kilogram-force per square millimeter | MPa | megapascal | 9.80665 |

| TO CONVERT | FROM | | TO | | MULTIPLY BY |
|---|---|---|---|---|---|
| | SYMBOL | NAME | SYMBOL | NAME | |
| mass | kgf s$^2$ / m | kilogram-force second squared per meter | kg | kilogram | 9.80665 |
| volume, capacity | kL | kiloliter (or, m$^3$) | L | liter | 1000 |
| velocity / speed | kph, km / h | kilometer per hour | m / s | meter per second | 0.277777778 |
| force | kp | kilopond (kilogram force) | N | newton | 9.80665 |
| energy | kW - h | kilowatt-hour | J | joule | 3600000 |
| energy | kW - h | kilowatt-hour | MJ | megajoule | 3.6 |
| force | kip | kip (1000 lbf) | N | newton | 4448.221615 |
| force | kip | kip (1000 lbf) | kN | kilonewton | 4.448221615 |
| pressure, stress | ksi | kip per square inch | kPa | kilopascal | 6894.757293 |
| pressure, stress | ksi | kip per square inch | Pa | pascal | 6894757.293 |
| pressure, stress | ksi | kip per square inch | MPa | megapascal | 6.894757293 |
| velocity / speed | kn | knot (nautical mile per hour) | m / s | meter per second | 0.514444444 |
| luminance | L | lambert | cd / m$^2$ | candela per square meter | 3183.098862 |
| energy per area | ly$_{th}$ | langley | J / m$^2$ | joule per square meter | 41840 |
| energy per area | ly $_{IT}$ | langley | J / m$^2$ | joule per square meter | 41868 |
| energy per area | ly $_{15}$ | langley | J / m$^2$ | joule per square meter | 41855 |
| energy per area | cal / cm$^2$ | langley | kJ / m$^2$ | kilojoule per square meter | 41.84 |
| length | l.y. | light year | m | meter | 9.46053E+15 |
| volume, capacity | L | liter | m$^3$ | cubic meter | 0.001 |
| illuminance | lm / ft$^2$ | lumen per square foot | lm / m$^2$ | lumen per square meter | 10.76391042 |
| magnetic flux | Mx | maxwell | Wb | weber | 0.00000001 |
| electrical conductance | mho | mho | S | siemens | 1 |
| length | μin | microinch | m | meter | 2.54E-08 |
| length | μin | microinch | μm | micrometer | 0.0254 |
| volume, capacity | μL | microliter (= 1 mm$^3$) | L | liter | 0.000001 |
| length | μ | micron | m | meter | 0.000001 |
| length | μ | micron | μm | micrometer | 1 |
| length | mil | mil (= 0.001 inch) | m | meter | 0.0000254 |
| length | mil | mil (= 0.001 inch) | mm | millimeter | 0.0254 |

| TO CONVERT | FROM | | TO | | MULTIPLY BY |
|---|---|---|---|---|---|
| | SYMBOL | NAME | SYMBOL | NAME | |
| angle, plane | mil | mil (angle) | ° | degree of angle | 0.05625 |
| angle, plane | mil | mil (angle) | rad | radian | 0.000981748 |
| length | mi | mile (international) | ft | foot | 5280 |
| length | mi | mile (international) | km | kilometer | 1.609344 |
| length | mi | mile (international) | m | meter | 1609.344 |
| velocity / speed | mph | mile (international) per hour | km / h | kilometer per hour | 1.609344 |
| length | nmi | mile (nautical) | m | meter | 1852 |
| length | mi | mile (U.S. survey) | km | kilometer | 1.609347219 |
| length | mi | mile (U.S. survey) | m | meter | 1609.347219 |
| velocity / speed | mph | mile (U.S. survey) per hour | m / s | meter per second | 0.447040894 |
| velocity / speed | mph | mile (U.S. survey) per hour | m / s | meter per second | 0.447040894 |
| length / volume | mpg | mile per gallon (U.S.) | km / L | kilometer per liter | 0.425143707 |
| length / volume | mpg | mile per gallon (U.S.) | m / m³ | meter per cubic meter | 425143.7074 |
| velocity / speed | mph | mile per hour | m / s | meter per second | 0.44704 |
| velocity / speed | mi / min | mile per minute | m / s | meter per second | 26.8224 |
| velocity / speed | mi / min | mile (U.S. survey) per minute | m / s | meter per second | 26.82245364 |
| velocity / speed | mi / s | mile per second | m / s | meter per second | 1609.344 |
| length | nmi | mile, nautical | m | meter | 1852 |
| fuel efficiency | mpg | miles per gallon (U.S.) | L / (100 km) | liter per hundred kilometer | formula: L / 100 km = 235.214583 / mpg |
| pressure, stress | mbar | millibar | hPa | hectopascal | 1 |
| pressure, stress | mbar | millibar | kPa | kilopascal | 1 |
| pressure, stress | mbar | millibar | Pa | pascal | 100 |
| volume, capacity | mL | milliliter | L | liter | 0.001 |
| pressure, stress | mm Hg (0 °C) | millimeter mercury (0°C) | Pa | pascal | 133.3221913 |
| pressure, stress | mm Hg | millimeter mercury) (conventional | Pa | pascal | 133.3224 |
| pressure, stress | mm H₂O | millimeter water (conventional) | Pa | pascal | 9.80665 |
| length | mμ | millimicron | m | meter | 0.000000001 |
| time | min | minute | s | second | 60 |
| angle, plane | ʹ | minute (arc) | ° | degree of angle | 0.016666667 |

| TO CONVERT | FROM | | TO | | MULTIPLY BY |
|---|---|---|---|---|---|
| | SYMBOL | NAME | SYMBOL | NAME | |
| angle, plane | ' | minute (arc) | rad | radian | 0.000290888 |
| angle, plane | min (arc) | minute (arc) | rad | radian | 0.000290888 |
| time | min | minute (sidereal) | s | second | 59.83617401 |
| magnetic field strength | Oe | oersted | A / m | ampere per meter | 79.57747155 |
| magnetomotive force | Oe - cm | oersted centimeter | A | ampere | 0.795774715 |
| resistance length | Ω - cm | ohm centimeter | Ω-m | ohm meter | 0.01 |
| resistance length | Ω - mil / ft | ohm circular mil per foot | Ω-m | ohm meter | 1.66243E-09 |
| resistance length | Ω - mil / ft | ohm circular mil per foot | Ω-mm² / m | ohm square millimeter per meter | 0.001662426 |
| mass | oz | ounce (avoirdupois) | g | gram | 28.34952313 |
| mass | oz | ounce (avoirdupois) | kg | kilogram | 0.028349523 |
| density, mass / volume | oz / in³ | ounce (avoirdupois) per cubic inch | kg / m³ | kilogram per cubic meter | 1729.994044 |
| density, mass / volume | oz / gal | ounce ) (avoirdupois) per gallon (U.S.) | kg / m³ | kilogram per cubic meter | 7.489151707 |
| mass | oz | ounce (troy or apothecary) | g | gram | 31.1034768 |
| mass | oz | ounce (troy or apothecary) | kg | kilogram | 0.031103477 |
| volume, capacity | fl oz | ounce (U.K. liquid) | m³ | cubic meter | 2.84131E-05 |
| volume, capacity | fl oz | ounce (U.K. liquid) | mL | milliliter | 28.4130625 |
| volume, capacity | oz | ounce (U.S. liquid) | m³ | cubic meter | 2.95735E-05 |
| volume, capacity | oz | ounce (U.S. liquid) | mL | milliliter | 29.57352956 |
| mass / area | oz / ft² | ounce (avoirdupois) per square foot | kg / m² | kilogram per square meter | 0.305151727 |
| mass / area | oz / in² | ounce (avoirdupois) per square inch | kg / m² | kilogram per square meter | 43.94184873 |
| mass / area | oz / yd² | ounce (avoirdupois) per square yard | kg / m² | kilogram per square meter | 0.033905747 |
| force | ozf | ounce-force | N | newton | 0.278013851 |
| moment of force, torque, bending moment | ozf in | ounce-force inch (torque) | N-m | newton meter | 0.007061552 |

| TO CONVERT | FROM | | TO | | MULTIPLY BY |
|---|---|---|---|---|---|
| | SYMBOL | NAME | SYMBOL | NAME | |
| moment of force, torque, bending moment | ozf in | ounce-force inch (torque) | mN-m | millinewton meter | 7.061551814 |
| mass per volume | oz / in$^3$ | ounce (avoirdupois) per cubic inch | kg / m$^3$ | kilogram per cubic meter | 1729.994044 |
| mass per volume | oz / gal (UK) | ounce (avoirdupois) per gallon (Imperial) | kg / m$^3$ | kilogram per cubic meter | 6.236023291 |
| mass per volume | oz / gal (UK) | ounce (avoirdupois) per gallon (Imperial) | g / L | gram per liter | 6.236023291 |
| mass per volume | oz / gal | ounce (avoirdupois) per gallon (U.S.) | kg / m$^3$ | kilogram per cubic meter | 7.489151707 |
| mass per volume | oz / gal | ounce (avoirdupois) per gallon (U.S.) | g / L | gram per liter | 7.489151707 |
| length | pc | parsec | m | meter | 3.08568E+16 |
| volume, capacity | pk | peck (U.S.) | m$^3$ | cubic meter | 0.008809768 |
| volume, capacity | pk | peck (U.S.) | L | liter | 8.809767542 |
| mass | dwt | pennyweight | gr | grain | 24 |
| mass | dwt | pennyweight | g | gram | 1.55517384 |
| mass | dwt | pennyweight | kg | kilogram | 0.001555174 |
| permeability | perm | perm (0 °C) | kg / (N-s) | kilogram per newton second | 5.72135E-11 |
| permeability | perm | perm (23 °C) | kg / (N-s) | kilogram per newton second | 5.74525E-11 |
| permeability | perm in | perm inch (0 °C) | kg / (Pa-s-m) | kilogram per pascal second meter | 1.45322E-12 |
| permeability | perm in | perm inch (23 °C) | kg / (Pa-s-m) | kilogram per pascal second meter | 1.45929E-12 |
| illuminance | ph | phot | lm / m$^2$ | lumen per square meter | 10000 |
| illuminance | ph | phot | lx | lux | 10000 |
| length | pi | pica (computer, ⅙ inch) | m | meter | 0.004233333 |
| length | pi | pica (computer, ⅙ inch) | mm | millimeter | 4.233333333 |
| length | pi | pica (printer's) | m | meter | 0.004217518 |
| length | pi | pica (printer's) | mm | millimeter | 4.2175176 |
| volume, capacity | pt | pint (Imperial) | m$^3$ | cubic meter | 0.000568261 |

| TO CONVERT | FROM | | TO | | MULTIPLY BY |
|---|---|---|---|---|---|
| | SYMBOL | NAME | SYMBOL | NAME | |
| volume, capacity | pt | pint (Imperial) | L | liter | 0.56826125 |
| volume, capacity | dry pt | pint (U.S. dry) | $m^3$ | cubic meter | 0.00055061 |
| volume, capacity | dry pt | pint (U.S. dry) | L | liter | 0.550610471 |
| volume, capacity | pt | pint (U.S. liquid) | $m^3$ | cubic meter | 0.000473176 |
| volume, capacity | pt | pint (U.S. liquid) | L | liter | 0.473176473 |
| length | p | point (computer, 1/72 inch) | m | meter | 0.000352778 |
| length | p | point (computer, 1/72 inch) | mm | millimeter | 0.352777778 |
| length | p | point (printer's) | m | meter | 0.00035146 |
| length | p | point (printer's) | mm | millimeter | 0.35146 |
| viscosity, dynamic | p | poise | dyn - s / $cm^2$ | dyne - second per centimeter squared | 1 |
| viscosity, dynamic | p | poise | Pa-s | pascal second | 0.1 |
| volume, capacity | pottle | pottle (U.S. liquid) | $m^3$ | cubic meter | 0.001892706 |
| volume, capacity | pottle | pottle (U.S. liquid) | L | liter | 1.892705892 |
| other | lbf ft / in | pound-force foot per inch | N-m / m | newton meter per meter | 53.37865938 |
| other | lbf in / in | pound-force inch per inch | N-m / m | newton meter per meter | 4.448221615 |
| mass | lb | pound (avoirdupois) | kg | kilogram | 0.45359237 |
| mass | lb | pound (troy or apothecary) | kg | kilogram | 0.373241722 |
| moment of inertia | lbm $ft^2$ | pound mass foot squared | kg - $m^2$ | kilogram meter squared | 0.04214011 |
| moment of inertia | lbm $in^2$ | pound mass inch squared | kg - $m^2$ | kilogram meter squared | 0.00029264 |
| density, mass / volume | lbm / $ft^3$ | pound mass per cubic foot | kg / $m^3$ | kilogram per cubic meter | 16.01846337 |
| density, mass / volume | lbm / $in^3$ | pound mass per cubic inch | kg / $m^3$ | kilogram per cubic meter | 27679.90471 |
| density, mass / volume | lbm / $yd^3$ | pound mass per cubic yard | kg / $m^3$ | kilogram per cubic meter | 0.593276421 |
| mass / length | lbn / ft | pound mass per foot | kg / m | kilogram per meter | 1.488163944 |
| viscosity, dynamic | lb / (ft - h) | pound per foot hour | Pa-s | pascal second | 0.000413379 |
| viscosity, dynamic | lb / (ft - s) | pound per foot second | Pa-s | pascal second | 1.488163944 |

# Common Conversions

| TO CONVERT | FROM | | TO | | MULTIPLY BY |
|---|---|---|---|---|---|
| | SYMBOL | NAME | SYMBOL | NAME | |
| density, mass / volume | lbm / gal | pound mass per gallon | kg / m3 | kilogram per cubic meter | 119.8264273 |
| density, mass / volume | lbm / gal | pound mass per gallon | kg / L | kilogram per liter | 0.119826427 |
| density, mass / volume | lbm / gal (UK) | pound mass per gallon (Imperial) | kg / m$^3$ | kilogram per cubic meter | 99.77637266 |
| density, mass / volume | lbm / gal (UK) | pound mass per gallon (Imperial) | kg / L | kilogram per liter | 0.099776373 |
| mass / energy | lb / (hp - h) | pound per horsepower hour | kg / J | kilogram per joule | 1.68966E-07 |
| mass / time | lb / h | pound per hour | kg / s | kilogram per second | 0.000125998 |
| mass / length | lb / in | pound per inch | kg / m | kilogram per meter | 17.85796732 |
| mass / time | lb / min | pound per minute | kg / s | kilogram per second | 0.007559873 |
| mass mer mole | lb / (lb - mol) | pound per pound mole | kg / mol | kilogram per mole | 0.001 |
| mass / time | lb / s | pound per second | kg / s | kilogram per second | 0.45359237 |
| mass / area | lb / ft$^2$ | pound per square foot | kg / m$^2$ | kilogram per square meter | 4.882427636 |
| mass / area | lb / in$^2$ | pound per square inch | kg / m$^2$ | kilogram per square meter | 703.0695796 |
| mass / length | lb / yard | pound per yard | kg / m | kilogram per meter | 0.496054648 |
| pressure, stress | pdl / ft$^2$ | poundal per square foot | Pa | pascal | 1.488163944 |
| viscosity, dynamic | pdl s / ft$^2$ | poundal second per square foot | Pa s | pascal second | 1.488163944 |
| force | lbf | pound-force | N | newton | 0.138254954 |
| force | lbf | pound-force | N | newton | 4.448221615 |
| moment of force, torque, bending moment | lbf - ft | pound-force foot (torque) | N-m | newton meter | 1.355817948 |
| moment of force, torque, bending moment | lbf - in | pound-force inch (torque) | N-m | newton meter | 0.112984829 |
| force / length | lbf / ft | pound-force per foot | N / m | newton per meter | 14.59390294 |
| force / length | lbf / in | pound-force per inch | N / m | newton per meter | 175.1268352 |
| thrust / mass | lbf / lb | pound-force per pound | N / kg | newton per kilogram | 9.80665 |

| TO CONVERT | FROM | | TO | | MULTIPLY BY |
|---|---|---|---|---|---|
| | SYMBOL | NAME | SYMBOL | NAME | |
| pressure, stress | psf | pound-force per square foot | Pa | pascal | 47.88025898 |
| pressure, stress | psi | pound-force per square inch | kPa | kilopascal | 6.894757293 |
| pressure, stress | psi | pound-force per square inch | Pa | pascal | 6894.757293 |
| viscosity, dynamic | $\mu'_e$, lbf - s / ft² | pound-force second per foot squared | Pa-s | pascal second | 47.88025898 |
| viscosity, dynamic | lbf - s / in² | pound-force second per square inch | Pa-s | pascal second | 6894.757293 |
| mass | lb-mol | pound-mole | mol | mole | 0.45359237 |
| energy | quad | quad | J | joule | 1.05506E+18 |
| volume, capacity | dry qt | quart (U.S. dry) | m³ | cubic meter | 0.001101221 |
| volume, capacity | dry qt | quart (U.S. dry) | L | liter | 1.101220943 |
| volume, capacity | qt | quart (U.S. liquid) | m³ | cubic meter | 0.000946353 |
| volume, capacity | qt | quart (U.S. liquid) | L | liter | 0.946352946 |
| absorbed dose | rad | rad | Gy | gray | 0.01 |
| absorbed dose | rad | rad | J / kg | joule per kilogram | 0.01 |
| absorbed dose rate | rad / s | rad per second | Gy / s | gray per second | 0.01 |
| temperature | °R | Rankine (interval) degree | K | kelvin | 0.555555556 |
| temperature | °R | Rankine (temperature) degree | K | kelvin | 0.555555556 |
| dose equivalent | rem | rem | Sv | sievert | 0.01 |
| acceleration, angular | r | revolution | rad / s² | radian per second squared | 6.283185307 |
| angle, plane | r | revolution | rad | radian | 6.283185307 |
| velocity, angular | rpm | revolution per minute | rad / s | radian per second | 0.104719755 |
| velocity, angular | r / s | revolution per second | rad / s | radian per second | 6.283185307 |
| per viscosity, dynamic | rhe | rhe | 1 / (Pa-s) | per pascal second | 10 |
| length | rd | rod | ft | foot, U.S. survey | 16.5 |
| length | rd | rod | m | meter | 5.029210058 |
| exposure (x and gamma rays) | R | roentgen | C / kg | coulomb per kilogram | 0.000258 |
| angle, plane | " | second | ° | degree of angle | 0.000277778 |
| angle, plane | " | second | rad | radian | 4.84814E-06 |

## Common Conversions

| TO CONVERT | FROM | | TO | | MULTIPLY BY |
|---|---|---|---|---|---|
| | SYMBOL | NAME | SYMBOL | NAME | |
| angle, plane | sec | second | rad | radian | 4.84814E-06 |
| time | s | second (sidereal) | s | second | 0.997269567 |
| time | | shake | ns | nanosecond | 10 |
| time | | shake | s | second | 0.00000001 |
| time | | shake | s | second | 0.00000001 |
| mass | slug | slug | kg | kilogram | 14.59390294 |
| density, mass / volume | slug / ft$^3$ | slug per cubic foot | kg / m$^3$ | kilogram per cubic meter | 515.3788184 |
| mass / length | slug / ft | slug per foot | kg / m | kilogram per meter | 47.88025898 |
| viscosity, dynamic | slug / (ft - s) | slug per foot second | Pa-s | pascal second | 47.88025898 |
| mass / area | slug / ft$^2$ | slug per square foot | kg / m$^2$ | kilogram per square meter | 157.0874638 |
| velocity / speed | c | speed of light in a vacuum | m / s | meter per second | 299792458 |
| area, plane | ft$^2$ | square foot | cm$^2$ | square centimeter | 929.0304 |
| diffusivity, thermal | ft$^2$ / h | square foot per hour | m$^2$ / s | square meter per second | 2.58064E-05 |
| diffusivity, thermal | ft$^2$ / s | square foot per second | m$^2$ / s | square meter per second | 0.09290304 |
| area, plane | in$^2$ | square inch | m$^2$ | square meter | 0.00064516 |
| area, plane | in$^2$ | square inch | cm$^2$ | square centimeter | 6.4516 |
| area, plane | mi$^2$ | square mile (International) | km$^2$ | square kilometer | 2.58998811 |
| area, plane | mi$^2$ | square mile (International) | m$^2$ | square meter | 2589988.11 |
| area, plane | mi$^2$ | square mile (U.S. survey) | km$^2$ | square kilometer | 2.58999847 |
| area, plane | mi$^2$ | square mile (U.S. survey) | m$^2$ | square meter | 2589998.47 |
| area, plane | yd$^2$ | square yard | m$^2$ | square meter | 0.83612736 |
| electric current | statA | statampere (ESU current, obsolete) | A | ampere | 3.33564E-10 |
| charge, electric, electrostatic, quantity of electricity | statC | statcoulomb (ESU charge, obsolete) | C | coulomb | 3.33564E-10 |
| electric capacitance | statF | statfarad (ESU capacitance, obsolete) | F | farad | 1.11265E-12 |
| inductance, electrical | statH | stathenry (ESU inductance, obsolete) | H | henry | 8.98755E+11 |

| TO CONVERT | FROM | | TO | | MULTIPLY BY |
|---|---|---|---|---|---|
| | SYMBOL | NAME | SYMBOL | NAME | |
| electrical conductance | ℧ | statmho (ESU conductance, obsolete) | S | siemens | 1.11265E-12 |
| electric resistance | statΩ | statohm (ESU resistance, obsolete) | Ω | ohm | 8.98755E+11 |
| electric potential difference, electromotive force | statV | statvolt (ESU voltage, obsolete) | V | volt | 299.792458 |
| volume, capacity | st | stere | $m^3$ | cubic meter | 1 |
| luminance | sb | stilb | $cd/cm^2$ | candela per square centimeter | 1 |
| luminance | sb | stilb | $cd/m^2$ | candela per square meter | 10000 |
| viscosity, kinematic | St | stokes | $cm^2/s$ | centimeter squared per second | 1 |
| viscosity, kinematic | St | stokes | $m^2/s$ | meter squared per second | 0.0001 |
| volume, capacity | tbs | tablespoon (U.S. dry or liquid) | $m^3$ | cubic meter | 1.478676E-05 |
| volume, capacity | tbs | tablespoon (U.S. dry or liquid) | $mL^3$ | milliliter | 14.78676478 |
| volume, capacity | tsp | teaspoon (U.S., dry or liquid) | $m^3$ | cubic meter | 4.928922E-06 |
| volume, capacity | tsp | teaspoon (U.S., dry or liquid) | $mL^3$ | milliliter | 4.928921594 |
| mass / length | tex | tex | kg / m | kilogram per meter | 0.000001 |
| energy | therm | therm (EEC) | J | joule | 105506000 |
| energy | therm | therm (U.S.) | J | joule | 105480400 |
| mass | AT | ton (assay) | g | gram | 29.16667 |
| mass | AT | ton (assay) | kg | kilogram | 0.02916667 |
| energy | $10^6$ kcal | ton (from energy equivalent of one ton TNT) | J | joule | 4184000000 |
| density, mass / volume | ton / $yd^3$ | ton (long) per cubic yard | $kg/m^3$ | kilogram per cubic meter | 1328.939184 |
| mass | t | ton (metric) | kg | kilogram | 1000 |
| mass | t | ton (metric) | Mg | megagram | 1 |
| mass | t | ton (metric), tonne | kg | kilogram | 1000 |
| volume, capacity | ton | ton (register) | $m^3$ | cubic meter | 2.831684659 |
| mass | ton | ton (short) | kg | kilogram | 907.18474 |

## Common Conversions

| TO CONVERT | FROM | | TO | | MULTIPLY BY |
|---|---|---|---|---|---|
| | SYMBOL | NAME | SYMBOL | NAME | |
| force | ton | ton-force (short) | N | newton | 8896.443231 |
| force | ton | ton-force (short) | kN | kilonewton | 8.896443231 |
| density, mass / volume | ton / yd$^3$ | ton (short) per cubic yard | kg / m$^3$ | kilogram per cubic meter | 1186.552843 |
| mass / time | ton / h | ton (short) per hour | kg / s | kilogram per second | 0.251995761 |
| energy | 10$^6$ kcal | ton of oil equivalent | J | joule | 41840000000 |
| heat flow rate | ton | ton of refrigeration (U.S.) | W | watt | 3516.852842 |
| mass | ton | ton, long (2240 pound) | kg | kilogram | 1016.046909 |
| mass | ton | ton, long (2240 pound) | lb | pound (avoirdupois) | 2240 |
| mass | t | tonne (metric) | kg | kilogram | 1000 |
| pressure, stress | Torr | torr (mm Hg 0 °C) | Pa | pascal | 133.3223684 |
| magnetic flux | | unit pole | Wb | weber | 1.25664E-07 |
| energy, electrical | W - h | watt hour | J | joule | 3600 |
| power density, power / area | W / cm$^2$ | watt per square centimeter | W / m$^2$ | watt per square meter | 10000 |
| power density, power / area | W / in$^2$ | watt per square inch | W / m$^2$ | watt per square meter | 1550.0031 |
| radiance | W / (m$^2$ - sr) | watt per square meter steradian | | | |
| radiant intensity | W / sr | watt per steradian | | | |
| energy | W s | watt second | J | joule | 1 |
| magnetic flux density, induction | Wb / m$^2$ | weber per square meter | T | tesla | 1 |
| length | yd | yard | ft | foot | 3 |
| length | yd | yard | m | meter | 0.9144 |
| time | yr | year (365-day) | s | second | 31536000 |
| time | yr | year (sidereal) | s | second | 31558150 |
| time | yr | year (tropical) | s | second | 31556930 |

# Bibliography

Adam, Stephen F. 1969. *Microwave theory and applications*. Englewood Cliffs, NJ: Prentice-Hall.
Agilent Technologies. 2003. *Agilent 34401A product overview*, publication number 5968-0162EN. Palo Alto, CA: Agilent Technologies.
———. 2003. *Agilent Technologies fact book*. Palo Alto, CA: Agilent Technologies Corporate Media Relations. www.agilent.com/about/newsroom/facts/agilentfactbook.pdf
———. 2003a. *Application Note 1449-1: Fundamentals of RF and microwave power measurements (Part 1): Introduction to power, history, definitions, international standards and traceability.* Publication number 5988-9213EN. Palo Alto, CA: Agilent Technologies.
———. 2003b. *Application Note 1449-2: Fundamentals of RF and microwave power measurements (Part 2)—Power sensors and instrumentation*. Publication number 5988-9214EN. Palo Alto, CA: Agilent Technologies.
———.2003c. *Application Note 1449-3: Fundamentals of RF and microwave power measurements (Part 3): Power measurement uncertainty per international guides*. Publication number 5988-9215EN. Palo Alto, CA: Agilent Technologies.
———. 2002a. "Power Measurement Basics." *Agilent 2002 back to basics seminar*. Publication number 5988-6641EN. Palo Alto, CA: Agilent Technologies.
———. 2002b. *Application Note 1382-7: VNA-based system tests the physical layer*. Publication number 5988-5075EN. Palo Alto, CA: Agilent Technologies.
———. 2000a. *Application Note 1287-1: Understanding the fundamental principles of vector network analysis*. Publication number 5965-7707E. Palo Alto, CA: Agilent Technologies.
———. 2000b. Application Note 1287-2: *Exploring the architectures of network analyzers.* Publication number 5965-7708E. Palo Alto, CA: Agilent Technologies.
———. *Application Note AN 154: S-Parameter design*. Publication number 5952-1087. Palo Alto, CA: Agilent Technologies.
———. 1996. 34401A Operating manual. Palo Alto, CA: Agilent Technologies.
———. Smith chart. www.educatorscorner.com/index.cgi?CONTENT_ID=2482
AIAG (Automotive Industry Action Group). 2003. *AIAG history highlights*. Southfield, MI: AIAG. www.aiag.org/about/history.pdf
ANSI/ASQC Z1.4-2003, *Sampling procedures and tables for inspection by attributes.* Milwaukee: ASQ Quality Press.
ANSI/ASQC Z1.9-2003, *Sampling procedures and tables for inspection by variables for percent nonconforming*. Milwaukee: ASQ Quality Press.
AIAG (Automotive Industry Action Group). 2002. MSA-3: *Measurement systems analysis (MSA)*, 3rd ed. Southfield MI: AIAG.
American Association for Laboratory Accreditation (A2LA). 2003. *Mandatory guidance on editorial principles for calibration scopes of accreditation*. Frederick, MD: A2LA.

ANSI/ISO 17025-1999, *American National Standard—General requirements for the competence of testing and calibration laboratories*. Milwaukee: ASQ Quality Press.

ANSI/ISO/ASQ Q9001-2000, *American National Standard—Quality management systems—Requirements*. Milwaukee: ASQ Quality Press.

ANSI/ISO/ASQ Q9000-2000, *Quality management systems—Requirements*. Milwaukee: ASQ Quality Press.

ANSI/NCSL. 1997. ANSI/NCSL Z540-2-1997, U. S. Guide to the expression of uncertainty in measurement. Boulder, CO: NCSL International. (Also called the GUM).

ANSI/ASQC M1-1996, *American National Standard for calibration systems*. Milwaukee: ASQC Quality Press.

American National Standards Institute (ANSI). New York. www.ansi.org

ANSI/NCSL Z540-1-1994, *American National Standard for calibration—Calibration laboratories and measuring and test equipment—general requirements*. Boulder, CO: National Conference of Standards Laboratories.

ANSI Z17.1-1973. *American National Standard for preferred numbers*. New York: ANSI.

American Society for Testing Materials (ASTM). 1999. ASTM E 691 – 1999, *Standard practice for conducting an interlaboratory study to determine the precision of a test method*. West Conshohocken, PA: ASTM.

———. 1990. ASTM E 29-90. *Standard practice for using significant digits in test data to determine conformance with specifications*. West Conshohocken, PA. ASTM.

———. 1995. ASTM E 1301 – 1995, *Standard guide for proficiency testing by interlaboratory comparisons*. West Conshohocken, PA: ASTM.

———. 1973. NSI/IEEE 474-1973. *Specifications and test methods for fixed and variable attenuators, DC to 40 GHz*. New York: Institute of Electrical and Electronic Engineers. (Available online at ts.nist.gov/ts/htdocs/230/233/calibrations/Electromagnetic/pubs/IEEE-474.pdf)

*The American Heritage Dictionary*. 1985. Boston, MA: Houghton-Mifflin Co.

Anthony, D.M. *Engineering metrology*. Bristol, UK: Pergamon Press.

Arden, Paul. 2003. *It's not how good you are, its how good you want to be: The world's best selling book*. New York: Phaidon Press.

ASQC Statistics Division. 1996. *Glossary and tables for statistical quality control*, 3rd ed. Milwaukee: ASQC Quality Press.

ASQ. 2003. Greg Watson examines software quality. Online video segment: www.asq.org/news/multimedia/board/watson4.html

———. 2003. Origins of the American Society for Quality. www.asq/org/join/about/history

Barker, Joel Arthur. 1993. *Paradigms: The business of discovering the future*. New York: HarperCollins.

Bentley, Robert E. 2000. *Uncertainty in measurement: the ISO guide with examples*. Sydney: National Measurement Laboratory CSIRO.

Bertermann, Ralph E. 2002. *Understanding current regulations and international standards; Calibration compliance in FDA regulated companies*. Mt. Prospect, IL: Lighthouse Training Group.

BIPM (International Bureau of Weights and Measures). 2000. The Convention of the Metre. www.bipm.fr/enus/1_Convention/foreword.html

———. 1998. *The International system of units (SI)*. 7th ed. Sevres, France: International Bureau of Weights and Measures. www1.bipm.org/en/publications/brochure

Bird, Malcolm. 2002. "A few small miracles give birth to an ISO quality management systems standard for the automotive industry." *ISO Bulletin* (August). www.iso.ch/iso/en/commcentre/isobulletin/articles/2002/pdf/automotiveind02-08.pdf

Blair, Thomas. 2003. Letter to author (Payne), 28 September 2003.

Box, George E.P., William G. Hunter, and J. Stuart Hunter. 1978. *Statistics for experimenters, an introduction to the design, data analysis and model building*. New York: Wiley & Sons.

Brassard, Michael and Dianne Ritter. 1994. *The memory jogger II*. Salem, NH: GOAL/QPC.

Brownlee, K.A. 1960. *Statistical theory and methodology in science and engineering*. New York: John Wiley & Sons.

BSR/ISO/ASQ Q10012:2003(E), *Measurement management systems—Requirements for measurement processes and measuring equipment*. Milwaukee. ASQ Quality Press.

Bucher, Jay L. 2000. *When your company needs a metrology program, but can't afford to build a calibration laboratory . . . What can you do?* Boulder, CO: National Conference of Standards Laboratories.

Busch, Ted, Roger Harlow, and Richard L. Thompson. 1998. *Fundamentals of dimensional metrology*, 3rd ed. Albany, NY: Delmar Publishers.

Calhoun, Richard, 1994. *Calibration & standards: DC to 40 GHz*. Louisville, KY: SS&S Inc.

Carr, Joseph J. 1996. *Elements of electronic instrumentation and measurement*. 3rd ed. Upper Saddle River, NJ: Prentice-Hall.

Central Limit theorum, www.animatedsoftware.com/statglos/sgcltheo.htm

Clarke, Kenneth K. and Donald T. Hess. 1990. "Phase measurement, traceability, and verification theory and practice." *IEEE Transactions on Instrumentation and Measurement*, 39 (February).

Columbia Encyclopedia. 2003. "Measurement" expanded Columbia electronic encyclopedia, Columbia University Press. www.historychannel.com/perl/print_book.pl?ID-100597 (Accessed June 2003).

Creech, Bill. 1994. *The five pillars of TQM: How to make total quality management work for you*. New York: McGraw-Hill.

Crosby, Phillip B. 1979. *Quality is free: The art of making quality certain*. New York: McGraw-Hill.

Daimler-Chrysler Corporation, Ford Motor Company, General Motors Corporation. *Measurement System Analysis (MSA)*, 3rd ed. Southfield, MI: Automotive Industry Action Group (AIAG).

Dandekar, A. V. and E. E. Hutchison. 2002. "The QuEST Forum—The future of telecom quality development." CD-ROM. Milwaukee: ASQ Quality Press.

www.dartmouth.edu/~ogehome/CQI/PDCA.html

Davidson, C. W. 1978. *Transmission lines for communications*. New York: Halsted Press.

Delaney, Helen. 2000. *Impact of conformity assessment on trade: American and European perspectives*. Proceedings of the SES 2000 conference, Baltimore. www.ses-standards.org/library/00proceedings/delaney.pdf

Delta Air Lines. 2003. *Electronic calibration laboratory glossary*. Atlanta: Delta Air Lines.

Deming, W. Edwards. 1993. *The new economics for industry, government, education*. Cambridge: MIT Press.

———. 1982. *Out of the crisis*. Cambridge: MIT Press.

deming.eng.clemson.edu/pub/tutorials/qctools/qct.htm

Doebelin, Ernest O. 1989. *Measurement systems, application and design*. 4th ed. New York: McGraw-Hill.

de Silva, G.M.S. *Basic metrology for ISO 9000 certification*. Oxford, UK: Reed Educational and Professional Publishing Ltd.

DeWitt, Peter. 2003. Telephone conversation with Graeme C. Payne, 12 September.

Driver, L. D., F. X. Ries, and G. Rebuelda. 1978. NBSIR 78-871, *NBS RF voltage comparator*. Boulder, CO: National Bureau of Standards. ts.nist.gov/ts/htdocs/230/233/calibrations/Electromagnetic/pubs/78-871.pdf

Electronic Industries Alliance (EIA). www.eia.org

Engen, Glenn. 1978. "Calibrating the Six-Port Reflectometer by Means of Sliding Terminations." *IEEE Transactions on Microwave Theory and Techniques* MTT-26 (December): 951–57. www.geocities.com/frank_wiedmann/publications.html

———. 1980. "A least-squares technique for use in the six-port measurement technique." *IEEE Transactions on Microwave Theory and Techniques* MTT-28 (December):1473–77.

———. 1977. "An improved circuit for implementing the six-port technique of microwave measurements." *IEEE Transactions on Microwave Theory and Techniques* MTT-25 (December):1080–83. www.geocities.com/frank_wiedmann/publications.html
www.geocities.com/frank_wiedmann/publications.html

———. 1977. "The six-port reflectometer: An alternative network analyzer." *IEEE Transactions on Microwave Theory and Techniques* MTT-25 (December):1075-80.
ts.nist.gov/ts/htdocs/230/233/calibrations/Electromagnetic/pubs/MTT25-12.pdf

Engen, Glenn F., and Cletus A. Hoer. 1979. "Thru-Reflect-Line: An improved technique for calibrating the dual six-port reflectometer automatic network analyzer." *IEEE Transactions on Microwave Theory and Techniques* MTT-27 (December):987–93.
ts.nist.gov/ts/htdocs/230/233/calibrations/Electromagnetic/pubs/MTT27-12.pdf

European Cooperation for Accreditation. 1999. EA-4/02, *Expression of the uncertainty of measurement in calibration*. Paris: European Cooperation for Accreditation.

Evans, James R., and William M. Lindsay. 1993. *The management and control of quality*, 2nd ed. St. Paul, MN: West Publishing Company.

Farago, Francis T. 1982. *Handbook of dimensional measurements*, 2nd ed., New York: Industrial Press, Inc.

Federal Aviation Administration. 2003. Advisory Circular 145–9 *Guide for developing and evaluating repair station and quality control manual*.

Federal Aviation Regulations (Title 14, United States Code of Federal Regulations). Washington, DC: US Government Printing Office.
www.access.gpo.gov/nara/cfr/cfrhtml_00/Title_14/14tab_00.html

FDA Backgrounder, May 3, 1999, Updated August 5, 2002; www.fda.gov/opacom/backgrounders/miles.html

FDA. Food and Drug Administration—Center for Devices and Radiological Health. *Medical device quality systems manual: A small entity compliance guide, 1st ed.* (Supersedes the *Medical device good manufacturing practices [GMP] manual*).
www.fda.gov/cdrh/ dsma/gmp_man.html

FDA (U.S. Food and Drug Administration). 2003. *Guidance for industry: Part 11 electronic records; Electronic signatures—Scope and application*. Rockville, MD: FDA.
www.fda.gov/cder/guidance/5667fnl.pdf

———. 2002. *General principles of software validation: Final guidance for industry and FDA staff*. Rockville, MD: FDA. www.fda.gov/cdrh/comp/guidance/938.pdf

———. 2001. 21 CFR 820.72, *Inspection, measuring and test equipment*. Washington, DC: FDA:145 – 146.

Feigenbaum, Armand V. 1991. *Total quality control*. 3rd ed. rev. (40th ann. ed.) New York: McGraw-Hill.

Fischer, Thomas C. and David Williams. 2000. *The United States, the European Union and the "globalization" of world trade: Allies or adversaries?* Westport, CT: Quorum Books: 24.

Fluke Corporation. 1994. *Calibration: philosophy in practice*. 2nd ed. Everett, WA: Fluke Corporation.

FS 1037C. 1996. Federal Standard 1037C, *Telecommunications: Glossary of telecommunication terms*. Washington: General Services Administration. www.its.bldrdoc.gov/fs-1037

Gaunt, Ken. 2003. "It May Be Time to 'Retool" Your Quality System." *Automotive Excellence* (Winter/Spring): 6–7.

Gillespie, Helen. 1994. What Is ISO 9000? *Scientific Computing and Automation*. (February).
www.limsource.com/library/limszine/applica_tions/apscaiso/apwhat94.html

Griffith, Gary. 1986. *Quality technician's handbook*, 3rd ed. Englewood Cliffs, NJ: Prentice Hall.

Grob, Bernard. 1977. *Basic electronics*. 4th ed. New York: McGraw-Hill.

Grun, Bernard. 1991. *The timetables of history*. New 3rd rev. ed. New York: Simon & Schuster.

GSA. 1996. FS 1037C. Federal Standard 1037C, *Telecommunications: Glossary of telecommunication terms*. Washington: General Services Administration. www.its.bldrdoc.gov/fs-1037

———. 1975. FS GGG-G-15C. *Federal Specification: Gage blocks and accessories (inch and metric).* U.S. General Services Administration. Washington DC: U.S. Government Printing Office.
Hailey, Victoria A. 2003. Personal correspondence with Graeme C. Payne, October 2003.
Harral, William M. 2003. What is ISO/TS 16949:2002? *Automotive Excellence* (Summer/Fall): 5–6.
Harris, Georgia L. 1995, *Western regional assurance program workbook.* Gaithersburg, MD: NIST, Office of Weights and Measures.
———. (editor and lead). 2003. NIST Special Publication 1001, *Basic mass metrology*, CD-ROM interactive. Gaithersburg, MD: NIST.
Harris, Georgia L., and Jose A. Torres. 2003. NIST IR 6969, *Selected laboratory and measurement practices, and procedures, to support basic mass calibration.* Gaithersburg, MD: NIST.
Hazarian, Emil. 2003. *"Elements of measurement techniques,"* MSQA Program, QAS 516 Measurement and Testing Techniques, Supplement. Dominguez Hills: California State University.
———. 2002. Mass Measurement Techniques IV. "Scales uncertainty." Tutorial. Anaheim: Measurement Science Conference.
———. 1994. *Techniques of mass measurement workbook.* Los Angeles: Technology Training, Inc.
———. 1992. *"Some characteristics of weighing devices"* Technical paper at Measurement Science Conference. Anaheim.
The Healthcare Metrology Committee. 1999. *Calibration control systems for the biomedical and pharmaceutical industry.* Recommended Practice RP-6. Boulder, CO: National Conference of Standards Laboratories.
Helfrick, Albert D. 2000. *Principles of avionics.* Leesburg, VA: Avionics Communications Inc.
Herscher, Bret A. 1989. "A Three-Port Method for Microwave Power Sensor Calibration," *Microwave Journal* (March):117–24.
Hess, Donald T., and Kenneth K. Clarke. 1995. "Phase standards: Design, construction, traceability and verification." *Cal Lab* (November–December).
Hewlett-Packard. 1989. *Test and measurement catalog: 1989.* Publication number 55959-8256D. Palo Alto, CA: Hewlett-Packard.
Hills, Graham. 2003. The effect of ISO/TS 16949:2002. *InsideStandards* (November).
Hirano, Hiroyuki. 1995. *5 pillars of the visual workplace: The sourcebook for 5S implementation.* Translated by Bruce Talbot. Shelton, CT: Productivity Press.
Hochberg, Adam. 2001. "Analysis: How Bell South's plan to get out of the pay phone industry will affect the lower-income and those in need of emergency assistance." National Public Radio, *All Things Considered.* Broadcast date 14 February 2001.
Hoer, Cletus A. 1979. "Performance of a Dual Six-Port Automatic Network Analyzer." *IEEE Transactions on Microwave Theory and Techniques* MTT-28 (December): 993–98. www.geocities.com/frank_wiedmann/publications.html
Hutchison, E. E. 2001. The road to TL 9000: From the Bell breakup to today. *Quality Progress* (June): 33–37.
IAS (International Accreditation Service). 2002. IAS Calibration Laboratory, *Accreditation program: definitions.* Whittier, CA: IAS.
International Aerospace Quality Group (IAQG). www.iaqs.saae.org/iaq
———. 2003. "Excerpts from charter of the International Aerospace Quality Group (IAQG)." www.iaqg.sae.org/iaqg/about_us/charter.htm
———. 2003. "IUPAC periodic table of the elements", version dated 7 November 2003. Research Triangle Park, NC: International Union of Pure and Applied Chemistry. www.iupac.org/reports/periodic_table
———. 2002. "The IAQG discusses industry initiatives at the 19th FAA/JAA international conference." www.sae.org/iaqg/about_us/news/june-04-2002.htm
Institute of Electrical and Electronic Engineers, and American National Standards Institute. 2002. IEEE. ASTM SI 10-2002. *American National Standard for use of the international system of units(SI): The modern metric system.* West Conshohocken, PA: ASTM.

Institute of Electrical and Electronic Engineers. 1998. *IEEE Standard for a smart transducer interface for sensors and actuators: Transducer to microprocessor communication protocols and Teds formats*. West Conshohocken, PA: ASTM.
International Electrotechnical Commission (IEC). www.iec.ch
International Organization for Standardization (ISO). Geneva, Switzerland. www.iso.ch
———. 2003a. "Where ISO 9000 came from and who is behind it." www.iso.org/iso/en/iso9000-14000/tour/wherfrom.html
———. 2003b. "The ISO 9000 family." www.iso.org/iso/en/iso9000-14000/iso9000/selection_use/iso9000family.html
———. 2003c. ISO 10012:2003, *Measurement management systems—Requirements for measurement processes and measuring equipment*. Geneva: ISO.
———. 2003a. "ISO Members." www.iso.ch/iso/en/aboutiso/isomembers
———. 2003b. "Overview." www.iso.ch/iso/en/aboutiso/introduction
ISO/TS 16949:2002 *Quality management systems—Particular requirements for the application of ISO 9001:2000 for automotive production and relevant service part organizations*. 2nd ed. Geneva: ISO.
ISO/ANSI/ASQ 9001-2000 *Quality standard*. Milwaukee: ASQ Quality Press.
ISO. 1997. ISO 9000-3:1997 *Quality management and quality assurance standards—Part 3: Guidelines for the application of ISO 9001:1994 to the development, supply, installation and maintenance of computer software*. Geneva: ISO.
———. 1997. ISO Guide 43-1 and 2, 2nd ed., *Proficiency testing by interlaboratory comparisons*. Geneva: ISO.
———. 1994. ISO 5725 Parts 1-4. Geneva: ISO.
———. 1993. *International vocabulary of basic and general terms of metrology (VIM)*. Geneva: ISO.
———. 3-1973 *Preferred numbers—Series of preferred numbers*. Geneva: ISO.
———. 17-1973 *Guide to use preferred numbers and of series of preferred numbers*. Geneva: ISO.
———. 497-1973, *Guide to the choice of series of preferred numbers and of series containing more rounded values of preferred numbers*. Geneva: ISO.
ISO/ANSI. 1973. *Preferred number standards*. Geneva: ISO.
International Telecommunications Union (ITC). www.itu.int
IUPAC. 2003. "IUPAC Periodic Table of the Elements", version dated 7 November 2003. Research Triangle Park, NC: International Union of Pure and Applied Chemistry. www.iupac.org/reports/periodic_table
Jones, Frank E., and Randall M. Schoonover. 2002. *Handbook of mass measurement*. Boca Raton, FL: CRC Press.
Juran, Joseph M. 1999. *Juran's quality handbook*. 4th ed. Wilton, CT: McGraw-Hill.
———. 1997. Early SQC: A historical supplement. *Quality Progress* 30 (September): 73-81.
———. 1995. "A history of managing for quality." *Quality Progress* 28 (August): 125-29.
———. 1974. *Juran's quality control handbook*. Wilton, CT: McGraw-Hill.
Keithley. 1998. *Low level measurement*. 5th ed. Cleveland, OH: Keithley Instruments.
Kempf, Mark. 2002. "TL-9000: Auditing the adders." Proceedings of ASQ's 56th Annual Quality Congress. CD-ROM. Milwaukee: ASQ.
Kershaw, Patrick. 1998. *Validation of calibration software by the calibration facility*. Measurement Science Conference. www.fluke.com/Download/Calibrators/pkmsc98.pdf
Kilian, Cecelia S. 1992. *The world of W. Edwards Deming*. Knoxville, TN: SPC Press.
Kimothi, S. K. 2002. *The uncertainty of measurements, physical and chemical metrology impact and analysis*. Milwaukee: ASQ Quality Press.
Kochsiek, M., and M. Gläser, (editors). 2000. *Comprehensive mass metrology*. Berlin. Wiley-VCH.
Kurtz, Max. 1991. *Handbook of applied mathematics for engineers and scientists*. New York: McGraw-Hill,
Kymal, Chad and Dave Watkins. 2003. How can you move from QS-9000 to ISO/TS 16949:2002? More key challenges of QS-9000 and ISO/TS 16949:2002 transition. *Automotive Excellence*. (Summer/Fall): 9-13.

Lamprecht, James L. 1996. *ISO 9000 implementation for small business.* Milwaukee: ASQC Quality Press.

Larsen, Neil T. 1976. "A New Self-Balancing DC-Substitution RF Power Meter." *IEEE Transactions on Instrumentation and Measurement.* vol. IM-25 (December):343-347.

Laverghetta, Thomas S. 1981. *Handbook of microwave testing.* Dedham, MA: Artech House.

Lombardi, Michael, Lisa Nelson, Andrew Novick, and Victor Zhang. 2001. "Time and frequency measurements using the Global Positioning system." *Cal Lab* (July-September): 26-33.

Lombardi, Michael. 1999. "Traceability in time and frequency metrology." *Cal Lab* (September-October): 33–40.

———. 1996. "An introduction to frequency calibration, part I." *Cal Lab* (January-February).

———. 1996. "An introduction to frequency calibration, part II." *Cal Lab* (March-April).

Luke, Robert. 2003. "Era ends for pay phones." *Atlanta Journal-Constitution.* (27 December).

Malcolm Baldrige National Quality Program. www.quality.nist.gov

Marash, Stanley A. 2003. What's good for defense . . . how security needs of the Cold War inspired the quality revolution. *Quality Digest* 23, (June): 18.

Marquardt, Donald W. 1997. Background and development of ISO 9000 standards. *The ISO 9000 Handbook,* 3rd ed., Edited by Robert W. Peach. New York: McGraw-Hill: 14–20.

McConnell, Steve. 1998. *Software project survival guide.* Redmond, WA: Microsoft Press.

Merriam-Webster Online Dictionary. 2003. "valid." www.m-w.com

Mettler-Toledo. 2000. *Glossary of weighing terms: A practical guide to the terminology of weighing.* With the assistance of Prof. Dr. M. Kochsiek, Physikalisch-Technische Bundesanstalt, PTB. English Translation by J. Penton. Berlin: Mettler-Toledo.

Microsoft Corporation. 2003. "1942: Aviation," *Microsoft Encarta 2004.*

———. 2003. "Chariots." In *Microsoft Encarta 2004.* Redmond, WA: Microsoft Corporation.

Millbrooke, Anne Marie. 1999. *Aviation history.* Englewood, CO: Jeppesen Sanderson Inc.

Mitard, Nathalie. 2001. From the cubit to optical inspection. *Quality in Manufacturing* (September/October). www.manufacturingcenter.com/qm/archives/0901/0901gaging_suppl.asp

Montgomery, Douglas C. *Introduction to statistical quality control,* 4th ed. Hoboken, NJ: John Wiley & Sons.

Nadal, Maria, 2003. *Colorimetry.* Gaithersburg, MD: NIST. www.nist.gov/public_affairs/update/upd20021216.htm#Colorimetry

NASA (National Aeronautics and Space Administration). 2003. "NASA Quality Management Systems." www.hq.nasa.gov/iso

———. 1994. NASA Reference Publication 1342, *Metrology-calibration and measurement processes guidelines.* Pasadena, CA: NASA.

Natrella, Mary G. 1963. *Experimental statistics* (NBS Handbook 91). Washington, DC: U.S. Government Printing Office. (Reprinted in 1966 with corrections.)

National Conference of Standards Laboratories. 1999. Recommended Practice RP-6. *Calibration control systems for the biomedical and pharmaceutical industry.* Boulder, CO: National Conference of Standards Laboratories.

NCSL International (NCSLI). 2003. "The Royal Egyptian Cubit." ww.ncsli.org/misc/cubit.cfm

———. 2000. *Laboratory design,* Recommended Practice, RP-7. Boulder, CO: National Conference of Standards Laboratories.

———. 1999. *Guide for interlaboratory comparisons* RP-15. Boulder: NCSL International.

———. 1999. Recommended Practice RP-6: *Calibration control systems for the biomedical and pharmaceutical industry.* Boulder, CO: NCSL International.

———. 1999. *Guide to selecting standards—Laboratory environments,* Recommended Practice. RP-14. Boulder, CO: National Conference of Standards Laboratories.

NCSL. 1995. *Handbook for the application of ANSI/NCSL Z540-1-1994.* Boulder, CO: NCSL.

NCSL. 1994. *NCSL glossary of metrology related terms.* Boulder, CO: NCSL.

———. 1990. Recommended Practice RP-3, *Calibration procedures.* Boulder, CO: NCSL.

———. 1989. Recommended Practice RP-9: *Calibration laboratory capability documentation guideline*. Boulder, CO: National Conference of Standards Laboratories.NCSL. 1999 *NCSL glossary of metrology related terms*, 2nd ed.. Boulder, CO: NCSL.
NBS Metric Information. 1976. Letter Circular LC 1071, *Factors for high-precision conversion*. Gaithersburg, MD: National Bureau of Standards.
NBS (National Bureau of Standards). 1975. *The international bureau of weights and measures 1875–1975*. NBS Special Publication 420. Washington, DC: U.S. Government Printing Office.
NIST (National Institute of Standards and Technology). 2003. Frequently asked questions about the Malcolm Baldrige National Quality Award. www.nist.gov/public_affairs/factsheet/baldfaqs.htm
———. *NIST Policy on traceability*. www.nist.gov/traceability
———. 2003. Calibration Services, Gaithersburg, MD: NIST. www.ts.nist.gov/ts/htdocs/230/233/calibrations/optical-rad/cal.op.htm
———. 2003. "NIST frequency measurement and analysis service." www.bldrdoc.gov/timefreq/service/fms.htm (viewed November 2003).
———. 2001. NIST Special Publication 330–2001 Edition, *The international system of units (SI)*. Gaithersburg, MD: U.S. Dept. of Commerce.
———. 1995. Special Publication 811, 1995 Edition, *Guide for the use of the international system of units (SI)*. Gaithersburg, MD: National Institute of Standards and Technology.
NIST/SEMATECH. 1999. "2.5.7.1 Degrees of Freedom," e-Handbook of Statistical Methods. Gaithersburg, MD: NIST and Austin, TX: International SEMATECH. www.itl.nist.gov/div898/handbook/mpc/section5/mpc571.htm
NPL. 2003. "Standard time and frequency transmissions." Teddington, England: National Physical Laboratory. www.npl.co.uk/time/time_trans.htm
National Telecommunications and Information Administration (NTIA). www.ntia.doc.gov
———. 1996. *United States frequency allocations chart*. National Telecommunications and Information Administration. Internet, www.ntia.doc.gov/osmhome/allochrt.pdf
Oldham, N. M., O. B. Land, and B. C. Waltrip. 1987. "Digitally Synthesized Power Calibration Source." *IEEE Transactions on Instrumentation and Measurement* IM-36 (June): 341–6.
Okes, Duke and Russell T. Wescott. 2001. *The certified quality manager handbook*. 2nd ed. Milwaukee: ASQ Quality Press.
Ontario Power Generation. *2003 hydroelectric generating stations: Niagara plant group*. www.opg.com/ops/Stations/OntarioPower.asp
Page, Stephen B. 1998. *Establishing a system of policies and procedures*. Mansfield, OH: BookMasters Inc.
Paton, Scott M. 2002. "Juran: A Lifetime of Quality." *Quality Digest* 22 (August): 19–23.
Pellegrino, Charles R., and Joshua Stoff. 1985. *Chariots for Apollo*. New York: Atheneum.
Pennella, C. Robert. 2004. *Managing the metrology system*. 3rd ed. Milwaukee: ASQ Quality Press.
Pinchard, Corinne. 2001. *Training a calibration technician . . . in a metrology department?* Boulder, CO: National Conference of Standards Laboratories.
Process Instruments. 2003. *History of Julie Research Laboratories*. www.procinst.com/jrlhistory.htm
projects.edtech.sandi.net/staffdev/tpss99/processguides/brainstorming.html
Powell, Robert C., and Anne C. Miller. 1987. "Determination of the reflection correction when using a symmetrical two-resistor power splitter to calibrate a power sensor." *IEEE Transactions on Instrumentation and Measurement* IM37 (June): 458–67.
Prowse, David D. 1995. *The calibration of balances*. Australia: Commonwealth Scientific and Industrial Research Organization.
QuEST Forum. questforum.asq.org
———. 2003. *QuEST Forum member companies*. questforum.asq.org/public/memcomp.shtml
Quest Geo Solutions. 2003. GPS tutorial. Liphook, UK: Quest Geo Solutions, Ltd. www.qgsl.com/gps_tutorial/gps.shtml

Saad, Theodore S., Robert C. Hansen, and Gershon J. Wheeler. 1971. *Microwave engineer's handbook.* vol. 1. Dedham, MA: Artech House.

Schulmeyers, G. Gordon and James I. McManus. 1996. *Total quality management for software.* Boston: International Thomson Computer Press.

Shepard, Alan, and Deke Slayton. 1994. *Moon shot: The inside story of America's race to the moon.* Atlanta, GA: Turner Publishing.

Shera, Brooks. 1998. "A GPS-Based Frequency Standard." *QST* (July): 37–44.

Shewhart, Walter A. 1980. *Economic control of quality of manufactured product.* New York: D. Van Nostrand, Co.; reprint, Milwaukee: ASQC Quality Press.

Shewhart, Walter A. 1939. *Statistical method from the viewpoint of quality control.* Washington DC: Graduate School of the Department of Agriculture. New York: Dover Publications, 1986.

Shrader, Robert L. 1980. *Electronic communication.* 4th ed. New York: McGraw-Hill.

Smith, Phillip H. 1995. *Electronic applications of the Smith chart in waveguide, circuit, and component analysis.* Norcross, GA: Noble Publishing.

Software Engineering Institute, Carnegie Mellon University. www.sei.cum.edu

———. 2003a. *Capability Maturity Model® (SW-CMM®) for Software,* July 21, 2003. www.sei.cmu.edu/cmm/cmm.sum.html

———. 2003b. *Process maturity profile—Software CMM® CBA IPI and SPA appraisal results—2003 mid-year update.* September 2003. www.sei.cmu.edu/sema/pdf/SW-CMM/2003sepSwCMM.pdf

Stecchini, Livio C. *A history of measures.* www.metrum.org.measures/indx.htm

Stein, Philip G. 2000. Don't whine—calibrate. Measure for Measure column, *Quality Progress* 33. (November): 85.

Stenbakken, Gerard N. 1984. "A wideband sampling wattmeter." *IEEE Transactions on power apparatus and systems,* vol. PAS-103, (October): 2919–2926.

Surveyors Historical Society. 1986. "The Changing Chains" *Backsights Magazine.* Reprinted from *Reflections* (Autumn), a publication of First American Title Insurance Company. www.surveyhistory.org/the_changing_chains.htm

Tague, Nancy R. 1995. *The quality toolbox.* Milwaukee: ASQC Quality Press.

Taylor, Barry N. 2001. *NIST Special publication 330, 2001 edition: The international system of units (SI).* Gaithersburg, MD: National Institute of Standards and Technology. physics.nist.gov/Pubs/SP330/sp330.pdf

———. 1995. *NIST Special publication 811, 1995 edition: Guide for the use of the international system of units (SI).* Gaithersburg, MD: National Institute of Standards and Technology. physics.nist.gov/Document/sp811.pdf

Tektronix. 2002. *Handheld battery operated Oscilloscope/DMM/Power Analyzers, THS700 series: Characteristics.* Beaverton, OR: Tektronix Inc.

Today in Science. 2003. *Lord Kelvin, physicist.* www.todayinsci.com/K/Kelvin_Lord/Kelvin_Lord.htm

TickIT. www.tickit.org

———. 2003. *TickIT certified organizations.* www.tickit.org/cert-org.htm

———. 2001. *The TickIT guide: executive overview.* ww.tickit.org/overview.pdf

Turgel, Raymond S., and N. Michael Oldham. 1978. "High-Precision audio-frequency phase calibration standard." *IEEE Transactions on Instrumentation and Measurement,* vol. IM-27 (December): 460–464.

UKAS Laboratory. Accommodation and Environment in the Measurement of Length, Angle and Form. www.ukas.com/Library/downloads/publications/LAB36.pdf.

U.S.A.F. 1997. HO ESAQR2P031-000-1, *Measurement and calibration handbook.* Keesler AFB, MO: U.S.A.F. Air Education and & Training Command.

U. S. Department of Defense. 2000. MIL-HDBK-1839A, *Calibration and measurement requirements.* Washington, DC: Government Printing Office.

U. S. Department of Defense. 2000. MIL-STD-1839C *Standard Practice for Calibration and Measurement Requirements*. Washington, DC: Government Printing Office.

U. S. Department of Defense. 1997. MIL-PRF-38793B *Performance Specification—Technical Manuals: Calibration procedures—Preparation*. Washington, DC: Government Printing Office.

U. S. Department of Defense. 1992. MIL-STD-1309D *Definitions of terms for testing, measurement and diagnostics*. Washington, DC: Government Printing Office.

U.S. Government. 2001. Code of Federal Regulations, Title 21, Volume 8, Part 820—*Quality system regulation, Sec. 820.72 Inspection, measuring, and test equipment*. Revised April 1, 2001. Rockville, MD. U.S. Government Printing Office.

U.S. Government Printing Office, National Archives and Records Administration. Code of Federal Regulations. www.access.gpo.gov/nara/cfr

U.S. Department of Labor, Occupational Safety & Health Administration, www.osha.gov/SLTC/radiationionizing

U. S. Navy. 2003. "United States Ship Constitution." Washington, DC: Department of the Navy www.ussconstitution.navy.mil/statpage.htm

———. 1978. NAVAIR 17-35QAL-9, Rev. 2, *Dimensional measurements*. Pomona, CA:U.S. Navy Metrology and Calibration Program, Metrology Engineering Center.

———. 1976. NAVAIR 17-35QAL-2 Rev. 3, *Physical measurements*. Pomona, CA: U.S. Navy Metrology and Calibration Program, Metrology Engineering Center.

———. 1966. *Why calibrate?* Washington, DC: Department of the Navy. Training film number MN-10105.

USNO. 2003. "GPS Time transfer performance." April 2003. Washington DC: U.S. Naval Observatory. tycho.usno.navy.mil/pub/gps/gpstt.txt

———. "GPS Time Transfer." Washington, DC: U.S. Naval Observatory. tycho.usno.navy.mil/gpstt.html

———. "USNO NAVSTAR Global Positioning System." Washington, DC: U.S. Naval Observatory.tycho.usno.navy.mil/gpsinfo.html

Wadsworth Jr., Harrison M. 1990. *Handbook of statistical methods for engineers and scientists*. New York: McGraw-Hill.

Wavetek Corp. 1998. *User's handbook for the model 9100 universal calibration system: Volume 2—Performance*, Issue 8.0.

Wells, Alexander T. 1998. *Air transportation: A management perspective*. Belmont, CA: Wadsworth Publishing Company.

Weidman, M. P. *Direct comparison transfer of microwave power sensor calibrations* (NIST Technical Note 1379.) Boulder, CO: NIST.
ts.nist.gov/ts/htdoc/230/233/calibrations/Electromagnetic/pubs/TN1379.pdf

White, Robert A. 1993. *Electronic test instruments*. Englewood Cliffs, NJ: Hewlett-Packard Professional Books.

Wiedmann, Frank, Bernard Huyart, Eric Bergeault, and Louis Jallet. 1999. "A New Robust Method for Six-Port Reflectometer Calibration." *IEEE Transactions on Instrumentation and Measurement* 48 (October): 927–31.
www.geocities.com/frank_wiedmann/publications.html

Wildi, Theodore. 2001. *Metric units and conversion charts*. Weinheim, Germany: John Wiley & Sons.

Wilson, Bruce A. 1996. *Dimensioning and tolerancing*. ASME Y 14.5. Chicago: Goodheart Willcox.

Winter, Mark. J. WebElements™ periodic table.

———. 1987. *Electronic test instruments: a user's sourcebook*. Indianapolis, IN: Howard W. Sams & Company.

Witte, Robert. 1987. *Electronic test instruments: a user's sourcebook*. Indianapolis, IN: Howard W. Sams & Company.

# Author Biographies

## KEITH BENNETT

Keith Bennett has been in the field of metrology for over 23 years and is proficient in multiple disciplines within the field. He spent 10 years in the United States Air Force working primarily in the areas of physical/dimensional, primary DC/low frequency, and RF/microwave. Keith's career progressed from calibration technician to quality process evaluator to instructor for the USAF Metrology School, Precision Measurement Equipment Laboratories (PMEL).

After the military, Keith spent the next 10 years working for Compaq Computer Corporation in its corporate metrology laboratory. His primary responsibilities at Compaq focused on analytical metrology, automated test and evaluation (ATE), and championing manufacturing support teams tasked in the calibration of in-line equipment used in making printed circuit boards. While at Compaq, Keith also held the position of radiation/laser safety officer.

Keith currently is the director of metrology for Transcat Calibration Services. He is primarily responsible for planning, initiating, and directing all activities associated with archiving and maintaining accreditation to ISO/IEC/17025 and Z-540 for 13 commercial calibration laboratories. His responsibilities also include development of new disciplines and improvement of all others.

Keith holds an associates degree in electronic engineering technology and is pursuing his bachelor's degree in engineering. Keith's outside activities include camping, hunting, and fishing.

## HERSHAL BREWER

Hershal Brewer is an accreditation assessor for the International Accreditation Service (IAS), performing assessments of calibration and testing laboratories to ISO/IEC 17025 and of inspection agencies to ISO/IEC 17020. He is the former corporate lead auditor and corporate metrologist for Newport Corporation. Hershal has presented at the National Conference of Standards Laboratories International (NCSLI), the International Conference on ISO 9000, the Measurement Science Conference, and the ASQ Audit Conference. He has been published in *Cal Lab Magazine*, in an NCSLI *Recommended Practice (RP)*, and in *Building Safety Journal*.

Hershal has extensive experience in metrology, especially in time/frequency, phase, RF/microwave, and lightwave/laser/fiber-optics. He has experience in other fields, including manufacturing, audio, program/project management, medical devices, and research and development. He holds CCT certificate number 10. Hershal is a member of Team Metrology, which is the team writing *The Metrology Handbook* under sponsorship from the ASQ Measurement Quality Division (MQD). He is the program chair for the MQD. Hershal is the primary author of the *Measurement Uncertainty Primer*, available through the International Code Council (ICC).

Hershal served four years in the Air Force, 11 years in the Navy, and is currently serving in the California National Guard. He has undergraduate degrees in electronics technology and business, and an MBA from the University of Redlands. Hershal lives in southern California.

## DAVID EUGENE BROWN

David Eugene Brown, BSME, MBA, ASQ CQE, CRE, CCT, has been employed in industry since 1962, holding positions from draftsman, tool engineer, machine designer, and senior product engineer to manager, director, and vice president of engineering. He holds 10 U.S. and five foreign patents; authored several publications; taught math, physics, SPC/SQC, Excel, ANOVA, and DOE at the community college level; and has conducted numerous seminars for educational and industrial organizations.

He is a senior member of both ASQ and ISA (Fellow, and two times past director of the ISA Process Measurement and Control Division), and an active member of several ASTM committees.

David also leads a consulting practice, the READ groups, inc., that provides statistical and engineering services and tools using Microsoft Excel. Several of these tools are provided in the CD for use with this text. READ also supplies fluid device and systems engineering to OEMs.

## JAY L. BUCHER

Jay L. Bucher is the manager of metrology services for Promega Corporation. He is also president of Bucherview Metrology Services, LLC. He started his career in metrology in 1971 by attending the U.S. Air Force's Precision Measurement Equipment Laboratory (PMEL) School in Denver, Colorado. After graduation, Jay held numerous positions in PMELs, including technician, QA manager, calibration lab chief, and flight chief. In 1994 he led a technical assistance team to upgrade the Indonesian Air Force's PMEL program by establishing its initial quality assurance and scheduling programs. After retiring from the U.S. Air Force, he developed and implemented all facets of an ISO 9001 and cGMP compliant program for Promega's metrology department.

Jay has been a member delegate for the National Conference of Standards Laboratories International (NCSLI) since 1997. He started a new NCSLI chapter for the southern Wisconsin area in 2000 and is its section coordinator. Jay has been an ASQ member since 2002 and passed ASQ's certified calibration technician (CCT) exam in June of 2003. He is an officer for ASQ's Measurement Quality Division. He has had several articles published by NCSLI, *Cal Lab Magazine,* and MQD's *The Standard.* Jay lives in DeForest, Wisconsin, with his wife and daughter.

## CHRISTOPHER GRACHANEN

Christopher Grachanen started his metrology career in 1979 as a USAF PMEL technician. He is presently Hewlett-Packard's (formerly Compaq Computer) manager of Houston Metrology group.

Chris spearheaded the development of ASQ's Certified Calibration Technician (CCT) program, is an editorial advisory for *Cal Lab Magazine*, is an officer of ASQ's Measurement Quality Division (MQD), is the author of three freeware metrology packages in use throughout the world, is the 1998 recipient of NCSL International's Dr. Allen V. Austin Award for Conference best paper, and is Test & Measurement World's (T&MW) 2004 Test Engineer of the Year.

Chris holds a bachelor's degree in technology and management from the University of Maryland, a bachelor's degree in electronics engineering from the Cooks Institute of Electronics Engineering, and an MBA from Regis University.

## EMIL HAZARIAN

Emil Hazarian has over 35 years experience in metrology, mechanical engineering, statistical process control, design of experiments, and quality assurance. He has taught metrology courses at Cal. State, Dominguez Hills; participated in various Professional Engineering Society technical sessions; and taught mass measurement techniques for Southern California Edison Co. (1994). In Romania; he taught metrology for regional metrologists (1972), mining industry and engineering metrology courses at the Technical School of Metrology (1974–1976), and taught introductory quality assurance classes for the Automation Co. in its nuclear and petrochemical activities (1981).

Between 1967–1978 Emil was working at the National Institute of Metrology, Bucharest, Romania, as a metrology technician, metrology engineer-scientist, and assistant manager for the Dimensional Primary Laboratory. Between 1984–2003, Emil worked for the County of Los Angeles, AC/Weights and Measures Department. He was the manager of the metrology laboratory and was responsible for overall operation including procedures, calibration, certification and accreditation of the metrology laboratory, covering legal metrology for Los Angeles County and scientific metrology for southwestern United States. In 2003 Emil joined the NSWC Corona, U.S. Navy, the biggest applied metrology organization in the world. He also operates a private consultancy, for clients including NASA/Jet Propulsion Laboratories, Boeing North America, U.S. Department of Commerce and Southern California Edison.

Emil received his bachelor's degree in electro-energetics engineering in 1965 and his master's degree in mechanical engineering in 1975, both from the Polytechnical Institute of Bucharest, Romania. He also earned a bachelor's degree in metrology from the Technical School of Metrology, Bucharest, Romania, 1967 and a master's degree in quality assurance from California State University, Dominguez Hills, California in 1992.

## GRAEME C. PAYNE

Graeme C. Payne is the president and principal consultant of GK Systems, Incorporated. Located near Atlanta, Georgia, he provides calibration engineering and support services such as development and preparation of calibration procedures, identification of calibration requirements, and consulting on measurement-related issues.

He has helped one client, the 10-person electronic calibration laboratory of a major airline, achieve and maintain registration to ISO 9001:2000-the only part of the company to be registered to any quality management standard.

Earlier, Graeme worked in both production and reference standards calibration laboratories with the Department of the Navy, as well as a Navy product testing laboratory. He has more than 19 years total experience in electronic calibration, specializing in DC/low frequency and RF/microwave areas. Other experience includes post-manufacturing final testing of products, quality management systems, and service in both the U.S. Marine Corps and U.S. Air Force.

Graeme is a Senior Member of ASQ and is certified as a Quality Engineer (CQE), Calibration Technician (CCT), and Quality Technician (CQT).

## DILIP SHAH

Dilip Shah has more than 25 years of industry experience in metrology, electronics, and instrumentation, measurement and computer applications of statistics in the quality assurance areas. He is currently a principal of $E = mc^3$ Solutions, a consulting practice that provides training and other solutions in ISO9000//TS16949, ISO17025, measurement and computer applications.

Dilip is an ASQ Certified Quality Engineer and is the past chairman of Akron-Canton (Ohio) ASQ section. He is the chairman of ASQ's Measurement Quality Division and also belongs to the ASQ's Statistics Division. His past participation with the Measurement Quality Division has been the development of an ASQ Certified Calibration Technician exam and, more recently, the publication of *The Metrology Handbook*. Dilip has been a member of the advisory board of the University of Akron Engineering and Science Technology Division since 1988. He conducts workshops and seminars in quality and metrology issues and participates actively in the metrology and measurement-related issues through various professional organizations. Dilip resides in Ohio and has two sons.

# Index

Note: Page numbers in italics refer to illustrations; a *t* following a page number refers to a table.

## A

abbreviations, 463–70
AC (alternating current), 325, 333–34, 334*t*
Access (software), 140
accreditation
    certificate of, 472
    criteria for, 472
    definition of, 472
    ISO/IEC 17025 Standard for, 27, 55, 126
    proficiency testing for, 203–4
    scope of, 481
    and traceability, 61–62
accreditation body, 472
accuracy of a measurement, 472
accuracy of a measuring instrument, 472
acids, in water, 406
acoustics, 261
acres, 6
acronyms, 463–70
Action procedures, 120–21
Adams, John Quincy, 3
addition, scientific notation for, 230–31
Advisory Circular 145-9 (FAA), 129
AECMA (European Association of Aerospace Companies), 124
aerospace manufacturing, 124
Agilent Metrology Forum, 138
Agilent Technologies, 201n.5
Air Commerce Act (U.S., 1926), 132n.42
Air Mail Act (U.S., 1934), 132n.42
Airline Deregulation Act (U.S., 1978), 128
airline industry safety, 448
airlines. *See* civil aviation
Alert procedures, 120
Allied Quality Assurance Procedures (AQAP), 29
alpha particles, 403, 404, *404*
alternating current (AC), 325, 333–34, 334*t*
alternating voltage (AV), 186, 325, 332–33, 332*t*

altitude (specification), 185
AM, 349, 363
American National Standards Institute. *See* ANSI
American Society for Nondestructive Testing (ASNT), 453–54
American Society for Quality (ASQ), 452–53, 455–62
American Society for Quality Control. *See* ASQC
American Society for Testing Materials, 379
American Society of Test Engineers (ASTE), 454
aminotriazole, 13
ampere, 240, 241
analysis of variance (ANOVA), 210, 211–12, 211*t*
Andromeda galaxy, 228
angle gage blocks, 391–92
angle squares, 391
angles
    conicity, 389
    definition of, 389
    instruments for measuring, 390–93
    mensuration of, 290–91
    slope, 389
    units of measure, 389–90, 390*t*
ANOVA (analysis of variance), 210, 211–12, 211*t*
ANSI (American National Standards Institute), 26
ANSI/ASQC M1-1996 Standard
    on calibration labels, 95
    on calibration procedures, 42
    on calibration records, 48
    on environmental controls, 103
    on quality systems, establishing, 19
    on traceability, 61
ANSI/ISO/IEC 17025 Standard
    on traceability, 61
    on training, 99
    on vendors/suppliers, 443–44
ANSI/ISO/IEC 17025-1999 Standard
    on audits, 89–90

on calibration certificates, 51
on calibration records, 47–48, 49
on customer service, 418, 421
on document control, 37
on environmental controls, 103
on quality systems, establishing, 19
on software for automated calibration, 137
ANSI/ISO/ASQ Q9001:2000 Standard, 103
ANSI/ISO/IEC 17025:2000 Standard, 109–11
ANSI/NCSL Z540-2-1997 Standard, 305, 318, 319, 461, 471, 476, 486
ANSI/NCSL Z540-1-1994 Standard, 19, 37
on environmental controls, 103
history of, 29
on traceability, 61
on training, 99
ANSI Z1.15 Standard, 19
Apollo 1 (spacecraft), 29–30
apparent (conventional) mass, 377
AQAP (Allied Quality Assurance Procedures), 29
Arden, Paul, 417–18
area, Babylonian measurement of, 4
ASNT (American Society for Nondestructive Testing), 453–54
ASQ (American Society for Quality), 452–53, 455–62
ASQ Code of Ethics, 413–14
ASQC (American Society for Quality Control), 28, 30
assessment, definition of, 472
Assize of Bread (Britain, 1202), 11–12
ASTE (American Society of Test Engineers), 454
astronomy, 9
atomic watches/clocks, 338, 339
attenuation, 356–57, 357$t$
attenuators, coaxial, 367, 367
audio frequencies, 326
audits
and cGMP, 91, 122
definition of, 89
external, 91
internal, 90, 477
of quality systems, 27
requirements for, 89–91
auto-correlation, 283
autocollimators, 394
automated calibration system/procedures, 141–44
Automotive Industry Action Group, 124
automotive manufacturing, 26, 114, 124–25
AV (alternating voltage), 186, 325, 332–33, 332$t$
aviation. *See* civil aviation

Avogadro project, 384
Avogadro's number, 227, 228
AVSQ 94 Standard, 125

# B

Babylonian measurement, 4–5
balances, 381, 382, 383
bar graphs, 264
barleycorn, 4
barretters, 352
bases, in water, 406
bases (scientific notation), 229, 253
battery, electric, 11
BEAM (Business Excellence Acceleration Model), 127
beer, as a unit of measurement, 4, 9
Bell, Alexander Graham, 260
bell curve (Gaussian distribution), 279, 305, 309
Bertermann, Ralph E., 118–19
best-fit line, 274–75
best measurement capability, 472
best practices, 417–27, *422*
beta particles, 403, 404, *404*
bias, 163, 472
bimodal distributions, 267, *268*
binomial distribution, 280
Biologics Control Act (U.S., 1902), 12
biotechnology industry safety, 448
BIPM (International Bureau of Weights and Measures), 61, 87, 373
Blackbody radiation law, 10
Boeing, 201n.2
bolometers, 352
brainstorming, 22
bread laws, 11–12
Brownlee, K. A., 318
BS 5179 Standard, 29
BS 5750 Standard, 29
BSR/ISO/ASQ Q10012:2003 Standard
on audits, 90
on calibration procedures, 41
on calibration records, 48
on environmental controls, 103
and the ISO 9000 family, 113, 115
vs. ISO 17025:1999 Standard, 118
on personnel organizational responsibilities, 429
on resource management, 441
on software for automated calibration, 137–38
on training, 100
on vendors/suppliers, 444
*See also* ISO 10012 Standard

bubble graphs, 264
budgeting, 441–42
Bureau International des Poids et Mesures (BIPM), 61, 87, 373
   *See also* International Bureau of Weights and Measures (BIPM)
Bureau of Chemistry and Soils, 12
Business Excellence Acceleration Model (BEAM), 127

## C

CAA (Civil Aviation Administration), 132–33n.42
   *See also* Federal Aviation Administration
CAB (Civil Aeronautics Board), 132–33n.42
cable TV, 347
calculators, 348
calendars, 3
Calibrate Before Use (CBU) stickers, 97
calibration
   accessories needed for, 437
   activities/providers of, definition of, 474
   vs. alignment/adjustment, 473–74
   automation system of, 141–44
   certificates of, 51–53, 52, 474
   circular, 62
   definition of, 162, 472–74
   intervals between calibrations, 67–70, 122–23, 130, 183
   labels/stickers of, 95–97
   of like items, 435–36
   by linearization, 173
   methods/techniques, 173–74
   by nulling, 174
   physical-dimensional, safety in, 449
   procedures for, 41–43, *43–44*, 199–201, 474
   programs of, 474
   records of, 37, 45–51, 62, 121
   reports on, 474
   results of, 177
   seals of, 474–75
   self-calibration, 481–82
   by spanning, 174
   by spot frequency, 174
   and traceability, 62–64, *63–64*
   valid results, 114, 132n.4
   by zeroing, 174
   *See also* calibration standards
*Calibration: Philosophy in Practice* (Fluke), 162
calibration engineers, 431–32
calibration rooms, temperature/humidity controls in, 104–6, 105*t*
calibration standards, 71–88
   CGPM, 8, 71, 87
   CIPM, 87
   definition of, 73, 86, 475
   intrinsic, 87
   for mass/weight, 378–80
   numerical values of quantities, 77–85
   realization techniques for, 87, 88*t*
   SI prefixes, 72*t*, 75–77, 80–81
   SI units, 10, 71–75, 72*t*, 78–80
   substituting, 195–201, 196–99*t*
   types of, 86*t*
   *See also* SI
calibration technicians
   definition of, 431
   knowledge/skills needed by, xvi
   tasks for, 431, xvi
   training for, 99–100
calibrators, 186–87
calipers, 316
candela, 240
cannons/cannonballs, 7–8
Capability Maturity Model (CMM), 130–31, 143
capacitance, 334–36, 335*t*
capacity, Babylonian measurement of, 4–5
carat format, 229
cause-and-effect diagrams, 21
CCT (Certified Calibration Technician), 455–56, xiii–xiv
centesimal grades, 390
central limit theorem, 268
central tendency, 266–67
certification vs. registration, 27
Certified Calibration Technician (CCT), 455–56, xiii–xiv
Certified Quality Engineer, 30
certified reference material, 87
cGMP (current good manufacturing practices), 118–23
   and audits, 91, 122
   Bertermann on, 118–19
   on calibration personnel, 119
   on calibration requirements, 119
   on documentation detail, 121, 122
   on environmental controls, 121–22
   on equipment calibration procedures, 120–21
   on inspection/measuring/test equipment, 119–20
   on overdue lists/inventory control, 122
   and public safety, 120–21
   on traceability, 121, 122–23

CGPM (General Conference on Weights and Measures), 8, 71, 87, 387
CGS, 200–201
chains, 399–400
checksheets, 20
chemical/electro-optical/analytical/radiation parameters, 323, 324t, 397–98, 397–409
Chi-squared-distribution, 279
Chrysler, 124
cigarette legislation, 15
CIPM (International Committee for Weights and Measures), 87, 133n.49
circles, mensuration of, 288, 293–94, 295
Civil Aeronautics Board (CAB), 132–33n.42
civil aviation
    death statistics, 133n.46
    government regulation in the U.S., 128–29, 132–33n.42
    industry-specific requirements in, 128–30
    life cycle of airliners, 195, 201n.2
    repair stations, 129
    safety in, 448
    traceability in, 129–30, 133n.49
Civil Aviation Administration (CAA), 132–33n.42
    *See also* Federal Aviation Administration
civil engineering, 131
CMM (Capability Maturity Model), 130–31, 143
Code of Ethics (ASQ), 413–14
code officials, 131
coefficients, 229
Collier, Peter, 12
colorimetry, 402–5
column graphs, 264
combined standard uncertainty, 475
combined uncertainty, 282
commerce, 9
common cause, 21
communication technology, 11
compasses, 399–400
competence, 62, 475
competition, response to, 26
competitive advantage, 26
complaints, 56, 425–27
computer industry safety, 449–50
computer software. *See* software
computers, home, 348
computers and automation, 137–44
concave lenses, *398*
concave mirrors, *398*
cones, 298–99
confidence interval, 475
confidentiality

ISO/IEC 17025 on, 111
    of measurement data, 172–73
    in proficiency testing, 213
conicity, 389
constants, 150–51t, 257
construction, 9
Consumer Bill of Rights (U.S.), 14
Contract Air Mail Act (U.S., 1925), 132n.42
control charts
    definition/function of, 21
    for MAPs, 215–16, 215–17t, 218, 219–20t, *221*, 221–22
    and measurement capabilities, 164–68, 165–66t, *167*
    for proficiency testing, 214
control limits, 166t, 215–17t, 216, *218*
control numbers, 37
Convention of the Metre treaty (1875), 8, 87, 373
conventional mass, 377
conventional reference conditions, 377
conversions
    common, 487–511
    for dimensional/mechanical parameters, 387
    metric-British, 375, 376t
    for SI units, 252–53, 252t, 256
convex lenses, *397*
corellogram, 283
correction of error, 476
corrective action, 476
correlation, 272–73
correlation coefficients, 318–19
cosines, 291
Council for Optical Radiation Measurements, 402
counts (of a digital display), 181
coverage factor, definition of, 476
craftsmen/artisans, 9, 10
cranberry inspection, 13
Crosby, Philip B., 35
cubits, 4, 5, 9
Cuomo, Paul, 401
currency, decimal, 7
current good manufacturing practices. *See* cGMP
customer confidentiality. *See* confidentiality
customer requirements, 26
customer satisfaction/complaints, 425–27
customer service, 418–21

# D

date due calibration, 95
Davy, Humphrey, 10

day, duration of, 340
days, measurement of, 3
dB/bit scale, 262
dB/Hz scale, 262
dB scale, 261
dBA scale, 261
dBC scale, 261
dBi scale, 262
dBm scale, 262
dBr scale, 262
dBu scale, 262
dBuV scale, 262
dBV scale, 262
dBW/K-Hz scale, 262
dBW scale, 262
DC/low frequency, 325–44
    AC/DC converters, 332–33
    alternating current, 325, 333–34, 334$t$
    alternating voltage, 186, 325, 332–33, 332$t$
    capacitance, 334–36, 335$t$
    DC, definition of, 325, 329
    DC, measuring/parameters, 329–30, 330$t$
    direct voltage (DV), 186, 187, 325, 327–29, 327$t$, 329$t$
    electrical power, 342–43, 343$t$
    inductance, 336–37, 336$t$
    low frequency, definition of, 325–26
    overview of, 323, 324$t$
    phase angle, 341–42, 341$t$
    resistance, 330–31, 331$t$
    specifications of, 186–87
    time interval, 337–40, 337$t$
    time of day, 340–41
decibel measures, 259–63
decimal markers, 227, 237
decimal system, 7
Defense Department. *See* DOD
deficiency, 476
degrees, 236, 341
degrees, sexagesimal, 390
degrees of freedom (DOF), 265–66, 318
Deming, W. Edwards, 21, 28, 30, 35
Deming Prize, 28
Department of Agriculture, 12
Department of Defense, 26, 130
departure value, 476
difference (proficiency testing), 213
differential voltmeters, 201n.3
diffusion, 405–6
digits, significant, 225–27
digits (ancient measurement), 5–6
digits (of a digital display), 180–81
dimensional/mechanical parameters, 323, 324$t$, 387–95, 390$t$

angles, instruments for measuring, 390–93
angles, units of measure, 389–90, 390$t$
classification of geometrical quantities, 387
conversion factors, 387
length measures, 388–89
levels, 393–95
diode sensors/meters, 352, 353–54
Dirac, Paul A. M., 11
direct current. *See* DC/low frequency
direct voltage (DV), 186, 187, 325, 327–29, 327$t$, 329$t$
distribution types/properties, 279–82, 305
division, scientific notation for, 231–32
Do Not Use, Out of Calibration labels, 96
document control, 37–38
document control administrators, 433
documentation and traceability, 62
DOD (Department of Defense), 26, 130
Dodge, Harold F., 27–28
DOF (degrees of freedom), 265–66, 318
*Dotterwich, U.S. v.*, 13
doughnut charts, 264
drift (specification), 164, 183
Drug Importation Act (U.S., 1848), 12
drug laws, 12–15
Durham-Humphrey Amendment (U.S., 1951), 13
Dutch, Duke, 400, 401
DV. *See* direct voltage

# E

E format, 229
EAQF 94 Standard, 125
Edgar the Peaceful, 6
EDM (Electronic Distance Meter), 400–401
Edward I, king of England, 6–7
Egyptian measurement, 3–4, 5–6, 9
EIA (Electronics Industries Alliance), 369n.7
electrical signals, 261–63
electricity, 11, 248$t$
electrolysis, 11
Electronic Distance Meter (EDM), 400–401
electronic encoders, 395
electronic field books, 401
electronic levels, 394
Electronics Industries Alliance (EIA), 369n.7
electronics industry safety, 448
electrons, 11
elevation, 185
ellipses, mensuration of, 288–89, 294, 295–96
ellipsoids, 297, 300–301
$E_n$ number, 213
end measures, 391

engineering. *See* civil engineering
environmental controls, 103–6, 105t, 121–22
equalvalence, 476
equations, quantity/numerical-value, 83–84
equipment
    manuals for, 41–43, *43–44*
    status of, 95–97
    substituting, 199–200
error
    correction of, 476
    definition of, 305, 476
    distributions of, 305 (*see also* uncertainty)
    maximum permissible for weights, 379–80
    random, 305, 480
    standard error of the mean, 271
    systematic, 305, 483
estimation, scientific notation for, 233
ethics, 413–14
European Association of Aerospace Companies (AECMA), 124
European Union (EU), 31
Excel, 138, 140, 263–64
exponents, 229–30

## F

F-distribution, 279
FAA (Advisory Circular 145-9), 129
FAA (Federal Aviation Administration), 26, 133n.42, 133n.49
Fahrenheit, Daniel Gabriel, 10
FAR (Federal Aviation Regulations), 129
Faraday, Michael, 11
FCC (Federal Communications Commission), 348
FDA (Food and Drug Administration)
    cranberry inspection by, 13
    drug research/marketing promoted by, 14
    establishment of, 12, 14
    laboratories consolidated, 15
    medical devices tested by, 14
    quality standards required by, 26
    *See also* cGMP
Federal Aviation Administration (FAA), 26, 133n.42, 133n.49
Federal Aviation Regulations (FAR), 129
Federal Communications Commission (FCC), 348
Federal Trade Commission, 13
Fines Enhancement Laws (U.S., 1984, 1987), 14
fission, 11
fixed distribution assumption, 265
fixed model assumption, 265
fixed variation assumption, 265
floor term, 182
flowcharts, 20, 55
Fluke, 333
FM, 349, 363
FMAS (Frequency Measurement and Analysis System), 339
Food, Drug, and Cosmetic (FDC) Act (U.S., 1938), 13
food, pH of, 406–7
Food Additives Amendment (U.S., 1958), 13
Food and Drug Administration. *See* FDA
Food and Drug Administration Act (U.S., 1988), 14
Food and Drug Administration Modernization Act (U.S., 1997), 15
Food and Drugs Act (U.S., 1906), 12
food laws, 11–15
Ford, 124
Franklin, Benjamin, 11
Frequency Measurement and Analysis System (FMAS), 339
fusion, 11

## G

gage blocks, 131, 391–92
gage R&R, 168–69, 169–70t, *171*, 476
gamma rays, 403, 404, *404*
Gaussian distribution (bell curve), 279, 305, 309
General Conference on Weights and Measures. *See* CGPM
General Motors, 124
*General Principles of Software Validation*, 139–40
geometrical dimensioning. *See* dimensional/mechanical parameters
George IV, king of Great Britain, 7
Global Positioning System. *See* GPS
global trade, 111–12
GLONASS, 338
GMT (Greenwich Mean Time), 340
GOAL/QPC *Memory Jogger II*, 165, 167–68
goniometers, 395
government contract requirements, 26
GPS (Global Positioning System), 338–41, 348, 402
grain, as a unit of measurement, 4
graphs, 264
GRAS (Substances generally recognized as safe), 13
gravitational system, 200
Great Pyramid of Giza, 5
Greenwich Mean Time (GMT), 340
Griffith, Gary, 168

Grun, Bernard, 9
guilds, 9–10
GUM, 476
Gunter, Edmund, 399

## H

Hadco Instruments (California), 400
Hall effect, 330
hard disk drive (HDD), 185
Hawthorne Works (Chicago), 27–28
heat, mechanical equivalent of, 10
Henry III, king of England, 6
hertz, 337
Hertz, Heinrich R., 11
Hewlett-Packard Company, 201n.5
high voltage/current industry safety, 448
Hills, Graham, 124
Hipot test, 476–77
histograms, 21, 264
history and philosophy of metrology/calibration, 3–15
    ancient measurement, 3–6, 9
    and the Industrial Revolution, 7, 10–11
    international standardization, 8–9, 71
    and international trade, 7–9, 87
    and military requirements, 7–9
    prehistoric metrology, xiii
    progress summary, 6–7
    quality standards history, 9–10
    uses/effects of metrology, 3, 11, xiii
housekeeping, 447–50
humidity, 104–6, 105t
hypergeometric distribution, 280

## I

IAQG (International Aerospace Quality Group), 124
IATF (International Automotive Task Force), 125
ICC (International Code Council), 131
IEC (International Electrotechnical Commission), 27, 369n.7
IEEE (Institute of Electrical and Electronic Engineers), 453
*If Japan Can . . . Why Can't We?* (NBC News), 30
IF substitution, 357
IIDN, 265
Imperial System of Weights and Measures (Great Britain), 7
IM&TE, 477
    *See also* calibration

indication (of a measuring instrument), 159
inductance, 336–37, 336t
Industrial Revolution, 7, 10–11
industry-specific requirements, 109–33
    in aerospace manufacturing, 124
    in automotive manufacturing (SAE AS9100A), 114, 124–25
    cGMP, 118–23
    in civil aviation, 128–30
    in civil engineering, 131
    in computer software (CMM), 130–31
    in computer software (ISO 9000-3 and TickIT), 123–24, 127–28
    for global trade, 111–12
    ISO 9000, standards based on, 123–24
    ISO 9000 Standard, 111–16
    ISO 9001:2000 Standard, 114–16
    ISO 10012 Standard, 117–18
    ISO 17025 Standard, 109–11
    ISO/TS 16949 Standard, 125–26
    in medical device manufacturing, 123
    QS-9000 Standard, 124–25
    in telecommunications (TL 9000), 126–27
information, 56, 78–79
infrared, 349
infrasonic frequencies, 326
input-output relationships, 35
insertion loss, 356–57, 357t
Institute of Electrical and Electronic Engineers (IEEE), 453
Instruments, Systems, and Automation Society (ISA), 453
insulation resistance test, 477
integrated systems, 35
integrating sphere, 405
interlaboratory comparisons
    definition of, 203, 477
    schemes, *204*, 204–5, 205–6t, 207–8, *207–8*
interlaboratory testing scheme, 208, *209*, 209–10t, 210
internal audits, 90, 477
International Aerospace Quality Group (IAQG), 124
international atomic time scale (TAI), 241, 337, 340
International Automotive Task Force (IATF), 125
International Bureau of Weights and Measures (BIPM), 61, 87, 373
International Code Council (ICC), 131
International Committee for Weights and Measures (CIPM), 87, 133n.49

International Electrotechnical Commission (IEC), 27, 369n.7
International Federation of Standardizing Associations (ISA), 35n.6
International Organization for Standardization. *See* ISO
International Organization of Legal Metrology, 373, 379
international standards, 86t, 112
  *See also specific standards*
International System of Units. *See* SI
International Telecommunications Union (ITU), 27, 348–49
International Temperature Scale (1927), 10
International Vocabulary of Basic and General Terms of Metrology. *See* VIM
Internet, 348
interpolation, 275, 276–79
inventory control, 122
ionizing radiation, 402–3
irrational number constants, 226
ISA (Instruments, Systems, and Automation Society), 453
ISA (International Federation of Standardizing Associations), 35n.6
ISO (International Organization for Standardization)
  members, 26 (*see also* ANSI)
  number rounding method, 235, 236
  origin of name, 35n.4, 478
  overview/scope, 26–27, 478
  preferred numbers, 234, 235, 236
ISO 9000 Standard
  family of standards of, 113–14
  in global trade, 111–12
  history of, 29, 31–35
  as industry-specific, 111–16
  industry-specific standards based on, 123–24
  third-party audits in, 113
ISO 9000-3 Standard, 127–28, 143
ISO 9000-3:1997 Standard, 127–28
ISO 9000:2000 Standard, 33, 112, 113
ISO 9001 Standard, 26
ISO 9001:1987 Standard, 56
ISO 9001:1994 Standard, 31–32
ISO 9001:2000 Standard, 31–35, 56, 113, 114–16, 162
ISO 9002:1994 Standard, 31
ISO 9003:1994 Standard, 31
ISO 9004:2000 Standard, 113
ISO 10005:1995 Standard, 113
ISO 10006:1995 Standard, 113
ISO 10012 Standard, 29, 117–18
ISO 10012-1 Standard, 115
ISO 10012-2 Standard, 115
ISO 10012:2003 Standard, 419
ISO 10013:1995 Standard, 113
ISO 10015:1999 Standard, 113
ISO 14001 Standard, 32
ISO 90003 Standard, 127
ISO/IEC 17025 Standard
  accreditation via, 27, 55, 126
  on confidentiality, 111
  as industry-specific, 109–11
  on measurement uncertainty, 110
  on personnel organizational responsibilities, 429
  on proficiency testing, 203–4
  on quality managers, 430–31
  on quality manuals, 56–57, 57–58
  on software for automated calibration, 138–39
  on technical managers, 430
ISO/IEC 17025:1999 Standard
  vs. BSR/ISO/ASQ Q10012:2003 Standard, 118
  importance of, 27
  as international standard, 109
ISO/TC176 Standard, 29, 31, 125
ISO/TR 10014:1998 Standard, 113
ISO/TS 16949 Committee, 26, 125–26
ISO/TS 16949:2002 Standard, 113, 114, 125–26
ITU (International Telecommunications Union), 27, 348–49

## J

JAMA (Japan Automobile Manufacturers Association), 125
JAN (Joint Army-Navy specifications system), 369n.7
Japanese quality standards, 28, 30
Jefferson, Thomas, 7
John, king of England, 6, 11–12
Joint Army-Navy specifications system (JAN), 369n.7
Josephson junction, 327–28
Julian dates, 49–50
Juran, Joseph H., 9, 27–28, 35, 164
JUSE (Union of Japanese Scientists and Engineers), 28

## K

Kefauver-Harris Amendments (U.S., 1962), 14
Kelsey, Frances, 13–14
kelvin, 240

Kelvin, Lord, 10, 323
Kennedy, John F., 14
Key Comparison Index, 133n.49
kilograms
    definitions of, 240, 376, 384–85
    international standardization of, 8
    prototypes of, 376
    and SI prefixes, 76
    as units of mass, 375, 376
Kimothi, S. K., 160
kurtosis, 272

## L

labels, calibration, 95–97
laboratory liaisons, 418–21
laboratory managers, 430
lasers, 11, 406
law of gravitation, 373–74
leading zeroes, 237
length, 285–88
    ancient measurement of, 4, 5–6
    definitions of, 3
    early British standards of, 8, 8
    international standardization of, 9
    measuring instruments for, 388–89
lenses, *397–98*
level of confidence, 478
levels (instruments), 393–95
life cycle of systems, 195, 201n.2
light, units of, 248–49*t*
light (optical radiation), 405–8
lightning, 11
like items, 93
Lincoln, Abraham, 12
line charts, 264
line graduated measures, 390–91
line power term, 184
linear interpolation methods, 276–77
linear regression, 274–75
linear relationships, 273–75
linear scale, 163
linearization, 173, 264
lines, mensuration of, 285–88
load term, 184
logarithms, 259–60, 263–64
logistical support personnel, 433
Lombardi, Michael, 337
LORAN-C, 338, 340
Loschmidt constant, 228
Louis XVI, king of France, 7
low frequency. *See* DC/low frequency

## M

M1-1996 Standard. *See* ANSI/ASQC M1-1996 Standard
Magna Carta, 6
magnetic fields, 11
magnetic moment, 228
magnetism, units of, 248*t*
Malcolm Baldrige National Quality Award (MBNQA), 30–31
management review, 478
managing a metrology department/calibration laboratory, 413–54
    best practices, 417–27, *422*
    budgeting/resource management, 441–42
    getting started, 413–15
    housekeeping/safety, 447–50
    personnel organizational responsibilities, 429–33
    process workflow, 435–40, *439*
    professional associations, 451–52 (*see also specific associations*)
    vendors/suppliers, 443–46
MAPs (measurement assurance programs)
    elements of, 214
    establishing, 215–17*t*, 218, 219–20*t*, *221*, 221–22
    objectives/importance of, 214–15, 222
Marash, Stanley, 29
mass comparators, 377, 382
mass/weight, 317, 373–85
    Avogadro project, 384
    balances, 381, 382, 383
    calibration standards, 378–80
    classification of weighing devices, 380–85
    concept of mass, 373
    definitions/conventions, 374, 377–78
    design of weights, 379
    equilibrium position, 381–84
    handling/storage/packing/shipping of weights, 380
    installation of weighing devices, 381
    international standardization of, 9
    mass comparators, 377, 382
    mass measuring technique, 381
    mass vs. weight, 374–75
    materials for manufacturing weights, 379
    maximum permissible error for weights, 379–80
    metric-British conversions, 375, 376*t*
    operator participation in weighing process, 381
    physical mass, new, 384
    scales, 377, 381–82, *382*

substitution method, 383–84
transposition method, 384
units of measures for mass, 375–78
Watt Balance approach, 385
weighing methods, 383
material handlers, 433
mathematics, 9, 149
*See also* scientific notation
MBNQA (Malcolm Baldrige National Quality Award), 30–31
MDL (Microwave Development Laboratories), 369n.6
mean/average, 266–67
mean solar day, 340
measurand, 157
measurement
data, 172–73
definition of, 157, 159, 478
fundamental assumptions of, 265
results of, 159
uncertainty of (*see* uncertainty)
units of, 149–50
*See also* measurement methods; measurement parameters; measurement systems; SI
measurement assurance programs. *See* MAPs
measurement capabilities, 162–71
best measurement capability, 472
bias, 163
gage R&R, 168–69, 169–70*t*, 171
linearity, 163
repeatability, 163–64
reproducibility, 163, 164
stability, 163, 164
statistical process control/control charts, 164–68, 165–66*t*, 167
measurement methods, 157–59
definition of, 157
differential, 158
direct, 158
indirect, 158
ratio, 158
reciprocity, 158
substitution, 158
transfer, 158
measurement parameters
chemical/electro-optical/analytical/radiation, 323, 324*t*, 397–98, 397–409
dimensional/mechanical, 323, 324*t*, 387–95, 390*t*
overview of, 323–24
physical measurements, 323, 324*t*

and SI units, 151–54*t*, 153
*See also* DC/low frequency; mass/weight; radio frequency and microwave
measurement signal, 159
measurement standards, 86, 86*t*
*See also* specific standards
measurement systems, 159–62, *161*
CGS, 200–201
definition of, 159, 478
gravitational, 200
MKS (metric), 200–201
*See also* SI
Measurement Systems Analysis (MSA) Reference Manual, 160
measurement uncertainty, 62, 110
*See also* uncertainty
measuring chains, 159
measuring instruments, 159
*See also specific instruments*
Meat Inspection Act (U.S., 1906), 12
mechanical parameters. *See* dimensional/mechanical parameters
mechanical protractors, 392
mechanics, units of, 249–51*t*
median, 267
Medical Device Amendments (U.S., 1976), 14
Menes, ruler of Egypt, 4
mensuration
angles, 290–91
circles, 288, 293–94, 295
definition of, 285
ellipses, 288–89, 294, 295–96
length, 285–88
perimeters, 294–96
plane areas, 291–93
surface area, 299–301
volume, 296–99
meter (length), 8, 240, 387
Metric Act (U.S., 1866), 7
metric system, 7, 200–201
metrics (measurements), 421–23, *422*
metrological confirmation, 117–18
metrological function, 117–18
metrologists, 432–33
metrology
definition of, 478
development of, 373
functions of, 373
overview of, 149–51, 150–54*t*, 153
*Metrology—Calibration and Measurement Processes Guidelines*, 159–60
microprocessors, 348

Microwave Development Laboratories (MDL), 369n.6
microwaves. *See* radio frequency and microwave
MIL-C-45662 Standard, 29
MIL-H-110 Standard, 29
MIL-HDBK-50 Standard, 29
MIL-Q-5923 Standard, 28
MIL-Q-9858 Standard, 28, 31
MIL-Q-21549B Standard, 28
MIL-STD-105 Standard, 28
MIL-STD-414 Standard, 28
MIL-STD-45662A Standard, 29
military specifications/standards, 28
millieme system, 390
minutes symbol, 236
mirrors, *398*
Mitard, Nathalie, xiii
MKS (metric system), 200–201
mobile operations, 478
mode, 267
modulation, 362–63, 362*t*
mole, 240, 241
monochromators, 405
monuments (surveying), 401
moon cycles, 3
Mouton, Gabriel, 7
multimeters, 187, 191–92
multiplication, scientific notation for, 231
muskets, 10

# N

Napoleon I, emperor of France, 7
NASA (National Aeronautics and Space Administration), 26
National Bureau of Standards (NBS), 61, 353
National Conference of Standards Laboratories (NCSL) International, 451–52, 479
National Institution of Standards and Technology. *See* NIST
national metrology institutes. *See* NMIs
National Physical Laboratory (NPL; U.K.), 87, 138
National Society of Professional Engineers (NSPE), 454
national standards, definition of, 86*t*
National Telecommunications and Information Administration (NTIA), 348–49
NATO (North Atlantic Treaty Organization), 29
natural (physical) constant, 478–79
navigation satellites, 338–39
NAVSTAR, 338

NBS (National Bureau of Standards), 61, 353
NCSL (National Conference of Standards Laboratories) International, 451–52, 479
NCSL RP-7 Standard, 104
NCSL RP-14 Standard, 104
NCSLI RP-1 Standard, 68–69
NCSLI RP-6 Standard
  on audits, 90
  on calibration labels, 95
  on calibration procedures, 41–42
  on calibration records, 48
  on environmental controls, 103–4
  on software for automated calibration, 138
  on traceability, 61
  on vendors/suppliers, 444–45
neutron mass, 228
Newton, Sir Isaac, 373
newtons, 375
Niagara Falls power plant, 195
NIST (National Institution of Standards and Technology; U.S.)
  color standards of, 402
  frequency system of, 339
  kilogram prototype at, 376
  physical constants worksheet, 257
  on SI units/prefixes, 73–85
  and traceability, 61, 64, 121, 133n.49
  UTC provided by, 341
  weights classified by, 378–79
NMIs (national metrology institutes), 61, 64, 73, 133n.49
No Calibration Required (NCR) labels, 95–96
noise, 328, 340, 363–64, 363*t*
Nondestructive Testing (NDT), 479
nonlinear interpolation methods, 277–79
nonroutine activities, 438
normal distribution, 279, 281, 305, 309
normalization. *See* calibration
North Atlantic Treaty Organization (NATO), 29
NPL (India), 87
NPL (National Physical Laboratory; U.K.), 87, 138
NRC (Nuclear Regulatory Commission), 26
NSPE (National Society of Professional Engineers), 454
NTIA (National Telecommunications and Information Administration), 348–49
Nuclear Regulatory Commission (NRC), 26
nulling, 174
numbers
  bases, 229, 253
  formatting issues, 236–37
  ISO preferred, 234, 235, 236

rounding methods, 235–36
  See also scientific notation
numerical values, 229

## O

OEM (original equipment manufacturer) procedures, 42, 445
offset, 479
Ohm, George Simon, 11
Ohm's law, 11, 330, 342
oil, as a unit of measurement, 4
Oldham, N. M., 341–42
on-site operations, 479
optical dividing heads, 394
optical protractors, 393
optical radiation (light), 405–8
optical rotary tables, 395
optics, 397–402
  concave lenses/mirrors, 398
  convex lenses, 397
  reflecting/refracting telescopes, 398
  in surveying, 399–402
organizations vs. suppliers, 32
Orphan Drug Act (U.S., 1983), 14
oscillosopes, 187–88
Out of Calibration labels/stickers, 96
out-of-tolerance standards/results, 64–65, 94, 177
overdue lists, 122
OWC (Owning Work Center), 420

## P

paper trails, 122–23
  See also traceability
parallelograms, 292
parameters. See measurement parameters
PARD (periodic and random deviation), 201n.9
Pareto charts, 20
part per million/billion/trillion, 83
PDCA (plan do check act) cycles, 21–22, 35
peak sensor meters, 355
percent difference (proficiency testing), 213
percent of hydrogen (pH), 406–8
percentage symbol, 82
Performance Test, 479
perimeter, 294–96
periodic and random deviation (PARD), 201n.9
perpendicular lines, 288
personnel organizational responsibilities, 429–33
pH (percent of hydrogen), 406–8
pharmaceutical, 47, 63, 203, 404, 444

industry safety, 448
phase angle, 341–42, 341t
phase comparison, 339
photoelectric effect, 11
physical constants, 150–51t, 257
physical-dimensional calibration, safety in, 449
physical measurements, 323, 324t
physics, 10, 11
Physikalisch-Technische Bundesanstalt (PTB; Germany), 87
pie charts, 264
Pinchard, Corinne, 100
plan do check act (PDCA) cycles, 21–22, 35
Planck, Max, 10
plane area, 291–93
platinum, 10
PM, 363
PM (preventive maintenance), 424–25
PMELs (Precision Measurement Equipment Laboratories), 71, 420
point-slope equation of a line, 287
points in a plane, distance between, 287
Poisson distribution, 280–81
policy, 56, 479
polygons, 392
potentiometry, 406–7
pounds, 375
power frequencies/meters. See radio frequency and microwave
power splitters, 355–56, 356
practical mil system, 390
precision, 480
Precision Measurement Equipment Laboratories (PMELs), 71, 420
predictability, 328
Prescription Drug Marketing Act (U.S., 1988), 14
Prescription Drug User Fee Act (U.S., 1992), 15
pressure, 316
preventive action, 480
preventive maintenance (PM), 424–25
primary standards, 86t
prisms, rectangular, 296, 299
procedures, 480
  See also specific procedures
"Procedures for the Appraisal of the Toxicity of Chemicals in Food," 13
process workflow, 435–40, 439
product realization, 32–33
products vs. services, 25, 34–35, 35n.1
professional associations, 451–52
  See also specific associations
proficiency testing, 203–14
  acceptability of data, 212–14

for accreditation, 203–4
confidentiality in, 213
definition of, 203, 480
interlaboratory testing scheme, 208, *209*, 209–10*t*, 210
international recognition of, 110
measurement comparison scheme, *204*, 204–5, 205–6*t*, 207–8, *207–8*
neutrality in, 213
split-sample testing scheme, 210, *211–12*, 211*t*
statistical control in, 214
protractors, 392–93
Proxmire Amendments (U.S., 1976), 14
PSWR, 359
 *See also* standing wave ratio
PTB (Physikalisch-Technische Bundesanstalt; Germany), 87
pyramids (Egypt), 4, 5, 9, 400
pyramids (geometry), 297–98
Pythagorean Theorem, 290

# Q

QCM (Quality Control Manual), 129
QHE (quantum Hall effect), 330
QS-9000 Standard, 124–25
QS9001-2000 Standard, 19, 37
*Quality Control Handbook* (Juran), 164
Quality Control Manual (QCM), 129
quality management system, 25
 *See also* quality systems
quality management system standards, 25
 *See also* quality standards
quality managers, 430–31
quality manuals, 55–57, 57–58, 480
quality policy, 56
quality standards
 in automotive manufacturing, 26
 in civil aviation, 26
 and competition, response to, 26
 and competitive advantage, 26
 and customer requirements, 26
 definition of, 25–26
 evolution of, 27
 and government contract requirements, 26
 history of, 9–10, 27–31
 importance of, 25–26
 laws/regulations requiring, 25
 voluntary, 25–26
quality systems, 19–23
 act on the difference, as a premise of, 19, 89
 audits of, 27
 brainstorming, 22

cause-and-effect diagrams, 21
check the results, as a premise of, 19, 89
checksheets, 20
control charts, 21
do what you say, as a premise of, 19, 89
flowcharts, 20
histograms, 21
input-output relationships in, 35
overview, 19–20
Pareto charts, 20
PDCA cycles, 21–22
process approach in, 32
process improvement techniques, 21–23
quality tools, 20–21, 23
record what you did, as a premise of, 19, 89
say what you do, as a premise of, 19, 41, 89
scatter diagrams, 21
 *See also* quality standards
quantum Hall effect (QHE), 330
quantum mechanics, 10
QuEST Forum, 126–27

# R

radar graphs, 264
radian, 240, 341, 389–90
radiation
 health risks from, 405
 ionizing, 402–3
 optical, 405–8
 sources of, 404–5
radio frequency and microwave, 323, 324*t*, 347–70
 attenuation/insertion loss, 356–57, 357*t*
 bands/range of frequencies, 349, 349*t*
 calibrating power sensors/meters, 355–56, *356*
 circuit boards, 347
 definitions/overview, 347–50
 diode sensors/meters, 352, 353–54
 infrared, definition of, 349
 microwaves, definition of, 349
 modulation, 362–63, 362*t*
 noise, 363–64, 363*t*
 peak sensor meters, 355
 radio waves, definition of, 348
 reflection coefficient/standing wave ratio, 358–61, 359*t*, *360*
 RF examples, 328
 RF power, 351–56, 351*t*, *356*
 RF voltage parameters, 361–62, 361*t*
 s-parameters, *367*, 367–69, *369*
 scalar network analysis, 364–66, *366*

spectrum analysis of, 364
thermistor sensors/meters, 352–53
thermocouples, 10, 352
vector network analysis, 365–67, *366*, 368–69
waveguides, 350, 350*t*, 369n.6
radio waves, 348
radiology, units of, 251*t*
railway tracks, spacing of, 9
raising to powers, scientific notation for, 232–33
random distribution assumption, 265
random error, 305, 480
random variation, 21
range display, 181
ratios, 259–63
recall, 94
rectangles, 291–92, 294
rectangular distribution, 281, 305, 309–10
rectangular prisms, 296, 299
reference material, 87, 482–83
reference standards, 86*t*
reflecting telescopes, *398*
reflection, 405, 406
reflection coefficient, 358–61, 359*t*, *360*
reflection-transmission test set, 367
refracting telescopes, *398*
refraction, 405–6
registration vs. certification, 27
relative uncertainty, 282
repair, 480–81
Repair Station Manual (RSM), 129
repeatability, 163–64
reported value, 481
reproducibility, 163, 164
residuals, 266
resistance, 11, 330–31, 331*t*
resolution, 172
resource management, 441–42
reverse traceability, 64–65, 94, 116
Reynard, Charles, 234
Reynard series, 234
RF (radio frequency). *See* radio frequency and microwave
rimailho millieme, 390
RMS (root mean square), 268
Roman Empire, standardization in, 9
Roman numerals, 83
Roosevelt, Theodore, 12
root mean square (RMS), 268
root sum square (RSS), 269
round robin (interlaboratory comparison), 477

royal cubit sticks, 5, 9
RSM (Repair Station Manual), 129
RSS (root sum square), 269
run charts, 264

**S**

s-parameter test set, 366–67
s-parameters, *367*, 367–69, *369*
SAE AS9100A Standard, 124
Safe Medical Devices Act (U.S., 1990), 15
safety, 447–50
sample standard deviation, 270
sample variance, 270
satellites, 11
scalar (value), 365
scalar network analyzer, 364–66, *366*
scales, 377, 381–82, *382*
scatter diagrams/plots, 21, 264
schedulers/scheduling, 93, 433
scientific notation
    for addition/subtraction, 230–31
    for division, 231–32
    for estimation, 233
    formats for, 228–30
    for multiplication, 231
    for raising to powers, 232–33
    vs. SI prefix system, 233–34
    significant digits, 225–27
    vs. standard notation, 227–28, 233–34
seasons, 3
second (time interval), 240, 241, 337, 340
second law of motion, 373–74
Second Law of Thermodynamics, 10
secondary standards, definition of, 86*t*
seconds symbol, 236
Seebeck, Thomas Johann, 10
SEI (Software Engineering Institute; Carnegie-Mellon University), 130
SEM (standard error of the mean), 271
semiconductors, 11
sensitivity coefficients, 318
service, 418
sexagesimal degrees, 390
Shewhart, Walter A., 21, 27–28, 35, 164–65
SI (International System of Units), 9, 27
    definition of, 478
    establishment of, 71, 150
    as the metric measurement system, 35n.7
    and MKS, 200–201
    numerical values of quantities, 77–85
    and traceability, 62
SI prefixes
    as calibration standards, 72*t*, 75–77, 80–81

currently recognized, 245t
examples of, 244
vs. scientific notation, 233–34
SI units
   base, 240–41
   base factors worksheet, 256
   as calibration standards, 10, 71–75, 72t, 78–80, 200
   by category, 247, 248–51t
   coherence of, 241
   conversion factors and their uses, 252–53, 252t
   conversion worksheet, 256
   converting between equivalent measurement units, 253, 255
   decimal basis of, 241, 253
   derived, 240–41, 242–51t, 244–47 (*see also* SI prefixes)
   from/to lists, 252–53
   from/to matrix table, 252t, 253
   and measurement parameters, 151–54t, 153
   multiples of base/derived, 245, 246t
   names of, 239–40
   NIST worksheet, 257
   and physical constants, 150–51t
   properties worksheet, 256–57
   realization techniques for, 87, 88t, 240–41
   references worksheet, 256
   symbols for, 239–40
   uncertainties for, 253, 254t
   units not to be used within SI unit system, 245–46, 246–47t
   *See also specific units*
significant digits, 225–27
sine bars/plates, 393
sines, 291
Sisson, John, 399
SJAC (Society of Japanese Aerospace Companies), 124
skewness, 271–72
slope, 389
slope-intercept equation of a line, 286
slope of a line, 286
slugs, 375
Smith, Phillip H., 359
Smith chart, 359–61, *360*
Society of Japanese Aerospace Companies (SJAC), 124
software
   acquisition of, 137–40
   for automated calibration, 137–44
   for calibration management, 104
   construction/coding of, 142
   COTS, 138, 139, 141, 143
   design of, 142
   and hardware, 141
   industry-specific requirements in, 123–24, 127–28, 130–31
   maintaining/changing of, 143
   quality planning for, 141
   requirements definition for, 141–42
   testing of, 142–43
   validation process, 141
   validation tasks, 141–43
   validation/verification of, 116, 138–40
Software Engineering Institute (SEI; Carnegie-Mellon University), 130
Software Support for Metrology, 138
sound pressure/power, 260–61
space, units of, 251t
space exploration, 11
spanning, 174
SPC. *See* statistical process control
specification limits, 178, 183
specifications, 177–93
   absolute, 184–85
   baseline, 179–82
   comparing instrument to measurement task, 192–93
   comparing two instruments, 188–92, 189–90t
   definition of, 482
   floor, 182
   forms of writing, 182–83
   IM&TE (*see* calibration)
   modifiers, 179, 183–84
   one-way, 178
   output, 180
   qualifiers, 179, 184–88
   scale, 180–81
   tables of, 185–88
   vs. tolerance, 177
   two-way, 178
spectrum analyzer, 364
spheres, 296, 300
spider graphs, 264
split-sample testing scheme, 210, *211–12*, 211t
spot frequency, 174
spread-spectrum technology, 348
SQC (statistical quality control), 28, 30
SS (sum square), 269
stability (direct voltage), 328
stability (measurement capability), 163, 164
stability (specification), 183
standard deviation, 269–70
standard error of the mean (SEM), 271
standard normal distribution, 279
standard operating procedure (SOP), 482

standard reference material, 482–83
standard uncertainty, 483
standard vs. scientific notation, 227–28, 233–34
standardization (self-calibration), 481–82
Standardize Before Use stickers, 97
standards (document), 482
    See also specific documents
standards (measurement)
    definition of, 73, 86, 482
    types of, 86*t*
    See also specific standards
standing wave ratio (SWR), 358–61, 359*t*, *360*
Stanford Applied Engineering (SAE) International, 124
Star Wars defense project, 406
State Plane Coordinates System, 401
statistical analysis, early, 27–28
statistical process control (SPC), 164–68, 165–66*t*, *167*, 214
statistical quality control (SQC), 28, 30
statistics, 265–83
    auto-correlation, 283
    bimodal distributions, 267, *268*
    central limit theorem, 268
    central tendency, 266–67
    correlation, 272–73
    degrees of freedom, 265–66
    distribution types/properties, 279–82
    formats of tabular data, 275
    fundamental measurement assumptions, 265
    interpolation, 275, 276–79
    kurtosis, 272
    linear interpolation methods, 276–77
    linear relationships, 273–75
    mean/average, 266–67
    median, 267
    mode, 267
    nonlinear interpolation methods, 277–79
    residuals, 266
    root mean square, 268
    root sum square, 269
    sample standard deviation, 270
    sample variance, 270
    skewness, 271–72
    standard deviation, 269–70
    standard error of the mean, 271
    sum square, 269
    variance, 269
    zero and span relationships, 275
Stein, Philip, 115
steradian, 240
Steven Simon, 7
stickers, calibration, 95–97

subcontractors, 32
Substances generally recognized as safe (GRAS), 13
substituting calibration standards, 195–201, 196–99*t*
substitution measurement method, 158
subtraction, scientific notation for, 230–31
Sulfanilamide, 13
sum square (SS), 269
Sumerian period, 4
superscript format, 229
suppliers, 32, 443–46
surface area, 299–301
surface graphs, 264
surveying, 9, 399–402
surveys, 425–27
SWR (standing wave ratio), 358–61, 359*t*, *360*
systematic error, 305, 483

**T**

t-distribution, 279
tabulation formats, 275
TAI (international atomic time scale), 241, 337, 340
tamper seals, 96
tangent bars, 393
tape measures, 399, 400
taper gages, 391
taxation, 9
Taylor, Barry, 241
Tea Importation Act (U.S., 1897), 15
Tea Tasters Repeal Act (U.S., 1996), 15
technical managers, 430
telecommunications, 126–27
telephone service, 347
telescopes, *398*
television service, 347
temperature
    absolute zero, 10
    control of, 104–6, 105*t*
    degree Celsius, and SI units, 76–77
    International Temperature Scale, 10
    standardization of measurement of, 10
    thermal noise, 364
    and uncertainty, 317
temperature term (specification), 184
terrestrial time (ephemeris time), 340
test accuracy ratio, 483
test uncertainty ratio (TUR), 123, 483–84
testing, 140
    See also proficiency testing; validation; verification
thalidomide, 13–14

thedolites, 395, 399, 400
thermal noise, 364
thermistor sensors/meters, 352–53
thermocouples, 10, 352
thermodynamics, 10
thermoelectric potentials, 328–29, 329t
thermometers, 10
Thomson, J. J., 11
three-D graphs, 264
TickIT, 128, 143
time, 3, 241, 251t
time interval, 337–40, 337t
time of day, 340–41
time term, 183–84
title law, 401
TL 9000, 126–27
tolerance, 177, 484
torque, 317
total stations, 401
traceability
    cGMP on, 121, 122–23
    in civil aviation, 129–30, 133n.49
    definition/scope of, 61, 65, 484
    elements of, 61–63
    and NIST, 61, 64, 121, 133n.49
    pyramid of, 63–64, *63–64*
    reverse, 64–65, 94, 116
    and TUR, 123
    *See also* calibration
traceability path, 111
training
    cost of, 100
    on document changes, 37
    in guilds, 9–10
    records, 91, 100
    requirements for, 99–100
transfer measurement method, 484
transfer standards, 86t, 484
transits, 399
trapezoids, 292
Treaty of the Meter (1875), 8, 87, 373
triangles, 290, 292–93, 295
triangular distribution, 281–82, 305, 309, 310–11
tropical year, 340
Troy pound, 8
true mass, 377
truncation of numbers, 235–36
TUR (test uncertainty ratio), 123, 483–84
    4:1 TUR, 51, 52
Turgel, R. S., 341–42
two-point slope-intercept relationship, 274
Type A evaluation of uncertainty, 104, 307–8, 307–8t, 484–85

Type B evaluation of uncertainty, 309, 485
typeface of unit symbols, 73

## U

U-shaped distribution, 282, 305, 309, 311
UKAS (United Kingdom Accreditation Service), 106
ultrasonic frequencies, 326
unbroken chain of comparisons, 61–63
uncertainty, 305–19
    budget, *311*, 311–12
    combined, 282, 312–13
    considerations, 316–18
    and correlation coefficients, 318–19
    definition of, 485
    and degrees of freedom, 318
    determining, 306–17
    expanded, 313
    identifying uncertainties, 306–7
    managing, 319
    normal distribution of, 281, 305, 309
    overlap of, 207–8, *208*
    process for determining, 306
    reasons for determining, 306
    rectangular distribution of, 281, 305, 309–10
    relative, 282
    reports, 313–14, *314–15*
    and sensitivity coefficients, 318
    standard, definition of, 483
    triangular distribution of, 281–82, 305, 309, 310–11
    Type A evaluation of, 104, 307–8, 307–8t, 484–85
    Type B evaluation of, 309, 485
    U-shaped distribution of, 282, 305, 309, 311
    *See also* measurement uncertainty
uncertainty budget, 485
*The Uncertainty of Measurements* (Kimothi), 160
*Understanding Current Regulations and International Standards* (Bertermann), 118–19
Union of Japanese Scientists and Engineers (JUSE), 28
United Kingdom Accreditation Service (UKAS), 106
units of measurement, 149–50
    *See also* SI
universal coordinated time (UTC), 241, 340–41
universal time (UT), 340
U.S. Coast and Geodetic Survey, 399
U.S. Coast Guard, 338
U.S. food and drug laws, 12–15
U.S. Marine Corps, 417

U.S. Metric Association (USMA), 9
U.S. Mint, 7
U.S. Naval Observatory (USNO), 340–41
U.S. Pharmacopeia, 12
*U.S. v. Dotterwich*, 13
USMA (U.S. Metric Association), 9
USNO (U.S. Naval Observatory), 340–41
*USS Constitution*, 7
UT (universal time), 340
UTC (universal coordinated time), 241, 340–41
UUC/UUT, 339, 485
UV rays, 405

## V

validation
    of automated calibration procedures, 143–44
    of COTS, 143
    definition of, 140, 485
    of software, 116, 138–40
    software validation process, 141
    software validation tasks, 141–43
    *See also* testing; verification
values of quantities, 77–85
variance, 269
VDA 6.1 Standard, 125
vector (value), 365
vector network analyzer (VNA), 365–67, *366*, 368–69
vendors, 443–46
verification, 140, 485
    *See also* testing; validation
Victoria, queen of Great Britain and Ireland, 8
VIM (International Vocabulary of Basic and General Terms of Metrology), 61
    on bias, 163
    definition of, 485
    on drift, 164
    on linear scale, 163
    on measurement methods, 157–59
    on measurement systems, 159
    on repeatability, 163–64
    on reproducibility, 164
    on resolution of measurement data, 172
    on stability, 164
    on units of measurement, 150
Vitamins and Minerals Amendments (U.S., 1976), 14
VNA (vector network analyzer), 365–67, *366*, 368–69
Volta, Alessandro, 11
voltaic pile, 11
Voltaire, 417
voltmeters, 316
voltmeters, differential, 201n.3
volts, 11
volume, Babylonian measurement of, 4
von Klitzing, Klaus, 330
von Klitzing constant, 330
VOR signals, 363
VSWR, 359, 406
    *See also* standing wave ratio

## W

waveguides, 350, 350*t*, 369n.6
War Department (U.S.), 28
warships, 7
Washington, George, 399
water molecules, 406
Watson, Greg, 127
Watt Balance approach, 385
watts, 343, 351
Weibull distribution, 280
weighing methods, 383
weight, 3, 5, 8, 374
    *See also* mass/weight
Western Electric, 27–28
Wetherill, Charles M., 12
Wheeler-Lea Act (U.S., 1938), 13
Whitney, Eli, 10
Wiley, Harvey W., 12
workflow, 435–40, *439*
work instruction, 485
working standards, 86*t*

## X

X rays, 403, 404
x-y points, distance between, 287

## Y

y-intercept of a line, 285–88
yard, 9
years, 3
Young, William J., 399

## Z

z-Score, 213
zener diodes, 328
zero and span relationships, 275
zeroes, leading, 237
zeroing, 174